THE BEEKEEPER'S HANDBOOK

THE BEEKEEPER'S HANDBOOK

FIFTH EDITION

Diana Sammataro
Alphonse Avitabile

Foreword by Dewey M. Caron

COMSTOCK PUBLISHING ASSOCIATES

an imprint of

Cornell University Press

ITHACA AND LONDON

First edition published 1978 by Peach Mountain Press
Second edition published 1986 by Macmillan Publishing
    Company
Third edition published 1998 by Cornell University Press.
    Fourth edition 2011. Fifth edition 2021.
Third edition printing, Cornell Paperbacks, 1998.
    Fourth edition 2011.

Printed in the United States of America

Library of Congress Cataloging-in-Publication Data
Names: Sammataro, Diana, author. | Avitabile, Alphonse,
    author. | Caron, Dewey M. (Dewey Maurice), writer of
    foreword.
Title: The beekeeper's handbook / Diana Sammataro and
    Alphonse Avitabile, foreword by Dewey M. Caron.
Description: 5th edition. | Ithaca [New York] : Comstock
    Publishing Associates, an imprint of Cornell University
    Press, 2021. | Includes bibliographical references and
    index.
Identifiers: LCCN 2020017573 |
    ISBN 9781501752612 (paperback)
Subjects: LCSH: Bee culture—Handbooks, manuals, etc. |
    LCGFT: Handbooks and manuals.
Classification: LCC SF523 .S35 2021 | DDC 638/.1—dc23
LC record available at https://lccn.loc.gov/2020017573

Contents

Foreword to the Fifth Edition

In my introduction to the fourth edition I wrote "Beekeeping is different things to different people . . . for some a business, or [for others] a way to supplement income from the 'daytime' job; for others, a pleasure, an intense learning experience, something to really delve into." Whatever your beekeeping scale, the fifth edition of *The Beekeeper's Handbook* is an important resource to help you beekeep.

Although we may not be fully aware of it, newly available information on honey bees appears virtually every month. Some of it is extremely significant, like the demonstration that mites are not bloodsuckers but rather are feeding on vitellogenin, or that flower resource competition of native bees and honey bees leads to improved and better seed set for a number of food crops. Others are perhaps less significant but nonetheless may help improve our husbandry of honey bees and/or our understanding of bees. The fifth edition of *The Beekeeper's Handbook* thoroughly updates and integrates this new material.

The loss of honey bee colonies over the wintering period and queen events and other issues that occur during the season may mean average losses of 50 percent of colonies within the starting, and subsequent, beekeeping years. The learning curve of how to keep bees healthy and alive is steep. Colonies are distinct families and, along with other animals we keep, face significant challenges. Our management choices can help improve colony survival. Experience improves beekeeping success, as does shepherding more colonies.

The road to successful maintenance of healthy

bees includes skillful management of the population of the single queen, the thousands of workers, and the hundreds of drones within the hive. Beekeepers seek to anticipate, rather than merely react to, what we find on colony inspection. The fifth edition is a valuable tool for reliable and current information toward flattening the learning curve and reducing annual losses.

The fifth edition updates the area of maladies, diseases, pests, and predators. Seasonal management continues as a real strength of the *Handbook*. Clear and concise information with options of how we might accomplish bee care are detailed. Useful management strategies are offered in step-by-step summaries. How you can provide appropriate care is fully covered.

Another strength of the *Handbook* is coverage of how bees obtain and process—and how we can then subsequently harvest—honey and other bee products. One beauty of keeping bees is in appreciating their balance to the environmental ecology. For example, bee colonies harvest flower resources in a nondestructive manner. Their fortune is related to resource availability and to the weather that enables them to leave their home to obtain those resources and develop a population sufficiently robust to successfully exploit them. The *Handbook* helps us understand and manage the rewards we can obtain from our colonies.

How might you get the greatest benefit from using the *Handbook*? Well, it can be read from cover to cover, but I doubt you will have the time or want to do this. Rather read it in sections—chapters. That will help you get a feeling for what your tool can do. Continued return to the tool will improve your familiarity and improve its value.

Mark sections that you want to return to at the appropriate season of the year or when confronted with an issue with your bees. I like sticky tabs sticking out from the margin. The table of contents and index will be a useful way to assist finding the relevant information you may need to review. There is so much here you might not immediately find useful, but over a season or two, returning to consult this tool will definitely help improve your beekeeping success.

Beekeeping is never the same. Seasons vary, the bees vary, and the variety of challenges will vary.

Use this tool along with another requisite tool—communication with other beekeepers. How are they keeping bees healthy? How are they doing in harvest of honey? Are they too having a difficult time with yellow jackets? Are they controlling mites, and if so, how? It is our responsibility to bee health that we know how many mites are in a colony and then seek to reduce mite numbers. High mite loads can significantly impact colony health and, if uncontrolled, negatively impact the health of neighboring colonies.

Remember, all beekeeping is local. Better beekeeping begins with better tools. You are off to a great start using the fifth edition of *The Beekeeper's Handbook*. Good bees and good luck.

—*Dewey M. Caron*
Portland, Oregon
2020

Preface and Acknowledgments

Beekeeping is a fascinating activity, and in this age of environmental problems, with reports of large losses of many insects and wildlife, it is important that beekeepers do their part to help sustain their bees and help them thrive. Bees are fascinating and have a unique social system that we are still studying. With the advent of new techniques and knowledge (molecular biology, including DNA, RNA, and protein interactions), scientists are still learning how bees interact with each other and their environment.

This, the fifth edition, updates some methods of beekeeping. The most important changes have been in the area of pests, predators, and diseases of bees. With the advent of predatory mites, beekeeping has changed forever from the days when Diana Sammataro's grandfather (George Weber) and his brothers kept bees in the 1900s. Then, beekeepers set up their hives and left them until they needed to add supers or take off the honey. Now we, as beekeepers, cannot keep bees without tending to them. We, as beekeepers, have been selecting for certain characteristics of our bees, such as gentleness, productivity, and pest/pathogen resistance. While we have changed the bees by these selections, we have also changed the pests/pathogens at the same time. Beekeepers are now responsible for keeping their bees healthy and thriving but also from becoming a nuisance.

We have updated information and added new ideas while still trying to keep the same theme of this book, which is not telling you *what* to do, but giving you options on different techniques that you can use in your beekeeping work. This leaves the decision of what to do in your hands; we just give you some pros and cons for the different techniques.

With the advent of the internet, more ideas and opinions are readily available. A word of caution, however: some of the information online may be incorrect or misleading. As always, check with other beekeepers in your club or region to see what works well in your area. Ask questions and try different techniques.

Also, keep current with new legal requirements in your area. Some cities forbid beekeeping, as bees still are considered a nuisance. With the spread of the Africanized honey bees, it is even more important to know your bees and requeen with docile queen lines. Keeping current with the new research, and knowing what your colonies are doing and how your bees are reacting to their environment, are all vital to being a good beekeeper.

A take-home message for any beekeeper: Remember, bees do not read the bee books; they do what they want. There is no one correct way to keep bees.

But have fun and eat lots of honey.

ACKNOWLEDGMENTS

I would like to acknowledges my parents, Joseph Michael Sammataro, an architect, and Nelva Margaret Weber, a landscape architect, who guided the many interests and curiosities of their daughter's childhood with gentle kindness and encouragement. I also remember my maternal grandfather, George Weber, who first introduced me to the world of bees at the tender age of twelve in Arrowsmith, Illinois. His two brothers, Fred and Harry, the Weber brothers, were commercial beekeepers in Blackfoot, Idaho, early in the 1900s.

I wish to mention Dr. Bert Martin, of Michigan State University, whose gentle encouragement gave me the necessary direction and creative outlet to learn more about bees. I managed to talk the Ann Arbor Adult Education staff into letting me "teach" a beekeeping course. This forced me to learn about beekeeping by reading all the available books, taking lots of notes, and drawing lots of sketches. The first edition was envisioned when Doug Truax of Peach Mountain Press in Dexter, Michigan, who was taking the class, suggested making the teaching notes into a book.

I am always grateful to Jan Propst (who created the original layout with its horizontal format design) that the idea of making the rough notes into a book became a reality.

It was providential that while working at the White Memorial Nature Center and Museum in Litchfield, Connecticut, I met Al Avitabile, professor of biology (ecology and evolutionary biology) at the University of Connecticut's Waterbury campus.

Over the years many beekeepers and bee researchers have been kind enough to express their enthusiasm and honest appreciation of this book. To all of you who personally shared opinions, comments, photos, and observations, your kind words have helped more than you will ever know. Thank you.

—*Diana Sammataro*

I wish to dedicate this book to Mr. Lenard Insogna for persuading me to pursue a degree in biology, to my parents who allowed me to spend most of my time in the woods and ponds near my home studying nature, and to my wife, Ruth, for her support throughout my studies of honey bees.

On a May day in 1965, students from my class at the University of Connecticut's Waterbury campus rushed to my office to report the presence of a swarm of termites clustered in a forsythia bush just outside my office. That swarm of termites turned out to be a swarm of honey bees that launched my study of honey bees. To my student Bruno West, who rushed from Waterbury to Bristol, Connecticut, to gather some used beekeeping equipment to house the aforementioned swarm, to Dr. Roger A. Morse, whose Cornell beekeeping correspondence course on the husbandry of bees provided the basis for how to care for honey bees, and to both him and Dr. Rolf Boch, who provided opportunities for me to carry out research on honey bees at Cornell University in Ithaca, New York, and at Canada's Central Experimental Farm in Ottawa, thank you all.

I would like to thank Peter Borst for his valuable research assistance during the preparation of this fifth edition.

—*Al Avitabile*

Contributors to this edition include Dr. R.W. Currie of the University of Manitoba and Dr. Andony Melathopoulos of Oregon State University, who provided information about indoor wintering of honey bees. Retired Drs. Gordon D. Waller and Gerald Loper, who worked at the USDA Carl Hayden Bee Lab in Tucson, Arizona, and Fred Terry of Oracle, Arizona, contributed information on how to keep bees in the Southwest. Dr. Malcolm Sanford, retired extension service entomologist from the University of Florida, added some good information on southern beekeeping. James Fischer, a beekeeper in New York City, and Toni Burnham of the [Washington] DC Beekeepers Alliance, wrote the appendix on extreme urban beekeeping.

The late Ann Harman helped tremendously by pointing out areas in the book that needed changing. Dr. Dewey Caron, always encouraging, added great information and guidance.

Thank you all.

THE BEEKEEPER'S HANDBOOK

Introduction

Beekeeping is an interesting, rewarding activity for hobby beekeepers as well as for individuals who make their living tending to their bee colonies. If you enjoy nature and its wonders, keeping honey bees will connect you with many parts of the natural world, including the flowering plants in your area, and the unique way bees have of communicating to their nest mates the location of food and raw materials provided by these plants. This method of communication, often referred to as the dance language, was discovered by the late Dr. Karl von Frisch; in 1973 he received the Nobel Prize for identifying this language, which continues to intrigue scientists, beekeepers, and other students and lovers of nature.

Flowering plants and the ancestor of today's bees coevolved. The flower provides the bee with its nutrient requirements, sugar in the form of nectar and proteins in the form of pollen, and the bee "rewards" the flower by transferring pollen to the female reproductive part of the flower, which culminates in the production of seed. Thus, the bee inadvertently plays a major role in assuring the perpetuation of flowering plants year after year, and in so doing perpetuates itself as well.

This fifth edition of *The Beekeeper's Handbook* provides options for you to care for your bees as you move from amateur status to knowledgeable beekeeper. Whether you intend to start keeping or are already engaged with honey bees, this book offers numerous illustrations and diagrams together with written explanations to assist in finding both success and satisfaction in this venture. We outline the various colony activities you are most likely to observe and describe the major management options available that are often required to manipulate the internal workings of a colony. We list the advantage and disadvantage of each important technique/method, to help you decide which option may work best in your situation. For example, when entering your apiary to work your colonies, you will be able to interpret to some extent the colonies' condition by observing the number of bees bringing in pollen loads and the number of bees flying in and out. In addition, many sections of this text are cross-referenced to point you to even more detailed information if needed. Remember, **there is no single way to manage bees**. You may read about a manipulation and tweak it to suit your situation or try something entirely different. Space is provided at the close of each chapter for you to make notations on your successes and failures that will serve as reminders the next time you are faced with the same observation and/or problem.

In this fifth edition of *The Beekeeper's Handbook*, the reference section has been updated to include additional relevant books, bee organizations, and internet resources. We have also included an updated glossary to help beginners understand bee vocabulary. A word of caution: with the advent of the age of the internet, with its instant information being just a keystroke away, it is easier than ever to find the information you are searching for. However, be careful to separate opinion from time-tested results. For example, on the internet you may find remedies for eliminating a given number of varroa mites from your colony—but you are not permitted to use some of those remedies, involving the introduction of chemicals into your hives, unless permitted by law.

The internet source might not have provided that important bit of information.

This book also provides information that in most instances will allow you to avoid the negative aspects of this husbandry. Although often considered a gentle art, beekeeping can be at times physically demanding and strenuous. The typical picture of a veiled beekeeper standing beside one of his or her colonies with a smoker in one hand and a hive tool in the other does not reveal the aching back, sweating brow, smoke-filled eyes, or, sometimes, the very painful stings that can come with the territory. When you are in the presence of fellow beekeepers, ask a lot of questions about their experiences and share your successes and failures with them.

A final note: bees do not read bee books. They simply go about the activities that sustain them. We have yet to tease out everything bees do to maintain their species. They have been independent of us and successful at it for millions of years. Beekeeping will have many rewards and eureka moments for you, as well as some disappointments. Nevertheless, welcome to the world of beekeeping. We wish you a successful journey.

LEGAL REQUIREMENTS

Most, if not all, of the 50 states that constitute the United States, along with the provinces of Canada and many other countries around the world, have laws and regulations requiring that all active honey bee colonies and their location(s) be registered with the office of the state entomologist or with the state's agricultural department. (These laws and regulations do not pertain to bees living in tree cavities or other "natural" or non-human-managed spaces.) In most cases, beekeepers are required to renew their registration every year. Regulations and laws governing the keeping of honey bees are not uniform from state to state or from country to country, but a good place to start to locate important information about laws and regulations pertaining to your honey bees is by contacting your department of agriculture.

All state apiary inspection laws are designed to protect beekeepers; the inspectors are your friends. For example, your state's entomologist or his or her agent will notify you when bee diseases such as American foulbrood may be prevalent in your area or when insect spraying may take place at or near your apiary sites, thus providing ample opportunities for you to take the necessary steps to keep your bees safe. In most countries, including the United States, all colonies must be housed in structures that will allow for inspection of the inside of the colony. All colonies should be housed in enclosures containing movable frames. In recent decades, several serious pests originally confined to very limited areas of the world are now found far beyond their original boundaries, namely the varroa mite, the small hive beetle, and the Africanized honey bee. As a consequence, some states and regions have had to enact special laws and regulations regarding the keeping and care of honey bees. For specific legal requirements and regulations, check with your state's department of agriculture.

General apiary laws and regulations usually include some of the following:

- Beekeepers may have to register the number of colonies and their locations with their state bee inspector.
- The director of agriculture or appointed deputies may be authorized to inspect colonies and treat, quarantine, disinfect, or destroy any diseased colony that presents a risk to other colonies in your apiary, as well as neighboring ones.
- Before you move your bees to another town or city within your state or to another state or province, you may have to clear the move with the bee inspector or someone authorized to issue you such a permit.

Because bees sting and are considered by some to be a nuisance, some cities and towns have restrictions or zoning laws as to how many colonies, if any, you are permitted to have in a given location. But with the worldwide declining bee population, many of these restrictions have been lifted. New York City now permits beekeepers to keep bee colonies, which are often located on the rooftops of buildings. If in doubt as to whether you can keep bees in your town or city, call an agent or zoning officer in your municipality for information before purchasing equipment and bees.

Failure to comply with your state, province, or country's apiary regulations may result in monetary penalties.

BEE-STING REACTIONS

An important question you must consider as a bee-keeper is your response to bee stings. Although most beekeepers may never exhibit a serious allergic reaction to a bee sting or multiple stings, the possibility of such a reaction should never be discounted. After a few years of keeping bees and being stung, or even in much less time, some individuals will develop a severe allergic reaction to bee venom, and on occasion also an allergy to bees, their body parts, and the hive equipment.

When you are stung, the bee's stinging apparatus pierces your flesh, and its venom is injected into the surrounding tissues and then transported by the fluids and cells in your veins and arteries throughout your body. Fortunately, for most beekeepers, only a localized reaction takes place—that is, pain, reddening, itching, and swelling at and around the site of the sting. At times, the swelling can be alarming, but it will usually subside over the course of a few days, as long as it remains in the general area of the sting site.

On the other hand, you may experience a more serious, potentially life-threatening reaction to bee stings. This is referred to as a systemic reaction, which is a true sign that you are allergic to bee venom, bees, their parts, and in general the hive equipment. A systemic or general reaction means that the entire body is reacting to the venom and its proteins. Signs of a systemic reaction may include those experienced in a localized reaction, as well as far more severe symptoms: itching of the extremities (hands and feet), an outbreak of hives all over the body, difficulty breathing, swelling away from the sting sites, sneezing, abdominal pain, vomiting, diarrhea, the loss of consciousness, and ultimately, perhaps even death.

These reactions are referred to as anaphylactic shock, a life-threatening manifestation of an allergic reaction caused by the release of histamines and other compounds in a person sensitized to an allergen, a substance that causes an allergic reaction to things such as bee venom and peanuts.

Note that the preceding is primarily about bee venom. And note clearly that we are not medical doctors; only they have the ability to fully diagnose and resolve an allergic reaction issue. Although anaphylactic shock is rare, it does occur. We cannot say whether the first sting or successive ones will trigger a reaction. There is no substitute for consulting with professionals in the medical field, in particular an allergist. Bee venom and its components have specific protein allergens that are different from those found in the venom of other stinging insects such as hornets, wasps, and yellow jackets. Because they are different, an individual may be allergic to a yellow jacket sting and may not be allergic to a honey bee sting.

We strongly advise anyone working with stinging insects, including honey bees, to seek the advice of members of the medical community. The percentage of individuals who become allergic to bee stings is very small, but for those individuals who are allergic, the allergic reaction requires an immediate visit to a doctor's office or a call for ambulance service. Death from an allergic reaction can occur in less than 20 minutes. Fewer than seven deaths are recorded each year in the United States as a result of bee stings (this number represents only reported deaths), far less than from other leading causes of death such as heart attacks or car accidents, but we should not minimize the possibility of a life-threatening allergic reaction to a bee sting. If there is ever any question about whether you are developing an allergy to bee stings, bees or their body parts, or bee products and equipment, you should not return to your bees until first seeking the advice of a physician or an allergist. A final note: members of your family who do not tend to bees may also become allergic simply because of their contact with venom and other mentioned allergens carried into the house on the beekeeper's clothing.

For more information on the effects of bee stings, see "What to Do When Stung" in Chapter 5, as well as the venom sections in the References and Appendix C, and check current medical websites. Finally, as a precaution, please seek the advice of a physician about the need to carry an EpiPen when you are around bees; it might be a wise thing to do.

CHAPTER 1

Understanding Bees

BEE ANCESTORS

Although the fossil record is incomplete, insect fossils first appear from about 300 million years ago, during the Carboniferous period, often divided into the Mississippian and Pennsylvanian periods of the Paleozoic era. The fossil record also indicates that the order Hymenoptera (phylum Arthropoda) evolved around 200 million years ago during the Triassic period of the Mesozoic era. However, fossil insects preserved in Permian rock, the last period of the Paleozoic era, display hymenopteran-like structures, including membranous wings and antlike and wasplike waists, both hallmarks of the order Hymenoptera.

Approximately 50 million years later, during the Jurassic period of the Mesozoic era, the order Hymenoptera is firmly established in the fossil record, preserved primarily in amber. Bees appear to have evolved from a predatory sphecid wasp ancestor about 100 million years ago, during the mid-Cretaceous period of the Cenozoic era. The switch from animal to plant protein, referred to as pollen, and the presence of branched hairs among other anatomical features, separates bees from wasps.

During the vast periods of time that followed, the flowering plants became more specialized and more dependent on mobile pollinators. Insect visitors such as the bees were very important, and they, and the plants they pollinated, coevolved structures to their mutual benefit as a result of this interdependence.

It was not until 66 million years ago during the Tertiary period of the Cenozoic era that stinging hymenopterans became abundant, and the family Apidae (bees), now numbering 20,000 or more known species, co-evolved with the dominant plant group angiosperm (flowering plants), with almost 300,000 known species. Flowers provided this group of hymenopterans with their source for protein (pollen) and their main source for carbohydrates (nectar). The angiosperm flowers evolved colors, shapes, and odors that attracted bees to them as well. The flowers benefited from this relationship by the fact that while the bee was collecting pollen and/or nectar from the flower, it also inadvertently transferred some of the pollen to the flower's stigma, resulting in the fertilization of the flower's ovules, producing seeds for subsequent generations. At the same time, the bee was evolving to take full advantage of the flower's bounty. Bees developed branching hairs on their bodies to trap the pollen, and large "baskets" (the *corbiculae*) on their third set of legs, which allowed for copious amounts of pollen to be carried back to the hive. Similarly, a large inflatable organ, the honey stomach or sac, allowed the bees to collect copious amounts of nectar/honeydew before returning to the colony. The honey sac and the corbiculae, the pollen-bearing structures, enabled the bees to sustain large colonies with numbers in the vicinity of 40,000 individuals per colony. In addition, a highly structured social order developed, with elaborate defense and communication systems to exploit the most rewarding floral sources.

In the order Hymenoptera, worldwide there are over 150,000 species in over 100 families, now grouped into 25 superfamilies and over 2000 genera.

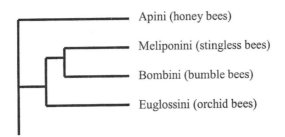

Apinae

Taxonomy of the Family Apidae. After S.A. Cameron. 1993. Multiple origins of advanced eusociality in bees inferred from mitochondrial DNA sequences. Proc. Natl. Acad. Sci. 90:8687–8691.

With current DNA analysis, some of these families have been moved around. Currently, the placement of the honey bee in the animal kingdom is:

- Kingdom: Animalia.
- Phylum: Arthropoda (many-jointed, segmented, chitinous invertebrates, including lobsters and crabs).
- Subphylum: Hexapoda.
- Class: Insecta (six-footed).
- Order: Hymenoptera (Hymen is the Greek god of marriage; hence the union of front and hind wings [*pteron*]).
- Suborder: Apocrita (ants, bees, and wasps).
- Superfamily: Apoidea.
- Family: Apidae (characterized by food exchange, pollen baskets, storage of honey and pollen); three subfamilies (see the figure on taxonomy).
- Tribe: Apinae (long tongues, nonparasitic, highly eusocial).
- Genus: *Apis* (bee, Linnaeus, 1758; native of the Old World, probably evolved in India and Southeast Asia).
- Species: *mellifera* (honey bearing), also called *mellifica* (honey maker), Western honey bee.

(Note: A fossil newly discovered in Nevada contains a now-extinct New World honey bee species, named *Apis nearctica*: www.sciencenews.org, July 2009.)

EVOLUTION OF SOCIAL STRUCTURE

Social structure is defined by the degree of community living. The true, highly specialized, or *eusocial*, societies are those of honey bees, ants, and termites. The eusocial society of honey bees is characterized by two traits: (1) individuals exhibit extreme task specialization while living in multigenerational family groups, and (2) the majority of individuals cooperate to aid a single reproductive member, the queen. The sophistication of the social structure of honey bees is indicated by a number of other characteristics as well:

- Longevity of the female parent (queen) coexisting with her offspring.
- Presence of two female castes: queens and worker bees.
- Siblings assisting in care of the brood.
- Progressive feeding of brood instead of mass feeding.
- Division of labor, whereby queen lays eggs and the workers perform other functions.
- Nest and shelter construction, storage of food.
- Swarming as a reproductive process.
- Perennial nature of colony.
- Communication among colony members.

Another description of a eusocial community of honey bees is one that consists of two female castes, a mother (the *queen*) and her daughters (sterile *workers*), that overlap at least two generations. Because hornet and wasp colonies do not overwinter in temperate climates as do honey bees, they are termed *semisocial* insects.

Most insects are solitary—they neither live together in communities nor share the labor of raising their young. The 12,000 species of insects that do live in communities are the ants, termites, wasps, and bees (and include some beetles, aphids, and thrips). The origin of social insects troubled even Darwin, who could not rationalize how a special, sterile caste—the workers—could pass on their genetic information if they could not produce offspring, for one of the goals of reproduction is to transmit one's genes to one's children. In other words, workers were displaying "altruism" toward their siblings at the expense of having their own children. Over the years, there have been several ways to explain the evolution of sociality. Following are two of the more popular theories, briefly explained: kin selection and parental manipulation

Kin selection:. In 1964, William Hamilton proposed a way around Darwin's dilemma with his theory of kin selection, which states that, in the case of honey bees, workers surrender their reproductive activities to their queen for the benefit of the colony. Instead, the workers care for their siblings, enabling them to pass on their genes by way of their queen.

Parental manipulation asserts that the mother gains net survival or reproductive success by tricking or manipulating others. The queen dominates her daughters by means of chemical signals called *pheromones*. Such signals reduce the reproductive potential of the daughters, forcing them to become slaves and tend their siblings instead of laying their own eggs.

The behavior of existing subsocial and primitively social insects can be accounted for by either or both of these hypotheses. Newer instrumentation and molecular ecology have made rapid advances in this area. Evolution of social behavior makes for fascinating reading, and you should look at other works on the complex social behavior of insects for a more complete understanding of the subject (see "Social Insects" in the References).

RACES OF BEES

According to Dr. Stanley S. Schneider (2015), at the University of North Carolina Charlotte, "although there are over 25 geographic races (subspecies) of *A. mellifera* adapted to a wide variety of habitats and climates, most research has concentrated on the 4–5 European races most commonly used in beekeeping in Europe and North America."

A *species* is a group of organisms that can only interbreed within their species, producing offspring that can do the same. Currently bees are grouped by size, such as the Dwarf honey bees (in the subgenus *Micrapis*) that include *A. florea* Fabricius, 1787, and *A. andreniformis* Smith, 1858, and Giant honey bees (subgenus *Megapis*), such as *A. dorsata* Fabricius, 1793. The Asian cavity-dwelling bees are *A. cerana* F. 1793, *A. koschevnikovi* Enderlein, *A. nigrocincta* Smith 1861 for the Eastern species, and the Western or the European honey bee *A. mellifera* Linnaeus, 1758, which contains 24 races. Some bee researchers have identified new races (subspecies) in *mellifera*, *cerana*, and *dorsata* (which in *cerana* include *A. nuluensis*, and in *A. dorsata* include *A. binghami, laboriosa*, and *breviligula*). *A. andreniformis, florea, koschevnikovi*, and *nigrocincta* all lack subspecies so far.

By races, we are referring to populations of the same species (e.g., *mellifera*) that originally occupied particular geographic regions with different climates, topography, and floral resources. In these areas, bees evolved characteristics that made them unique from other races within the same species. F. Ruttner, in *Biogeography and Taxonomy of Honeybees* (1988), divided bees into four groups: (1) African, (2) Near Eastern, (3) Central Mediterranean and southeastern European, and (4) Western Mediterranean and northwestern European. From the European groups came the Italian, Carniolan, and German black bees; the Near East group includes the Caucasian bees. These four races provide the raw materials from which modern hybrid bees are derived.

The first of the four European races of honey bees imported to North America from England was the German dark bee. These bees arrived at the English colony in Jamestown, Virginia, in 1621. Bees were needed for honey production, for making beeswax candles, and to pollinate fruit trees and other crops. Native Americans witnessed honey bees for the first time and called them "the white man's fly." Then, in 1859, the first Italian queens were imported to America. This variety was quickly recognized as superior to the German, which is aggressive, has a shorter tongue, and has low resistance to bee diseases. All these negative factors led to the diminished use of the German black, and now few beekeepers in North America have these bees. Honey bees spread quickly in the New World.

Today, the Italian honey bee is the most widely distributed race in the Western Hemisphere. The other two popular races, the Carniolan and the Caucasian, were brought to the United States circa 1883 and 1905 respectively; as with the Italians, they are frequently crossbred, interbred, and inbred for disease/pest resistance, hardiness, and gentleness.

Importation of live adult bees into the United States was halted in 1922 because of the danger of introducing bee diseases and pests that did not already exist here. This action was precipitated by the discovery of tracheal mites on the Isle of Wight (UK) in 1919. However, because of a shortage of available bee colonies in the United States in 2006 and 2007, California began to import packaged bees from Australia to help pollinate its almond crop. This practice has been discontinued.

South America did not have such restrictions, and in the 1950s, Brazil imported African honey bees

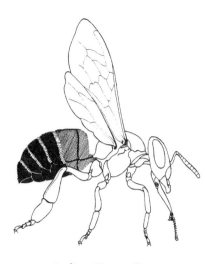

Italian Honey Bee

(*A. m. scutellata* Lepeletier 1836, one of many African races) to improve breeding stock. The accidental release in 1956 of the volatile bee known as the Africanized honey bee (labeled the "killer bee" by the press) has led to the spread of this subspecies throughout all of South America, all of Central America, and Mexico. In 1990, swarms of the Africanized bee crossed the border into Brownsville, Texas, and became established in the southern United States and parts of California. (Check the internet for the most recent updates.) How many states will ultimately be occupied by this bee is disputed, but there is general consensus that the southern half of the United States will have them, as either year-round or summer residents.

Although the most common honey bee in America is the Italian, you may be interested in experimenting with other bee races or hybrids. If you raise your own queens and do not control the drone source, crossbreeding could result in inferior queens. It is important to pay attention to the quality of both the queen mother and the drone fathers of the daughter queens you are rearing (see "Queen Rearing" in Chapter 10). Some beekeepers try to maintain only one race of bees in any one apiary, believing that pure strains are more resistant to diseases. An overview of the most commonly available races of honey bees now used in the United States is provided below:

The **Italian honey bee** (*Apis mellifera ligustica* Spinola 1806) originated in the Apennine Peninsula (the boot) of Italy. This race, introduced into the United States circa 1859, has several color types. Today, the Italian honey bee is the most widely distributed race in the Western Hemisphere. In general, the Italians are yellow with dark brown bands on their abdomen; the "goldens" have five bands, while the "leathers" have three. They are known for laying a solid brood pattern, producing lots of bees late into the fall, and making a good surplus of honey. On the other hand, they forage at shorter distances and therefore tend to rob nearby colonies of their honey stores, especially during dearth periods. They also drift frequently, because they orient by color rather than by object placement.

Advantages

- Good, compact brood pattern, making a strong workforce for collecting lots of nectar and pollen.
- Excellent foragers.
- Light color makes queen easy to locate.
- Moderate tendency to swarm.
- Moderate propolizers, so hive furniture is not glued together too much.
- Resistant to European foulbrood disease.
- Relatively gentle and calm, making them easy to work.
- Moderate to high cleaning (hygienic) behavior.
- Readily builds comb cells; white cappings common.

Disadvantages

- Can build lots of brace and burr comb; Italians have a slightly smaller cell size.
- Poor orientation to home hive; drifts to other hives, spreading diseases/pests and causing uneven colony populations.
- Can be bothersome by persistently flying at beekeeper when worked.
- Short-distance foragers, which have a tendency to rob weaker colonies, creating a robbing frenzy in the apiary, particularly during periods of dearth when there is little or no available floral nectar.
- Can be susceptible to many diseases and pests.
- Slow to build populations in spring, not good for early honeyflow.
- Brood rearing continues after the main honeyflow has ceased, sometimes late into the fall; as a consequence, bees may enter the winter period with too many adult bees and insufficient stores of surplus honey, resulting in starvation.

The **Carniolan honey bee** (*Apis mellifera carnica* Pollmann 1879) was originally brought from Yugo-

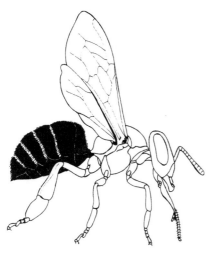

Carniolan Honey Bee

slavia and Austria, where the winters are cold and the honeyflows variable. They were introduced to the United States circa 1883 and are popular in northern areas. Carniolans have a grayish-black-brown body with light hairs; the drones and queens are dark in color. In general, they were bred for fast buildup when the spring flow starts, and to shut down brood production early in the fall. They are known for their gentle disposition and low propolis and brace comb production, but can swarm if not given ample expansion room. Currently the New World Carniolans are found in the United States and were developed and improved by Sue Cobey (University of California, Davis).

Advantages

- Build strong populations.
- Brood rearing decreases if available forage is diminished, thus conserving honey stores.
- Exceptionally gentle; less prone to sting and easier to work.
- Few brood diseases, so less medication may be needed.
- Economic honey consumers, therefore over-winters on smaller honey/pollen stores.
- Little robbing instinct, as they are long-distance foragers and are object oriented.
- Can have very white wax cappings, making comb honey sections attractive to customers.
- Little brace comb and propolis, making hive manipulations easier.
- Overwinters well; queen stops laying in fall, and a smaller number of bees overwinter on fewer stores.
- By comparison to other races, forages earlier in

the morning, on cool, wet days, and later into the afternoon.

Disadvantages

- Tend to swarm unless given enough room.
- Slow to begin brood rearing in the spring. As a consequence, colonies cannot be effective pollinators of early spring crops.
- Strong brood population depends on ample supply of pollen; can be slow to build up in summer if forage is late or unavailable.
- Dark queen difficult to locate, making requeening operations slower.

The **Caucasian honey bee** (*Apis mellifera caucasia* Gorbacher 1916) is originally from the high valleys of the central Caucuses near the Black Sea, where the climate ranges from humid subtropical to cool temperate. Caucasian bees are black with gray or brown spots and short gray hairs. They have the longest tongue compared to the other three races described here, which could make them superior pollinators of some crops. The drones are dark, with dark hairs on the thorax. These bees were introduced into the United States from Russia, circa 1905. In general, they are gentle bees, with low swarming instincts, and are good in areas of marginal forage or long honey flows. However, sources and breeders of these bees are currently difficult to find.

Advantages

- Build strong populations, but slow to start in the spring; not good for early spring crops.
- Gentle and calm on comb, making them easier to work.
- Have a long tongue so can exploit more species of flowers.
- Little tendency to swarm, resulting in strong colonies.
- Forage at lower temperatures, earlier in the day and on cool, wet days.
- Overwinters well, shutting down brood production in the fall; conserves stores.

Disadvantages

- Maximum propolizers, making hive manipulations difficult unless collecting propolis for sale.
- Can have wet wax cappings over honey, making comb honey less attractive to consumers.

- Can sting persistently when aroused, making inspections difficult.
- Late starters in spring brood rearing; not good for early spring pollination.
- Dark queen difficult to find.
- Can drift and rob.
- More susceptible to nosema disease; may require more medication.
- Difficult to find breeders.

Hybrid Bees and Select Lines

In addition to these races, there are hybrid bees, which can be crosses between the races or between selected strains within a race. Some common hybrids are the Starline (four-way Italian cross), Midnite (Caucasians × Carniolan), and Buckfast bee lines. Many queen breeders have their own variations of these races as well, to meet the needs of their customers.

The **Buckfast Hybrid** was a product of Brother Adam (1898–1996) from the Buckfast Abbey in the United Kingdom. He crossed many races of bees (primarily Anatolians with Italians and Carniolans) in search of a superior breed that would be tolerant of tracheal mites, work well in the microclimate of the area, and be gentle and productive. This hybrid is reported to have high cleaning instincts and disease resistance and to possess good overwintering abilities. Currently, this hybrid is difficult to find.

The **Starline Hybrid** was a four-way Italian cross, combining several Italian stocks. They were known for their gentle behavior and large brood numbers, providing a large workforce to exploit many nec-

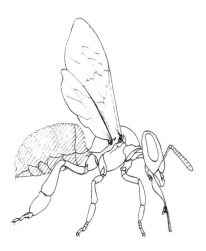

Cordovan

tar resources and therefore ideal for commercial beekeepers. Currently this line is not commonly available.

The **Cordovan** line is a color mutation, originally developed by Dr. Bud Cale to serve as a genetic marker in his Starline queen breeding program. They were used to trace behavior and kinship relationships for research purposes (now done with molecular markers and single-drone inseminated queens). Some queen breeders select Cordovan for hardiness, production, or disease resistance. They are easily identified because the black body color comes out red in Italians or a purple-bronze in Caucasian and Carniolan bees; this latter is called a Purple Cordovan. They are known for their gentle behavior and pretty color, excellent for showcase observation hives.

Russian Bees, a carefully introduced varroa-tolerant line from the Primorsky Krai region of northeastern Russia, were imported and isolated on a Louisiana island by Dr. Tom Rinderer in 1995; they were also bred at the USDA lab in Louisiana. This Russian stock is now available from selected breeders.

SMR (suppressed mite reproduction), now called **VSH (varroa sensitive hygiene)** and **Hygienic bees**, are new hybrids and crosses that are resistant to varroa mites. The VSH and the Hygienic lines were bred for cleanliness (pulling out diseased brood) and resistance to varroa mites. The VSH line was developed by Dr. John Harbo from the Baton Rouge USDA-ARS lab, and the Hygienic line by Dr. Marla Spivak from the University of Minnesota. Hygienic bees were originally recognized and developed at Ohio State University by Dr. Walter Rothenbuhler (1920–2002), who was breeding them to remove diseased (foulbrood) larvae; now they also will remove larvae or pupae infected with mites.

Current queen breeders can be found advertising in bee journals. All these lines that are being developed should be tried with caution, as they may not perform as advertised, because of poor breeding, your hive location, or the region where you live. If you are planning to requeen colonies, try 5 to 10 queens from one breeder first, to evaluate how they will do in your area; also select different breeders to determine which give the best results.

Advantages

- Tolerant of tracheal/and or varroa mites (VSH, Hygienic, Russian, Buckfast).

- Low swarming instinct.
- Minimal propolizers.
- Disease resistant (Hygienic lines), including chalkbrood.

Disadvantages

- Some lines build populations slowly in spring unless good honeyflow is in progress; not good for early spring pollination.
- Offspring queens from hybrid mother may not be like the original queen; daughter queens or their progeny may not have desired characteristics.
- Poorly mated or open-mated queens may not perform as advertised.
- Super-hygienic lines such as VSH may not be able to build up an adequate population of bees. In the process of removing the mites, they remove the brood—the future adult bees—to which the mites are attached.
- Requeening every other year may be necessary to ensure the colony is headed by hybrid queen and was not superseded.

A final note: when trying some new queen lines, **go slowly**. Start with 5 to 10 queens and observe how they compare to each other (queens vary within a line) and to the other queen lines in your apiary. Take notes and compare with other beekeepers in your area.

ANATOMY OF A BEE

The general anatomy of the honey bee is similar to that of other members of the class Insecta, with the exception of certain organs and structures that are unique to honey bees that allow them to become specialists within their niche. Bees share the three basic anatomical structures found in all insects, namely, a head, a thorax, and an abdomen (see the illustration of external anatomy on page 12). They have, along with many other insects, the following: an exoskeleton made up of hardened plates that are connected to one another by membranes; a respiratory system consisting of tracheae; a ventrally located spinal cord; an open circulatory system with a heart absent of veins and arteries, with the exception of a short aorta; a muscular system for locomotion, for both walking and flying; a reproductive system; a nervous system; and an excretory system.

The surface of the bee's head contains several important structures: It has five eyes, two of which are compound and three of which are simple eyes known as the *ocelli*. Attached to the head are the antennae, which are used to detect odors and measure flight speed. The head also contains the tongue, or *proboscis*, which is employed to obtain liquids by lapping and sucking up fluids such as water, nectar, honeydew, and honey. It also contains the jaws, or *mandibles*, which are used for crunching pollen and shaping beeswax into honeycomb (hexagonal cells), queen cups, and queen cells; they are also used for biting intruders and removing debris. The mandibles of queens contain the mandibular glands that produce an essential pheromone often referred to as *queen substance.*

The middle section of the bee, the thorax—the segment between the head and the abdomen—contains the flight muscles for the wings. Other muscles within the thorax control the three pairs of legs, which not only assist the bee in locomotion but are covered with hairs that are used for grooming and for the gathering and transferring of pollen to the third set of legs. It is important to note that the first pair of legs is also equipped with an antenna cleaner; the antennae can be pulled through the notch and brush components of the cleaner to keep them free of debris.

The respiratory system consists of three components: the *spiracles* (openings in the body wall that connect to the tracheae), the *tracheae* (or air tubes), and the air sacs, which branch off the tracheae. There are total of 20 spiracles, 10 on each of the lateral sides of the bee. Six are located on the thorax and 14 on the abdomen. The spiracles with the largest openings are the first set, located between the *prothorax* and the *mesothorax.* These spiracles are protected by dense hairs. In spite of these hairs, the large size of the openings creates an entry point for the tracheal mites. In the branching tracheae, the tracheal mites find refuge, feed on circulating blood, and carry out their reproductive activities (see "Tracheal Mites or Acarine Disease" in Chapter 14; this disorder was first observed in 1904 on the British Isle of Wright). The exchange of oxygen and carbon dioxide takes place in the air sacs.

The abdomen of the bee contains many important organs. Externally, it is armor-plated with scalelike dorsal segments called *tergites* and ventral segments

External Anatomy of a Worker Honey Bee

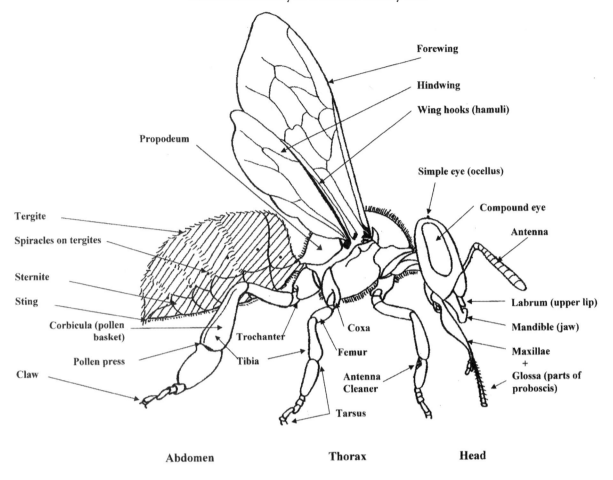

Forewing

Hindwing

Wing hooks (hamuli)

Simple eye (ocellus)

Compound eye

Antenna

Propodeum

Tergite

Spiracles on tergites

Sternite

Sting

Corbicula (pollen basket)

Pollen press

Claw

Trochanter

Tibia

Coxa

Femur

Antenna Cleaner

Tarsus

Labrum (upper lip)

Mandible (jaw)

Maxillae + Glossa (parts of proboscis)

Abdomen   **Thorax**   **Head**

called *sternites*. These armored plates protect the bee from predators and from desiccation. The bee's *sting* and associated glands possessed by the two female casts—the female workers and the queen—are located internally, just anterior to the abdomen's tip. The scent gland (the *Nassonoff gland*) is located near the dorsal tip of the abdomen just above the sting apparatus, and the wax glands are on the ventral part of the abdomen.

Both queens and worker bees have ovaries located in their abdomens. The queen has a storage sac called the *spermatheca* located in her abdomen for housing drone semen. In addition, there are pheromone-producing glands in the queen's abdomen and in the sting complex. The queen has no wax glands (see Appendix A for more information on internal anatomy and Appendix B on pheromones). The drone's abdomen contains the male reproductive organs, but there is no sting complex; the sting is a modified ovipositor, which would not be found on drones. Their abdomen has no wax glands. On occasion, drones may have both male and female structures, including the sting apparatus. These rare *gynandromorphs* may be able to sting you!

Bee Vision

Bees have five eyes—three simple eyes (*ocelli*) and two compound eyes. Each ocellus is a thick, biconvex lens or cornea and reacts only to light intensities. The compound eyes are composed of thousands of individual light-sensitive cells called *ommatidia* (singular, *ommatidium*). It is with the compound eyes that bees perceive color, light, and directional information from the sun's rays.

The color range of bee vision has been shown to include the violet, blue, blue-green, yellow, and orange colors, as well as ultraviolet light, which is invisible to humans. Because bees compete with each other for available pollinators, flowers that depend on bee pollination are within these color ranges. Those plants that succeeded in attracting bees with their color, nectar, and pollen gained an edge over other plants during their evolutionary development.

The structures and arrangement of the ommatidia permit polarized light to pass through certain parts of each ommatidium at any given instance. The sun's position and the bee's direction are the factors determining which section of the ommatidia will receive full, partial, or shaded regions. This pattern serves as a kind of compass to the bee, giving directional information. The bee is able to monitor these shifting patterns continually as it flies and, if necessary, adjust its course.

Antennae

Most of the tactile (touch) and olfactory (smell) receptors of bees are located on the antennal segments. These receptors are in the form of seta or hair tactoreceptors and plates and recessed chemoreceptors. These sensory organs help guide bees both inside and outside the hive and enable them to differentiate between hive, floral, and pheromone odors. If the antennae are cut off, the bee will not be able to negotiate within and outside the colony, will be unable to make comb, and will soon die. Once odors or other tactile stimulation are detected, signals are transmitted down the nerve cord to the brain. There are about 3000 plate organs on each antenna of the queen, 3600 to 6000 on each of the worker's, and 30,000 on the drone's.

Pollen/Propolis-Collecting Structures

The third pair of legs of a worker bee are highly specialized for collecting and transporting large amounts of pollen or propolis. Pollen is the male sex cell of flowers and the source of protein, fats, and minerals for bees. Propolis is a sticky substance created when resin exuded by the buds of trees and shrubs is mixed with a bee's wax and saliva. Propolis contains antimicrobial properties. Bees use this propolis to plug extra openings in a beehive, and they spread it inside both the cavity walls and the entrance of their dwelling. Its sticky quality makes it difficult for beekeepers to separate different parts of the hive's furniture without a hive tool when undertaking a hive inspection.

Bees actively collect pollen by scraping it off the anthers of flowers with their mandibles and with their first two pairs of legs. As the pollen is removed from the flowers, the bee adds a small amount of honey/nectar regurgitated from its honey stomach to the pollen, which accomplishes two things: the honey/nectar makes the pollen sticky, thus making it more easily adhere to the bee's body part, and it adds beneficial bacteria to the pollen. Furthermore, during the collection process, additional pollen grains adhere to the bee's body by the action of static electricity. As previously stated, the collection of resins, when mixed with bee saliva and wax, becomes the substance we refer to as propolis; it is transferred from the middle pair of legs of the bee directly to the corbiculae. Upon returning to the hive, this material is so sticky that the bee carrying the propolis cannot unload it at the colony without the assistance of other bees.

In order to harvest copious amounts of pollen from the anthers of flowers, bees evolved special anatomical structures, particularly on their third pair of legs. A bee begins gathering pollen by scraping it off the anthers of flowers with its forelegs; the pollen grains become sticky when mixed with honey. Additional pollen falls on the body hairs of the bee. The forelegs remove the pollen grains from the bee's head and from parts of the thorax.

The bee then takes to the air, and while hovering, the pollen is transferred from its forelegs to its middle legs. These middle legs then move the pollen to the combs (rows of hairs that resemble the teeth of human hair combs, not to be confused with wax comb) located on the basitarsae of the bee's third pair of legs. The average pollen load carried by a bee ranges from 0.000226 ounces (6.4 mg) to as high as 0.000263 ounces (7.44 mg).

The anatomical parts of the third set of legs begin with the coxa, which is attached to the thorax. Moving distally from the coxa, the remaining parts are the trochanter, the femur, the tibia, and the tarsus. Our interest here is to concentrate on the specialized anatomical structures located on the tibia and that section of the tarsus referred to as the *basitarsus*, which receives the pollen grains from the first two pairs of legs.

On the medial section, or inner surface, of each basitarsus are those rows of hairs that resemble the teeth on a human hair comb. Bees use their second pair of legs to transfer pollen onto these combs/hairs. On the medial and distal part of each tibia is a row of stiff hairs, referred to as *rakes*. When the combs on the basitarsus are loaded with grains of pollen, the rake apparatus of the left leg scrapes the pollen grains off the combs of the right leg. As the right leg moves upward encountering the rake apparatus of the left leg, the pollen grains raked off these combs are deposited onto a structure on the left leg called the press. Subsequently, a similar action takes place with the right leg.

The press consists of two somewhat flat surfaces that are located at the distal part of the tibia and the proximate part of the basitarsus. Once the rake has deposited the pollen grains scraped off the combs into the gap between these two surfaces, the gap closes and the pollen is squeezed onto a slightly concaved area on the lateral surface of the distal part of the tibia called the pollen basket, technically, the corbicula (see Appendix A). These actions repeat themselves until the pollen grains accumulate into a relatively large, ball-shaped mass on each corbicula. As noted, a similar activity is taking place on the other leg, and when the pollen balls on each corbicula reach the maximum weight the bee can carry, she returns to the colony with her bounty.

When the bee carrying a load of pollen returns to the colony, it backs into a cell and with its first two pairs of legs dislodges the pollen pellets into the cell. The same bee or other bees enter the cell head first and press and flatten the pellets by pushing against them with their mandibles. In general, bees fill only the lower third to half of a cell with packed pollen. As the pollen was being collected and while bees are packing it, they continue to add beneficial microbes and enzymes to the pollen. Once the packed pollen has filled a third to one half of the cell, the bees apply a thin film of honey, covering the exposed surface of the pollen; this action aids in preservation of the pollen. Through the action of fermentation, the packed pollen is worked on first by bacteria, then by fungi, yeasts, and molds and is eventually converted into the nutritious product referred to as *bee bread*. Beneficial yeasts and molds in the colony can be compromised when the collected pollen has been gathered from fields that were sprayed with fungicides. In such cases, the colony can be compromised, because the organisms needed for the conversion of the raw pollen to bee bread have been killed by the fungicide.

Occasionally, the bee bread may contain toxins such as pesticides; in such cases, the bees will seal the bee bread with propolis. Now the bee bread is referred to as *entombed bee bread*.

Often mice or shrews will enter a bee colony in the fall; if one of these creatures should die in the colony, because of its weight, the pall bearer bees will be unable to remove it from the colony, so instead they will remove the hairs from the dead creature and coat it in propolis, presumably to seal it from bacterial invasion. Propolis has other uses as well. For example, it is collected and sold by beekeepers as another hive product as it is used frequently in salves and hand lotions. Not all propolis is the same; some types have higher antimicrobial properties than others, depending on the source from which they were derived. Keep current by reading updated articles.

Wings

Honey bees have two pairs of wings, which fold together over their abdomen when the bee is not flying. In preparation for flight, and during flight, the wings are extended horizontally. As the wings are extended, the forewings are drawn over the front part of the hindwing, and the hooks of the hindwings, called the *hamuli* (singular *hamulus*), catch the marginal folds of the front wings, connecting the two pairs of wings together. The wings are strengthened by tubular thickenings called *veins*. Honey bee wings have a distinct pattern by having only four major tubular veins; most other insects, such as dragonflies, have many more. Wing vein patterns in Hymenoptera are unique and are used to help identify this group of insects.

Honey Stomach, Sac, or Crop

The esophagus of the bee begins at the back of the mouth and continues through the thorax, terminating in the anterior part of the bee's abdomen, where it expands to form the honey stomach, or sac or crop. Worker bees collecting nectar, honeydew, or water temporarily store these substances in their honey stomachs. The walls of the honey stomach are pleated and invaginated, allowing it to expand, maximizing the amount of these liquids they can gather on any single forging trip.

We now know that beneficial bacteria (e.g., *Lactobacillus* spp.) and enzymes reside in the honey stomach. When bees return to the hive and pass on the contents of their honey stomachs to receiving bees, these liquids include enzymes and bacteria. Pollen-collecting bees add honey/nectar to the pollen as it is being gathered. As a result, the incoming pollen loads include beneficial microbes as well. These are important in the conversion of nectar and pollen into honey and bee bread, respectively, and may even provide the bees some protection against pathogens.

A muscular valve at the posterior end of the crop called the *proventriculus* prevents the contents of the honey stomach from continuing forward into the bee's intestine. The bee can control the opening and closing of this valve via its nervous system; it can later open the valve to pass on food from the honey stomach to the intestines.

When the foragers return to the colony with the liquid contents of their honey stomachs, particularly nectar or honeydew, they transfer these liquids to the processing/receiving bees. The receiving bees then add additional enzymes to these liquids. This transfer from the foragers to the processing bees allows the foragers to return quickly to their foraging duties. The processing bees suspend these liquids from their proboscises. This activity, along with the air circulating in the hive created by the fanning of the bees' wings, accelerates the evaporation of water from the nectar or the honeydew; this results in concentrating the sugar content of these liquids, giving us the product we call honey or honeydew honey. At some point in this process, the more highly concentrated sugary solution is deposited by the processing bees into cells, where further evaporation takes place. Ideally, when the water content of the nectar or honeydew has been reduced to no more than 18 percent, the cells are filled to the brim and then sealed with a wax capping. (This process can be viewed as being somewhat similar to industrial or down-home canning of a large variety of products to preserve them for future use.) One can differentiate between the cappings covering larvae/pupae and those covering honey or honeydew: cappings covering larvae/pupae are convex and those covering honey or honeydew are slightly concave.

The Sting

Stinging insects, or *aculates*, belong to the order Hymenoptera, which includes both social and soli-tary bees and wasps. The more defensive species of stinging insects are the hornets and the yellow jackets (both of the Vespidae family); less volatile are the bumble bees (Bombidae) and the honey bees (Apidae). The venoms of each of these stinging insects are chemically distinct. Thus, a beekeeper who is allergic to yellow jacket venom will not necessarily develop an allergy to honey bee venom or the venom of other stinging insects.

The sting complex found in many members of the order Hymenoptera evolved from an ovipositor, an anatomical structure used by some female insects to deposit their eggs. Given that females are the egg layers, it is clear that the sting apparatus would be part of the anatomy of females. This is why the drones—the male bees—lack the sting complex. The exception is when a drone is part female and part male; in such rare cases this drone, referred to as a gynandromorph, would be capable of stinging. Queens generally use their stings only to dispatch rival queens. The entire stinging apparatus consists of a poison sac (sometimes called the *acid gland*), an alkali (or Dufour) gland associated alarm substances, and the mechanical equipment (muscles and hardened plates) of the sting (see Appendix C for information on sting reaction and anatomy).

The recurve barbs of the sting's lancets, which resemble the anterior ends of a series of fishhooks, prevent the bee from withdrawing or retracting its sting. Once the lancets have pierced the beekeeper's skin, as the stinging bee attempts to fly off, its entire sting complex, including the venom sac, is torn from the bee's abdomen, and the muscles attached to the lancets continue to work their way deeper into the beekeeper's flesh. Muscle pumps near the base of the now-detached sac force more venom into the wound for about a minute. Alarm odors are released at the sting site, inducing other workers to sting near the same site. To minimize the amount of venom received, it is important to remove the sting **promptly** by scraping or flicking it off with your fingernail. Since the sting site is now "tagged" with alarm odors, apply smoke to the site to mask these odors. Bees usually die shortly after stinging but occasionally may live for hours or even days. Africanized bees have the same venom as European bees but are more volatile and respond quickly to the release of the alarm odors. (For more information on the compounds in bee venom, see "Bee Venom" in Chapter 12.) **Caution:** Before applying smoke to the site, be sure your smoker is not discharging hot flames.

LIFE STAGES OF A BEE

There are three types of bees in a colony, divided into two female castes (workers and queens) and the males, or drones (see Appendix A for information on the morphology of the different bees). Under normal circumstances, the queen lays all the eggs in a hive. If the queen is lost and the bees are unable to rear a new one, workers will sometimes lay unfertilized eggs. Unfertilized eggs produce drones; fertilized eggs give rise to female bees. A fertilized egg is formed by the union of sperm and egg; each contains the haploid (16) number of chromosomes, and when united, the diploid number of chromosomes (32) is restored. Diet determines whether a female larva will become a worker or a queen. Since drones emerge from unfertilized eggs, they have only a haploid number of chromosomes. The egg (whether fertilized or not) contains the genetic material and a large yolk reservoir, which provides food for the developing embryo. The nucleus and the component parts of the cell that include the cell membrane, cytoplasm, and organelles of the fertilized or unfertilized egg, now referred to as the zygote, will undergo cell division, a process known as mitosis, and divide into 2 cells, then 4, 8, 16, and so on. As the cells continue to divide, they will disperse throughout the egg's yolk. As the cells continue to multiply by mitotic division, the yolk part of the egg provides the essential nutrients to fuel this multiplication.

These rapidly multiplying undifferentiated cells at first are scattered throughout the egg's yolk. Eventually, most of these cells migrate to form a layer of cells along the margins of the egg's vitelline membrane, forming the *blastoderm*. These cells then form the so-called *germ band*; this band then differentiates into the three distinct kinds of germ cells: the ectoderm, the mesoderm, and the endoderm. These three categories of cells will give rise to the tissues and organs that will eventually produce the adult honey bee.

Many insects, including honey bees, have four distinct stages of development, namely, the egg, larva, pupa, and adult. This passage from egg to adult is referred to as *complete metamorphosis*. Other well-known insects that pass through four stages of development are butterflies, moths, and beetles. Insects such as grasshoppers skip the pupa stage; this type of insect development is referred to as *incomplete metamorphosis*. The eggs of honey bees are incubated in the nursery region, called the *broodnest* area, of the comb at 91.4 to 96.8°F (33°–36°C). Honey bee larvae emerge three days after the eggs are laid as the vitelline and chorion membranes encapsulating the egg break down, releasing the larvae.

The life cycle of honey bees includes the need to maintain large colonies to help ensure winter survival and have a sufficient number of bees to collect and store their winter food. It also requires cadres of bees to carry out many other important colony tasks as well as a highly prolific queen capable during the spring and at least part of the summer of laying approximately 2000 eggs per day.

During all four of the developmental stages, ranging from eggs to adults, some of the developing larvae, pupae, and adults may die or become diseased. Under such circumstances, healthy adult bees will usually remove and discard them. This trait of removing dead or diseased members of the colony is crucial hygienic behavior, and bee colonies that have such traits play an essential role in colony health.

During periods of dearth where there are no available external food sources and very little if any reserves of food within the colony, bees will eat existing eggs, larvae, and pupae. Note well that, although bees may have evolved from a carnivorous sphecid wasp ancestor, and, during this evolution, turned to a vegetarian diet consisting of honey to fulfill their carbohydrate requirement and bee bread to fulfill their protein, fat, and mineral requirements, under certain circumstances, bees will revert to their carnivorous ancestral roots and devour the colony's larvae and pupae.

Once the larvae have emerged (pearly white in color), these wormlike grubs become eating machines, with a huge digestive system consisting of a mouth, salivary glands, midgut, hindgut, and a closed excretory system. In addition, the larvae have spiracles (breathing apparatus) and silk glands for spinning cocoons when they are about to enter the pupal stage.

Larvae lying in the bottoms of their cells look like white, C-shaped worms. Each larva is fed between 150 and 800 times per day and will gain around 900 times the egg weight by the fifth day (see Table 1-1 on development time; Appendix E on weight of bees). About 33 percent of the larval weight (dry

Table 1-1 Average Development Time of a European Honey Bee

	Egg[a]	Larva	Pupa	Total	Adult Life Span	Weight[b]
Queen	Fertilized 3 days	4.6 days	7.5 days	15 to 17 days	2-5 years	178-292 mg
Worker	Fertilized 3 days	6.0 days	12.0 days	19 to 22 days	15-38 days summer	81-151 mg
					140-320 d winter[c]	
Drone	Unfertilized 3 days	6.3 days	14.5 days	24 to 25 days	8 weeks	196-225 mg

Sources: M.L. Winston. 1987. Biology of the honey bee. Cambridge, MA: Harvard University Press. E. Crane. 1990. Bees and beekeeping: Science, practice, and world resources. Ithaca, NY: Comstock.

Note: Average time between metamorphic stages, in days at 93°F (33.9°C). Conversion: 1 mg = 0.000035 oz.; 1 mm = 0.004 in.

[a] Egg dimensions: Worker and queen eggs weigh 0.12–0.22 mg, are 1.3–1.8 mm long, and take 48–144 hours to hatch, with an average of 72 hours.

[b] Weight at emergence. Weights of emerging adults vary depending on cell size, number of nurse bees, colony population, food availability and type, and season of the year.

[c] Workers in winter have well-developed hypopharyngeal glands and more fat bodies, which may enable them to live longer.

weight) is made up of *fat bodies*, tissues that are utilized in the pupal stage. The membrane (skin) surrounding a rapidly growing larva does not expand as the larva grows larger in size. As a consequence, the larva sheds its skin, or *molts*, six times during its development; these molts are often referred to as *instar stages*. The first four instars take place during the first four days of larval life, the fifth takes place in the pre-pupa stage, and the final one occurs just before it emerges as an adult bee. The technical term for organisms that shed their exoskeletons as they increase in size is *ecdysis*.

Two different diets are fed to larvae destined to female worker bees. For the first two or so days of their larval life, these larvae are lavishly fed glandular secretions from both the mandibular and hypopharyngeal glands of young nurse bees. This early diet is referred to as *royal jelly*. Both of these glands are located in the head region of worker bees. The secretions from the hypopharyngeal glands are clear in color and contain primarily protein molecules, whereas the secretions from the mandibular glands are white in color and contain primarily lipid compounds.

On the third day of larval life, larvae selected to become workers no longer receive secretions from the mandibular glands of nurse bees. Instead, they are now fed on secretions from the hypopharyngeal glands as well as honey and bee bread, when it is required rather than lavishly. This less nutritious diet is referred to as *worker jelly*. This diet change (which contains fewer proteins, lipids, minerals, vitamins, and sugars) and the switch from mass feeding to progressive feeding appears to be the salient factor for their differentiation into female worker bees as opposed to becoming female queen bees. Thus, all female bees at "birth" are potentially either workers or queens, but their diet determines their destiny. It is presumed the larvae destined to become drones receive a diet similar to the workers.

On the other hand, female larvae destined to become queens are fed *royal jelly* (see below) throughout their larval life, receiving equal amounts of secretions from the hypopharyngeal and mandibular glands. Not only are they provided these glandular secretions from the nurse bees, but they receive them in abundance. In fact they are fed so lavishly that even after they enter the pupal state, there often remains a plug of their food at the bottom of their cells.

At the end of day 5 (see illustration on Developmental Stages of Honey Bees), the cell, which until this time is called *uncapped* or *open brood*, is capped with a waxlike cover and is now called *capped* or *sealed brood*. This wax covering consists of old wax, propolis, and other components. The larva molts into a prepupa, defecates, and spins a cocoon with silk produced from the thoracic salivary glands. Normal hive temperatures of about 95°F (35°C) are necessary for normal development; if colder, the development time can be delayed by several days.

Developmental Stages of Honey Bees

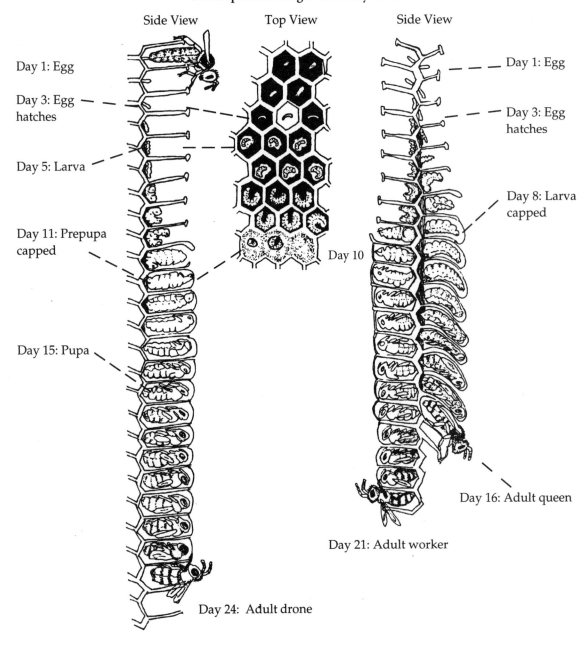

Side View

Top View

Side View

Day 1: Egg

Day 3: Egg hatches

Day 5: Larva

Day 11: Prepupa capped

Day 15: Pupa

Day 10

Day 24: Adult drone

Day 1: Egg

Day 3: Egg hatches

Day 8: Larva capped

Day 16: Adult queen

Day 21: Adult worker

The next stage is called the pupal stage, where massive internal and external morphological changes take place. Recognizable parts of the adult bee form—the legs, wings, and abdomen—and all the internal organs and muscles develop. Most internal changes happen in eight to nine days. The pupae are full of *fat bodies*, carried over from the larval stage. Fat bodies, cell-like organelles, or tissues, serve as food storage sites for lipids, proteins, and glycogen, a stored form of sugar, and have an abundance of mitochondria. These organelles are where the high-

energy molecules adenosine triphosphates (ATP) are produced. For example, when ATP breaks down to adenosine diphosphates (ADT), the energy released causes a bee's flight muscles to contract.

THE WORKER

The most numerous members of a bee colony are the workers, the sterile female caste incapable of laying fertile eggs; in a colony headed by a well-mated and healthy queen the worker populations can reach numbers of 40,000 or more by mid or late summer. The workers are smaller than the drones and have shorter abdomens than the queen.

In the late pupal stages of a worker bee's development, the exoskeleton, or cuticle, she will possess as an adult will gradually darken and begin to exhibit the colors of an adult honey bee. After a final molt, or sixth instar, she is ready to emerge from her cell. However, before she can emerge, she has to remove the wax capping, or cover, just above her head. She accomplishes this task by cutting through the cap with her mandibles and then relocating the wax pieces near the edges of her cell where other bees collect and reuse them to cap additional cells containing uncapped larvae.

Once the cell is uncapped, she remains in her cell for several more hours; during this period, nurse bees feed her honey mixed with beneficial microbes and bee bread. Even though this young bee has the anatomy of an adult bee, her exoskeleton is not fully hardened. In fact, it remains relatively soft, and the hairs on her body are light in color and matted down. In this condition, the young bee is referred to as a *teneral*, or *callow*, bee. Usually a bee in this state is unable to sting because her exoskeleton has not fully hardened.

By consuming adequate amounts of bee bread, over the next four or five days after emerging from their cells, these young bees will develop healthy glands, particularly their mandibular and hypopharyngeal glands. They will also build up a reservoir of fat bodies. With this bee bread diet, they will not only ensure healthy glandular development but also live a full six weeks during the summer months, the average life span of a bee. On occasion, they will also beg for food from the nurse bees.

These young teneral bees will soon begin the first of many tasks that they will perform during their life span of approximately six weeks. Their glandular development, their genetics, and the needs of the colony will prioritize which activities they will perform as they progress from house bees to foragers (see Chapter 2). Generally, workers from one to three weeks of age remain within the hive, where under ideal circumstances they perform a series of tasks related to their age, often referred to as *polyethism*. However, it is important to note that although as they age they progress from one task to the next, in the absence of those who normally carry out a specific task,

it is known that bees can both revert back to previous tasks or progress to more complicated tasks.

The following tasks, which are listed in the order they are undertaken as a bee ages, have been observed and reported by many scientists. It begins with the newly emerged teneral bees:

- Clean cells of their debris, including fecal material, cocoons, and exoskeletons.
- Place wax cappings over older larvae.
- Feed and care for developing larvae.
- Tend to the queen, forming a retinue around her, feeding and grooming her, and transferring her pheromones to other bees.
- Accept nectar from incoming foragers, curing and storing it.
- Pack pollen.
- Construct comb.
- Cap cells containing older larvae and cells filled with honey.
- Remove dead bees.
- Ventilate and circulate the air in the hive.
- Guard the hive entrance and exit, and patrol areas throughout the hive for intruders.
- Take brief orientation flights (also referred to as *play flights*) to familiarize themselves with landmarks near their colony.
- After the third week, the bees become foragers and scouts. The foragers seek nectar, honeydew, water, pollen, and resins, and when a colony swarms, many of the foragers become scouts who search for new home sites for the swarm.

Once these worker bees have spent their first three weeks on primarily in-house duties, their glands that produced larval food and wax have atrophied. These bees then move away from the warm broodnest, where the eggs, larvae, and pupae are located, and relocate to combs essentially absent of brood. Here they come in contact with returning foragers and are eventually recruited to locate food sources or other products needed by the colony.

As foragers, they will collect honeydew, nectar, pollen, water, and resins. Foraging activities take a heavy toll on workers; most of them die after three weeks of outdoor duties, and some do even sooner. During the winter, however, many workers survive for several months. For a more complicated breakdown of worker activities, see Chapter 2.

Relative Cell Sizes

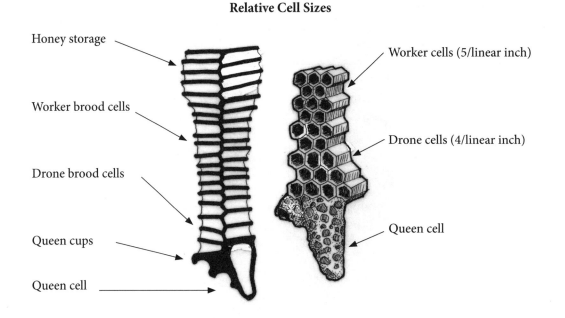

Honey storage

Worker brood cells

Drone brood cells

Queen cups

Queen cell

Worker cells (5/linear inch)

Drone cells (4/linear inch)

Queen cell

THE QUEEN

Bee colonies are usually *monogynous*—that is, they have only one egg producer, the queen. The measure of a queen's length from her head to the tip of her abdomen is approximately three-quarters of an inch (2 cm), twice the length of a worker bee, and her blimp-shaped abdomen—usually without color bands—distinguishes her from both workers and drones. In addition, her wings cover only two-thirds of the dorsal section of her long abdomen, whereas the wings of both drones and worker bees cover nearly the entire dorsal section of their abdomens. (The most common mistake by beginning beekeepers when trying to locate a queen is mistaking a drone for the queen because the drone stands out, being larger than worker bees.) If a queen has been removed or accidentally killed by a beekeeper, has been injured, or is elderly, worker bees can rear a new queen from larvae up to three days of age, but note that very young larvae less than a day old and those fed lavish amounts of royal jelly make the best queens. Bees also rear queens when the bees are preparing to swarm, a phenomenon known as *colony reproduction*. In this instance, the old

Worker, Queen, and Drone Bees

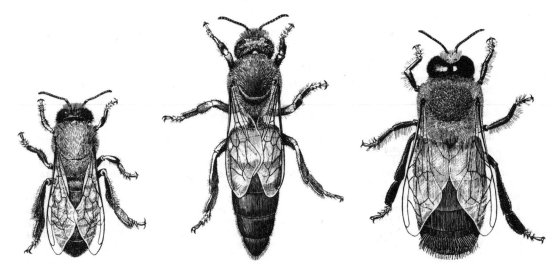

Source: E.C. Martin et al. 1980. Beekeeping in the United States. USDA Ag. Handbook 335.

queen departs with the swarm; prior to their departure, the bees have reared a new queen to leave behind.

The ability to find the queen is important if her presence in the colony needs to be confirmed or if you wish to replace her. *Requeening*, or replacing an old queen with a new one, is successfully accomplished when the existing queen is located and removed from the colony. New beekeepers need to gain the facility to find the queen among the workers and drones; once you have accomplished this, you are truly a beekeeper.

Any larva hatching from a fertilized egg is a female bee. This fact simplifies the raising of queens for commercial purposes and gives worker bees a wide latitude in selecting larvae to become new queens (see Chapter 10).

The pathway the female larvae follow is directly connected to the food they receive during their larval life. Worker bees often initiate queen rearing by constructing special cup-shaped cells. These *queen cups* are usually located on the lower edges of combs; after the queen has deposited an egg in them, they are then called *queen cells*. The presence of cups or larvae in queen cells does not necessarily lead to the production of queens.

When queen cells are needed, worker bees clean the cups, or, in the case where this preparation hasn't been made, worker cells can be modified to become queen cells. Larvae in these cells are mass-fed royal jelly—a thick, white, creamy material created from the combination of secretions from both the mandibular and hypopharyngeal glands of nurse bees—during their entire larval development. The larvae are fed more of this royal jelly than they are able to consume; this is clearly evident in that a plug of this food remains at the bottom of queen cells after they have emerged. This diet, supplemented with the juvenile hormone secreted by developing queen larvae, produces queens.

Whether cells containing queen larvae begin as cups or as worker cells, as the larvae grow, these cells are enlarged and elongated by worker bees, gradually taking on a peanut-shell–like appearance. The openings of drone and worker cells lie horizontally but are inclined slightly upward on the comb; cells that cradle the queens hang vertically (see illustration showing relative cell sizes).

Isn't it interesting that young worker bees play such an important role in a colony by selecting the next generation of queens? Remember, female larvae selected to be workers begin larval life on a diet similar to that of queen larvae, but after two days they are weaned from it and thereafter receive worker jelly, which consists primarily of proteins mixed with honey and bee bread.

Queen Cell Production by Bees

Three conditions trigger queen rearing by honey bees: (1) The colony is making preparations for swarming, (2) the queen's physiological and behavioral activities are substandard, or (3) the queen is lost or dies. In each case, the purpose is to replace the existing or resident queen in the colony.

Reproductive swarming is a process whereby honey bee colonies duplicate themselves. The existing queen will depart the colony with about half the workers and a few hundred drones, to form a new colony. Colonies preparing to swarm will begin this process by constructing a great number (from 10 to 40) of queen cups on the lower edge of combs. Newly constructed cups are light yellow in color and hang vertically from the lower edges of the honeycomb. These cups become queen cells once the queen deposits eggs in them and larvae begin to grow. If the cells are found during the swarming season—a period when colonies are casting swarms—they are called *swarm cells*.

The second condition that leads to queen cell construction occurs when bees prepare to replace a queen that is substandard; this type of replacement is called *supersedure*. Workers begin this process by either constructing queen cups or by modifying existing worker cells containing young larvae. In this case, these supersedure cells are few in number and can be found in the brood area. Because the queen cells are usually constructed of old wax, these cells will be brown to dark brown in color.

This replacement is triggered when the queen's physiological or behavioral activities, or both, decline —for example, her egg production is declining or her pheromone levels are reduced (usually an aging queen) or she is injured. Worker bees are able to recognize these conditions and will rear queens to replace the resident one. After the new queen emerges, mates, and begins to lay eggs, she may coexist with her ailing mother. In time, however, only the replacement queen will be found.

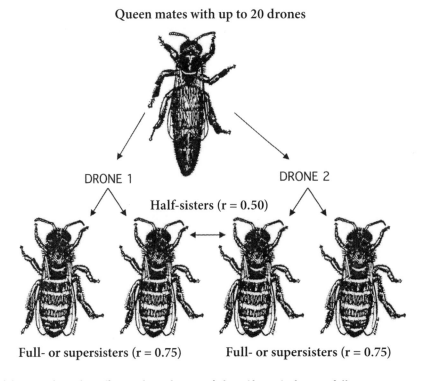

Queen mates with up to 20 drones

DRONE 1   DRONE 2

Half-sisters (r = 0.50)

Full- or supersisters (r = 0.75)   Full- or supersisters (r = 0.75)

If the queen's worker offspring have the same fathers (drones), they are full- or supersisters;
if they have different fathers, they are half-sisters.

The last condition of queen replacement, or *emergency*, occurs when the queen is absent from the colony. This can be from natural causes, beekeeper clumsiness, disease, or predation. Occasionally, the queen will fall off the comb during hive inspection and is unable to return to the hive or is crushed between two frames as they are being removed or replaced (a process called *rolling the queen*). In such cases, unless by good fortune there are queen cells already present, it will be necessary for bees to convert worker cells containing eggs or larvae into queen cells. Worker bees will "select" young larvae in worker cells and modify them into queen cells; emergency cells are found on the face of the comb, but in small numbers.

Virgin Queens

While still in the queen cells, virgin queens will often *pipe* or *quack* to one another. After emerging, a virgin queen may toot, or call to other virgins. She then begins to search for and partially destroy any other queen cells, leaving the workers to discard the pupae or larvae inside. Some cells may contain queens ready to emerge, in which case the emerged queen will par-tially open these cells and sting the occupants. While performing these tasks, she may also encounter other emerged queens; fighting ensues, and ultimately only one virgin queen survives.

About six days after emerging, the queen will leave the hive on a mating flight; if weather is inclement, this flight will be delayed until more favorable weather appears. During her flight, the queen's pheromones attract male bees from *drone-congregating areas* (DCAs). Research by Dr. David Tarpy at North Carolina State University indicates she may mate with up to 14 or more drones in succession during a single mating flight or over a series of flights taken over a period of days. When her sperm sac (*spermatheca*) is filled, she will never leave the colony again, unless accompanying a swarm. Three days or so after mating, the now bigger and heavier queen will begin to lay eggs. The queen continues to lay eggs the rest of her life, pausing for a month or so in late fall. The general consensus is that a healthy young queen is capable of laying close to 2000 eggs per day during the spring and early summer, assuming there are sufficient numbers of vacant cells and worker bees to care for the larvae that will follow. Over her lifetime, a good queen will lay 200,000 eggs per year.

Genetic Traits

Because the queen mates in the open, the beekeeper has limited control over which drones will inseminate her. Those few that do mate with her may be from several apiaries or from "wild" or *feral*, colonies. Now because of the presence of parasitic mites and other predators, most feral colonies have been killed off, drastically reducing the number of wild drones.

As a consequence of this random mating pattern, the queen's spermatheca may contain semen from genetically different drones. Her worker and queen progeny, therefore, will consist of individuals that are not necessarily genetically alike (that is, they will be half sisters, full sisters, or supersisters; see illustration). The drones, hatching from unfertilized eggs (called *parthenogenesis*), are all full brothers, because the queen will lay genetically similar drone eggs whether she has been inseminated or not. Only when the queen has been instrumentally inseminated with semen from recorded drone stock (or from a single drone) will a colony's workers be of known origin. Poorly mated queens, either because of low sperm, or weak drones from mite parasitism or other problems, will not produce strong colonies. Researchers have discovered that multiply mated queens, in comparison to queens that have mated with only a few drones, produce colonies that survive the winter in better condition, are more resistant to disease, and swarm less.

Because the queen is the sole egg producer, she is responsible for the genetic traits of a colony. If a colony has undesirable traits, requeening should change the hive's genetic makeup and therefore its character.

Queens should be of superior stock to optimize desirable (or minimize undesirable) characteristics, such as:

- Temperament (how calm the bees are on the comb).
- Handling ease (how easily aroused to sting).
- Industry (how early in the morning bees are out foraging).
- Production (how much honey is collected).
- Propolizing tendency (excessive use of propolis).
- Burr-comb building (building excessive comb between frames).
- Pollen hoarding (some strains are excessive pollen collectors).
- Plant preferences (mixed or single-source pollen loads).
- Tongue length and nectar-carrying capacity (can be measured).
- Honey hoarding (some strains store more honey).
- Whiteness of honey cappings (compared to "wet" cappings).
- Conservation of stores (bees manage their stores well).
- Total hive population (large or moderate population).
- Brood pattern (compact or scattered).
- Swarming tendency (colony's tendency to cast frequent swarms even when their queen is less than two years of age).
- Winter survival (colony survives the winter and emerges in the spring with sufficient numbers to develop into a robust colony).
- Hygienic behavior (removing dead or diseased brood).
- Disease resistance (few or no diseases).
- Mite tolerance (overall fewer mites).

THE DRONE

Because drones are larger and plumper than worker bees, beginning beekeepers often mistake them for queens. They can be distinguished from queens in several ways. The abdomen of a queen tapers to a point, whereas the posterior section of the drone's abdomen is blunt or rounded; the entire abdomen of a drone is almost barrel-like in shape, making him chunky in appearance. The number of drones per colony may be in the hundreds to thousands, usually around 15 percent of the colony's total population. However, we wish to note that prior to the onset of winter, most if not all the drones are evicted from their colony, and during periods of dearth even drone larvae and pupae, as well as adults, are removed from their colony. However, at times there may be an excess of drones because the queen has depleted all the sperm in her spematheca and therefore is capable of laying only unfertilized eggs. Another factor that can influence the number of drones in a colony is that drones returning from drone-congregating areas tend to drift into other colonies.

Drone larvae hatch from unfertilized eggs, which under normal conditions are laid by a mated queen in hexagonal wax cells similar to, but larger than, worker cells (see diagram of relative cell sizes, p. 20). On the fourth day of their larval life, drones are fed a diet of modified worker jelly, which contains large quantities of bee bread and honey.

After six and a half days of feeding, the cells of drone larvae are capped with wax. The capped drone cells are dome shaped, like a bullet's head, and are readily distinguished from the slightly convex shape of the capped worker cells. Remember, capped hexagonal cells that are pitched some seventeen degrees upward from a horizontal plane are either drone or worker cells. Cells that are shaped similar in appearance to the shell of a peanut and are suspended vertically are queen cells.

Newly emerged adult drones are fed by workers for two to three days and then will beg, from nurse bees, food that contains a mixture of pollen, honey, and brood food. Older drones feed themselves from the honey stores. Adult drones have no sting (remember, the sting is a modified female egg-laying structure) and have very short tongues that are unsuitable for gathering nectar (see Appendix A). Drones never collect food, secrete wax, or feed the young. Their sole known function is to mate with virgin or newly mated queens; think of them as flying gametes (sperm cells).

Drones first leave the hive, about six days after emerging, on warm, windless, and sunny afternoons. As they get older, the drones congregate and hover in designated areas in anticipation that a queen will fly into this area for reproductive purposes. A queen may mate on average with 14 or more drones during her nuptial flights. Those drones that successfully copulate with a given queen die soon after they have transferred their sperm to her.

Whenever there is a dearth of nectar (when no food is being collected), worker bees expel drone brood and adult drones from the colony. During the summer, you can see workers dragging drones in various stages of metamorphosis out of their cells and dropping them in front of the hive. Normally in the fall, all adult drones and any remaining drone brood are gradually evicted from the hive. The evicted drones die of starvation and/or exposure to the elements. Queenless hives and those with laying workers or drone-laying or failing queens usually retain drones longer.

Drone Layers

An unmated queen can lay only unfertilized eggs. A failing queen is one that has mated but is no longer capable of normal egg-laying activity, laying all or nearly all unfertilized eggs. Failing queens may result from sperm deficiency, physiological impairment, disease, mite infestation, or old age. Some workers of hopelessly queenless colonies (those unable to rear another queen) undergo ovary development and start to lay eggs. These eggs are, of course, all unfertilized. Unfertilized eggs that are laid by healthy, mated, unmated, or failing queens, or by *laying workers*, will produce mature drones, capable of mating.

Unlike a mated queen that lays unfertilized eggs in drone cells, a failing or unmated queen will often deposit such eggs in worker cells. Laying workers usually place their unfertilized eggs in worker cells as well, but though these unfertilized eggs are laid in worker cells, they will hatch into drone larvae, and as they near the pupal stage, the cappings will have the characteristic dome shape found on regular drone cells. The presence of scattered worker cells with drone cappings is usually a clear indication that the colony is queenless and has a cadre of laying workers carrying out the functions of a queen. Attempts to requeen a queenless colony where laying workers have taken over are usually futile.

On further inspection, you may find that each uncapped cell within a scattered brood pattern contains not one, but several eggs. These eggs, instead of being deposited at the bottom of the cell as is characteristic of eggs laid by queens, adhere to the cell walls. This is a result of the worker's abdomen not being long enough to reach the cell bottom. If you find these patterns in your hive, read about what to do in "Laying Workers" in Chapter 11.

The presence of clusters of occupied drone cells in the spring, summer, and early fall in a *queenright* colony (where a healthy, mated queen is present) is a normal part of the colony cycle. Because drones attract varroa mites, many beekeepers use this fact to trap the mites. They add at least one frame of drone-size comb in each hive body to attract female varroa mites to lay their eggs. Once the drone cells are capped, the frame is frozen to kill the mites (see "Varroa Mite" in Chapter 14).

As previously noted, it takes drones twenty-four days to complete their metamorphosis and only twenty-one days for worker bees. Female varroa mites are aware of the difference between these two maturation periods. As a consequence, if given a choice, a female mite known as a *foundress* will select a cell containing a developing drone in place of a cell containing a developing worker in which to deposit her eggs because more of her offspring will have sufficient time to reach maturity before the drone departs from its cell. On average, approximately 2.8 mites plus their mother will exit a drone cell. In contrast, only 1.8 mature offspring plus their mother will exit a worker cell.

Given this biology, researchers and beekeepers deliberately place deep frames of comb imprinted with only drone cells or combs with fully developed drone cells into a colony. Counting both sides of a comb, the total number of drone cells equals 4,272 cells. Hypothetically, if each cell has a female mite and all the cells produce 2.8 mature young female mites, the number of new mites in the colony would be 11,961.6.

Adding that to the 4,272 foundress mites, it would yield a total of 16,233.6 mites. WOW!

Now, should the beekeeper using drone comb to control the mite population fail to pull this frame out of the colony before the drones and mites emerge, the beekeeper would have unwittingly placed the colony in grave danger. On the other hand, if the beekeeper pulls the frame after the drone cells are all capped, it would eliminate 4,272 foundress mites. Using drone frames can help control the mite population by removing the frame in a timely manner.

The use of drone comb as a technique to get rid of varroa mites is questionable; there is always some payback with this approach. It forces the queen bee to shift from primarily laying worker eggs to laying an extraordinary number of drone eggs—that's 4,272 unfertilized eggs—and the worker bees have to feed and care for all 4,272 drone larvae and cap their cells with wax. This requires an expenditure of great effort by the queen and takes the colony out of its more natural reproductive rhythm, shifting workers to caring for these drone eggs, larvae, and pupae.

 **Notes**

CHAPTER 2

Colony Activities

COLONY LIFE

The worker bees carry out the broadest range of chores necessary to maintain and promote the colony's well-being. Drones and queens have a far more restricted range of duties; however, their duties are equally key to honey bee survival.

You should now be able to distinguish between the two female castes, the workers and queen. *Caste* in social insects is applied to individuals of the same sex that differ both in behavior and in morphology, which deals with form and structure. Drones, the male bees, are not members of a caste, since all drones exhibit the same morphology and behavior. Now we will concentrate on the worker caste, which is responsible for doing many of the tasks necessary to maintain the colony unit.

A colony of individuals that do not themselves reproduce but work as a unit to undertake important duties to maintain the colony and protect and rear reproductive members is called a *superorganism*. Such is an ant, a termite, and a honey bee colony, where the individual workers have particular but complex duties to protect and maintain the reproductive queen and drones.

Division of Labor

The activities of worker bees can be divided into two major categories—those that take place primarily inside the colony, and those that take place primarily outside. There is some overlap, but in general terms, inside bees are younger, and outside bees are older. Furthermore, it is important to note that during the foraging season, the array of chores by worker bees follows a somewhat dictated age progression. This means that when a day-old bee first emerges from her brood cell, the work that she does throughout her lifetime is roughly related to her age.

During the summer season, the array of chores by worker bees follows a progression generally dictated by the bees' age. But age is only one factor: hormones, genetics, and pupal temperature also play a role, in addition to the time of year and colony conditions. Following the summer solstice, in northern latitudes, as the amount of daylight begins to decrease, the queen begins, slowly at first, to lay fewer eggs per day; by September and October, her egg laying declines more markedly. One of the consequences of the decrease in the number of eggs laid per day is that newly emerged bees that would shortly assume the role of nurse bees have fewer and fewer larvae to feed; this results in a buildup of the glucolipoprotein vitellogenin in the fat bodies located in these bees' hypopharyngeal glands and in their abdomens. These nurse bees, referred to as "winter bees," live through the winter months, whereas the "summer bees" have a much shorter lifespan: usually only six weeks. By having extended longevity, many of these winter bees are the ones responsible for feeding the larvae that will begin to emerge in small numbers in mid-January. The survival of colonies during the winter months is to a great extent dependent on having a large cadre of winter bees that have not been compromised by varroa mite–vectored infections. This is why it is essential to have low mite counts in your colonies in the late summer and fall.

In northern latitudes, this progression in age-related duties is often interrupted or curtailed during the period when the queen's egg laying declines in the fall, which is followed by a period of complete or near complete absence of egg laying. The absence of eggs means that bees within the colony become progressively older, and therefore the duties related to age are no longer operative. So by the time foraging begins in the spring, most of the bees are older than six weeks. And once spring is under way, the initial succession of age-related duties falls upon bees that have survived the winter. The ability of these older bees to switch jobs is a good example of the plasticity of workers.

A functional knowledge of bee biology and an understanding of the activities performed by the members of the colony will assist you in becoming a better manager of your bees. You should be able to recognize which activities, or the absence thereof, signify the colony's overall status. It is important also for you to recognize the different labors of worker bees and when (or if) they are being performed. For instance, if none of the foraging bees returning to the hive are carrying pollen loads, this could be an indication that there is no brood in the colony. Obviously the lack of brood may indicate that the queen is absent or she has been superseded and the new queen has not commenced egg laying.

INSIDE ACTIVITIES

Upon emerging from its capped cell, a young worker begins to perform the first in a sequence of many tasks she will carry out during her short but highly productive six-week life span, the average life span of spring, summer, and early fall bees that are known collectively as summer bees. Over the next few days, the age of the bee, its glandular development, and the environmental conditions of the colony will rule bee activities. The organizational factors determining worker bee activities is called *age polyethism* (in which the same individual passes through different forms of specialization as it grows older). These tasks include cleaning cells, capping and caring for the brood, tending and feeding the queen, receiving nectar from foragers, making honey, removing the trash, packing pollen, building comb, hive ventilation, guard duty, thermoregulation, and finally orientation and foraging flights. When a colony swarms, foragers assume the role of scouts, which will play the vital roles of searching, discovering, and leading the swarm to a new homesite.

Researchers are able to record and document age polyethism in bee colonies by using a glass-walled observation hive containing frames of brood, adult workers, drones, and a laying queen. The researchers mark several hundred newly hatched bees on the dorsal side of their thorax with paint and then introduce them into the observation hive. Over the next six weeks, the activities of the painted bees are noted, and bees under normal conditions progress from one specific activity to another. Such studies continue to this day, and each new study adds more clarity to our knowledge of age polyethism.

The progression of work "assignments," and the overall decision making of which bee does which job, suggest there is some command structure within this superorganism. However, research has shown that in fact each worker appears to be making individual decisions, and stimulates or recruits others to follow along, in activities such as comb building, comb use, patrolling, or even swarming. Although most young workers progress from one specific task to another, there is a great deal of plasticity in the jobs and ages of bees. This enables the bee colony to thrive during difficult circumstances or in changeable environments. For example, if the majority of foragers have been eliminated by pesticide poisoning, younger bees who have not reached normal foraging age are able to step up and replace the deceased foragers. The reverse is also true, where older forager bees can once again feed larvae or secrete wax if a majority of the younger bees are eliminated. But age is only one factor in dictating the activity of workers; hormones and genetics also are factors.

Unlike other insects, honey bees that age have higher levels of a hormone called *juvenile hormone* (JH), which is normally higher in young insects. In honey bees, JH levels increase during the worker bee's life and could be one of the reasons that older bees learn better. Learning is needed in the outside world, to navigate, to find food, and to cope with the dangerous outside environment. In addition, the genetic makeup of individual worker supersisters (those that have the same father) will also influence when and how long they do particular tasks. Recent work has also suggested that the temperature at which the pupae are incubated affects the activities the adults perform.

Cross Section of Honey Bee Frame

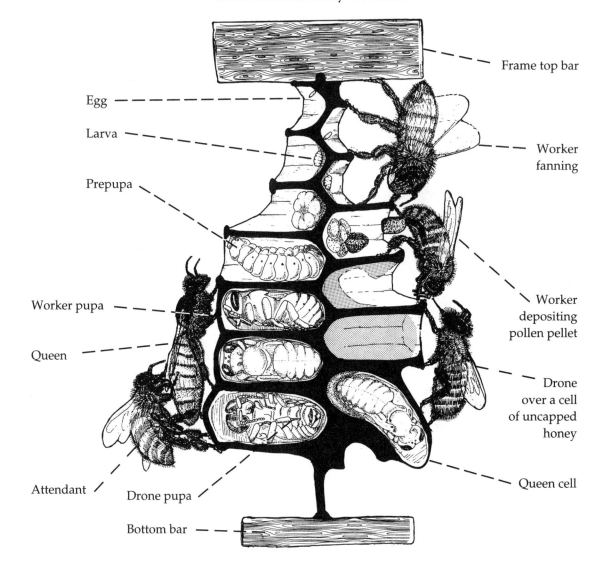

Frame top bar

Egg

Larva

Prepupa

Worker pupa

Queen

Attendant

Drone pupa

Bottom bar

Worker fanning

Worker depositing pollen pellet

Drone over a cell of uncapped honey

Queen cell

As we already mentioned, the age of the worker, her glandular development, and the needs of the colony will trigger which chores will acquire her attention. The development and maturation of her *exocrine* glands (a gland that releases secretions externally), namely the mandibular, hypopharyngeal, post cerebral, thoracic, and wax glands, as well as the conditions both within and outside the hive, will provide the stimulus needed as she progresses from one task to another. Three weeks into the six-week life span, due to atrophy, hypopharyngeal and mandibular glands no longer produce brood food; the bees graduate to foraging activities, collecting nectar, pollen, resins, and water.

Shortly after the winter solstice, in northern climates, the queen resumes egg laying. At first, she lays a few eggs per day; gradually, more and more eggs are laid until she reaches levels close to 2000 per day. As a consequence, over the next several months, more and more young bees will continue to emerge and the distribution of age-related activities will be reestablished. When this occurs, workers will progress from one inside task to another as mentioned above. Generally speaking, when you inspect a colony or observation hive, the age of various bees can be determined by the task or duties they are engaged in; for example, a bee returning to the colony with a load of pollen is a forager and at least three weeks old.

Cell Cleaning and Capping

When workers emerge from their cells and are only a few hours old, they begin cleaning chores. Cell cleaning prepares the cells for egg deposition by the queen or for receiving pollen or nectar. Cleaning involves the removal of debris and the remains of the cocoon. The cocoon, which is silk thread left by the previous occupant, was spun by the larva before entering the pupal stage. Bees are unable to remove the entire cocoon, and therefore, over a period of years, the interior of the cells diminish in size. As a consequence, emerging bees are smaller in size over time; it is for this reason that some beekeepers practice comb rotation to replace old comb with new. Cell cleaning also involves the removal of fecal matter, deposited in the cell before the larva starts to spin the silken case.

Once the queen perceives the cells to be clean, she will deposit eggs in these cells (one egg per cell). Recent studies indicate that since the inside of the hive is in darkness and the queen is not equipped with a miner's light, the cleaning bees leave a marker (pheromone) that signals that a cell is prepared to receive eggs (or nectar or pollen).

Surfaces that are difficult to clean are coated with fresh wax and/or *propolis*, a resinous substance collected from buds or the stems of trees and carried back in the pollen baskets. Often referred to as bee glue, it is a dark-reddish to brown resin, sticky when warm, brittle when cold. It is used to strengthen the combs and to cover any foreign matter that cannot be removed (e.g., a dead mouse). Propolis has antifungal and antibacterial properties that provide bees with some defense against pathogens. On the other hand, propolis makes it necessary to use a hive tool, in order to pry apart the components of a bee hive.

These same young workers also progress to capping the brood cells that are near the end of their larval stage. These cells are capped with beeswax mixed with some propolis, giving the convex caps the characteristic brownish color; honey cells with concave cappings are light in color, since they contain no propolis. Capped brood cells are easy to distinguish from capped honey by their color.

Cleaning or *hygienic behavior* is an inherited activity that helps maintain the health of the colony, since workers rapidly remove dying or dead brood or mites from cells. Research has confirmed that colonies with workers prone to good hygiene are more likely to be free from diseases and pests, and have reduced mite levels. This hygienic behavior was observed and researched early in the twentieth century by Dr. Walter Rothenbulher, who placed frames containing sealed brood in a freezer for 24 hours, thus killing the brood, after which he returned the frames to the hives. These frames were checked after 24 hours to determine how quickly the bees had uncapped and removed the dead brood. In some colonies, the removal of the brood was rapid; in others very little had been removed. Those colonies that rapidly evicted the dead brood had fewer bee diseases, particularly American foulbrood.

This hygienic behavior has been recently revisited by Dr. Marla Spivak (University of Minnesota) and other researchers who are looking for ways to reduce varroa mite infestation. A super-hygienic line named Varroa Sensitive Hygiene (VSH, originally SMR) was developed by Drs. John Harbo and Jeff Harris. This line removes fertile varroa mites from brood larvae, actually uncapping the cells to remove those mites that are actively laying eggs. (For more information on testing for this behavior see Chapter 10 and "Varroa Mite" in Chapter 14). It has been shown that colonies with strong genetic tendencies toward housecleaning are less likely to succumb to a variety of bee maladies, and it is a good policy to promote such behavior by raising queens with these genetic attributes.

In summary, hygienic workers take on the task of keeping the hive clean and can be seen:

- Removing dead or dying brood and adults from the hive. These workers are called *undertaker bees* and compose about 1 percent of the worker population.
- Removing debris such as grass and leaves, as well as pieces of old comb and cappings.
- Keeping the bottom board clean of debris and dead bees.
- Removing granulated honey or dry sugar and moldy pollen.
- Coating the insides of the hive and wax cells with bee glue or propolis.
- Propolizing cracks and movable hive parts, including frames, bottom board, and inner cover; some races use more propolis than others, making a propolis "gate" at the entrance; see Chapter 1, "Races of Bees."

● Removing healthy brood—usually drone brood and adults—when the colony is starving.

Under certain circumstances, workers remove healthy brood during nectar dearths or when the hive is lacking in food reserves. Cannibalism is not uncommon under these conditions, and usually the drone brood is the first to be pulled, followed by worker brood. Later, adult drones are evicted in order to preserve the core of the colony—the workers and the queen.

Tending to the Brood

From cleaning and capping cells, young workers 5 to 15 days old move on to their next task, which includes feeding the brood (worker, drone, and queen larvae) as well as the adult queen. Young worker bees in areas containing uncapped brood are primarily involved in feeding the larvae. During this activity they are referred to as *nurse bees*. At this point in their adult life the nurse bees' hypopharyngeal and mandibular glands, located in their heads, are fully developed. Many work-related activities are correlated with the maturation of these glands. Nurse bees require a large supply of bee bread (fermented pollen) in order to provide a portion of the raw materials their hypopharyngeal and mandibular glands need to produce brood food that they feed to developing larvae. Recall that queen larvae are fed this rich brood food throughout their larval life. This is why when beekeepers are raising queens, combs filled with bee bread are placed in proximity of queen cells; it ensures that nurse bees have copious amounts of bee bread to supply the glands with ingredients to produce royal jelly.

The combined substance produced by these glands is referred to as *royal jelly*. All young queen, worker, and drone larvae are lavishly fed this diet, which is also referred to as *mass provisioning*. When viewing nurse bees in an observation hive, the bees' heads can be seen poking into the cells for a few seconds to evaluate how much food remains and make sure the amount of food continues to be abundant at the bottom of these cells.

Queen larvae remain on a diet of royal jelly throughout their larval life; workers' and drones' diets change. After the third day of larval life, they are no longer provided with royal jelly. Their diets now

consist primarily of honey/nectar on an as-needed basis. This is referred to as *progressive provisioning*. Brood food, therefore, falls into two categories, either glandular (massive provisioning) or essentially non-glandular (progressive provisioning). It is important to note these diets are still under study and further research is required to fully understand all of the components that are included in these diets. For instance, whether or not pollen or bee bread is fed directly to worker and drone larvae remains an open question.

Over the course of brood tending, one bee will rear two to three larvae. Assuming it takes one bee to rear three larvae, and on a given comb we have 1800 newly hatched larvae, these larvae will require the attention of 600 nurse bees over the course of their development. This serves to demonstrate the time and effort required to rear brood by bee colonies, from the time the nurse bees place royal jelly in cells containing eggs that are about to hatch, until the brood is capped.

Tending the Queen

In addition to caring for the brood, nurse bees tend to the adult queen and continue to maintain her on a diet of royal jelly, the very same diet that reared her during larval life. The ingredients in royal jelly and the amount of royal jelly available to her serve

Queen and Retinue of Workers

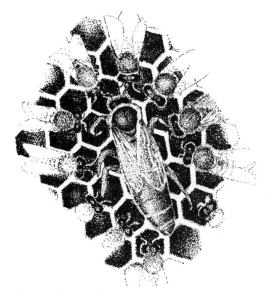

Drawing by Jan Propst.

to nourish her as well as to stimulate her ovaries in order to maximize egg production.

Again, using an observation hive, researchers watching the activities of nurse bees surrounding the queen noticed that while some were feeding the queen, others were making tactile contact with her by licking and touching her with their antennae and forelegs. This group of bees (between six and ten) surrounding the queen are referred to as *attendant bees*; they form a *retinue* or circle around the queen (see the illustration of the queen's retinue). By attending and caring for the queen in this manner, they enable her to concentrate on her most important function: laying eggs and lots of them!

At first the activities of the retinue may have been simply interpreted as a group of bees feeding and making tactile contact with the queen. However, further observations of the activities of the retinue revealed a great deal more. It was noted that the retinue remained with the queen for very brief intervals—between thirty seconds to nearly a minute. Then, individual members of the court dispersed, and as one or more took their leave, other nurse bees replaced them. This replacement activity keeps the retinue intact. Researchers noted that as the departing bees moved within the colony, they encountered other bees that initiated tactile contact with the former retinue members. These encounters are not like "idle handshakes" but instead play a major role in maintaining colony stability.

During these contacts, the *queen substance* or pheromone secreted by her mandibular glands is spread to other members of the colony. The transfer of food between bees is known as *trophallaxis*. After bees depart from the queen's court, they may exchange food with other bees they encounter. It is speculated that during these encounters they are transmitting the queen's pheromones to other workers. In fact, these former members of the queen's court can be called pheromone distributors. Remarkably, each departing attendant within the next 30 minutes will be physically contacted by up to 56 other nest mates. As a consequence of this mode of pheromone distribution, thousands of bees get samples of the queen's pheromones. The effect is not only to notify other members of the colony that the queen is present, but also to provide information as to the amount or level of her pheromone output. Thus, a colony of forty thousand or more bees, most of which may

never have direct contact with the queen, are fully aware of her presence!

As long as her pheromone levels are interpreted as sufficient, the colony remains in a stable condition. When the queen's pheromones are in actual decline or perceived to be so, the colony undertakes certain activities to correct this change. Reduced queen pheromones may trigger other activities by the workers, such as swarm preparation or queen supersedure. There are several scenarios wherein the queen's pheromones are (1) in decline (the queen is old or diseased), (2) perceived to be so, or (3) totally absent.

If the colony grows too large, some of the workers may receive only a limited amount of the queen pheromone; if this occurs between mid-spring or early summer in northern latitudes, bees will construct queen cups that lead to queen cells and then the eventual emergence of a virgin queen or queens; see "Swarming" and "Queen Supersedure" in Chapter 11. If the bees are preparing for swarming, before the new queen(s) emerge, the colony is likely to cast its first swarm (*primary swarm*), which is usually accompanied by the old queen. Thus the original colony rids itself of its mother and replaces her with her daughter whose pheromone levels return the colony to a stable unit.

If, on the other hand, the queen is perceived to be of low quality, due to disease, poor mating, or other reasons, a colony will replace the mother queen not by swarming but by supersedure. The outcome is similar in that the new queen replaces the old queen. Sometimes both the old and new queens (mother/daughter) remain members of the same colony. It is likely that the old queen's pheromones are nonexistent, and therefore the young queen, her replacement, does not seek to eliminate her mother.

If a queen is accidentally killed or lost, within 24 hours the entire colony will know that it is queenless. Occasionally a queen dies for any number of reasons. For example, while a beekeeper is inspecting a comb, the queen could fall to the ground and be crushed when the beekeeper inadvertently steps on her. Bees will then construct *emergency queen cells* to rear a replacement.

In another instance, a queen dies and the bees are unable to replace her because there are no eggs or young larvae available to rear a new queen. In this case, a number of worker bees undergo ovary maturation because of the absence of worker brood and queen pheromones, which, when present, suppresses ovary development. Workers that are able to lay eggs are referred as *laying workers*. However, they are unable to mate, and any eggs they lay are haploid and will produce only male bees (drones). Such a colony, unless discovered early and successfully requeened, is doomed.

Comb Building

The wax comb (and surrounding cavity) is the nest and abode of the honey bee. Although comb is referred to as honeycomb, which implies that it is comb containing honey, it has many more functions than solely to store honey. The hexagonal cells are used to store either bee bread or the nectar that is being processed into honey. Comb is also the place for depositing water droplets needed to cool the hive when rising temperatures would cause problems for the colony. In addition, the cells serve as a depository for the queen's eggs, as incubators, and cradling and transforming sites as the developing bees move through four metamorphic stages (eggs, larvae, pupae, and adult). It is also a communication platform, especially the area of comb in the vicinity of the hive entrance. This area is referred to as the *dance floor*, where returning scout bees perform the important round and wagtail dances. The purpose of these dances is to recruit other bees to visit nectar, honeydew, pollen, resin, and water sources. Furthermore, vibrations produced by the dancers travel to more distant areas of the honeycomb, where they are detected by other bees that then move to the dance floor to receive information from the dancers. Honeycomb also provides the colony with its particular odor, and defends the colony (through the agency of propolis) against pathogens.

A prerequisite to the construction of honeycomb by bees is that they first must locate themselves, or be located by humans, within a suitable cavity. An ideal cavity includes the following elements: low interior light, a dry interior, protection from adverse weather, sufficient capacity to hold a reasonably large bee populations, ample room for future combs (for brood rearing and honey and pollen storage), and an entrance/exit that allows easy passage for the bees while restricting predators or parasites.

Other than in human-made bee boxes, the comb is usually confined to where the light levels are low,

such as a cavity located in a tree, in a space between the walls of a building, or anywhere in a cave. Some nests, referred to as *exposed comb nests*, can be found in the open, almost always in a deeply shaded area. In the desert Southwest, for example, where trees are not abundant, feral bees will nest in small caves formed by rock outcrops. Once bees move into a cavity, either on their own (as in the case of swarms that have not been captured), or are placed into one by a beekeeper, the process of comb building begins. Occasionally, honey bees will construct comb outside of a cavity; such undertakings are suicidal in northern climates. As we have previously noted, many work-related activities of bees parallels the maturation of specific bee glands. In the case of wax glands, they become fully developed when worker bees are between 12 and 18 days of age, and it is bees of this age that not only extrude the *wax scales* (the building blocks for honeycomb), but also are the ones that become directly involved in the construction of the comb. The wax for honeycomb is produced by four pairs of wax glands, located beneath the ventral segments (*sternites*) of a worker bee's abdomen. The wax is secreted as a liquid from each of the paired glands, and as it comes in contact with air, it solidifies into thin oval scales. The ventral surface of the sternites is referred to as the *wax mirrors*. The hardened wax flakes are referred to as *wax scales.*

Each wax scale is removed from its wax mirror with special hairs located on the bees' hind legs. From there, each scale is passed forward by the next two pairs of legs and on to the bee's mouth. Here the bee's mandibles (jaws) masticate or knead the scale and mix it with a secretion from the mandibular glands. The end result is a pliable product that bees can employ to construct comb from scratch or augment existing comb. Bees in the process of producing wax can be observed suspended from one another in formations that resemble strings of beads or chains. These formations are referred to as *festooning* (see illustration of worker bees festooning on p. 33). The function of festooning still is not entirely understood. Workers in a festoon will stay there for a time and then move off to feed brood or do other tasks, thus allowing the wax glands to recharge.

During the construction process, bees are also warming the wax to more than 109°F (43°C) in order to construct the prefect hexagonal cells. When you see a series of parallel combs constructed by bees,

Worker Bees Festooning

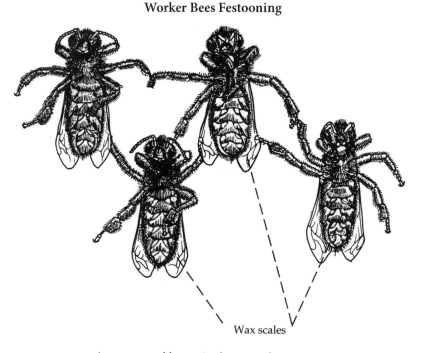

Wax scales

Appearance of festooning bees on a frame

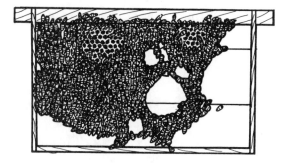

it is difficult to believe that so many combs can be produced from tiny wax scales. Most hive bodies have eight or ten frames of honey comb. Each comb, having two sides, is separated from an adjacent facing comb by a space equal to the width of two bees. The width of two bees on opposite facing combs with their backs to each other measures ⅜ of an inch (8–10 mm). This spacing is referred to as the *bee space* and permits bees on the surfaces of opposite-facing combs to go about their respective chores without bumping into each other. Any space greater than ⅜ of an inch will be filled in by bees with additional comb and/or burr comb.

The discovery of the bee space in 1851 by Reverend Lorraine Lorenzo Langstroth, a graduate of Yale's Divinity College, revolutionized beekeeping because beekeepers are now able to both withdraw and insert comb into a bee hive without damaging the nest. As a result of this discovery, beekeepers can examine any given comb with minimum disturbance to the bees on combs adjacent to the one being withdrawn. Prior to this invention, bees were kept in skeps, logs, and boxes that were without movable frames; in order to remove combs of honey or any other combs, the beekeeper was obligated to tear them from the nest, which caused a lot of destruction to the nest and to the bees.

The cells of the honeycomb do not lie on a completely horizontal plane: the openings are slanted slightly upward by 9° to 13°. We may speculate that this inclination probably prevents nectar, which has a low viscosity, from spilling out and also serves to concentrate the brood food at the base of the cells holding the young larvae. The cell walls are 0.003 inch (0.07 mm) thick, and the combs hang vertically. Bees have special sensory hairs in the joints of their legs that enable them to detect gravity, and they measure the cell thickness by using sensory hairs on the tips of their antennae.

The production of beeswax is economically very expensive; it takes 50 pounds of honey (22.7 kg) to produce 10 pounds (4.5 kg) of wax. Therefore, bees need to engorge large quantities of honey, nectar, or sugar syrup and pollen, in the form of bee bread, to stimulate wax production. Your hive scale plays an important role, because as it begins to display weight gains, it is informing you that a honeyflow is in progress. This is the ideal time to provide bees with foundation, which will be converted by the bees into wax comb. This is precisely why the best time to have foundation "drawn" into honeycomb is during a honeyflow and/or by feeding sugar syrup to bees or by capturing a swarm.

As a colony prepares to cast a swarm, worker bees will engorge on excess amounts of honey/nectar and store the same in their honey sacs. This activity serves three vital purposes: (1) it provides a reserve of food in the bees' honey sacs while the swarm is clustered and scouts are searching for a home site; (2) it stimulates their wax glands, which will be needed to construct honeycomb at the new homesite; (3) the remaining honey/nectar can be regurgitated into the new honeycomb, providing food for the new colony.

Beekeepers often hive swarms on foundation so the bees will "draw out the foundation" by adding wax to it, resulting in new, white comb. For the same reasons, swarms can be used to produce comb honey: the cappings over the honey cells will be a clean white color, pleasing to the eye and giving the product a more edible appearance to those who eat comb honey.

Honeycombs, once constructed, consist primarily of hexagonal cells; each cell is connected to its neighbors to form a large comb containing thousands of cells on each side. The hexagonal cells are of two sizes: the smaller worker cells and the larger drone cells.

An exception to the hexagonal-shaped cells of honeycombs is the cells intended to rear queens. These cells are fabricated of wax, suspend from the comb, and have the shape of a miniature inverted cup (resembling the cap on an acorn) and are referred to at this stage as *queen cups*. Once an egg has been placed in these cups by the queen, these cells are then referred to as *queen cells*. As the hatched queen larvae begin to grow, these cells are elongated with the addi-

tion of more wax and eventually look somewhat like a peanut shell. In some cases queen cups are never used for their intended purpose; or after a virgin queen emerges from the queen cell, bees remove the wax from these structures and use it elsewhere.

One caveat to the formation of queen cells is that in cases of supersedure, or the need to replace the queen in an emergency (e.g., she dies unexpectedly), queen cells are shaped from the hexagonal cells and slowly assume the shape and form of regular queen cells. To begin raising a queen during supersedure or an emergency by starting with a hexagonal cell means that no initial preparations were made to replace a queen. Therefore, the hive lacked queen cups or queen cells with eggs or larvae in them. However, if hexagonal cells containing fertilized eggs or larvae less than three days of age were present, the bees have the capacity to convert these cells into queen cells.

When inspecting a hive, you will see that the wax cappings over honey cells tend to be light in color, whereas the cappings over the sealed brood tend to be tan to brown in color. The reason for this is that the wax cappings over brood is mixed with propolis.

Nest Homeostasis

The maintenance of temperatures inside the colony at a constant level despite external conditions is termed *homeostasis*. Colony homeostasis is maintained by cooperative living or social behavior and is found in all social insects (ants, termites, wasps, and bumble bees) and some mammals (naked mole rats). The ability of a colony to survive temperature extremes or times of dearth is an obvious advantage over living a solitary life (which includes most other insects).

The ability of bees to organize and execute the many tasks needed for colony survival, including nest design and construction, is indeed a marvel of the natural world.

Food Exchange, Handling, and Hive Odor

Bees within a hive exchange honey or nectar. Foragers returning from the field pass food to the hive bees, who then pass it to other bees. This food exchange, called *trophallaxis*, not only communicates what the colony is receiving from the foragers, but also indicates the availability and quality of incoming food.

A returning forager will offer a drop of nectar to

Beneficial Microbes in Honey Bees

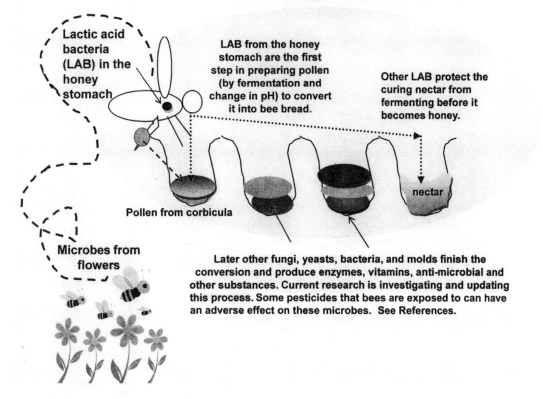

Lactic acid bacteria (LAB) in the honey stomach

LAB from the honey stomach are the first step in preparing pollen (by fermentation and change in pH) to convert it into bee bread.

Other LAB protect the curing nectar from fermenting before it becomes honey.

nectar

Pollen from corbicula

Microbes from flowers

Later other fungi, yeasts, bacteria, and molds finish the conversion and produce enzymes, vitamins, anti-microbial and other substances. Current research is investigating and updating this process. Some pesticides that bees are exposed to can have an adverse effect on these microbes. See References.

two or three house bees, who will then move off to a quiet corner to work the nectar by extending their tongues and exposing the droplet to the warm air. This helps cure the nectar into honey by evaporating some of the 80 percent water that is in nectar. As the water content is reduced, the nectar is placed in cells, where further curing takes place. This ripening is finished in the honey cells and will take from one to five days to complete, depending on the amount of humidity in the air, the water content of the nectar, ventilation, and the amount of nectar being cured. Through this ripening process, the final product, honey, will have a water content of less than 18 percent.

House bees also process the pollen pellets deposited in cells by foragers. They pack the pellets down with their mandibles and forelegs and moisten them with honey and saliva, which contains enzymes and beneficial microbes. When a cell is partially filled, the packed pollen is sealed with a thin film of honey, resulting in the product referred to as *bee bread*.

An additional function of food transmission is the spread of the hive's unique odor. Each colony has its own characteristic odor, which may aid the bees of one hive in distinguishing bees from other hives (such as robbing bees) and foreign queens (see "Queen Introduction" in Chapter 10). To keep foreign bees out, guard bees patrol the colony entrance and challenge any incoming bees that may be intruders. Guard bees are older workers that have very high concentrations of the alarm pheromones.

Some foragers will return with their honey sacs full of water and some will return with them full of nectar. The needs of the colony will then dictate which of these two sets of foragers will be the first to be relieved of the contents of their honey sacs by the receiving bees. For example, if the temperature within the colony is rising above normal levels, foragers returning with water will have their loads emptied before foragers carrying nectar, permitting most of the water foragers to quickly return to gather more water. Some of these water foragers will move to the dance floor and recruit more bees to the water source. This is an example of how quickly a colony can redirect its activities to fulfill, in this case, an urgent need for water.

Ventilation

Bees can often be seen fanning their wings on the extended deck of the bottom board with their heads facing toward the hive entrance. In this position, warm air is pulled out of the hive. This fanning also takes place inside the hive, as well as on the portion of the bottom board within the hive that is hidden from view. The best time to observe fanning behavior is on warm days when abundant amounts of nectar are being collected. Workers of all ages perform this task, but many young bees, less than 18 days old, are often fanning, especially on hot days.

Fanning circulates air though the hive and helps to:

- Regulate the hive's humidity at a constant 50 percent.
- Reduce or eliminate accumulations of gases, such as carbon dioxide (CO_2).
- Regulate brood temperature.
- Evaporate water carried into the hive to reduce internal temperatures (see the diagram on temperatures on p. 118).
- Evaporate excess moisture from unripened honey (nectar with a high percentage of water); as this moisture evaporates, it too, will cool or humidify the hive.
- Keeps the wax combs from melting as temperatures increase.

Bees also fan their wings to disperse pheromones. One pheromone of interest is the orientation pheromone that assists in guiding bees to a specific location. This pheromone, produced by the *Nasonov gland*, is often referred to as the *scent gland*, which is located on the dorsal tip of a worker bee's abdomen. When the gland is exposed, a mixture of volatile chemicals is released. Some of these chemicals include geranol, neurolic acid, geranic acid, and two citral compounds. When 50 or more bees are "scenting," a beekeeper can actually detect the lemony-scented citral compounds in this pheromone. One can witness this fanning pheromone-releasing ac-

tivity when shaking in front of the entrance either a package of bees, a swarm cluster off a branch, or one or more frames of bees.

When a colony casts a swarm and bees begin to land on some object such as a tree branch, bees will release pheromones from their Nasonov glands to guide bees still in the air to this branch, where they will gather together to form a swarm cluster. When the swarm takes flight again to head to a new homesite, many bees will settle around the entrance and start scenting. This action guides other bees still in flight to the opening to their new home.

This type of scent fanning is commonly seen:

- When a swarm or package of bees is emptied at the entrance of a bee hive.
- When bees are shaken off a frame or otherwise disoriented.
- When a hive is opened that is queenless or has a virgin or newly mated queen.
- When a swarm begins cluster formation.
- As a swarm enters a natural homesite such as a tree cavity.

Guarding the Colony

Worker bees between the ages of 12 and 23 days will defend their colony by flying at and often stinging an intruder. Each bee does this guard duty for only a few hours (or days) in her entire life. Guard bees can be recognized by their posture, standing on their last two pairs of legs with their antennae held forward, their front legs raised, and their mandibles opened; this seemingly menacing stance may serve to warn potential intruders that they are unwelcome. These guard bees often inspect incoming bees, and if an incoming bee does not smell or behave correctly, it might not be admitted. Stray or foreign drones, young workers, and foragers carrying a full load of food are generally allowed to enter. During strong nectar flows, foreign bees pass easily into a hive; during a dearth, however, guard bees closely inspect strange bees.

If a large animal approaches the colony, some of the guards will often fly out to challenge the intruder. This defensive action should not be interpreted as meanness or aggression. When an intruder approaches and enters or begins to open a hive, some bees raise their abdomens, begin fanning, and thereby disperse the *alarm odor* that is released by a gland at the base of the sting. This pheromone has an odor similar to that of banana oil and incites other worker bees to defend the colony. Once some of the attacking bees sting clothing or skin, some alarm odor remains at the site, tagging the victim. Thus tagged, the victim may become the target of further defensive acts as long as the odor remains on the clothing or skin.

Many factors influence the temper of a colony (see "Bee Temperament" in Chapter 5).

OUTSIDE ACTIVITIES

Flight

Except for occasional orientation flights, worker bees generally remain within the colony for the first three weeks of their adult lives, cleaning, feeding, building comb, ripening honey, and packing pollen. These routines are more or less discontinued at the end of the third week as bees turn to tasks that require flight. An ability to recognize the different types of flying activity will enable you to interpret activities at or near the hive entrance.

Orientation flight. Bees on orientation flights familiarize themselves with landmarks surrounding their hive as well as void feces. These bees hover near the hive entrance for very short periods. A single flight will last only five minutes; successive flights over the next few days will last longer. This is a common sight in the late afternoon, when young drones and workers are hovering in front of the entrance.

Foraging flight. This is the final task of a worker bee. Foraging bees fly out and away from their hives in search of nectar, honeydew, pollen, resins, and water. Because their brood food and wax glands have atrophied, these bees look smaller, and the edges of their wings are often torn and ragged. The characteristic patterns of returning with pollen, or flying straight into the hive, or onto the extended deck of the bottom board, will distinguish them from orienting bees. Foragers will have traveled about 500 miles (800 km) before they die (see the illustration on forage areas on p. 38).

Robbing flight. Unlike orientation flights, which are short in duration, robbing activity is a form of foraging. On first approaching a target hive, the robbers sway to and fro in front of the hive in a manner somewhat similar to a figure eight. Once a hive has been invaded, other robbing bees are "recruited" to this activity when the robbers return to their home and dance the location of the new food source.

Robbing often takes place during a dearth, when little or no nectar is available; abandoned and weakened colonies are the first targets for robbing bees. Beekeepers who expose combs, take off honey supers (see p. 135), or attempt to feed weak colonies at this time may inadvertently initiate robbing. You must minimize your activities in the beeyard during a dearth, or you can start a robbing frenzy (see "Robbing" in Chapter 11).

Cleansing flight. Individual bees usually release their body wastes or fecal material outside the colony during flight. Evidence of this activity is the yellow or brownish spots or streaks found on the snow, plants, cars, or laundry hung out to dry. (The so-called yellow rain discovered in Southeast Asia during the Vietnam War turned out to be bee excrement instead of chemical warfare.) As weather permits, when bees are able to exit their colonies on a daily basis, one will find little or no evidence of this activity. However, during periods of prolonged weather-related confinement— during the winter months, for example, or during wet and cold weather—or even from confinement in a package, rather than defecating in the hive, bees usually retain their fecal matter; however, as soon as they can exit the hive, it is released, usually while in flight. In these cases, the yellow or brownish droppings will be easily seen because of the large number of bees that are defecating. In such circumstances, the flights are referred to as *cleansing flights*. Such flights usually occur in the spring, as the young bees are first able to leave and orient to the homesite. If bees are confined for too long, or they have dysentery or are infected with nosema, defecation may take place inside the hive or just outside on the hive furniture. If you see dark stains on the outside of the hive, your bees may not make it through the winter.

Foraging and Communication

The gathering of food in the form of pollen and nectar/honeydew for both larvae and adult consumption, as well as the gathering of food for storage, requires a high degree of cooperation and communication among all members of the colony. Haphazard or random searches for food by foragers would be too time and energy consuming. Even if moderately success-

Forage Areas for Honey Bees

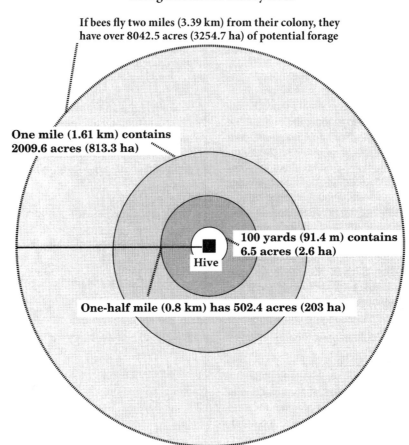

If bees fly two miles (3.39 km) from their colony, they have over 8042.5 acres (3254.7 ha) of potential forage

One mile (1.61 km) contains 2009.6 acres (813.3 ha)

100 yards (91.4 m) contains 6.5 acres (2.6 ha)

Hive

One-half mile (0.8 km) has 502.4 acres (203 ha)

ful, random searching would not be able to sustain honey bee colonies with large populations that have many mouths to feed, populations that also require an abundance of stored food for winter survival.

When foraging, a worker bee orients herself in her flights to and from collecting locations according to various external stimuli:

- The sun's position and polarized light.
- Landmarks, both horizontal and vertical.
- Ultraviolet light, which enables her to see the sun on cloudy days.

The honey bee colonies have evolved a communication system that markedly increases their efficiency of food-gathering activities. A worker bee is able to inform other bees about the location of a food source through a series of body movements called *dances*. Dr. Karl von Frisch, an Austrian scientist, won the Nobel Prize in 1973 for his research on the function of these dances. He and his colleagues ob-

served that the movement of the dancer on the surface of honeycomb could be outlined as circles and figure eights. From the figure-eight dance, bees learn both the distance to and the direction of where to find food, water, and resins. During the dancing, the dancer rubs her wings together, causing vibrations on the comb, which may provide additional information to recruited bees. Other clues may include particular samples of the provision being sought and its associated odors, as well as other chemical signals the dancers release. This information, as well as scent markers at the source, minimizes the time and energy needed by the recruits to locate the provisions.

Several dance configurations have been recognized, but the two most easily identified are the *round dance* and the *wagtail dance*. The round dance communicates that the provisions are no farther than 300 feet (100 m) from the hive in any direction. Bees being recruited to places beyond this distance obtain direction and distance information from the waggle, wagtail dance, or figure-eight dance, which resembles the

Dance Language of Honey Bees

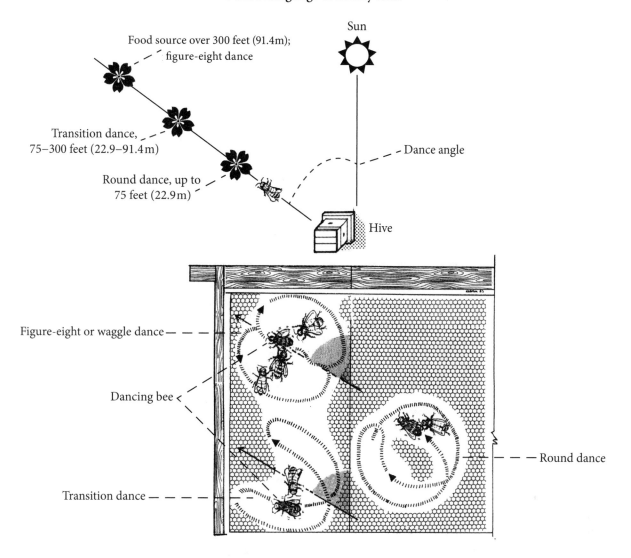

Sun

Food source over 300 feet (91.4m); figure-eight dance

Transition dance, 75–300 feet (22.9–91.4m)

Round dance, up to 75 feet (22.9m)

Dance angle

Hive

Figure-eight or waggle dance

Dancing bee

Transition dance

Round dance

number eight. The wagtail dance includes three components: comb vibration, frequency of waggles, and the number of waggles (see the illustration on dance language). The waggle vibrations transmitted to the surface of the comb are felt by other bees attending the dancer, especially if the dance is performed on uncapped cells. This vibration is thought to be picked up by the recruited workers via special organs on their legs. The middle section of the figure-eight wagtail dance is referred to as the *straight run*. Distance is communicated by the number of straight runs or waggles in the dance per 15 seconds; the greater the number of runs, the shorter the distance to the food source. Direction is telegraphed by the compass direction of the straight run in relationship to the sun.

A transition or sickle dance is done by some races when distances to the food are between the circle and wagtail dance. Workers have been observed doing other dances, and these are still being studied.

Other Behaviors

Washboard Movement

Beekeepers can often observe bees, usually in the late afternoon, on the front wall of the hive with their heads pointed toward the entrance. These bees are standing on the second and third pair of legs and seem to be scraping the surface of the hive with their mandibles and front legs, as if to clean it. As they scrape, their bodies rock back and forth in a motion similar to scrubbing clothes on a washboard. This is called the *washboard movement*. The exact purpose of this activity is not currently understood, although it occurs in only very populous colonies.

Bee Bearding

The phenomenon known as *bee bearding* refers to the clustering or hanging of older bees outside the front face of the hive during very hot weather, giving the impression of a beard. This behavior is common on hot, humid days when temperatures reach the high 80s and 90s (27–33°C). This activity is often viewed as an indication that bees are preparing to swarm; in some instances, this view may be correct. In northern climates there are two swarm seasons: an early prolific season that takes place from late April to mid-July, and a second but less prolific season from mid-August into the month of September. In one type of bee bearding, bees hang off the bottom board in a conical formation waiting for the swarm to issue from the hive, at which time these loitering bees will depart with the swarm. In the other case, due to high ambient temperatures, the bees cluster on the front side of the hive in a beard-like formation.

Yet, bee bearding takes place even outside the swarming season, or when temperatures are far greater than those from mid-spring to early summer. In most cases, a portion of the bees in a hive preparing to swarm hang from the bottom board in a conical cluster rather than spreading themselves over the front-facing portion of the hive in a beard-like formation.

Bee beard forms any time bees need to cool the inside of the hive. When summer is in full swing, the numbers of adult of bees are at their highest, which may lead to overcrowding in the colony. In addition to the overcrowding, other issues impact the colony's ability to continue its foraging activities because most of the cells are plugged with ripening nectar, capped honey, and bee bread. And the act of foragers bringing water to cool the hive may not be sufficient. As a consequence, to moderate the internal temperature of the hive, many bees vacate the interior of the hive and move to the outside in order to keep the brood from being killed by excessive heat and to prevent the possibility of the combs melting down.

When bearding occurs, and weather forecasts indicate a sustained period of high temperatures, the beekeeper needs to take some action to ameliorate the situation. Increase the ventilation by adding supers to reduce crowding. Screened bottom boards will also help resolve the problem. Propping open the outer cover will provide additional ventilation. In extremely warm climates, hives should be located in areas that receive shading during the afternoon. Assisting the bees in these ways will reduce the number of bees needed to collect water, and the bearding bees will be able to return to more productive activities. In addition, the following tips were outlined by Khalil Hamdan (*BeeWorld*, 2010) and can be useful:

- Nothing beats the screened bottom board for ventilation.
- Increase the entrance space. Larger entrances are good for hot weather.
- Place shade boards to shade the hives, especially if the sun is beating on them.
- If you have 10 frames, consider going to 9 to allow more space for ventilation.
- Provide an upper entrance to improve ventilation in warm humid conditions during the summer.
- Place a ventilation box with screened openings on top of the inner cover, then place the outer cover on top of the ventilation box. This allows for good airflow through the hive. Ventilation boxes can be made out of unused honey supers: drill holes in the sides on a slant so rain cannot enter, and screen them on the inside to prevent robbing. Sliding a super back slightly to improve airflow is not a recommended approach. This may encourage robbing and allow rain to get inside the hive.
- Paint hive bodies with white paint or some other light color to help reduce overheating. White reflects the sun's heat.
- Use an inner cover with either a wider opening or two openings.
- Provide a source of water for the bees in a partially shaded position near the apiary (within a half mile or less), or use Boardman feeders. On a hot day, a strong colony will use more than one-quarter of a gallon (1 liter) of water to cool the hive and to prevent overheating.
- Remove a few frames of honey or harvest supers when sealed and add new boxes of frames to provide space and ventilation.

Check the References for more information under "Bees and Beekeeping."

 **Notes**

CHAPTER 3

Beekeeping Equipment

Throughout the history of beekeeping, a variety of hives have been employed by beekeepers, including straw skeps used by Virgil, Mediterranean-style clay tubes, Top Bar hives, and Warre hives. Here we will concentrate on the modern Langstroth hive. The shape, size, and number of pieces to this hive may vary, but in all such hives, the ⅜-inch Bee Space—the distance between two facing combs or combs facing the walls of a hive—is maintained. A common type of Langstroth hive consists of a bottom board and two deep hive bodies, each containing 8 or 10 deep frames. One of these deep hive bodies sits on the bottom board and the other one above it; these two chambers are often referred to as the broodnest, the area of the hive where the queen lays her eggs and the subsequent larvae bees develop and finally emerge from their pupae stage as adults. Usually, shallow supers that are only half the depth of the deep boxes are placed above the second deep hive body. They, too, contain frames; it is in these boxes (supers) where bees usually store their surplus honey. Both an inner and outer cover are placed above this stack of supers. A configuration of this kind is designed to house an active honey bee colony. The depth of the hive bodies (supers) varies. (See diagram on p. 44 for terminology of basic hive parts.)

BASIC HIVE PARTS

Outer Cover

There are two basic types of outer covers for hives, a *telescoping* one and a *migratory* one; both serve as the roof of the hive. The telescoping cover is usually made of wood, with a metal sheet, usually of aluminum, covering the top. As the name implies, it "telescopes" over the rim of the inner cover and uppermost hive body. However, these covers can blow off on windy days, so it is always prudent to place a heavy object on top of the cover to secure it. If you live in an area with long, hot summers, some type of insulating material can be placed between the metal sheet and the wooden roof (newspapers work well) during construction of the cover. The illustration of a beehive shows a telescoping outer cover.

The flat California or migratory cover does not telescope over the sides and is used in drier areas; it will not last long in extended wet weather. The lack of sides telescoping over the edge of the hive allows beekeepers to strap hives together to move them to different locations for pollination services and to particular or abundant nectar sources. It also permits four colonies to be tightly packed together so they are flush against one another; therefore, when strapped to pallets, they can be easily loaded on and off trucks. Whichever outer cover you use, place a heavy weight on top of it to prevent it from being blown off by strong winds. If a colony loses its roof, exposure to robbing, heat, snow, rain, and cold winds could weaken or kill the bees.

Inner Cover

The *inner cover* is a wooden, Masonite, or plastic board with ½-inch (13 mm) rims and an oblong hole in its center. Some inner covers come with an additional half circle hole or square opening notched in the middle of one of the two shorter rims of the cover. This opening can be closed off with grass or leaves when not in use. It can be opened to allow hot air to escape from the hive and/or serve as second opening for bees to enter and exit. This aperture, as well as the oblong hole in the center of the cover, helps to vent moist air, especially in the winter. When migratory covers are employed, inner covers are absent.

If a device called the Porter Bee Escape is placed in the oblong center hole, the inner cover becomes an *escape board*. By placing this board below a honey super or supers, the worker bees passing through the one-way bee escape are unable to reenter the honey supers above, and thus the boxes are cleared of bees; see "Removing Bees from Honey Supers" in Chapter 9 for other removal methods.

Whenever it becomes necessary to feed a colony, you can invert a jar or pail of syrup over the oval hole of the inner cover. Bees can collect the food without venturing outside. Place an empty box on top of the inner cover to enclose the feeder, and set the outer cover on top of that box (see Chapter 7). This will protect access to the feeder by robber bees. By placing several shims under the feeder pail, bees will have access to more openings in the feeder pail's lid.

Shallow or Honey Supers

The shallow supers, which are usually used for honey storage, come in various depths and with frames and foundation of corresponding size. In the United States, three different honey supers are used:

- 4 13/16 inches (12.2 cm), referred to as the comb honey or section super.
- 5 11/16 inches (14.5 cm), or shallow super.
- 6 5/8 inches (16.8 cm), or medium super (also called Dadant, Illinois, or western super); this can also be used as a brood box.

The different super sizes are illustrated in "Super Sizes" in Chapter 9.

The number of honey supers per hive will vary during a honeyflow, depending on the amount of honey collected and the number of supers you have. Although supers are designed to hold 10 frames (see below), this configuration is generally used only in the brood supers. Experienced beekeepers put 8 or 9 frames of empty drawn combs in 10-frame honey supers, spacing them so they are evenly separated from one another. Bees will draw out (extend) these combs. When they are filled with honey, because

they were extended beyond the framing surrounding them, removing the capping is much easier as you prepare the frames for extracting the honey.

Note: you must have the bees draw comb with 10 frames first before reducing to 8 or 9 frames.

Queen Excluder

A queen excluder is a perforated plastic or metal sheet or a metal- or wood-framed metal grill. *Queen excluders*, as the name implies, exclude the queen from going beyond where the device is placed. Because of its small openings, only the worker bees can squeeze through; the larger drones and queens cannot. The device is placed on top of the broodnest to prevent the queen from entering the honey supers above. It is also used on two-queen colonies or for any other reasons that require the exclusion of the queen. Drones are often trapped by excluders, and if they die in large numbers, they can clog it with their bodies. This can be prevented by leaving (or boring) an escape hole in the honey supers and by eliminating adult drone or drone larvae from the supers. Be careful, as robber bees could also enter; so be alert.

Deep Super or Brood Chamber

The deep or standard brood box is 9⅝ inches (24.4 cm) in depth. Ten full-depth frames (9⅛ inches) are used in this box. The deep box is typically used in the United States to contain brood and winter stores. The general understanding is that since these deep, heavy boxes are not lifted or moved much, they are used as brood chambers; but to collect honey, the shallower supers keep the weight more manageable. At one time, bigger boxes called the Jumbo and the Modified Dadant deep supers were common. Commercial beekeepers tend to use only one size of box for both brood and honey production.

Some bee supply companies carry the smaller 8-frame hive equipment, which has the same vertical dimensions as the other boxes but holds 8 instead of 10 frames. This equipment, when full of honey, is lighter in overall weight and has smaller dimensions, and some beekeepers say that the bees do better in it; again, chat with beekeepers who have this equipment to see if they like it. The only problem is that this equipment cannot be mixed with the standard 10-frame bodies.

A Beehive

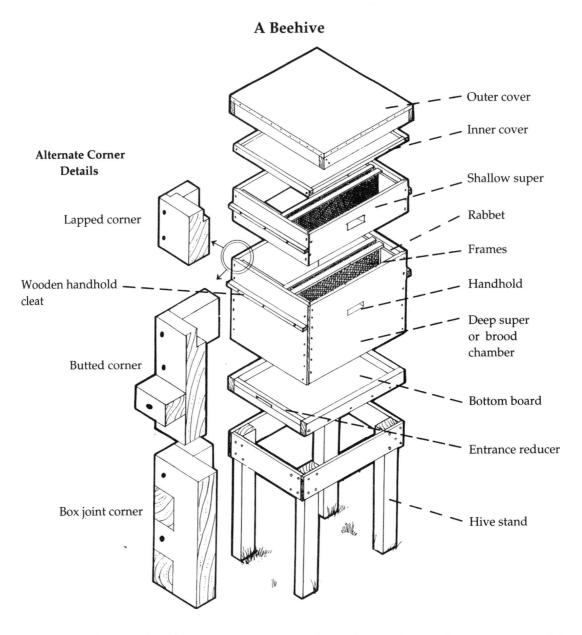

Alternate Corner Details

Lapped corner

Wooden handhold cleat

Butted corner

Box joint corner

Outer cover

Inner cover

Shallow super

Rabbet

Frames

Handhold

Deep super or brood chamber

Bottom board

Entrance reducer

Hive stand

Bottom Board

The floor of a hive is a wooden structure called a *bottom board*. Hive bodies are stacked on top of the bottom board, which should be placed on a firm and level foundation to keep honey-heavy hives from tipping over. Never place the bottom board directly on the ground, as it will rot quickly; use a hive stand, bricks, cinder blocks, or flat rocks (see the illustration of the beehive). Treating the bottom board with wood preservatives may extend its life, but many chemicals not only are toxic to bees but also can contaminate honey and wax. Bottom boards are a good candidate for paraffin dipping (see Appendix D and the "Equipment" section in the References).

Some bottom boards have two rim heights—a short winter rim and a deeper summer rim—and these are called *reversible bottom boards*. Many beekeepers, instead of reversing the bottom board (which is difficult to do), use an entrance reducer to restrict the hive opening in the fall. Plastic bottom boards are available, but they sometimes buckle if the hive is very heavy, rendering them useless. Commercial operations often use handmade pallets, on which bottom boards are permanently attached. This system serves as a convenient way to move bees (via forklifts) and provides a solid and safe hive stand.

Reducing the entrance of weak colonies enables guard bees to better protect the hive from robbing bees and other insect predators. However, the most important reason for reducing the entrance in the late summer (end of August in northern climates) is to keep mice and other rodents from taking up residence in a bee colony. Mice seek sheltered areas to spend the winter, and bee hives are very suitable to their needs. If you fail to place mouse guards at your hive entrance, you will likely find in the spring a nest made of dry grass and leaves and the size of a basketball inside your hive, with a family of mice nestled in it.

In addition to the unsanitary conditions created by the mice, the odor of their urine and feces and the chewed-away sections of honey comb to make room for their nest will require you to remove the nest and mice and replace damaged combs. By being proactive and installing a mouse guard, you will not have to witness the destruction mice can accomplish in a bee colony and thus may even save the colony. Metal mice guards are now readily available and are excellent for this purpose; check bee suppliers or talk to other beekeepers to see what they recommend.

Screened Bottom Boards

With the invasion of parasitic varroa mites and the small hive beetle, new equipment is available as a passive control method. The screened bottom board is an effective and noninvasive way to control for these mites, but it needs to be used with other mite-control options because it does not keep mite populations from reaching peak levels. However, the screened board is a means to determine the mite population in your colonies when it is used in conjunction with a sticky board to collect and count mites. Check the current research on these boards and speak with other beekeepers about using them. Also, special bottom boards or traps for the small hive beetle are available and should be researched if you have this pest in your area (see Chapter 14 for more information). Check bee supply catalogs.

Commercial Hive Stand or Alighting Board

Most bee supply companies sell an on-the-ground hive stand and alighting board combination. Although it is sufficient for most situations, this type of hive stand is not recommended, because unless it is made of cypress (or rot-proof material such as plastic), it will rot in a few years. It will also not protect

bees from ants, mice, or other predators. Therefore, most beekeepers eventually make their own hive stands from materials at hand in order to keep the bee hives elevated. For more information see "Hive Stands" in Chapter 4.

Bee Space

Bees do not space natural honeycombs at random in a wild colony or, for that matter, in the wooden beehive. They adhere to a strict code and do not construct comb in spaces less than ⅜ inch (9.5 mm). This space or gap (often referred to as the *bee space*) is found between the frames of combs and the hive's walls, and is present between the top of the frames and the inner cover, the bottom bars of the frames and the bottom board. This observation was first published by the Philadelphia minister L.L. Langstroth over 100 years ago. It was the basis on which he designed the prototype beehive used today. The ⅜-inch space enables beekeepers to remove frames without having to cut the combs from the walls and covers. The space between any two adjacent frames is the equivalent of two bee spaces, which allows bees on the vertical surfaces of self-facing combs to move about without colliding with one another (see illustration of the bee space).

By utilizing this natural spacing, beekeepers are assured that the bees will not attach comb to the walls or to other sections of comb, and that the frames can easily be withdrawn and reinserted. If the bee space is violated by the beekeeper, by failing to maintain the

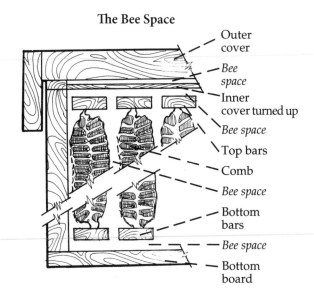

The Bee Space

Outer cover
Bee space
Inner cover turned up
Bee space
Top bars
Comb
Bee space
Bottom bars
Bee space
Bottom board

proper number of frames within the hive body (e.g., forgetting to return a frame after examining it), the bees will fill the gap by adding comb. By doing so, the bees restore the proper spacing of the combs (i.e., the bee space). Bees may produce an entirely new comb to fill the gap, or build less-than-satisfactory replacement. In any case, the beekeeper will be obligated to remove it (cut out this comb) in order to restore the proper number of frames and spacing. This could be a messy task, especially if there is brood and/or honey in the comb.

Any space less than the bee space will be filled with propolis or bee glue. This helps to keep cold air drafts out of the hive and kill any microbes that may live in these tiny spaces. See more about propolis in Chapter 12.

In the United States it is illegal to use beehives with fixed honeycombs such as straw skeps or log gums. Up until the mid-1800s, colonies of bees were kept in skeps or straw domes. To collect honey, the bees were killed, and the honey and wax combs were cut out and separated. Today, all bee equipment must have removable frames so bees can be inspected for diseases. If you have any questions, check with the state bee inspector or other beekeepers in your local group.

Frames

Bees in the wild attach honeycomb to the ceiling and often to the walls of a cavity. Such combs can be removed only by cutting or breaking them from their attachments. In modern beekeeping equipment, hive bodies contain *frames*, which are rectangular structures made of wood or plastic, each holding *foundation*, which is a sheet of beeswax or wax with a plastic base, embossed with either hexagonal worker or drone cells. When the foundation has been drawn out and completed with finished wax cells it is called *drawn comb*. There are now one-piece plastic frames with embossed plastic "foundation" available in several colors; darker colors make it easier to see eggs; again, check bee supply companies.

The vertical parts of frames, called *side bars*, are designed to space the frames in the hive; see the illustration showing frame sizes. Thus, when bees have fully drawn wax into honeycomb, the natural "bee space" is created between combs. This space allows the bees to move freely from comb to comb and the queen to lay her eggs.

As illustrated, frames come in many sizes and styles (European countries have more sizes, so be careful if you are ordering hive furniture from international sources). The top bars of frames will have either a *wedge* that is removed and then nailed back once the foundation is in place, or grooves that allow you to slide the foundation in. The bottom bar may be either solid, grooved, or two pieces. Before the foundation is set in, you should string the frame with wire. This will add support for the foundation, which is especially important for frames that will be placed in a honey extractor. Because the extractor spins frames at high speeds, stress is placed on the comb and may cause it to buckle or break. The wire support will prevent breakage in most cases. It is also important to make sure the foundation is perfectly flat in the frame; otherwise uneven or warped comb will result.

A simple jig can be made or purchased to facilitate wiring frames. It is a good idea to hammer eyelets in the holes on the wooden side bars before wiring; otherwise the taut wire strung through these holes may cut into the wood, which can cause the wire to slacken. The wire used for stringing frames is no. 28 tinned, although any thin wire that won't rust will do. Some beekeepers use 40-pound fishing line or monofilament instead of wire, in which case there is no need to embed the line in the foundation. However, there are reports that bees can cut the monofilament, especially if there is no strong honeyflow when these frames are installed.

If you decide to use plastic-based foundation, or plastic sheets sprayed with beeswax or plastic molded frames that already have the hexagonal honeycomb, wiring is not necessary. Check with your bee supply dealer or with other beekeepers to compare notes on which is the best to use in your area (see "Foundation" in this chapter).

With wired frames, the wire is embedded into the foundation by lightly heating the wire. Two tools can be used for this purpose. One is called the *spur embedder*, which is heated in boiling water and then rolled over the wire. The heat melts the wax around the wire; when hardened, the wax holds the wire-foundation combination in place. Another tool, called an *electric embedder*, heats the entire wire with

**Frame Sizes
(inches)**

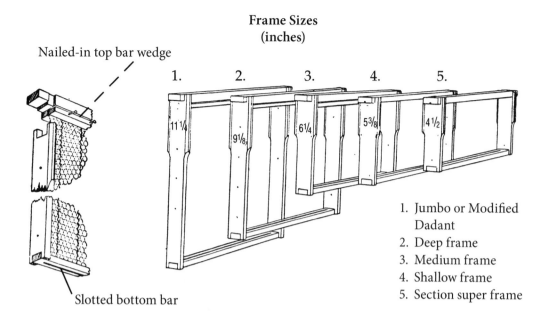

Nailed-in top bar wedge

Slotted bottom bar

1. Jumbo or Modified
   Dadant
2. Deep frame
3. Medium frame
4. Shallow frame
5. Section super frame

an electric current, which melts the surrounding wax enough for it to become embedded.

When the frames are fitted with sheets of wax foundation, start with 10 frames in the hive so the bees will draw out the foundation into even combs. Some beekeepers later remove one frame from either an 8-frame or 10-frame box, allowing for easier manipulation of the remaining frames; this can be applied to the brood boxes and/or the honey supers. To evenly space these 9 frames, use special spacers or follower boards, or merely space the 9 frames evenly by hand. If you do not space the frames evenly, bees will fill in any larger gaps with additional comb or expand the comb to fit the empty space. Make sure to add all 10 frames of foundation to a new box so the bees will draw out the comb correctly. This is especially true for the plastic foundation.

Empty frames—those that contain no foundation or comb—should not be placed in a colony, because the comb may be attached in an unsatisfactory way, such as at right angles, which would fasten all the frames together and make their removal and inspection impossible. Some beekeepers will use washed hardware cloth or plastic screen instead of foundation or, for reasons of economy, will put only a small strip of foundation (wax or plastic) attached to the top bar of a frame. Although straight, even comb may be achieved by this method, the practice of putting in small foundation strips is not recommended:

with deep frames, the bees may construct drone-size comb rather than worker cells, or fill the empty space with crooked comb, and it will not be wired, further weakening the comb.

With the advent of the varroa mite, some beekeepers practice mite control by encouraging drone production, because this mite prefers to infest drone larvae. You can now purchase either drone foundation or plastic drone frames. Another technique is to remove a deep frame and in its place substitute a shallow frame. Bees will draw drone comb beneath the bottom bar. Once the varroa mites are attracted to the drone brood, the infested pupae can be removed and frozen to reduce the population of mites (see "Varroa Mite" in Chapter 14). **Make sure** to label the top bar of this frame so you can find it easily. The drone comb can always be removed in the fall; this method should be used only for a few colonies; otherwise, buy drone foundation and, again, label those frames.

A word about plastic frames and plastic foundation: Some suppliers sell one-piece frames. While beekeepers report good success with such frames, and they may help with mite control, others do not use them at all. If you wish to experiment, try plastic frame units on a small scale; they may or may not work well in your area. Remember, if you find foulbrood in your colonies and need to burn frames, the plastic frames will not burn cleanly, and it may be illegal to burn

Frame Parts and Foundation

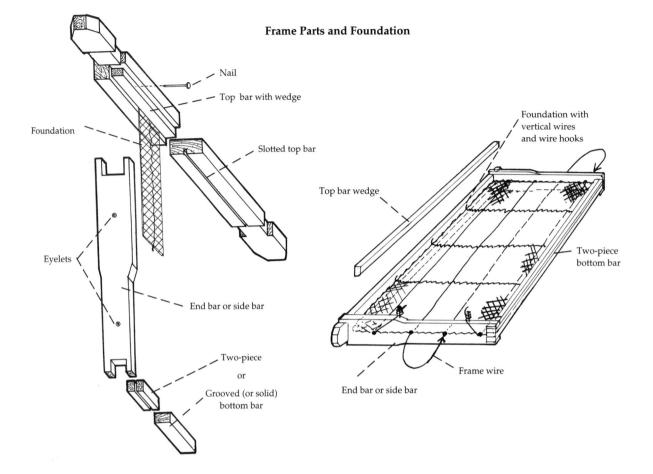

them because of the toxic gases released. You will then have to find another way to destroy or otherwise sterilize plastic equipment. Irradiation chambers exist in some areas for sterilizing diseased equipment.

Foundation

Foundation is a thin sheet of beeswax, or wax with a plastic base, embossed with the hexagonal shape of honeycomb cells. Sheets of plastic embossed with the hexagonal comb are also available; they come with or without a wax coating. They are sold with worker-size or drone-size cells.

A foundation sheet is set in the rectangular frame and wired or pinned in place. When it is put into a hive of bees that are gorged with honey, nectar, or syrup (a state that induces worker bees to produce wax), the bees will use the foundation as a base and draw out the walls of the cells. They do this by adding secreted wax to the base, then by drawing up the sides of the cell walls. When completely filled out, the frame is said to contain *drawn comb*.

The kinds and combinations of foundation can be confusing. Basically, there are three different categories:

● 100 percent beeswax (with or without vertical wires).
● Plastic sheets without beeswax.
● Plastic embossed sheets with beeswax sprayed on top.

Different thicknesses of wax foundation also exist, but basically the thicker sheets of 100 percent beeswax with inlaid vertical wires are used for brood frames and in the extracting supers. The thinner 100 percent unwired beeswax foundation is used in honey supers for producing cut-comb, round, or section comb honey. (Unwired frames should **not** be placed in a honey extractor.) Although most foundation sold today contains worker-size cells, drone-size-cell foundation, now in plastic, is also available for varroa mite control (see below). Drone foundation is also used to increase drone populations in queen mating yards to ensure there are enough drones to mate with virgin or newly mated queens (see Chapter 10). It is also used in honey supers.

To set the foundation firmly in the frame, especially in the brood frames, horizontal wires (or monofilament fishing line, 40-lb. weight) should be first strung across the wooden frame (see the illustration of frame parts and foundation on p. 48). Some wax foundation also has the sides encased with metal for better support and to keep the sheet from bowing. Use wired foundation, deep or shallow, in these wired frames that will go through a high-speed extractor. When a wired frame contains unwired foundation, the resulting comb will break apart when it is run through an extractor, and pieces of wax will "blow out," leaving large pieces of comb in the extractor. The embedded wires also help keep the comb from sagging, keep the cells from stretching in warm temperatures, and strengthen the honey-filled frames against breakage during extracting. Distorted cells in sagging combs are unsuitable for raising brood and decrease the number of cells in which the queen can lay eggs, which results in spotty brood patterns. Such frames should be replaced with new foundation or newly drawn comb.

The plastic-core foundation sheets are easy to install and require no wiring. Drawbacks include the bees not drawing new comb on the plastic sheet, should the wax separate from it. Bees prefer the 100 percent beeswax foundation.

Smaller worker-size-cell foundation has become the focus in controlling varroa mites. Some beekeepers have tried the smaller, 900-count foundation, but converting bees to the smaller cell size will take several generations, and most bees don't like to work the small foundation. Africanized bees, however, work this smaller comb very readily. The most recent research has found that small cell size does **not** reduce the varroa mite populations.

Foundation can be stored for a long time but should be kept sealed and away from heat or freezing temperatures. If kept in large plastic bags in a cool, dry place, the wax sheets will remain fresh and soft. Never store foundation in hot or sunny areas, as the wax will melt. The plastic bags will protect it from wax moth predation. Be aware that foundation may now contain pesticide contaminants. Check "Beeswax" in Chapter 12 for more information.

Note: Check with federal or state bee labs to see where you can get your wax tested for pesticides. All old wax will have some pesticides in it.

OF SUPERS AND HIVE BODIES

Some beekeepers use only standard or deep supers; others use the shallower supers for housing both the brood and the honey. If the standard 10-frame deep hive bodies are used, lifting off the honey will be very strenuous, because a deep super full of bees, brood, wax, wooden frames, and honey can weigh over 100 pounds (45 kg).

If, on the other hand, you choose only the medium supers to house your colony, the weight will be less, as one full 10-frame medium super weighs about 65 pounds (29 kg). Finding the queen in the broodnest, however, becomes much more time consuming and disruptive to the colony. If two medium supers are used as the broodnest, the queen can move more rapidly from frame to frame when you are in the hive than if she were on deep frames; in this case, it may take you longer to locate her, and thus you will be keeping the colony exposed longer.

The number of hive bodies needed to house bees varies throughout the season; in general the broodnest occupies the equivalent of two deep supers. In regions with cold weather, the extra honey supers are removed, and the bees winter in two deep and one or more shallow supers; the top deep and the shallow are used for winter stores of honey and bee bread. But this can vary; other combinations are one deep and one shallow, two deeps, or sometimes even three deeps or three shallows (see "Wintering" in Chapter 8). Whatever configuration is used, an ample supply of provisions must remain or be supplied before the onset of cold weather.

It has been traditional to paint the exterior of hive bodies white to reflect the sun's heat in the summer months, thus keeping the colony cooler, especially in southern or warmer regions of the United States. Remember to paint the metal top of the outer cover white; this will keep it cool to the touch during the hot summer months. Although white is most favored in southern climates, beekeepers in northern areas might consider painting hive bodies darker shades to retain the heat longer. For hives located in highly visible sites where vandals or thieves may be a problem, other colors can be used to camouflage hive equipment.

Whatever color is used, the sides and rims of the wooden hive parts should be painted to extend the life of the equipment. Because bees produce moisture as part of their metabolic activity, a latex paint would be least likely to blister as the moisture leaks out; lead-based or other toxic paints should never be used. Paint the bottom and sides of the outer cover and bottom boards (especially underneath) to extend their life, and make sure to paint underneath the ridge of the handholds.

To save money, you can buy off-color latex paints from hardware or paint stores; by mixing the paints together, you can create your own unique hive colors. It is unnecessary to paint or coat any interior hive parts with any substance. Some beekeepers paint the rabbets (see the diagram of a beehive), or apply a thin layer of petroleum jelly (such as Vaseline), to reduce propolizing in this area.

Some pieces of equipment, notably the inner and outer covers and the bottom boards, are available in plastic, but their suitability compared to wooden equipment can be disappointing. Visit with other beekeepers to obtain information about using plastic equipment in your area.

Beekeepers who live in damp and humid regions should consider other additional measures to extend the life of the hive furniture. Immersing wooden parts in boiling paraffin is an excellent way to preserve them (see Appendix D, along with the internet and the References for more options). This method works well but takes significant physical effort to heat the wax and can be dangerous, thus requiring great care. It is economical if you have a lot of equipment to dip; or you could combine your efforts with other beekeepers. Many wood preservatives used for lumber can be very toxic to bees and should not be used.

If your hives are in an area where loss through theft is a concern (or if you keep many colonies in various locations), brand the wooden hive parts with individual identification. However, branding your name or symbol on hive parts merely alerts the thief that eradication of such marks is necessary. A less expensive alternative is to leave your name and address under the tin roof of an outer cover, or, with a permanent marker, write your name **inside** your boxes; even if the bees cover your name with propolis, it can be easily scraped away. Or you can add or leave out an extra nail, or otherwise individualize equipment to assist in proof of ownership in the event of theft. If you have just a few hives, the extra time may not be worth the effort. New microchip technology now exists, and this option should be explored; check recent bee journals for the latest in identification technology. You can now locate your apiaries using GPS as well, so experiment and chat with others about the newest techniques.

FOR THE BEGINNER

To the beginning beekeeper, the plethora of equipment available from the bee catalogs may seem somewhat confusing. The basic equipment listed in the "Beginner's List" below provides a good starting point.

It is not a good idea to keep just one hive of bees, because if the colony fails to develop properly, or the queen becomes injured or dies, you will lose an entire year's experience at working and learning the art of beekeeping. Two to five hives is a better number for the beginner, or go in with a partner, each getting two or three hives.

While used hive bodies and frames are less expensive than new equipment, they could be contaminated with brood diseases that are not readily apparent. If equipment is questionable, it should be sterilized. The most economical way to sterilize hive bodies, bottom boards, and inner and outer covers is to scorch them with a propane torch, but this is not always reliable or effective. Frames are too difficult to sterilize and should be discarded.

A more effective sterilization technique is to use a boiling paraffin dip. This should be done only with hive bodies, not frames (see above); "fumigation chambers," where beekeepers could take diseased equipment to be sterilized, are generally no longer available. Check local beekeeping groups to see if they have one.

Warning: The use of boiling paraffin is very dangerous, and unless you have experience using heated liquids, solids, or gases, we urge you not to venture in this direction.

Other, more expensive equipment, such as honey extractors, can be shared by several beekeepers on a cooperative basis. Some small-scale beekeepers take their honey supers to a sideline or commercial beekeeper for extraction. Work out a fee beforehand, usually a percentage of the honey extracted. The joy of harvesting your own honey is part of the lure of becoming a beekeeper, but the actual process of extracting is a messy and sticky affair. Letting someone else who has the right equipment and setup do the extraction is a better alternative for most beginners.

Beginning beekeepers are cautioned not to buy every gadget on the market. When in doubt about the usefulness of a particular piece of equipment, seek the advice of more experienced beekeepers.

Beginner's List

Most hive parts come disassembled, and some require painting. More durable hive bodies are made from cypress, which has a longer life than pine or other softwood. Check prices and available equipment from various bee supply houses or from your local bee club. Generally, when you are first starting, it is best to buy all new equipment from one source. Hive dimensions vary just enough to cause problems when you mix and match hive furniture from different suppliers.

The current trend is to purchase hive furniture completely assembled and sometimes painted. This is certainly a time saver, but it will cost more, and you will lose out on the fun and education of putting together your own equipment.

While different bee supply companies have a variety of beginner's kits, a complete kit for one hive should include:

- 1–2 standard deep hive bodies, inner and outer covers, bottom boards, and frames (unassembled or assembled). Many beginners favor the medium, eight-frame boxes.
- 10–20 sheets of foundation to match size of hive body.
- 1 smoker.
- 1 veil or jacket with veil.
- 1 hive tool.
- Gloves.
- Entrance reducer.
- Bee brush.
- Feeder.
- Beginner's book and assembly instructions.
- 1–2 medium (6⅝ inches) or shallow supers and frames.
- 10–20 sheets of wired or unwired foundation, medium or shallow.

- Queen excluder.
- Smoker fuel.
- Subscription to a bee journal.

Total cost, $400–$700

Optional Equipment

- Paint (latex).
- Mouse guard.
- Bee suit, zip-on veil (or helmet and veil separate).
- Division board or top feeder.
- Bee escape or escape board.
- Roll of duct tape.
- Uncapping knife, electric.
- Extractor (two frames, hand powered).
- Jars for honey and labels.
- Another good beekeeping book.
- Year's subscription to another bee journal.
- Bee club membership.

Total cost, $400–$600

The prices listed are approximate for 2020, and they will vary somewhat depending on make, supplier, and so on. Shop around to get the best deals.

BEE WEAR

There is no stigma to wearing protective equipment; it will save you and your visitors a lot of worry if you are properly attired to look through your colonies without the fear of being stung. You will learn to gauge bee temper, handle equipment properly, and understand basic bee management by working bees. Listed below are the essentials.

Veil

A bee veil is a must. Although you will see photos that show beekeepers working without veils, such practice is discouraged. Stings on eyes, lips, scalp, or inside the nose or ear canal are extremely painful; it is downright foolish to risk them. All sensible beekeepers wear veils. Veils can be purchased separately or attached to helmets.

There are several kinds of veils:

- Wire-mesh veils (square, folding, or round) worn with a helmet.
- Sheer veils (tulle or nylon) worn with a helmet.

- Alexander-type veils, worn without a helmet, held in place by an elastic head strap.
- Veils attached to coveralls, with or without helmets.
- One-piece bug bag that fits over the entire torso and arms.

Check with other members of your bee club and try on different veils and suits to determine which type you like best.

Helmets

Helmets are worn to support veils, to keep the veils away from your face, and for shade. They are available in plastic, cloth, or woven mesh styles, with or without attached veil material. Some helmets have ventilation holes or louvers, but remember, if you live in an area that has the smaller Africanized bees, they can get through these holes. You will have to tape over or stuff cotton in these holes to keep them out.

Most helmets have bands inside that allow you to adjust them to your head size, but none fit very well. Helmets have a tendency to slip to one side or onto your face at inappropriate moments (such as when you bend over or are looking up at a swarm) and can create problems for the inexperienced beekeeper.

Felt or straw hats are sometimes used as helmets but are not usually strong enough to hold up to rough use. Veils and helmets are typically worn with a coverall; if veils are attached to the coverall with a zipper, the outfit becomes a *bee suit*. Jackets are also popular with beekeepers.

Coveralls and Bee Suits

Bees are less likely to sting people wearing light-colored attire. They are more prone to sting dark, furry objects (such as skunks and bears). Wearing lightweight, breathable material is preferable, because the best part of the day for working with bees is usually also the hottest part of the day. However, fabric that does not "breathe" well, such as nylon, may cause you to sweat more than usual, and heavy perspiration may aggravate the bees. Rule of thumb: fabric light in color and weight is best.

The easiest, most foolproof clothing to wear is a one-piece bee suit (with a zip-on veil), either jacket length or full length. **The ventilated mesh jackets and coveralls are rapidly becoming the favorites; again, check bee supply companies**. Bee suits come in handy if you are working a lot of colonies. Even though bees are able to sting through them, they rarely do. These suits keep your clothes cleaner and protect them from wax, honey, and propolis (the latter is difficult to remove thoroughly from clothing). If you are purchasing a bee suit, buy the best one you can afford, for cheaper ones tend to wear out quickly. The suits should incorporate all the features discussed in this section. Bee suits or coveralls, whether purchased or homemade, should be made of white cotton or a white polyester-cotton blend and have pockets and pouches to carry hive tools, matches, and the like. All wrist and ankle cuffs should have elastic in them to fit tightly, with no gaps. Some suits have zippers on the legs, which make getting into the suit easier, especially if you are wearing boots. Make sure the pockets of purchased bee suits or coveralls do not come with openings that access your inside pockets. Bees could get in.

If you are not wearing a bee suit, turn up your shirt collar before putting on a bee veil and close the trouser and sleeve cuffs with tape or elastic to keep bees from getting beneath clothing. You can also wear gauntlets. If using a jacket-style suit, some beekeepers tuck their trousers into their shoes or socks. If clothing is not closed tightly, bees can crawl inside unnoticed, and when a bee is pressed between clothing and skin, it will sting. Once a bee gets inside the clothing, you may attempt to release it, but it is easier to crush the bee before she stings you.

If bee suits do not come with a helmet, wear a visor cap or some other hat underneath to keep the sun off your face. Remember, veil mesh does not shade your face from sunburn. Sunscreen is always recommended.

Some bee supply companies sell child-size suits. Check online or in bee journals for other styles of suits. If you still can't decide, have your local bee club put on a "beauty" contest and show off the various bee wear. Wash your bee suit periodically to prevent the accumulation of dirt, disease spores, and venom from stings.

Gloves

Many old-time beekeepers disdain using bee gloves, but for the beginner it is a good idea to start with them. Again, there should be no stigma to wearing gloves while working with bees. As you gain experience, you will probably not wear your gloves, at least most of the time.

Gloves sold today at bee supply outlets are made from cloth or canvas, plastic-coated canvas, leather, or some kind of plastic. Wearing gloves that do not fit well will make handling frames more awkward and may even invite more stings than not wearing gloves at all (yes, bees can sting through leather). All bee supply companies carry men's sizes, but the smaller women's sizes may be harder to find; check the internet or other bee supply houses. Buy gloves that fit snugly. Even dishwashing gloves work well, and they can be replaced easily. A double layer (two pairs) of medical latex gloves, now widely available, does not interfere with dexterity and can reduce or even eliminate the risk of stings; however, your hands will sweat a lot. These gloves are disposable, thus reducing the risk of moving spores and other diseases among colonies. Work with other beekeepers to observe how they use (or don't use) gloves; you can learn much by watching others work their bees.

Gloves are a great help in keeping wax, honey, and propolis off your hands, but wash the gloves periodically, especially after working with diseased bees. One disadvantage of gloves is that they may retain the alarm odor long after bees sting them, which is another reason to wash them regularly. Most times, however, gloves are not necessary, and you should not make a habit of using them—you can get stung more wearing gloves than bare-handed. And never work someone else's bees with your used gloves, for they may contain foulbrood spores. In a pinch, latex or rubber gloves are a great alternative.

Some people wear leather gloves when applying miticide strips for varroa control. These are **not** recommended. Chemical-resistant rubber gloves must be used when applying miticide strips (see "Varroa Mite" in Chapter 14).

If you wear a black plastic or leather watch band, remove it during bare-handed apiary work, because such bands seem to incite bees to sting your wrists.

HARDWARE

Hive Tools

Several types and styles of hive tools are available from bee suppliers, including Teflon-coated tools.

The scraper nail-puller type can usually be found in most hardware stores (see the illustration of hive tools) and comes in 7-inch or 10-inch lengths—buy both. Hive tools are an invaluable aid to the beekeeper when prying apart hive bodies and frames that have been propolized.

It is a good idea to have several hive tools on hand, because they are easy to misplace. Hive tools should be periodically heat sterilized and scoured clean of excess propolis and wax; use steel wool and sandpaper. Paint your hive tools with bright colors (or stripes) such as a neon red, orange, or yellow, but do not use blue, green, or brown (which are camouflage colors). Painting your tools bright colors will help keep them from being lost in the grass and is a way to identify your tools from someone else's. The ends should be sharpened at least once a year, and the sharp corner edges blunted.

Try several kinds of hive tools and see what kind beekeepers like to use in your area.

Smoker

The second most important piece of equipment you will need, besides the veil, is the smoker. The smoker is a metal cylinder, in which a fire is lit, with attached bellows. Start the smoker by following these steps: (1) ignite some paper, (2) drop it to the bottom of the smoker, (3) pump air through the smoker by way of the bellows, and add more fuel once the lower layer is well lit. Smoke blown from the nozzle (see the illustration of the smoker) is directed into the hive and between the frames to encourage bees to gorge honey. Once engorged, bees are more docile and less prone to sting. Also, smoke masks the alarm pheromone bees release when disturbed.

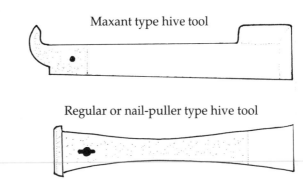

Maxant type hive tool

Regular or nail-puller type hive tool

Hive Tools

Bee Smoker

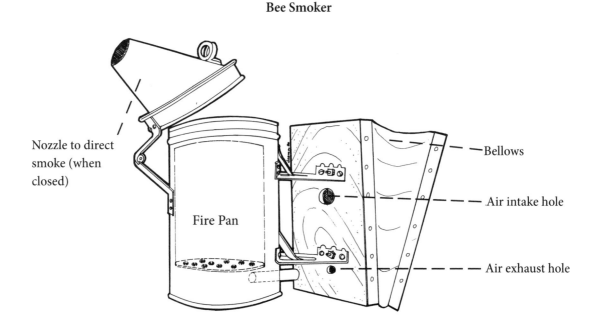

Nozzle to direct smoke (when closed)

Fire Pan

Bellows

Air intake hole

Air exhaust hole

Smokers come in 4 × 7 inch and 4 × 10 inch sizes. The small size generally does not hold enough fuel for more than a two-colony inspection trip; when purchasing a smoker, get the largest available size. For long life, buy a stainless steel smoker with an attached heat shield and front hook. See Chapter 5 for more information on lighting and using your smoker.

OTHER EQUIPMENT

Frame Grip

When held in one hand, this spring-loaded metal tool enables you to grip the center of the frame's top bar and pull out the frame in one motion. This is especially helpful to those who find it difficult to pull out frames. The frame must first be loosened with a hive tool, to break the propolis seal. However, if the grip is not used correctly, you can scrape or injure clinging bees or the queen if she is on the frame. You must lift the frame slowly, carefully, and straight up out of the hive.

Bee Brush

This soft bristle brush is used to clear bees off a frame, especially frames with honey that you want to extract. It is not a practical way to remove bees from many honey frames. It does come in handy, though, when you need to exchange frames (such as honey) from strong colonies to weaker ones (without the

clinging bees and mites). In a pinch, you can use a handful of grass, soft evergreens, ferns, feathers, or leaves for brushing bees from a frame. Remember, moving frames around the apiary will also move diseases and mites around, so be careful not to spread problems.

Top Bar Hives

There is a recent trend to "go organic," which now includes using the African-style top bar hives instead of conventional equipment. The top bar hive (TBH), also called the Kenyan top bar hive (KTBH), was designed to replace fixed comb log hives commonly hung in trees and is used in many parts of Africa. The KTBH has sloping sides and wide top bars or slats that fit together, forming a roof or top of the hive. The sloping sides of the hive box allow the bees to naturally draw comb, and they normally will not attach the comb to the sloping sides. Top bar hives became popular first in developing countries because the materials needed for their construction are inexpensive (local lumber), and assembling the parts can be done with low-tech tools. They can be made as long (up to 25 top bars) or as short as needed, and they were designed to hang, not rest on the ground.

The use of this hive has become increasingly popular. Why? Many beekeepers feel it is more natural for bees to construct their own honeycomb rather than being given wax foundation enclosed in a frame. When bees construct comb without foundation, they are making it to fit the natural size of the bees. An even more important advantage is that with bees making their own wax, the comb will not contain the contaminants now found in beeswax, even in purchased foundation. This type of hive uses little if any commercial foundation and forces the bees to draw fresh wax, thus creating a very clean and uncontaminated wax. There are many websites on the top bar hive, so be sure to look them up (see References, "Equipment").

Make sure if you use this equipment, it is legal to do so.

Advantages

- Materials for construction are inexpensive and often can be obtained at a local sawmill.
- The hive is a single horizontal unit, eliminating the need to lift off hive bodies that are stacked one on top of the other (such activity is necessary for removing honey and carrying out normal hive inspections).
- Reduces storage requirements; no empty honey supers and unused hive bodies to put away.
- Can be made to hang off the ground, which protects colony from ants, skunks, and other ground pests. All the commercial ones come with legs, and those who make their own put legs on them.
- Can be adjusted to fit honey supers.
- No queen excluder is needed.
- Honeycombs can be used as cut comb honey or chunk comb honey.

- The top bars are movable; hence, the hive conforms with laws that require movable frames.

Disadvantages

- Honey from comb cannot be extracted in the same manner as with conventional frames.
- Extraction of honey from these combs requires them to be crushed.
- The top bar needs to be inserted with a small strip of wax foundation; from this base the bees will construct the rest of the comb, which could become distorted and make removing frames difficult. Also, this wax could be very weak and break.
- Occasionally, comb may be attached to the side walls of the hive.
- During strong honeyflows you cannot add supers to a top bar hive (unless you have made a modified KTBH); this means you have to remove the filled combs, and additional top bars need to be added to continue to take advantage of any major honeyflow.
- Unlikely to be used in commercial operations.
- Can be difficult to overwinter colonies in cold climates.

There are good websites on the subject; check the References under "Equipment." Before trying this, check with other beekeepers who are using this method to see if it is a good option for you.

 **Notes**

CHAPTER 4

Obtaining and Preparing for Bees

Whether you are about to become a beekeeper, or are a novice or an established apiarist, there are a number of options available for obtaining your first bees, or augmenting the number of colonies you now wish to maintain. Here are the options:

- Buy package bees.
- Buy or produce your own nucleus colony or established colonies.
- Retrieve colonies nesting in natural or man-made cavities or from bait boxes designed to capture swarms.
- Collect wild bee colonies established on exposed combs.
- Capture a swarm that is still in a cluster formation.

BUYING PACKAGE BEES

Package bees come primarily from the southern states, California, and Hawaii and are shipped in the spring by mail or picked up by dealers and trucked to their final destination. To order packages, look for advertisements in the bee journals, such as the *American Bee Journal*, or *Bee Culture* magazine. Also, check your state beekeeping organization's newsletter or web page. Bee supply dealers and the internet will provide other package producers as well. Most national bee organizations have a web page from which you can order bees.

Call several package bee dealers to compare prices and delivery dates, and order packages early in the winter months (December and January) to obtain the desired number of packages and choice of shipping dates. Request delivery of your packages three to four weeks before the dandelions and fruit trees bloom in your area; this will enhance the chances that the bees will have enough time to develop into a strong colony with sufficient food reserves to carry them through their first winter. Depending on a multitude of factors (such as a prolific queen, an abundance of flowering plants, long periods of ideal weather for foraging, and the health of the bees), it is possible to obtain some surplus honey your first season.

All necessary bee equipment should be ordered and fully assembled well in advance of the arrival dates of your packages. For more complete information on packages see Chapter 6.

You can also contact local dealers or beekeepers within your own state who purchase packages and queens for resale. These local dealers are usually experienced beekeepers who will not only sell packages but may also have available for sale nucleus colonies, established hives, and an assortment of bee equipment. By attending your state or regional beekeeping organizations, you will meet dealers from your area (and other beekeepers).

Generally, a 3-pound (1.4 kg) package of bees contains approximately 10,500 bees. Most of the bees will be workers; there will also be a single young, mated queen and a small number of drones (for comparison, an established colony in the early spring can contain 10,000-15,000 bees). This will provide a sufficient number of bees needed to begin a new colony. The approximate cost of a single three-pound package in 2020 is between US$130 and $150. Prices are subject to change and will reflect current conditions. Shop around online or with your local club to compare prices.

Advantages

- Easier for beginners to work (fewer bees than in an established hive).
- More adult bees than in a nucleus colony.
- The bees in the package have been inspected and are certified to be in good health and from healthy stock.
- Free of brood diseases (American and European foulbrood).
- A minimum number of varroa mites.
- Replacement queens should be easy to obtain.
- Available in two-, three-, four-, and five-pound units (there are approximately 3500 bees per pound), with or without queens.
- Provides an opportunity to observe the development of a package into a colony and on to its maturity.

Disadvantages

- Queen could be injured or have died in her cage (the queen cage); or the queen might have been poorly mated, or failed to mate.
- Drifting is common (bees fly into other hives or become lost), especially during installation.
- Success is partly dependent on the weather; if it is very cold, bees may not be able to access the feeder of sugar syrup.
- Upon installation of the package, there are only adult bees. The first new bees will emerge 21 days after the queen commences laying eggs.
- Bees may be carrying viruses.
- Package bees placed on foundation require gallons of sugar water supplied by the beekeeper over their early development. (This action will prevent the bees from starving, and the sugar water will fuel their wax glands for wax production.)
- Bees may require a protein supplement if there are insufficient flowering plants to meet their protein requirements (see Chapter 7).
- Cold weather or sustained periods of rain may prevent the bees from foraging. Unless the bees are provided with sugar syrup and pollen supplement, the colony's growth and development will be impaired.

- Varroa mites and small hive beetles may be inadvertently included in the package (see Chapter 13).

On occasion, you will notice that the bees in a package are not clustered and calm; instead, they are scurrying about, and they look wet. It is likely they have regurgitated the contents of their honey stomachs. In addition, there will be a layer an inch or two deep of dead, wet bees on the floor of the package. Do not accept a package from a local dealer that has many dead, wet bees. If the package was received through the mail, notify the sender immediately (see "When the Packages Arrive," in Chapter 6). In most cases, the live bees in this package will not survive.

Bees may become overheated because of high temperatures during shipment, or were unable to settle quietly in the package and instead became agitated and overheated. In an attempt to cool down, the bees will regurgitate the contents of their honey stomachs. In a normal colony when the temperature exceeds 95°F (35°C), bees collect water, return to the hive, regurgitate it, and by fanning their wings induce evaporative cooling. Thus, the soaking-wet bees in the package may result from the bees regurgitating stomach contents on one another in an effort to cool down.

Often when swarms are collected in makeshift cardboard or wooden boxes with inadequate ventilation, the same phenomenon is witnessed. Such a swarm has been referred to as a "cooked swarm." On the other hand, some dead but dry-appearing bees at the base of the package should be expected.

BUY OR PRODUCE NUCLEUS COLONIES

Nucleus hives (or *nucs*) consist of from three to five frames in a small, nucleus hive box. A nucleus colony is a smaller version of a normal-size colony. A nuc gives the new beekeeper the opportunity to witness most, if not all, of the components of a standard hive. Nucs contain adult bees, comb(s) with eggs and brood, usually a laying queen, and various amounts of stored honey and pollen. Some nucs contain a virgin queen, or a queen cell, or are left to raise their own queen (not recommended). Nucs and nuc equipment can be purchased from dealers or beekeepers.

Advantages

- Usually produced and cared for by an experienced beekeeper.
- May be available locally, and the seller (usually a beekeeper) is available to provide information, assistance, and answer questions. Easy to transfer the nuc frames into a standard hive body.
- A full complement of adult bees likely, and a somewhat equal distribution of eggs, larvae, and pupae. This holds true when nucs are purchased in the spring and summer months.
- Will develop into a full-blown colony much faster; it is a miniature colony in itself with all the components of a large colony. A nuc contains a queen, worker bees, and drones, as well as eggs, larvae, pupae, and continuously emerging young bees, plus honey and bee bread. (In contrast, a package contains only adult bees; assuming the queen is released from her cage shortly after the package has been installed, it will take at least 21 days for the first compliment of young bees to emerge.)
- Easy to transport and install.

Disadvantages

- Nuc boxes and frames, if not new or relatively new, may contain disease spores and/or pesticide residue.
- Queen could be old (more than a year old) or of poor quality and stock.
- Bees may require additional feeding and medication.
- The nuc will likely have varroa mites and/or small hive beetles.
- In order to transfer the nuc into a full-size hive, additional equipment will be required (i.e., bottom board, two deep hive bodies, shallow supers, frames, and an inner and outer cover).

Installing a Nuc

If you plan to purchase a nuc, be sure the bees are disease free (ask the seller for a certificate). All colonies, however, have varroa mites, and the nuc will likely have any number of these mites. Small hive beetles may also be present. Ask the sellers if they have checked and treated the nuc for pests, diseases, or mites. No matter the case, it may be prudent to treat for mites before or after the nuc has been installed into regular hive equipment. In some states, the nucs need to be inspected by the state's bee inspector before they can be sold. If the frames in the nuc are dark brown or black in color, it is likely the nuc seller is using nucs as a convenient way to dispose of old frames. On the other hand, you should not expect your nucs to contain brand-new drawn combs (which will be white to light yellow in color). Ask questions before you purchase.

Usually, since a nuc colony is in a small hive box, the bee inspector will need to certify it to be a healthy colony. The contents of a nuc (the bees and the frames) will eventually have to be transferred into a normal-size hive body (deep super). Whatever the number of frames in the nuc (three, four, or five), the remaining space in the larger hive body needs to be filled with additional frames of drawn comb or foundation, or a combination of both. Depending on the time of the year and how much foundation is used to complete the complement of frames, it may be appropriate to feed the bees sugar syrup and apply a miticide treatment after the mite load has been determined (see Chapter 13).

As the colony continues to grow, add hive bodies or supers as needed. A nucleus colony will develop faster than a package for reasons already stated, provided that the queen is young, healthy, well mated, and has a good pedigree. Although all colonies will have varroa, tracheal mites are no longer found.

Installing the frames from the nuc into a deep hive body is an easy task; simply remove the frames from the nuc and place them (in the same order as they were in the nuc) into the center of a deep hive body. Fill the remaining space with wax foundation or drawn comb. If you are adding wax foundation, it may be prudent to feed the bees syrup (see Chapter 7). As the colony grows, you will have to provide additional room by adding more hive bodies to the colony, and, depending on many factors, you may have to continue to feed this colony as it grows, especially if there is a dearth and/or you are adding undrawn combs.

Nucs are an easy way to transition into beekeeping. Small colony units are easy to handle, and the nuc will contain all or most of the elements you would observe in a normal colony. Naturally, as with all methods or approaches to transitioning into being a beekeeper, there are associated risks. However, if the nuc producer is a reliable individual and can provide you with a young queen and disease-free bees, you are off to an excellent start.

BUYING A SPLIT OR AN ESTABLISHED COLONY

Established colonies in one or two deep hive bodies, with or without shallow supers when purchased in the spring, summer, and early fall, contain adult bees (workers and drones), brood of all ages, eggs, a laying queen, and various amounts of stored honey and bee bread. These colonies are usually purchased in at least two deep supers (each containing 8, 9, or 10 frames) and may also include any number of shallow supers. Retiring beekeepers often sell complete colonies and extra equipment. If you need to have expert help, call your local bee inspector to assess the health and condition of such colonies. You may be buying more trouble than you can afford. Before moving any bees and used equipment, check and comply with all legal requirements (see "Moving an Established Colony Over Three Miles" in Chapter 11). An alternative to purchasing colonies is to divide or split some or all of your existing colonies. As with nucs, **these colonies should be inspected and certified free of diseases and pests**.

Advantages

- Hive is fully assembled.
- The seller is available for assistance and information.
- Contains a full complement of bees.
- An established populous hive usually produces surplus honey the first year.

Disadvantages

- Equipment could be of different types and sizes (different size brood chambers, shallow supers, and frames).
- Combs could be old and may have to be replaced.
- May require monitoring of the mite load and the presence of small hive beetles. Bees will have some level of varroa mite infestation. Sampling for these pests should be done to ascertain whether a treatment is required.
- The hive, including supers, frames, bottom board, inner cover, and outer cover, may be old and need to be replaced.

- Old combs have reduced cell size and may harbor pesticide residue, foulbrood, or nosema spores or viruses, thus requiring replacement. Some pesticide residue in the wax may result from the application of miticides into bee colonies.
- Queen's age unknown, unless marked and records of her age provided; also may be of inferior stock.
- Depending on time of year, colony may be preparing to swarm or supersede the queen.
- Honey may be crystallized (if so, it cannot be extracted from comb).
- If the hive was purchased from a *bee haver* (one who does not attend bees consistently) or by a non-beekeeper relative, the frames may be heavily propolized (stuck together), or the wooden equipment may be rotten.
- Large established healthy colonies would be very populous and thus difficult to work, especially for a beginner. Moving established hives can be a challenge; see "Moving an Established Colony Over Three Miles" in Chapter 11. Such hives may be heavy and require a minimum of two strong adults to lift them. All hive bodies, the bottom board, and the cover need to be tightly secured before being moved. If anything comes apart during any phase of the operation, the situation will become problematic.

A final note concerning the purchasing of established or nucleus colonies: obtain from the seller assurance that the bees are disease free and have a low mite count. Most states have bee inspectors, so request their assistance in evaluating the bees, or ask a fellow beekeeper to accompany you in order to inspect the hive and the equipment before making a purchase. In addition, comply with all state regulations dealing with moving bees from one site to another, especially if you are crossing state lines.

If dividing your own colonies (making splits), the advantage is that you are familiar with the condition of the bees and equipment, and you can save money. The problem with making splits is that you will need additional equipment. In addition, the queenless portion of the split requires a queen. The options are to allow the queenless portion to rear its own queen or to provide it with a queen or a queen cell.

For more information on splitting or moving colonies see Chapter 11.

RETRIEVING FERAL COLONIES AND SWARMS

Feral Colonies

A colony of bees is a composite of thousands of workers, usually one queen, and a limited number of drones. Before the advent of human-produced enclosures for bees (beehives), bees occupied cavities in buildings, often between their outer and inner walls, or a hollow section of a tree or some other cavity.

In general, the method of retrieving a wild colony, once the cavity is opened, is to cut the combs from their attachment and transfer them to empty frames (frames without wax); cut the combs if necessary to fit into the empty frames and fasten them to these frames with elastic bands (or wire or string). Place each frame in an empty hive body, with a bottom board and an inner and outer cover.

During the removal of the wild colony's nest and its complement of bees, many bees will become airborne. With some luck, by leaving the new hive body near the site from which the bees were removed, and if the queen was on one of the removed honey combs, it is very likely that by nightfall all or most of the bees will have retreated into the hive where the queen is located.

Now, secure the hive and screen the entrance for transportation to your apiary. The hive body with the cut-out combs will be difficult to work (inspect). Instead of trying to deal with it, place a second hive body above the first and, to entice the bees to move upward into the second box, bait it with a frame of open brood. Once the bees have moved into it and the queen has also relocated in the new box, reverse the hive bodies, putting the new one on the bottom board. Next, place a bee escape above the new hive body, and within a few days all the bees in the box with distorted combs will move down into the first hive body. Now you can remove the original body.

The hive body with the distorted combs can now be used as a bait hive to capture swarms; you can also melt the comb down for its wax. Often in the process of removing the bees and transferring the cut combs, the queen will be lost or killed. If this is the case, the bees will rear a new queen providing there are eggs and/or brood less than three days of age. Another hitch in this process is when the queen is somewhere in the area of the original colony and not in the hive body. In this scenario, the bees in the box may depart before nightfall and reassemble at the queen's location. From all this information, you can safely conclude that removing bees from cavities can be a daunting task.

Bee colonies inhabiting cavities in buildings and trees can often be removed and collected by a beekeeper, once the landowner grants them permission to do so. Usually a landowner wants the bees removed rather than having them exterminated. However, removing bees from buildings and tree cavities is time consuming, and you should not undertake such an endeavor without some carpentry experience. If you are going to attempt this activity for the first time, it is advisable that you be accompanied and assisted by someone with previous experience in this activity.

In addition, property owners should be informed that once the bees have been removed, it will be their responsibility to take care and pay for all necessary repairs, unless other terms have been previously agreed upon.

Often, a more time-consuming way of removing the bees from buildings and tree cavities is to use a cone-shaped screen attached to the wall or the trunk of a tree around the entrance, guiding the bees into an empty movable hive body (see illustration). This method is a slower way of retrieving the bees, but it eliminates removing a section of a building or felling a tree.

Many prudent beekeepers who engage in the activity of bee removals carry insurance to protect themselves and also free the property owner from liability issues.

Honey bee colonies occupying cavities in trees and buildings are not as common as they were prior to 1986/87. The reason for this change is a parasite, previously not found in North America, which was discovered in a bee colony in Florida. This parasite, the varroa mite, has eliminated most unmanaged bee colonies throughout their present range, which is in most areas of the world where bees are maintained. Today when "feral" bee colonies are found, they in all likelihood resulted from a swarm that exited a beekeeper's colony and subsequently occupied a "natural" cavity.

Several beekeeping books and an experienced beekeeper should be consulted before deciding on any method of removing bees from trees or buildings; see the section on "Bees, Beekeeping, and Bee Management" in the References. You can also check with local professional bee removers and accompany them (maybe you could help and learn the tricks of the trade).

A word of caution: If you reside in areas known to harbor the Africanized honey bee (AHB), you should consider the possibility that bees in a building, a cave, a tree cavity, or a bait hive or swarm may be Africanized. Therefore, before undertaking any of the procedures described above, be extremely careful and have a vehicle nearby in which to take shelter. If you do live in an AHB area, work with a beekeeper who is familiar with them before attempting it yourself. Check online maps for the spread of the AHB.

Advantages

- Retrieving feral colonies can be interesting and educational.
- It may be a source of free bees to augment weak hives, make nucs, or start new hives.
- You may collect extra wax and honey from removed combs.

Disadvantages

- Could require a great deal of labor with little reward.
- Bees could be diseased and/or carry a heavy mite load and would have to be monitored and checked periodically for signs of disease or parasitic mites.
- In the process of removal, the queen could be lost, injured, or accidentally killed.
- Bees could be of inferior stock.
- Queen is often difficult to find and capture.
- Owner of a structure from which bees were removed may expect beekeeper to repair dwelling.
- Bees could be Africanized.
- Beekeeper may be liable for others getting stung.
- After the bees have been removed and repairs made, if the cavity is not eliminated or plugged, bees from another swarm may reoccupy it, and the owner may attempt to hold the beekeeper responsible.
- Transferring comb from a cavity onto frames without comb is time consuming, requiring cutting the combs to fit into the frames and fastening these combs to the frames with elastic bands or wire. This arrangement is only temporary, and eventually these frames will have to be removed and replaced.

Obtaining Bees from a Wall

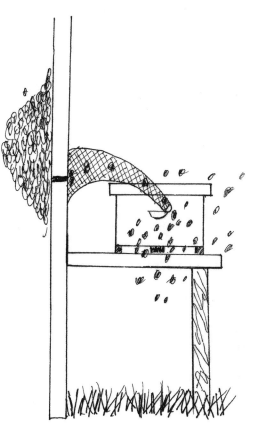

COLLECTING SWARMS

A swarm of bees is a portion of a colony that has exited a hive and temporarily clusters on some object, such as a tree branch, light post, or side of a building. Swarms are normally gentle and easy to collect, if they can be accessed without risk to life and limb. Given that adult bees carry the varroa mites with them, swarms will also contain these mites; thus after the swarm is hived and the queen begins egg laying, it would be appropriate to determine the varroa mite level in the new colony. If mite load is above an accepted threshold, the colony should be treated with a miticide.

Once a swarm has settled (a minimum of bees flying around), scout bees fly off the cluster and begin searching for a suitable homesite (a cavity) for the swarm. Capturing (retrieving) swarms was a common way for beekeepers to obtain bees free of charge. The swarm, once retrieved, may be used to start a new colony or be added to a weak one in order to strengthen it. Swarms either have exited from someone's colony or from a cavity in a tree or elsewhere.

A swarm normally remains at its cluster site for a short period, normally no more than two or three days, and as soon as there is agreement on a suitable cavity, the swarm will break cluster and fly to its new home. When a swarm alights on someone's property, most property owners do not want to have the bees killed and are grateful when a beekeeper shows up to collect them. Even municipalities are grateful to have the assistance of a beekeeper when a swarm clusters, for example, in a park.

Not only is retrieving a swarm a learning experience, but helping people who have a swarm on their property goes a long way in educating them and the public as to the important role played by honey bees. Today, because of liability issues and to avoid injury, we do not encourage taking risks in order to retrieve a swarm that, for example, may be clustered on a branch over 30 feet (about 9 m) above ground level.

Swarms can be a joy to collect under the correct conditions, especially when they are clustered in easy reach from the ground. Unfortunately, since the varroa mite was discovered in Florida in 1987, there are fewer "wild" colonies left to cast swarms.

Usually when a swarm alights, property owners seek an exterminator or (preferably) a beekeeper to remove them. Most swarms are calm and not prone to sting and therefore can be retrieved with a minimum of effort—again, provided that they are within easy reach. Occasionally, a swarm will exhibit aggressive behavior; therefore, as a precautionary step, you should always wear a bee veil and have a working smoker when collecting swarms. One additional caution: when you receive a swarm call, ask the caller to describe exactly why he or she thinks it is a swarm. People unfamiliar with honey bees and who see bees in a flowering bush (for example) may think they are witnessing a swarm. Or they may be seeing nests of various other stinging insects, such as white face hornets or yellow jackets. A trip to a swarm location may be time consuming, and to arrive there only to discover what the caller thought was a swarm turns out to be something else can be frustrating and time wasting.

The major problem with swarms is that not all of them are within easy reach, and it often seems that it is the larger ones that are beyond reach, creating a temptation to retrieve them at all costs. The risk to life and limb far outweighs any given swarm's value. Please be careful.

Before embarking on a swarm call, it would also be prudent to bring along some bait hives with pheromone lures, in the event that the swarm is out of reach. The possibility exists that scouts from the swarm searching for a homesite might select one of the bait hives for their homestead. Come back in a few days to see if they have occupied the bait hive.

Once the swarm is collected, either in a bait hive or in your swarm box, return to your apiary; it would be prudent to assess the colony for varroa mites three or four days later. If the mite counts reach the threshold levels for treating mites, then a miticide treatment should be applied, with follow-up mite counts as required. Swarms usually issue from strong, healthy colonies; however, you can't be certain that the hived swarm is not harboring disease spores of American foulbrood or other diseases. By monitoring the colony over the next four weeks, signs of the disease, if present, will be manifested. Consult with other beekeepers and/or your state's bee inspector as to the steps you need to follow should the disease's presence be confirmed. Another way to help prevent the occurrence of American foulbrood is to always hive a swarm on comb foundation. For a more complete discussion of swarms and how to retrieve them see Chapter 11.

Advantages

- If the swarm can be retrieved with ease, this is a great way of obtaining bees at no cost.
- Swarms captured in late spring may have ample time to develop into full-fledged colonies. An old saw in support of this says, "A swarm in May is worth a load of hay; a swarm in June is worth a silver spoon; but a swarm in July is not worth a fly." However, a late swarm (July) and even a fall swarm may have value when it is united to a moderately strong colony that has ample stores; by uniting the swarm to it, the swarm's ability to survive the winter will be increased. (Of course, swarming in southern or southwestern states takes place earlier in the year or even year-round.)
- Educational and interesting, swarms usually draw onlookers and on occasion the news media, providing the beekeeper with the opportunity to educate the public about honey bees. Media stories publicizing the activity of the beekeeper usually generate additional swarm calls from people wishing to rid their property of swarms.

Disadvantages

- Swarms will have mite loads and may carry diseases and viruses.
- Occasionally a swarm is unable to locate a new homesite, and its food reserves run out. Such a swarm is referred to as a "dry swarm" (the bees have used up all their reserves from their honey stomachs) and can become defensive.
- Another circumstance in which swarms exhibit aggressive behavior is when individuals, out of fear or malice, throw rocks or other objects at the swarm. Under these circumstances, the bees will release an alarm odor (pheromone); the odor permeates the area, and when someone approaches such swarms, the likelihood of being stung is real.
- The swarm could be Africanized bees.

BAIT BOXES

As soon as observers of bee behavior learned that honey bees required a cavity for their abode, individuals worldwide made an assortment of bait hives for the purpose of capturing swarms. Bait boxes furnished free bee colonies and hive products to these bee "trappers." The use of bee pheromones in these boxes helps lure scout bees to them. You can also leave out a nucleus box with a frame of dry comb to attract a roving swarm. These techniques work best when the box is placed off the ground, in a tree crotch, or otherwise elevated. Bees flying in and out of a bait box is not a guarantee that a swarm has taken up occupancy. It may just be scout bees inspecting the box. However, if bees are entering the box carrying pollen, this will serve as a clear sign there is now a bee colony present inside.

Advantages

- Bait boxes are relatively easy to construct or obtain.
- Bait boxes placed in your own apiary may reward you with swarms that may otherwise move elsewhere, possibly to another beekeeper's bait boxes.

Disadvantages

- Monitoring is needed to coincide with the prime swarm season.
- Bees need to be transferred into standard hive equipment, and if they have already made natural comb, this could be difficult.

- Bait boxes need to be securely fastened to trees, buildings, or poles.
- Bait boxes will also attract hornets and wasps, as well as Africanized bees. Caution is advised when checking bait boxes.

COLONIES THAT HAVE ESTABLISHED RESIDENCY ON EXPOSED COMBS

Occasionally, a swarm of bees fails to move from its cluster site to a new homesite (cavity). Such a swarm remains anchored to its cluster site and there initiates comb construction. With the presence of comb, bees begin collecting nectar, pollen, resins, and water, and the queen commences egg laying. The swarm reverts to a "normal" colony again, except that the nest is exposed to the elements. Two explanations have been proposed for this behavior: (1) scout bees fail to find an adequate homesite, or (2) inclement weather, accompanied by a reduction in daylight levels, stimulates bees to initiate comb construction. Additionally, Africanized bees are known to do this; if you live in areas where they live, you need to use caution.

Once foragers start to collect food, the swarm remains wedded to its cluster site. Over time, as the colony's population increases, more comb is constructed. The problem is that bees on exposed comb are unable to survive the winter months in northern latitudes.

In general, the method of retrieving a wild colony is similar to collecting bees from a cavity, outlined above, under "Feral Colonies."

Again, if you have any questions or problems, consult with local beekeepers in your group or the state bee inspector. Collections like this go best with a few other (experienced) folks who can help out.

Advantages

- As long as the exposed comb is within easy reach, retrieval should not be difficult.
- This activity is educational and interesting.

Disadvantages

- Bees could be diseased and will have a complement of varroa mites.
- Exposed comb may be entangled in bushes and shrubs, making it difficult to remove.
- The queen may be difficult to find and could be lost, injured, or accidentally killed in the process.

Ideal Apiary Site

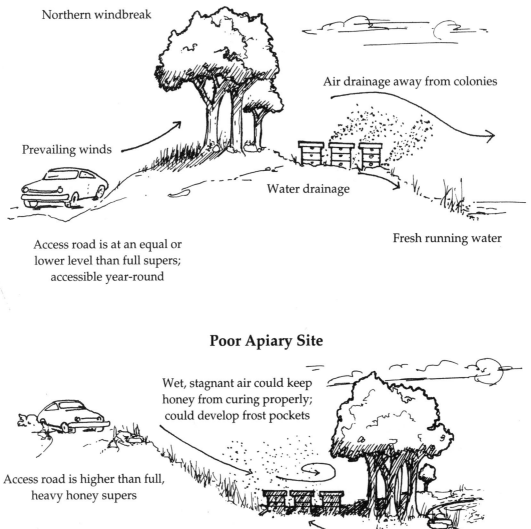

Northern windbreak

Air drainage away from colonies

Prevailing winds

Water drainage

Access road is at an equal or
lower level than full supers;
accessible year-round

Fresh running water

Poor Apiary Site

Wet, stagnant air could keep
honey from curing properly;
could develop frost pockets

Access road is higher than full,
heavy honey supers

Floodplain; colonies could be washed away

Drawing by J. Propst

- The process may require special bee equipment to properly hive the colony because the existing comb will have to be removed and fit into a standard hive body.
- Requires medication.
- The colony is not as gentle as a "normal" swarm.
- Bees may be Africanized.

THE APIARY

In this day of urbanization, changed farming practices, and public awareness of AHBs, or "killer bees," it is becoming increasingly difficult to obtain *apiary* (beeyard) sites. Many who keep bees as a hobby put bees in their own backyards, much to the concern of their neighbors; but this need not be a deterrent if you practice a "Good Neighbor Policy," as we discuss in this chapter. Maybe you can find a farmer or a friend with some rural property willing to take bees in exchange for some bottles of honey. In the case of a farmer, you can offer honey, as well as reassurance that having bees on the farm will immensely benefit any crops grown there that require pollination. Keep in mind that all apiary locations may require registration with appropriate town, city, and state agencies. Check with your state department of agriculture.

If possible, look for an apiary site that is close to fresh water, such as a lake, pond, or stream. In the absence of such, water **must** be provided. You can provide water in hive feeders or a shallow pan with water and some floating devices, so the bees will not drown. Recent studies indicate bees seem to prefer gathering "dirty" water, likely for its mineral contents. Bees have even been seen on cattle salt licks; this can be dangerous, as these blocks can contain insecticides. Cattle watering tanks are also favorite sites, in dry and arid areas, but the cattle owners may not like having their animals harassed by bees, especially if there are horses; bees and horses **do not** get along.

Water should be provided early in the spring so the bees can become habituated to this specific water site. In dry climates, a big problem is that bees will visit neighboring swimming pools and birdbaths. Bees appear to be attracted to some of the chemicals added to the water in swimming pools. Obviously, bee visits to pools will annoy your neighbors and often lead them to seek recourse from town officials or their lawyers.

Choose an apiary site that optimizes the following conditions (see illustration of ideal and poor apiary sites):

- Easy, year-round vehicle access.
- Near dependable nectar and pollen sources (within a 2-mile [3.2 km] radius).
- On upper sides of slopes to improve air drainage away from hives.
- Along the edges of open fields.
- With a northern windbreak for winter protection and noontime summer shade to keep hives cool.
- Near owner or neighbors who like bees (with clear commitments on both sides).
- With ground level and unobstructed enough so that the beekeeper does not have to be constantly looking down in order to avoid tripping.
- At least 3 miles (4.8 km) from other beeyards, to diminish the spread of diseases, dissemination of mites, and robbing (and killing of bees by other bees as robbing is in progress).

Thanks to modern technology and mapping search engines such as Google Earth, you may be able to learn what is in the flight range of your apiary, including water, food, factories, or parks. You could also zoom in to determine areas where bees might obtain good forage, such as swampland, fields, wild meadows, or other planted crops or orchards (which could get sprayed). In addition to searching for apiary sites, you may be able to locate other apiaries (depending on how recent the mapping is) that could be close to your apiary.

Sites to avoid include:

- Bottomlands where air tends to stagnate.
- Fire-prone regions or flood areas (check soils or land-use maps).
- Areas where vandalism and thievery may be prevalent.
- Areas where the flight path of bees will be an annoyance (public parks).
- Areas where black bears are prevalent (in recent years bears attacking beehives have become a common concern).

Bears are now relatively common throughout many states, and unless a state eliminates all its bee colonies or all its bears, there may be no way to avoid bear problems. In states or counties where bears are prevalent, the beekeeper has two alternatives: surround the hives with an electric fence, or move to an area known not to have bears.

Identification

You should post your name, postal address, email address, and phone numbers at each of your *outyards* (an apiary that is not located at your residence) in the event someone needs to reach you about your bees. A notebook with the names, addresses, telephone numbers, etc., of each property owner and where your outyards are located should be kept at your residence if you or others need to contact them. In the event that you can no longer tend to your outyards and you have to designate someone else to work your bees, they will be able to locate your yards.

The property owner may need to reach you for several different reasons: for example, a swarm has landed nearby; a bear has overturned your hive or has been seen lurking near your hives; suspicious individuals have been seen observing the hives; or the owner is about to sell the property and the new owners want the bees removed (or believe they can lay claim to them). The need for written agreements between you and the property owner that clearly state that the bees are *your* property will protect you should the owner, kin, or new owners claim the bees.

The property owner may also wish to be protected from liability and request that your insurance include them as being additionally insured on your policy. This process is now very routine and usually adds no additional cost to your homeowners policy. Certainly the fact that you can present this additional assurance to the property owners may influence their decision as to whether to grant you permission to place your hives on their property. Should the property owner inform you that he or she needs more time to make a decision, leave a jar of honey that contains your name, address, and telephone number. By this act, you are providing the owner with a way to get in touch with you; in a litigious society, where the gentleman's handshake is a thing of the past, everyone needs agreements in writing. Requisite information can also be enclosed in a waterproof plastic bag stapled to a hive or a nearby post. The posting will also enable the bee inspector, or anyone else, to get in touch with you.

Good Neighbor Policy

To keep on good terms with your neighbors, landowners, or other people likely to come in contact with your bees, here are some simple rules:

- First, make sure it is legal to keep bees at a particular location, especially if inside the city limits. Most municipalities have zoning and planning commissions that can provide you with information concerning honey bees.
- Keep your bees out of sight by planting tall shrubs in front to hide the beehives and to force bees to fly higher.
- Place hives behind evergreens, a stockade fence, or a stone wall, and paint the hives in colors that blend in with their surroundings; locate them in less traveled areas.
- Use a gentle race of bees, and keep them gentle by requeening.
- Learn to keep your bees calm; if possible, work them only during sunny, warm weather and during a nectar flow, and don't start a robbing incident.

- Erect a sturdy fence to keep out curious children and pets.
- Have no more than 2–3 hives at a home property site and not more than 20–30 in an out-apiary.
- Don't work hives if the neighbors are outside working in their yard (mowing or raking leaves) or holding outdoor events.
- Prevent your bees from swarming so as not to alarm your neighbors.
- Provide and maintain a water source so your bees are not in the nearby pool or birdbath or hummingbird feeder (use a bird or livestock fountain or Boardman feeder). A four-gallon plastic fount normally used to provide water for fowl can be used as an outside source of water for bees. Watering should begin as early in the year as possible, otherwise by the time you provide the bees with water, they may have discovered other sources, and as a consequence it may be too late to redirect them away. Do not underestimate the fact that bees require water!
- Give honey freely to next-door neighbors and a magazine or book explaining how beneficial bees are to people.

HIVE SCALE

A hive scale is a device that is placed under a strong colony and from which accurate records of weight gains and losses can be made. These scales can be a valuable piece of equipment. At times, marked changes can be recorded on a daily basis (for example, when a colony is strong and a heavy honeyflow is in progress); substantial weight gains can be witnessed in a single day. A substantial weight gain could be a half pound to several pounds (0.2 to 0.9 kg) daily gain. At other times, a hive scale may record subtle changes, or no changes.

It is important to have at least one scale per beeyard. Scales provide essential data to the beekeeper. If the hive scale indicates that a colony is gaining substantial weight over a successive period of days, it is a strong indication that a major honeyflow is in progress; the hive gains weight because of the incoming nectar. In such cases, the hives will need to have honey supers placed above the brood chambers as soon as possible. If the hives already have honey supers on them, you should reverse them, placing the ones containing more honey above the ones the bees have yet to work. If you plan to produce comb honey or cut comb honey, now is the time to put on supers dedicated to these purposes; at the same time, the brood chambers should be reduced from two to one, in order to crowd the bees. This will force them to work in this particular setup of honey supers (see Chapter 9 on comb honey production). If you are not proactive when a honeyflow is in progress, the colony can become "honeybound," which will reduce the space available to the queen for egg laying and may prompt the bees to initiate swarm preparations.

This is a good time to check what flowering plants or trees are in bloom in your area. If the scale shows that the hive has grown daily heavier, it means a strong honeyflow is in progress and the hives can be *supered*—that is, have extra honey supers (shallows or mediums) placed on top of the broodnest.

The scale often used in the past to weigh hives was a farmer's grain scale, but now several scales especially designed for bee colonies have become available. These scales may be obtained from companies or dealers that sell bee equipment, as well as some internet sites. Ask other beekeepers for suggestions, and check the "Equipment" section in the References for more information.

When a honeyflow is in progress, you should be alerted to do certain tasks, depending on the season:

- Add frames and/or supers full of foundation, as the worker bees' wax glands are stimulated during honeyflows. Timing supering to honeyflows almost guarantees that the foundation will be drawn into full honeycomb, so take advantage of such opportunities.
- Interchange the locations of weak and strong hives (see "Prevention and Control of Swarming" in Chapter 11) as long as the weaker colonies are not diseased or mite infested.
- Check hives for swarming preparations, especially during or shortly after a spring honeyflow (see "Swarming" in Chapter 11).
- Requeen or make splits when there is a good honeyflow in progress.
- Divide strong colonies to increase the number of colonies in your apiary.

Whenever the scale shows a hive gaining weight, check and note which flowers are in bloom in order to anticipate nectar flows in future years. Occasion-ally, hives may gain weight when no major nectar plants are in bloom. If you have been keeping good records, and this weight gain fails to correspond to previously recorded ones, and the temperatures are extremely hot outside, suspect that bees are harvesting water. Water is avidly collected in hot weather to cool the hives inside by evaporating droplets of water stored in the comb. If a colony becomes overheated, brood and bees can die, and comb could melt. Good beekeepers become good detectives.

Another occasion when hives gain weight and it isn't hot (and there is no honeyflow in progress) is when *honeydew* (a sugary liquid excreted by insects that are feeding on plant sap) is available. Don't forget: the possibility exists that a honeyflow, the availability of honeydew, and the need for water may occur simultaneously. Weight gains may also be attributed to the times of rapid population expansion in a colony (remember, 3500 bees = 1 pound or 0.45 kg).

On the other hand, when the scale records no change or shows the colony is losing significant weight, it's time to open up the colonies to see what is taking place. Bees could have swarmed, absconded, died, or been infested with pests; if such eventualities are caught early, you can help the colony survive.

During late fall, winter, and early spring, a slow but steady decline in weight should be anticipated because, with the exception of water, no other bee products are available. In these cases, colonies should be checked because they may be almost out of food and near the brink of starvation. Colonies in this condition need to be provided with supplemental food (sugar syrup, fondant, candy, dry granulated sugar, pollen, or pollen supplement). Naturally, a colony will show a weight loss when it casts a swarm, is diseased, is queenless, contains laying workers, is weakened by parasites or other maladies, or is being robbed.

NASA is interested in tracking climate change as it relates to the flowering times of bee plants and weight gains or losses in an apiary; for information, check out their website, https://honeybeenet.gsfc.nasa.gov/Sites.htm.

HIVE STANDS

The amount of bending and lifting that a beekeeper must do while working a colony can be minimized when the colony is placed on a stand about 18 inches (46 cm) above the ground (see illustration on p. 69).

Working bees should not require keepers to be a martyr to their hobby or vocation.

Hive stands, in addition to saving your back, will keep the hives dry, extend the life of the bottom board, and help keep the entrance clear of weeds. If stands are at least 18 inches high, they will discourage animal pests such as skunks, raccoons, and even coyotes (and coatis in areas where they live) from bothering the colonies. Additionally, wood that is continuously wet or damp, such as a bottom board, can quickly rot, and pests, such as carpenter ants and termites, are likely to nest in damp bottom boards (see "Minor Insect Enemies" in Chapter 14). Other pests, such as skunks and mice, have less easy access to hives that are placed on some sort of hive stand. Hive stands are also good in drier areas where fire ants, rodents, and snakes are a year-round problem. If ants are a continuous problem, paint the legs of a hive stand with oil or grease. Snakes have even been seen overwintering on top of the inner cover of hives that are on the ground.

In areas where dampness is not a problem, stands are sometimes constructed low to the ground. Low stands, especially if they help to create a dead air space underneath the hive, could provide extra insulation and enhance the wintering success of the colony.

When working a colony on a stand, set the hive bodies that are temporarily removed onto an empty super.

Types of Hive Stands

Make sure the colonies are level or slightly tipped forward, so water will not collect inside the hive; this is especially important in areas that experience a lot of rain. It is also important that the hives will not tip over in high winds when loaded with honey, or if the ground is saturated with a lot of rain.

Hives can be kept off the ground by placing them on any one or a combination of these materials:

- Cinder blocks covered with tar paper or shingles or painted with waterproof paint.
- Bricks or drain tiles.
- Wooden railroad ties, pallets, or 2 × 4 lumber, painted or covered with tar paper.
- Wooden stands made of durable lumber (but be careful, some wood preservatives are harmful to bees).

- Permanent cement platforms, metal stands, or even flat, level rocks.
- Pallets, unless with solid tops, are not recommended; you can easily twist an ankle or be easily injured especially when the platform for the pallet is rotten or weak. In addition, skunks and other pests have been known to live underneath these pallets—not a good thing for the bees.
- "Superhorse" or long work bench (see illustration on p. 70).

RECORD KEEPING

Diligent record keeping is the very key to success in beekeeping and likely in all occupations, whether you are raising bees for honey production, for pollination services, or for research. Maintaining accurate records will allow you to refine and tweak your activities and reduce or eliminate repeating previous mistakes or misinterpretation of honey bee activities. Relying on your memory, or by placing bricks in different positions on the hive's outer cover, will only work for a short time. When you are able to review detailed, written records about each hive, then and only then will you begin to see a bigger and clearer picture of the activities in your beeyards. The information gleaned from records will help you garner a true understanding of the activities of your bees and how environmental conditions may play a role in their success or failure.

Even if you just kept records of the daily temperatures, the blooming periods of flowers in your area, and the precipitation amounts each time it rained, over three or four consecutive years, you will become a better manager of your bees. For instance, you will be able to anticipate with greater accuracy when to add supers in advance of major nectar flows, and how the amount of rainfall or lack thereof affected the outcome of an anticipated flow. Let us assume you live in an area were the goldenrod is the last major flowering plant, which normally produces the last nectar in your area, and you and your bees are dependent on this flow to supply them with their winter stores. In the eventuality of a year of low precipitation and the flow fails, you will likely to need to begin feeding your bees to build up their stores well in advance of the cold weather.

Meticulous records of your bees are absolutely necessary if you desire to continually upgrade your stock. Although upgrading your stock should be your

High Hive Stand

Parts List

High stand (pressure-treated lumber):
- (4) 1 × 4 × 16.5 inches (42.9 cm)
- (4) 2 × 4 × 18 inches (46 cm)
- (2) 1 × 4 × 24 inches (61 cm)

Low stand (pressure-treated lumber):
- (2) 6 × 20 inches (51 cm)
- (2) 6 × 48 inches (122 cm)

32 inches (81.3 cm)

20 inches (50.8 cm)

Low Hive Stand

continuous goal, and its limitations, keeping records will help. For example, tracking your queens, when and from where they originated, is very important (see more in Chapter 10). The goal of most beekeepers is to have gentle bees that overwinter well, remain (within reason) disease free, and produce surplus honey, whenever weather and floral conditions permit (see "The Hive Diary" in Chapter 5). With reference to limitations, keep in mind whether you raise your own queens or purchase them; their mating habit makes it extremely difficult to obtain queens that are genetic duplicates of your best queens.

Financial records should also be kept for income tax and loan purposes and to determine the amount of income lost or gained in a season. During any given year you will have a certain amount of expenses, such as the purchasing of equipment, bees, vehicles, and the time employed in caring for the bees and assembling the equipment. All represent monetary expenditures. On the other hand, income represents the money you take in from pollination services, sales of bee equipment, sales of bee products (honey, pollen, propolis), and other services related to beekeeping (swarm or colony removal, if you were paid). If the expenses exceed the income, you have sustained a loss; if the income is greater than the expenses, you have made a profit.

On the actual beekeeping side, records of the con-

"Superhorse"

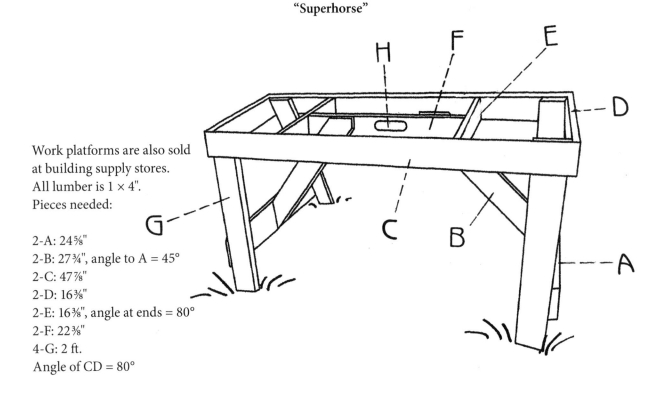

Work platforms are also sold at building supply stores. All lumber is 1 × 4".

Pieces needed:

2-A: 24⅝"
2-B: 27¾", angle to A = 45°
2-C: 47⅞"
2-D: 16⅜"
2-E: 16⅜", angle at ends = 80°
2-F: 22⅜"
4-G: 2 ft.
Angle of CD = 80°

dition of each hive will help guide you as to when queens need to be replaced, which colonies have mite infestation and the treatments provided, the times of major and minor honeyflows, the weight changes in hives during mild and not so mild winters, to name a few. A diary of the blooming times of major nectar- and pollen-producing plants (and crops) will help you anticipate the time important sources of forage will become available. This information makes it easier to plan wisely the activities necessary for successful beekeeping.

In summary, keep records for each apiary, if not each colony, and note such things as:

- Dates of all beekeeping purchases.
- Queen origin and age (marked, clipped, or unmarked).
- Equipment bought, destroyed, stolen, or sold.
- Mileage to apiary.
- Dead or stolen colonies, queens, or packages.
- Vendors where queens, packages, or equipment was purchased.
- Colonies killed by pesticide sprays.
- Medications for beekeeper and for bees.
- Organizational memberships; journal subscriptions.

- Lectures, talks, shows, and fairs attended or entered, with associated fees.
- Books and conference fees.
- Equipment for selling honey (labels, bottles, etc.).
- Amount of honey extracted, bottled, and sold.
- Amount of comb honey packaged and sold.

HIVE ORIENTATION

In most apiaries, the hives are placed in rows or are paired on hive stands in rows. The hives in a pair (on a shared hive stand) should be 6 to 8 inches apart (15–20 cm) and successive pairs spread 5 to 8 feet (1.5–2.4 m) apart. This minimizes vibrations and jostling while working a colony but keeps the hives close enough together to make your work more efficient.

When the hives are in long rows, there is a tendency for some bees to drift to the wrong hives. This drifting may be due to prevailing winds, which continually push returning bees toward the end of the rows. Drifting is undesirable and may lead to misinterpretation of hive scale readings. But the more critical problem that results from drifting is that it promotes the spread of diseases, mite dissemination, and robbing out of weaker colonies.

For all these reasons, placing colonies in long rows

Hive Inspection Sheet courtesy of Dadant & Sons, www.dadant.com

~ HIVE INSPECTION SHEET ~

Dadant & Sons, Inc.
51 S. 2nd Street, Hamilton, IL 62341
1-888-922-1293 • www.dadant.com
Since 1863

• Chico, CA (877) 332-3268 • Fresno, CA (877) 432-3268 •
• Paris, TX (877) 632-3268 • Sioux City, IA (877) 732-3268 • Watertown, WI (877) 232-3268 •
• Albion, MI (877) 932-3268 • Waverly, NY (877) 532-3268 • Chatham, VA (800) 220-8325 •
• Frankfort, KY (888) 932-3268 • High Springs, FL (877) 832-3268 •

Hive Obtained From: _____
Date Established: _____
Yard #: _____
Hive ID: _____
Date: _____
Weather Condition: _____
Who Worked Hive: _____
Next Inspection Due: _____

HIVE TEMPERAMENT
○ Calm ○ Nervous ○ Aggressive ○ Time to requeen

LOCATED QUEEN ○ No ○ Yes
(Marked? ○ No ○ Yes – Color_____
○ Replace queen – Date_____

LAYING PATTERN
○ Beautiful (Solid & Uniform)
○ Mediocre (Intermittent or random)
○ Poor (Spotty)

EGGS PRESENT ○ No ○ Yes Comments:_____

POPULATION ○ Heavy ○ Moderate ○ Low
○ Added deep body
○ Split hive (*new hive #_____*)
○ Swarming imminent – needs monitoring

EXCESSIVE DRONE CELLS ○ No ○ Yes
Drone Population Estimate:
○ Low: 30 ○ Ave.: 30 to 100 ○ High: 100+

QUEEN CELLS ○ No ○ Yes
Along frame bottom: #_____
Converted worker cell: #_____

DISEASE/PESTS ○ No ○ Yes
○ Chalkbrood ○ Nosema ○ Varroa Mites
○ Tracheal Mites ○ EFB ○ AFB ○ Small Hive Beetle
○ Other:_____

MEDICATIONS
Added – Date _____
○ Apistan ○ Apiguard ○ Mite Away II
○ Tylan ○ Fumagilin-B ○ Terramycin
○ Other:_____

Removed – Date _____
○ Apistan ○ Apiguard ○ Mite Away II
○ Tylan ○ Fumagilin-B ○ Terramycin
○ Other:_____

INTEGRATED PEST MANAGEMENT (IPM)
○ Screened bottom board
○ Powdered sugar mite drop

EARLY SPRING INSPECTION
○ Reversed brood box(es): #____ Deep___ Med.___ Shallow
○ Cleaned Bottom Board
SPRING FEEDING/BUILD-UP
○ Brood Builder: ____dry ____wet ____patties
○ MegaBee: ____dry ____wet ____patties
○ Fructose: _____
○ Pollen Sub _____
○ Sugar Syrup (1/1 ratio)_____
○ Other _____

SPRING/SUMMER HONEY FLOW PREPARATION
○ Added queen excluder
○ Added super(s): #____Deep____Med.____Shallow
○ Added ventilated inner cover
○ Added pollen trap (optional if sufficient stores)

HONEY REMOVAL/EXTRACTION
_____ # supers removed
_____ pounds honey extracted
_____ pounds comb honey
○ Remove excluder (if done with honey production)
○ Begin varroa control medication:_____

FOOD STORES/WINTER PREPARATION

	Honey	Pollen
High (Everywhere)	○	○
Average	○	○
Low	○	○
Near Brood	○	○

○ Fed Hive –
  ○ Sugar Syrup (2/1 ratio)_____
  ○ Other:_____
○ Added entrance reducer
○ Colony configuration: #____ Brood Boxes #____ Supers

HIVE CONDITION
○ Normal ○ Brace comb ○ Burr comb ○ Excessive Propolis
○ Normal odor ○ Foul odor ○ Equip. Damage
○ Replace Equipment – What:_____
○ Other:_____

Type of foundation: ○ Duragilt ○ Plasticell ○ EZ Frame
                    ○ Wired ○ Med. Brood ○ Cut Comb
Comb Condition or age:_____
○ Replace Foundation

Hive Orientations That Reduce Drifting

Horseshoe configuration   "S" configuration

Hive Orientation That Promotes Drifting

Hives lined up in a straight line disorient bees. Those at the end (dark boxes) will be stronger than those in the center (light boxes).

should be avoided whenever possible. In addition, during periods of nectar dearth, bees from strong colonies will drift into weaker ones and are likely to seize the opportunity to rob honey. Then finding their way back to their own colony, these robbers will recruit other bees to continue the robbing, and the weaker colonies can die. We cannot emphasize enough that the old practice of placing colonies in rows, no matter how aesthetic or convenient to work, needs to be curtailed.

Reduce drifting by placing the hives in a horseshoe configuration (entrances facing in or out), or by putting up a windbreak, or by shortening or staggering the rows (see illustration of hive orientation). If hives must be placed in a row, alternate the entrances front and back along the row to reduce drifting. If this is inconvenient (because you can't drive "behind" the colonies), paint the hives or bottom boards different colors or patterns, or place landmarks near the hives, such as rocks or bushes. In areas where there is a flat horizon, bees may not be able to find their way back to a colony because of the lack of vertical landmarks.

Consider erecting a snow fence or planting trees or shrubs to help the bees orient themselves.

RESOURCES

Numerous books, pamphlets, journals, web pages, and organizations are now available to beekeepers, on the internet and elsewhere. In addition, talking and working with other beekeepers can be an important way to learn more about the art and science of apiculture. For the beginner, professional, or part-time beekeeper, and for those who wish to learn more about beekeeping, local or state groups and university extension offices usually offer workshops, classes, or seasonal meetings so beekeepers can improve techniques or share experiences. Another reason for attending meetings will be the opportunity to hear from experienced beekeepers and individuals involved in honey bee research, as they present topics of interest to all beekeepers. To learn of bee organizations in your area, ask a local beekeeper, check meeting notices in the beekeeping magazines, or check your local newspaper or websites for dates, times, and location of meetings.

And don't forget state, national, and international conventions; they provide an opportunity to increase your knowledge as well as form new friendships with others in your profession. With changes in bee culture, especially with new pests and pathogens being discovered, it is essential to keep current. Take advantage of your computer and explore reputable (university) websites for more information on honey bees.

Notes

CHAPTER 5

Working Bees

It is important to know what you are going to look for and do before opening a hive. Be prepared so you can keep the amount of time spent in each colony to a minimum—no more than fifteen minutes, unless you are doing some specific task that requires more time. Each time a colony is examined, the foraging activities of the worker bees are disrupted, and it may be a few hours before normal foraging resumes. During a major honeyflow, this disruption could result in a measurable decrease in the quantity of nectar/honeydew collected.

It has been estimated that an average of 150 bees are killed every time a hive is worked. Bees that are killed or injured release their alarm pheromone, which may excite other bees to become more defensive. Careful handling of the bees and the hive furniture can minimize the bees' release of the alarm signal and could reduce the number of stings you receive.

Avoid quick movements when working with the bees, and do not jar the frames or other equipment. By proceeding slowly and gently, you allow time for the bees to move out of the way. Although killing some bees is unavoidable, the beekeeper who works slowly but precisely can keep the number of squashed bees to a minimum. Remember, work in slow motion.

WHEN TO EXAMINE A HIVE

A precise timetable for checking hives cannot be given, because conditions vary from colony to colony throughout the year, and some hives will require more attention than others. During the winter months, one needs to be aware of that fact that a colony's food reserves may be near depletion. On a warm day, when temperatures are between 40° and 50°F (4.4–10°C), one can gauge the weight of a colony by tilting it forward from the back side. If this tilting is accomplished with a minimum of effort, it is likely that winter stores are depleted and emergency feeding is in order. This procedure avoids opening the colony for a more thorough examination, which should be avoided because it may cause the winter cluster to move; the cluster may not reassemble in its original location where it was in contact with the remaining honey stores. Here are some general guidelines explaining when to open your hives:

- In the spring, when temperatures first reach over 55°F (12.8°C); briefly check the general conditions and determine if the colony has an adequate food supply and a laying queen. You can do this in cooler temperatures, but don't keep the colony open too long (less than five minutes).
- After the first fruit bloom, to check hives for growth, strength, and parasites, particularly the varroa mite, as well as small hive beetles and diseases such as American foulbrood.
- Check for signs that bees are making swarm preparations such as queen cups and queen cells.
- After a major honeyflow, to remove or add supers.
- Periodically after a honeyflow, to check condition of queen and brood.
- Before the winter season sets in.

Certain changes to a hive will require that you follow up to see how that colony responded to your previous visit. For example, check a hive:

- Two days after a package has been installed, to ascertain if the queen has been released from her cage.
- Five to seven days after checking to see if the queen has been released, to see if that queen is laying eggs. Look for eggs and/or larvae.
- Seven days after queen introduction.
- Seven days after dividing a colony.
- Whenever pesticide damage, disease, mite infestation, queenlessness, or similar conditions are suspected.

A hive should **not** be examined:

- During a major honeyflow, unless absolutely necessary—for example, if disease is suspected, to requeen, if pesticides sprays are imminent, or to add or take off supers.
- On a very windy or cold winter day.
- When it is raining.
- At night.

BEFORE GOING TO THE APIARY

The following equipment and supplies should be available before departing for the apiary. Although some of these items will not be needed during every trip to the apiary, it may be prudent to keep them near at hand, such as in the vehicle tool box or in the apiary shed. It is safest to house the smoker in a separate nonflammable container at all times, such as a metal can with a lid. All other bee supplies and tools can be stored in the back of a truck in a special box.

If you are going out for a quick look at several colonies, your checklist should include:

- Two hive tools.
- Matches, lighter, or propane torch.
- Dry fuel for smoker kept in a waterproof container, such as a small tin garbage can or plastic bag.
- A container of water to wash sticky hands, quench thirst, or to extinguish smokers.
- Newspaper to unite hives, start smokers, or to wrap comb samples for disease identification.
- Hive diary and extra markers, pens, and pencils.
- Duct tape and screen to close holes and cracks.
- Bottle of water for you.

- A nonflammable metal container for the hot smoker.

If you have many colonies (over 50) and are going out to visit several apiaries for the entire day, you may want to include these items:

- Can or jar of fresh syrup or dry sugar for emergency feeding, and extra feeders.
- Plastic spray bottle for emergency swarm capture (to dampen the bee's wings and keep them from flying off).
- Sample jars or plastic lunch bags to collect bee samples for mite or disease testing.
- Extra frames, in case you find broken ones.
- Extra hive bodies, outer and inner covers, and solid or screened bottom boards.
- Division screen or empty hive, for emergency colony division (e.g., should many queen cells or queens be found).
- Container to collect scrapings of wax or propolis; this keeps you from throwing away valuable hive products and spreading disease.
- Queen excluders.
- Old towels or other heavy fabric (check fabric or upholstery stores), to protect uncovered supers from robbing bees (if you are taking off honey). You can find these in fabric stores, or you can use plastic or nylon covers.
- String, rope, or hive straps to move colonies, or for other uses.
- Bee medication, if you are using it.
- Empty queen cages, for superseded queens, loose queens, or for emergency queen introduction.
- Marking paint (model paint) or markers sold by bee supply dealers.
- First-aid kit, including a sting kit or other medications for the beekeeper.
- Pruning clippers, saw, or sickle to keep weeds and brush under control.
- Skunk/mouse guards, entrance reducers.
- Hammer and nails for repairs.
- Spray can of herbicide or other spray to kill poison ivy or other noxious weeds growing on or around your hives. (Soapy water can kill some plants.)
- Your lunch, with extra water.
- Extra gloves, veil, and maybe even a bee suit.
- Extra smoker.
- Robbing screens.

THE HIVE DIARY

Methods of keeping track of the condition of each colony vary. Some beekeepers use a system of bricks or stones placed on top of the hives in some code to tell the queen's age, swarming tendencies, or the like. But because the stones can be removed by bee inspectors or others, or the code forgotten, other methods giving more precise information should be used.

A sheet of paper stapled or tacked to the underside of the outer cover is a good place to keep records. However, this is only a temporary solution, because the bees will chew their way through the paper, and all your notes will be lost; you can stall this process by placing your notes in a plastic sandwich bag and staple or tack it on the cover or outside the hive. Another technique is to write on the inside outer cover, or outside the hive with a grease pencil or permanent marker. Number your hives and be sure to note important events in your hive diary (such as if the colony swarmed or is queenless).

A better way to keep records is to use some sort of a hive diary that can be filled out each time you work your hives. Or you can purchase some inexpensive calendars and keep track of when you medicated, requeened, supered, etc., for each apiary. Later, you can enter the information into a spreadsheet in your computer. Some folks take their laptops, cell phones, or other electronic device and enter information at the apiary and then send it back to their computer; just don't get honey on your devices. Other beekeepers make special bar codes for each colony and keep track of them in this manner. No matter which method you use, be sure to keep good notes in some form. A good diary is also useful for tax purposes, as discussed in "Record Keeping," in Chapter 4.

Diaries are important if you plan to rear queens and select breeder queens from your best colonies. Queen characteristics should be recorded so you can tell whether or not a queen is up to your standards (see "Breeder Queens" in Chapter 10). In addition, by referring to the diary before going to the apiary, you will be less likely to forget any needed supplies or equipment. Every time a particular hive or group of hives is worked, note down some pertinent facts in the hive diary; add others as you go along. Note down:

- Layout of apiary, including windbreaks, water source, GPS and compass directions (so other people can find your apiaries if you are incapacitated).
- Date.
- Weather conditions (wind, temperature, humidity, etc.) the day you worked bees; if you want to be very fancy, get a weather station for your apiary.
- Colony strength—can be evaluated by the number of frames containing opened and sealed brood and frames covered by adult bees.
- Characteristics of the colony—defensive, gentle, productive.
- Swarming activity—include both the date the swarm was discovered and, if possible, the origin of the colony that cast the swarm. If you inspect a colony and discover sealed and opened queen cells, it is likely that this is the colony that cast the swarm. If the queen with the swarm was previously marked, then the origin of the swarm would be revealed.
- Manipulations made that day—reversing, supering, medicating, etc.
- Effects of last manipulation and time elapsed—after requeening, package installation, or other activities.
- Hive weight gained or lost since last visit.
- Amount of honey harvested for each colony—in case you want to raise queens from a good honey producer.
- Requeening schedule—age and origin of queen.
- Disease, pest, or mite record.
- Wintering ability—amount of stores consumed.
- Medication—what type, when, and for what reason.
- Mite controls used.
- Number of stings received and reaction.

DRESSING FOR THE JOB

If you absolutely do not want to get stung (although bees can sting through leather gloves) you can lessen your chances considerably by dressing appropriately for the job, much of which depends on reading the temper of the bees and dressing accordingly (see "Bee Temperament" in this chapter). First, bees are attracted to some odors, so do not wear strong perfumes or hair sprays. Then, depending on the degree of armor you want to put on (and the outside temperature), use the checklist below to make sure you or your friends, curiosity seekers, or other visitors are adequately dressed to work or observe you working bees (see also "Bee Wear" in Chapter 3).

Checklist:

- Bee veil, a necessity; getting stung on the face is dangerous and painful. **Always wear a veil**. Protect your eyes.
- Some sort of hat or helmet to keep off the sun, usually worn under the veil.
- A light shirt, bee suit, or coveralls to cover up gaps through which bees can and will crawl. A bee suit is best, as it keeps your other clothes clean; propolis and wax stains will not come out in the wash. Avoid floppy shirtsleeves or loose material that could get trapped between hive furniture.
- Long pants, with socks over the cuffs or some other closure (duct tape, bicycle straps, or a long bee suit with elastic cuffs). The same goes for the shirt—elastic closures around the wrists are important to keep bees from crawling up your arms.
- Boots or sturdy walking shoes; many apiaries have poison ivy and other "goodies," not to mention wet grass and brambles.
- Gloves are optional but handy to have if the bees decide to get fractious. They are essential to have if you are doing work that will irritate bees, such as taking off honey or setting upright a colony turned over by bears or preyed on by skunks.

If you do find yourself in a stinging situation, here are some simple rules:

- Do not swat at bees as they fly toward you or land on you. Bees will readily respond to quick movements, leading to a good possibility of being stung.
- If you are without your veil, a condition that must be avoided at all costs, and are wearing glasses, push them close against your face to protect your eyes, cover your mouth, keep your head down, and walk calmly into shrubbery or shady forest canopy. The best cover, if available, is an enclosure such as a vehicle or shed.
- At times, bees may manage to get inside your veil. First, seek refuge in a vehicle or in a safe area away from the bee yard. Then remove the veil to free the bees. Bees that get into your hair will usually work their way down to your scalp and then sting you. Try to avoid such a situation by keeping the strings or zipper of your veil secured.
- Never smash or pinch a bee that has landed on

you. If you do, the bee will release her alarm odor, and that odor will entice other bees to sting you as well. In short order, you will have to curtail your work and retreat in haste from the apiary.
- If attacked by Africanized bees that are very aggressive and tenacious, immediately take refuge in an enclosure. It is impossible to outrun them. Jumping into a body of water will not deter them; they will hover over the water waiting for you to resurface.

SMOKING BEES

The use of smoke while working bees is essential. No hive should be opened or examined without first smoking the bees. A few periodic puffs of smoke will help keep the bees under control, but bees that are oversmoked might become irritated. In a pinch, pipe or cigarette smoke can be blown in, but smokers work better.

Smoke works in several ways to keep bees from stinging. When bees are smoked, they seek out and engorge on honey or nectar in the hive, and bees with full stomachs are less prone to flying or stinging. In addition to it causing large groups of bees to seek nectar and honey, smoke also serves to redirect their attention from the fact that their colony is being invaded. When the hive is first opened, guard bees—which are sensitive to hive manipulations—release an alarm pheromone to alert other bees. When many bees are releasing this pheromone, you too can detect this alarm odor (especially if you open a hive in winter); it smells like banana oil. The alarm pheromone causes the bees to react defensively to protect their home from intruders. Smoke may reduce the sensitivity of the bee's receptors to the alarm pheromone; if they can't detect it, they won't release or react to it.

The following procedure works well: Before opening a hive for inspection, direct a few puffs of smoke into the entrance, then proceed to partially lift the outer cover and puff some smoke into the space between the outer and inner cover. Then re-secure the outer cover, wait ten seconds, remove the outer cover, and puff additional smoke into the oblong hole. As you remove the inner cover, add some smoke between the inner cover and the top bars of the frames. Now you can begin your examination. Continue to use smoke as needed. Remember, the judicious use of smoke will the keep the queen and bees from scurrying about, making it easier to search and find the queen. The use of the smoke will allow you to carry out a smooth inspection of the colony.

Smoke can also be used to drive bees away from or toward an area within the hive. Additionally, it can be used to mask the alarm pheromone after you have been stung. Because the gland that releases the alarm pheromone is at the base of the sting, some of this pheromone marks the area when you are stung. Other bees that detect this signal may also sting the tagged area. Hands, clothing, and bee gloves that have been stung should also be smoked (and washed regularly) to clean off the alarm odor, dirt, venom, honey, sweat, and disease spores.

Purchase the larger smokers—they are easier to light than the smaller ones, burn longer, and are less likely to fail when needed most. Smart beekeepers carry a lidded five-gallon pail in which dried fuel, extra matches, and hive tools are kept. Your smoker should be cleaned of soot when it builds up in the nozzle and thoroughly scrubbed with steel wool and hot soapy water, especially after working a diseased (especially foulbrood) colony. Keep your smoker in a fireproof container. **Never** place a smoker with live embers in a vehicle or in the open bed of a truck.

Lighting the Smoker

You should become thoroughly familiar with the smoker before using it at the apiary. It is a good idea to practice lighting it a few times to get the hang of it; nothing is worse than having your smoker go out when you are in the middle of examining a colony. All beekeepers have their favorite fuel and may use it exclusively. The best fuel to use is the fuel that works best for you and is readily available in your locale.

Some commonly used fuels are:

- Straw or dried grass mixed with something else (e.g., wood chips or bark).
- Dry, rotted, or punk wood produces good smoke but can burn too hot; use with other fuels.
- Sumac bobs.
- Brown, dry evergreen (pine) needles and cones that are old and open.
- Cedar bark or other bark chips or bark mulch.
- Peanut or other nut shells.
- Rice hulls.
- Twigs and pine needles.
- Wood shavings, sawdust mixture.

- Rags, 100 percent undyed cotton only.
- Dried horse or cow dung, which burns well without much odor.
- Cotton stuffing.
- Woodstove pellets.

To cool the smoke, put a thin layer of green grass on top of your burning fuel. This layer catches the ash and keeps your bees from getting burned.

Some companies sell smoker fuel as well, so check around. Some smoke was found to be toxic to bees, including dried corncobs and leaf tobacco; so be careful that you are burning clean, nontoxic fuel. In addition, synthetic materials, petroleum starters, or rags treated with pesticides should **never** be used, since they may also give off a toxic smoke. Newspaper and corrugated cardboard boxes should not be used as the only fuel, as the ash is too big and could burn the bees; also the cardboard may be treated with pesticides or other toxic compounds (e.g., glue). Some beekeepers use binder twine (from hay bales), but since this is usually treated with rot-retarding chemicals, which may be toxic to bees, we do not recommend its use. Generally, any organic material can be used, but be careful not to mix in dried poison ivy vines with your smoker fuel; this can be found in woodchips from country roadside tree trimmings (sometimes the workers leave the chips or give them away).

Here is how to light a smoker:

1. First clean the smoker and scrape out any clogging soot from inside the nozzle.
2. Use a match or lighter and ignite a small piece of newspaper or punk dry wood.
3. Drop a small amount of the blazing fuel to the bottom of the smoker.
4. Puff the smoker bellows and slowly add unburned material; puffing in air will help ignite the fuel as you pack the smoker. Do not over-pack, and make sure the fuel is well lit and smoking well. **Be careful not to get burned!**
5. If your fuel is damp, add bits of beeswax, or start your fuel with a small hand-held blowtorch or BBQ lighter.
6. Puff the bellows frequently until you are sure the smoker is ready to use.
7. Once it is going, put a handful of grass or green leaves on top of the fuel to cool the smoke and

catch hot ashes; make sure you don't put out your fire.

8. Do not pack the fuel cylinder too tightly, and keep filling it periodically with fresh fuel. A well-stocked, large smoker should last 30–45 minutes.

After finishing work in the apiary:

1. Place the hive tool(s) in the opened smoker and puff a blaze to sterilize the tools. This will make them very hot, so be careful when removing them; douse them with water to cool before storing them.

2. Empty the remaining fuel and ashes onto dirt or pavement and drench them with water (which you carried in your truck). Always have on hand some water to drench old, smoldering smoker fuel. Some beekeepers stuff a cork or green grass into the nozzle of the smoker to suffocate the fire if they have no water on hand. Another way to extinguish a smoker is to open the lid and place a sheet of newspaper that has been folded several times over the throat of the smoker and close the lid tightly. This also will help keep your nozzle from filling with soot.

3. Make sure the fire is out and the smoker is cool before putting it away, and never leave a lighted smoker in a vehicle, even the back of a truck. It can be fanned into a blazing fire just a short drive down the road.

EXAMINING A COLONY

The general method used by most beekeepers to open and examine a hive is outlined below. The procedure may vary somewhat, depending on the number of boxes on the hive and the purpose of the examination. Work with many different beekeepers and pick up their good habits on working bees:

1. Approach the hive from the side or back.
2. Do not stand in front of the hive at any time, since the flight path of outgoing and incoming bees will be blocked.
3. Puff some smoke into the entrance (be sure it gets inside) and wait 30 seconds so the bees can begin to gorge on honey.
4. Gently pry or take off the outer cover and direct a few puffs of smoke through the oblong hole of

the inner cover; again wait 30 seconds for the bees to gorge honey. Then gently pry off the inner cover, puffing some smoke underneath. If an inner cover is not used on the hive, puff some smoke under the outer cover as you take it off and wait 30 seconds.

5. Lay the inner cover at the entrance so clinging bees can reenter the hive; do not block the entrance.

6. After the covers have been removed, smoke the bees down from the top bars of the frames; smoke must be used judiciously—too much will cause the bees to run in every direction, making your work more difficult and decreasing the likelihood of finding the queen. In addition, too much smoke can damage brood cappings. Smoke bees just enough to make them move; experience will teach you what amount is right.

7. Use the outer cover (underside up) or a spare hive body or stand as a base for stacking supers as they are removed from the hive (see illustration on examining a colony).

8. As you begin to remove frames and set the hive boxes aside, avoid quick body movements or jarring or bumping the equipment; such actions tend to increase the defensive posture of bees. Slow, deliberate actions and the gentle handling of equipment will keep bees calm.

9. Throughout the examination, smoke the bees as needed to keep them out of your way and to keep them from getting squashed.

10. The purpose of the examination will dictate whether to first remove all supers above the bottom ones (to inspect the broodnest) or whether to work from the top down (to see if nectar or honey is being collected) during your inspection. Most of the time your visits will be related to an inspection of the broodnest, which is where you usually find the queen, eggs, larvae, sealed brood, and majority of adult bees. During the spring and summer, the broodnest is usually found in the second deep hive body.

11. Each time a super or hive body is pried off, puff a bit of smoke onto the frames below and to the bottom of the one you are moving.

12. Even if the colony has a robust population of bees, it is still best to work from the upper boxes down to the lower ones. If you use your smoker judiciously, you will not drive bees down to the

Examining a Colony

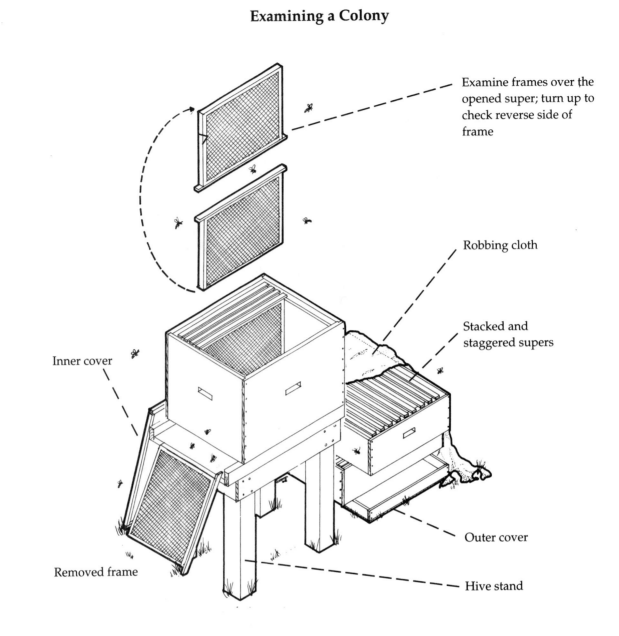

Examine frames over the opened super; turn up to check reverse side of frame

Robbing cloth

Stacked and staggered supers

Inner cover

Outer cover

Removed frame

Hive stand

lower boxes. Of course, if you do smoke each box excessively, to get away from that smoke the bees will migrate down to the lower boxes in great numbers, making your inspection of the lower boxes increasingly difficult.

Examining Frames

Now you have made it to the hive body, where you want to start examining the bees. Here's how you should proceed:

1. Whenever you decide to begin your examination, smoke the bees off the top bars and down

between the frames. Before removing frames, choose the one closest to the hive wall and push all other frames away from it with the hive tool; this will create sufficient space for easy removal of the frame. Avoid removing frames from the center of the hive first, as the queen may be crushed in the process of pulling out the frame.

2. Once the first frame is removed, you have created more space to remove subsequent frames. Lean the removed frame against the bottom hive body or some other object, out of the sun and where it won't be kicked or jarred, or place it in an empty hive body or on a frame holder. Frame

Reading the Frames

For each frame you inspect, quickly check for these items:

- Sealed brood: It should be compact (very few open cells) and in a somewhat concentric semicircle, with the upper reaches of the semicircle near center of the top bar of the frame. This pattern will vary somewhat from frame to frame. If there are many opened cells within this semicircle, it may mean the queen's eggs were not viable, or were removed before they hatched, or the larvae or the pupae were removed because of diseases or as a consequence of being infested by the varroa mite.

- Ratio of eggs to open larvae to capped pupae: A ratio of approximately 1:2:4 is ideal. This means there are twice as many larvae and four times as many capped pupae as there are eggs. It is an indication that the queen is laying continuously and there is sufficient food for the nurse bees to care for the larvae and an adequate number of other bees to maintain hive temperatures at the level required for compete metamorphic development (egg, larva, pupa, adult).

- No eggs found: If no eggs are found in the open cells, you can estimate how long ago the queen stopped laying by opening up some capped worker brood. Young pupae with white eyes will emerge in about seven days; if the eyes are purple, they will emerge in two to three days.

- Queen cells: If you find supersedure queen cells, usually few in number and usually found on the face of the combs, the queen is failing and the bees are in the process of replacing her. However, if you find open queen cells, containing queen larvae along the bottom edges of the frames, or in places where combs are distorted, and this finding coincides with either of the two swarm seasons, the colony is likely to swarm within the next seven days. Usually at this time, there are very few eggs, lots of sealed brood, and the colony is crowded with adult worker bees. On the other hand, if you find sealed queen cells, the colony may have already swarmed; or if you find holes in the queen cells, a virgin has emerged and has stung her sister queens while still in their cells. In this scenario, the colony has already cast its first swarm and, depending on the timing, several after-swarms.

- Other observations: Note changes in the behavior of a colony since your last visit, especially if the bees are more volatile; this could indicate lack of forage, pesticide use, pests, mites, queenlessness, or disease. Observe the amount of incoming nectar and pollen in case bees are starving (no honey, dead brood on bottom board) or are becoming honeybound (honey filling all available space, even into the brood combs). Also note the physical condition of combs and frames, including any wax moth damage and uneven comb or foundation; fix any broken frames.

holders are good since they keep the frames off the ground where they can be stepped on.

3. As any given frame is being examined, hold it above the hive body from which it was withdrawn; if the queen falls off the frame, she will drop back into the hive body rather than on the ground where she may be stepped on or be unable to find her way back into the colony.

4. Continue to examine each adjacent frame until your objective is completed.

5. Frames should be returned to their original positions and spacing in order to maintain the integrity of the broodnest. At times, there will be a need to replace old or broken frames or to swap frames of brood or honey from one colony to another. If that is the case, please note that, in the end, the broodnest should be reassembled as closely as possible to the original arrangement. For example, if you remove frame 6 from the box that contains young larvae and swap it into the 10th position in the box, bees may not be able to adequately care for those larvae. This disruption disturbs the organizational rhythm in the broodnest. If frames with brood and eggs are separated from the broodnest and placed elsewhere, those frames might become chilled, because the bees will have a hard time maintaining the proper temperatures in a scattered broodnest. This can result in brood chilling, making them more susceptible to chalkbrood disease; if chilled too long, the

whole frame of brood could die. Don't forget to return the initial removed frame before replacing the upper supers or hive covers.

6. If while working bees you see bees landing on the top bars, or bees fighting on uncovered frames, uncovered supers or boxes, or at the entrance, robbing may be in progress. This happens when there is no honeyflow. You must quickly cover the exposed equipment with a robbing cloth (wet cloth or blanket) or, better yet, curtail hive examinations for the day (see "Robbing" in Chapter 11).

7. When you are replacing the hive boxes, the bees in the one below will be milling on the top bars and rims; smoke the bees down so they will not get crushed as you replace the hive furniture.

8. Whenever possible, scrape excess propolis and *burr comb* (comb not in the proper place) from the frames with a hive tool. These materials should be placed in a closed container; the wax can later be melted down in a solar extractor (see "Beeswax" in Chapter 12). Never discard propolis and wax around the apiary or anywhere else. Not only will this material attract pests such as skunks, bears, and small hive beetles; it could also promote robbing and can transmit diseases. Remember, wax and propolis are also marketable products.

WORKING STRONG COLONIES

What is a strong colony? Perhaps one with a high population of bees dispersed throughout a hive that consists of three or more deep hive bodies, and above these, three shallow supers filled with honey. A strong colony may also be described as one with a large population of bees, two deep hive bodies, and three or four shallow supers chock full of honey.

From time to time, upon inspection of a strong colony, you'll find that the hive configuration needs to be brought to a more manageable arrangement. Before visiting such a colony, check your diary notes from a previous visit to help you determine what equipment you will need.

Let's discuss a hypothetical colony with a population of 40,000 bees housed in three deep hive bodies and three shallow hive supers filled with honey. After checking your diary notes, you determine that you should bring the following: three queen excluders; one bee escape board or Porter bee escape, or, in lieu of a bee escape, a fume board or bee blower; one or two extra deep hive bodies that contain their complement of frames; two shallow supers with frames; two bottom boards; and two inner and two outer covers.

When you arrive at the apiary, you observe the high traffic of bees flying in and out of the colony. The fact that it has a very high bee population tells you it may require a second trip in order to make all the necessary adjustments.

Now, begin your inspection. Using the smoker, apply several puffs of smoke into the entrance. From this point on, each time you remove a piece of hive furniture, smoke will be required. A hive tool will also be needed to break the propolis seal between the outer and inner cover, between the inner cover and the first super below the inner cover, and for every additional separation of hive furniture, as you work your way down to the bottom board.

Now, tilt the outer cover enough to puff some smoke beneath it, then lower it back in place. A minute or two later, remove the outer cover, invert it, and set it either on the ground, on a superhorse, or on an empty deep hive body, and then puff some smoke into the oblong hole located in the center of the inner cover. Then, as you remove the inner cover, apply a small amount of smoke over the top bars of the frames in the third honey super and set the inner cover aside. Now remove the third honey super, and as you tilt it back, apply smoke over the frames of the super below it. Set the third super either on the inverted outer cover, on a superhorse, or on top of an empty deep super and continue working downward. As you remove the second honey super, puff some smoke over the top of the frames of the honey super below it and set the second honey super on top of the third honey super and repeat this activity to remove the first honey super.

We will assume that on this day there is a honeyflow in progress, so robbing activity will not be a factor. As a precaution, drape a clean wet bed sheet (if you brought one along) over the honey supers.

Now you are confronted with three deep hive bodies chock full of adult bees, brood, and a queen. If part of your inspection is to locate the queen, I wish you good luck because you may have to expended an inordinate amount of time searching for her. But there is an option that involves separating the third deep hive body from the second, and then the second from the first. Now, STOP. Put a queen excluder over the first deep hive body, replace the second deep hive body and put a queen excluder on top of it. Replace the third deep hive body, place a queen excluder above it, and then add one empty shallow super above the third queen excluder. Next, place a Porter bee escape into the oblong hole of the inner cover, or replace this inner cover with an escape board, and then stack the three honey supers above the inner cover or the bee escape board. Add a second inner cover over the third honey super and then add the outer cover. Return in four days, which will allow the bees to recover from the previous manipulations. Remove the honey supers if they were not previously removed, and place them in your vehicle. Because you had installed either the Porter bee escape or an escape board, these supers should be free of bees—unless the queen is in one of the these supers.

Another option would have been to use the Bee Blower to blow the bees out of the honey supers, and then load the supers into your vehicle. This procedure eliminates the need to restack the honey supers.

Once you know the queen is not in the honey supers, begin your search for her. The deep hive body that contains eggs in the broodnest will be the one containing the queen. Once you have located her, carry out several manipulations. You may wish to make a split of this colony to reduce it from three deep hive bodies to two by removing one of the deeps and placing it on a separate bottom board, with additional deep supers optional at this point. You need to introduce a queen or queen cell into the split or have the split rear its own queen. You also need to add one inner and one outer cover to complete the hive configuration. This split can either remain in this apiary or be moved to another.

Please note that there are more ways than what is described above to rearrange a colony that is challenging to inspect. One of the intriguing aspects of bee husbandry is that there is more than one approach to solving problems that develop in the apiary.

WHAT TO LOOK FOR

In the spring the colony must build up strength in order to achieve the peak population of 40,000 or more; such numbers are needed to secure a good honey crop.

You should be able to verify that:

- A queen and/or eggs or young larvae are present (see above, "Examining Frames").
- There are adequate food stores, including bee bread and honey. If such stores are inadequate, you must provide frames of both honey **and** bee bread that have been kept on reserve or removed from another colony; honey provides the bees with their carbohydrate requirements, and bee bread with their protein, fats, and minerals. If you have no honey in this equation, you will need to substitute sugar syrup, sugar cakes, fondant, or simply dry sugar. If you have no bee bread, you will need to provide them with a pollen substitute.
- The brood pattern is compact for both uncapped (larvae) and capped (pupae) brood; if not present, determine the cause.

Also check for and take measures to correct the following adverse conditions:

- Queenlessness: add a queen or queen cell, or unite queenless colony.
- Queen cups and/or queen cells (supersedure or swarm cells): manage appropriately.
- Presence of a failing or a drone-laying queen: replace queen.
- Presence of laying workers: unite with a queen-right colony.
- Leaking feeders: replace and check to see if bees are not harmed by leaking syrup. If too much syrup leaks out, robbing could be started.
- Crowded conditions: add extra honey supers and/or brood chambers.
- Overheated conditions: provide shade or additional ventilation.
- Diseases, mites, and other pests: treat accordingly.
- Robbing activities: reduce size of entrance, add a robbing screen, seal cracks and other openings.
- Bottom board filled with bees, debris, wax moths, or propolis: check condition of colony, then clean off or replace bottom board.
- Wet, damp, or rotting equipment or carpenter ant infestation: replace.
- Dwindling populations: if in two colonies that are both disease free and have low mite count, join the two colonies.
- Broken combs or frames: replace.
- Cracked or broken equipment: replace.

- Obstructions in front of the entrance (grass, weeds): clear and mow.

Finding the Queen

The queen's presence and her reproductive state can be determined indirectly, without finding her. If you find brood frames with a concentrated pattern of capped worker cells, frames mostly filled with eggs or larvae (uncapped brood), or a combination of both, her presence and quality are indicated. The colony is healthy, headed by a fertile and productive queen.

If it is necessary to find her, use as little smoke as possible, open the hive gently (as outlined above in "Examining a Colony"), and remove the outermost frame. She will seldom be found on frames with just honey and pollen or on frames with capped brood; she will **most likely** be found on or near frames containing eggs and uncapped larvae.

The queen can often be spotted in the midst of her encircling attendants or retinue. When a queen moves slowly along the frame from cell to cell, the other bees will begin to disperse, but the circle will re-form when she pauses (see illustration of Queen and Court).

If the queen must be found—whether before re-queening, to kill her before uniting colonies, to mark her, or just to satisfy the need to see her—but cannot be located within 15 minutes or without disrupting

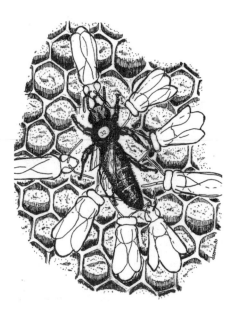

Queen and Court

the entire hive, it may be helpful to use the following method:

- Place queen excluders between the two brood chambers.
- Five days later, the queen will be in the hive body whose frames contain eggs. Because all eggs hatch in three days, the brood chamber from which she was excluded will have no eggs.

If you do not see the queen or eggs, or the colony is not behaving normally, refer to sections of Chapter 11 for potential problems of your colony, or have another beekeeper or bee inspector check it.

BEE TEMPERAMENT
Good Disposition

To encounter fewer defensive foragers and minimize the likelihood of being stung, work the hive on days when most field bees are foraging.

Generally, it is best to work bees:

- In the spring, when populations are lower and an early honeyflow is in progress.
- At the end of a good honeyflow.
- On warm, sunny, calm days.
- When populations are low, as with package bees.
- When bees are well gorged with food, as with a swarm or package bees that have been fed.
- Between late morning and early afternoon (roughly between 10 a.m. and 1 p.m., depending on season and time zone).

Irritable Disposition

Bees are more prone to sting when most of the foragers are in the hive. Conditions outside the hive (usually weather) are the reason for the foragers not being out. Other factors may also cause bees to become defensive:

- Queen temperament, genetically passed on to her offspring.
- The effects of pesticides.
- Disturbance by skunks, bears, or other pests.
- Poor honeyflow, when there is little food coming in.
- Autumn, after the honeyflow has ceased.

- Impending thunderstorm.
- Cool, wet, cloudy days.
- Hot, sultry, humid days.
- Windy days.
- Early morning or late afternoon or evening.
- Improper handling, resulting in killing many bees.
- Jarring of the hive or a hive part.
- Disease/mite infestation.
- Examining without using smoke.
- Removal of honey or leaving supers or frames exposed during a dearth, stimulating robbing activities.
- Reaction to pungent hair oils, lotions, deodorants, or perfumes.
- Queenlessness.
- Presence of laying workers

To minimize your getting stung, remember these rules:

- Work in slow motion; avoiding rapid, jerky movements around the bees.
- Don't swat at flying bees; ignore them, and they may ignore you.
- Don't drop, bang, or bump hive parts, as vibrations upset bees.
- Don't stand in front of the flight entrance of the colony; stand to one side or at the back. If you are working from the back, however, some colonies strongly object to your reaching over the frames to remove them.

UNEXPECTED OCCURRENCES

When working with bees, situations may sometimes arise for which you are not prepared. These are some of the more common unexpected happenings:

Bees get in your veil. Slowly retreat to an area where you can remove the veil and release the trapped bees, ideally in a sheltered location such as a vehicle or shed. In the absence of a shelter, slowly walk behind a bush or tree. If bees are following you, puff smoke in their direction. When bees are no longer pursuing you and you are in a safe location, remove your veil to release the trapped bees. Usually, they will fly off once they are released. Put your veil back on and return to the apiary. However, if the bees in the apiary demonstrated excessive aggressiveness, better to forgo any further work until another day.

Note well that if instead of following this advice you kill the bees inside your veil by pinching or squashing them, it is likely they will release their alarm odor and more bees will fly against the veil, looking for other ways to get inside. Veil strings should be tightened snuggly around your waist and veils with zippers completely closed to prevent bees from finding their way in. Holes in your veil will also serve as a port of entry.

Your smoker goes out. Cover exposed supers with extra outer cover(s) or cloth to prevent robbing and relight the smoker.

The queen flies away. During a package installation, some beekeepers deliberately or inadvertently release the queen from the queen cage. Unless you have worked with bees for many years, you should avoid the direct release of a queen. Queens that have been held captive in queen cages have lost a third of their body weight and are now capable of flying; in most cases, they will be unable to orient back to their intended colony. Virgin queens, queens in queen cages, and queens preparing to leave with a swarm are all light enough to fly. Laying queens are unable to sustain flight and will usually flutter to the ground. In the event a queen takes flight, you can take action to retrieve her. Near the entrance to the colony or on the top bars of frames of an opened colony, shake bees from either a frame or a package. These bees will begin scenting, and with some luck their activity may reorient the queen back to her colony. Or, look around the area in which you are working; if she has landed nearby, other bees may have formed a cluster around her. Retrieve the cluster and place it into the colony. When ordering a breeder queen, which can be quite expensive, some beekeepers request that one of her wings be clipped to avoid losing her. However, it is generally believed queens with clipped wings may not perform well over long periods (see "Swarming" in Chapter 11).

The queen is balled. If the queen is surrounded by a "ball" of workers, they may be trying to kill her. This can happen when she is released directly into a colony of package bees or when introduced into a hive that is being requeened. Balling may also happen when requeening a colony that already has a queen or when the hive is roughly handled. In these cases the bees consider the queen to be "foreign" and commence to surround or *ball* her. Balling will either suffocate or kill a queen by raising her body temperature too high, thus "cooking" her, or by tearing her apart. Do one of the following:

- Cover the hive quickly and hope for the best (not usually the best choice).
- Break up the ball with smoke or water and cage the queen; reintroduce her using the indirect release method discussed in Chapter 6.
- Break up the ball and spray the queen with syrup, then place her on a frame of uncapped brood.

You are chased by many bees. Blow smoke both on yourself and at the pursuing bees. Be sure your smoker is not discharging flames, which can happen; otherwise your clothing may catch fire. Walk slowly behind a bush or tree, or around your vehicle; if the bees are not Africanized they may end their pursuit. Bees are sensitive to quick movement and are less inclined to pursue objects moving at a slower pace.

The colony is exceptionally defensive. Close the hive as quickly as possible and wait for another day. Try to determine the reason for the bees' unusual behavior. Check other colonies in the same apiary to see if they exhibit the same behavior. A skunk or bear may be bothering your hives (see "Animal Pests" in Chapter 14). If a particular colony exhibits a defensive posture every time the hive is inspected, it may be a defensive trait, or it may have become Africanized (i.e., occupied by Africanized bees). In such cases, requeening the colony may be your only option. **Warning:** attempts to requeen a colony with Africanized bees should be left to someone who is familiar with them (see "Requeening Defensive Colonies" in Chapter 10).

BEE STINGS

What to Do When Stung

If a worker pierces your skin with the barbed lancets of her sting, she cannot withdraw them once they are embedded (see "Bee Venom" in Chapter 12 and Appendix C on sting anatomy). As the bee struggles to free herself, the poison sac attached to the lancets is ripped from her abdomen. This means that the bee will ultimately die; and having left most of her sting in your tissue, she will obviously not be able to sting again before she dies.

Other stinging insects have either smooth lancets or lancets with ineffectual barbs; they can therefore withdraw that portion of the sting and repeatedly reinsert it. The queen honey bee, hornets, and yellow jackets (superfamily Vespoidea) have such a sting. Queens rarely sting beekeepers even when handled—they use their sting against rival queens.

Scrape off the sting with a fingernail or hive tool as soon as possible to minimize the amount of venom pumped into the wound. Start to scrape the skin with your nail about an inch away from the sting and continue scraping through the sting, which will pull out easily. **Speed of removal, not method, is important to reduce the amount of venom injected and thus minimize the swelling and itching afterward.**

Because an alarm pheromone is associated with the sting, other bees are likely to sting in that vicinity. Apply smoke to the area of the sting to mask the alarm odor.

Treatment of Bee Stings

Local Reaction

Bee venom contains enzymes (hyaluronidase) and peptides (melittin) that cause the pain (see table under "Bee Venom," in Chapter 12). For local reactions, there is very little an individual can do except to relieve the itching. Since the sting barbs are so tiny and the puncture so small, no treatment will be effective in reducing the amount of venom other than the prompt, proper removal of the sting structure.

Every beekeeper has a favorite treatment for bee stings. Although the treatment does not *cure* the sting, it does give a different sensation to the area and thus takes one's mind off the momentary pain. The following items are often used to relieve the pain and itching of bee stings:

- Bee-sting treatment kits.
- Vinegar.
- Raw onions rubbed on the area.
- Toothpaste.
- Honey.
- Juice from the wild balsam, jewelweed, or touch-me-not (*Impatiens pallida*).
- Baking soda.
- Ammonia.
- Meat tenderizer, as a paste.
- Mud.
- Hemorrhoid treatment cream.
- Ice packs.

The best treatment is ice. These treatments work best if applied immediately after being stung, but immediate application of these items is usually impossible if you get stung away from home or through a bee suit. To give relief to the itching red welt that appears following a bee sting, apply calamine lotion or other insect-bite or poison ivy preparation, or hot water.

Systemic Reaction

A good summary of bee stings and allergy is in J.M. Graham's *The Hive and the Honey Bee* (Dadant & Sons, 1992), as well as on some online sites. In general, if you break out in a rash (hives) or have difficulty breathing after being stung by a honey bee, you are probably having an allergic reaction to bee venom.

In such cases, call for emergency medical help or take the person who has been stung to the hospital immediately! Time is critical.

Medication for an allergic reaction to bee stings can be obtained only by prescription. The drugs commonly prescribed are an antihistamine and epinephrine (adrenaline). Consult your doctor if there are any questions.

Here are some examples:

- Injected: EpiPen (www.epipen.com) or other insect sting kits are available with a prescription and include a syringe filled with epinephrine (adrenaline), with instructions for it to be injected under the skin (subcutaneously). Read instructions carefully (some kits may require refrigeration) and become familiar with how to administer the medication. Do **not** administer to others.
- Oral: Some over-the-counter antihistamines are helpful; your doctor may prescribe more powerful ones.
- Aerosol: An aerosol bronchial applicator, as for asthma sufferers, will offer quick relief of breathlessness as a result of a bee sting. Epinephrine inhalers are also effective. The dosage of two puffs should be repeated after fifteen minutes.

Although the above information provides an outline of what might be done for systemic or general allergic reactions to bee stings, exact and precise medical information should be strictly adhered to. No one should attempt to self-diagnose a response to bee stings or to prescribe medications. **Seek the advice of a physician.**

If you develop an allergy to bee stings and wish to still keep bees, the only other alternative is to go through an immunotherapy session or venom allergy shots. It is normally expensive, complicated, and inconvenient but does usually eliminate future systemic reactions. Consult your doctor or allergist for more details (see Appendix C, and "Venom" section in the References).

 **Notes**

CHAPTER 6

Package Bees

Many first-time beekeepers obtain their bees by purchasing a three-pound (1361 g) package of bees, which usually contains about 10,500 bees with one queen. The first bee packages were available for purchase well over 100 years ago; the only marked change is that instead of being constructed solely of wood and metal window screen, today's are also constructed of plastic. The design of bee packages takes into consideration the need for the package to be sturdy, while still providing the bees sufficient ventilation while in transit. Earlier packages were constructed with wooden tops, bottoms, and narrow ends, while the wider sides consisted of metal screening. Recently, a package made entirely of thick plastic screening on all sides has come into use. This package is typically referred to as a Bee Bus.

Package bees are prepared by honey bee package producers, usually from southern states or California or Hawaii. Bees are shaken off of frames into a funnel and slide, tumble, and fall down into the package. Once the requested number of bees (measured in pounds) is in the package, a queen cage containing a mated, laying queen, is added. A can of sugar syrup is inserted into the circular opening, and a lid is then placed over the can, securing it in place. A three-pound package of bees (3500 bees per pound), including queen, is also a great way to draw out foundation, as the bees are usually young, with well-developed wax glands.

The package is now ready to be shipped. It is safe to say that most beginning beekeepers started their first colony by installing a package of bees into a hive; queenless packages are also available to increase the populations of weaker, existing colonies.

Many packages are produced from a single production colony, and if that colony's queen was added to any given package, the production colony would become queenless. Instead of using the production colony's queen, many package producers also raise young, mated queens to accompany bees in packages. These introduced queens are foreign to the bees in the package and are placed into their own small *queen cage*, so the package bees cannot attack and kill her. While the package is in transit, the bees slowly become "acquainted" with this new queen, as her pheromones begin to be recognized and accepted by the bees in the package.

The ends of each wooden queen cage have openings. One is used to place the queen into the cage, and once she is inside, three to five worker bees from her mating box may be added. Then the opening is closed with a small cork. At the opposite end of the queen cage, a *candy plug* is used to seal off the hole, and another small cork is pushed into place. The purpose of the candy plug is to delay the release of the queen, thereby providing the bees more time to adjust to, and then adopt her as their own. Once the cork is removed from the candy end, the bees outside will begin to consume the candy plug, and slowly they create a passageway into the cage. They continue eating away at the candy until the queen can fit through and join the bees in the newly installed hive.

Plastic queen cages are now available and used by bee suppliers. The idea of keeping the queen confined with candy to delay her release is basically the same; just the design is different. The bees continue eating away at the candy until the queen can fit through and join the bees in the newly

Package of Bees

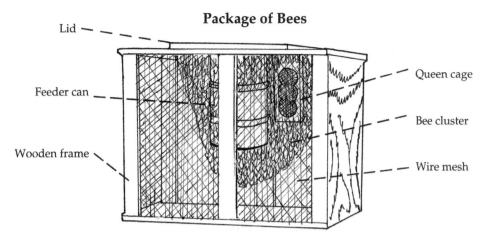

Lid

Feeder can

Wooden frame

Queen cage

Bee cluster

Wire mesh

installed hive (see "Designs of Queen Cages," in Chapter 10).

There are many different methods to install bees from packages into hives. One of the major differences is whether you choose to delay the release, or directly release, the queen from her cage during the installation of the package. In the three *indirect-release methods* presented here, the queen remains caged, and the bees are allowed access to the candy plug, which they must remove in order to release her. These methods simply delay the queen from being freed to mix with the other bees for a few more hours, or days, and increase the likelihood of their accepting her.

In the *direct-release method*, either the screen top or the cork opposite the candy plug in the wooden queen cage is removed, and the queen is released. If a plastic queen cage, such as the JZs BZs queen cage, is used, the plastic cap covering the opening used to insert the queen is removed, to allow the queen to exit.

A word of extreme caution here: Caged queens will have lost considerable body weight during confinement, and can fly. If you are not careful as to where you release her, the queen may fly off, never to be seen again. The best way to safeguard her from flying off, once you remove the screen, the cork, or plastic cap, is to quickly insert the cage into the hive entrance, where she may join the other bees. There is no guarantee that she will not turn around and fly out the entrance, unless you temporarily (and completely) block it. If you ordered a clipped queen, the queen cannot fly.

When the queen is released directly, the bees may still not be fully acquainted with her, and as a consequence they may form a tight ball of bees around her and, by raising her body temperature, attempt to suffocate and kill her. This process, called *balling the queen*, may result in the queen's death or permanent injury; see "Unexpected Occurrences" in Chapter 5. Combinations and variations of the direct-release and indirect-release methods are covered in "Installing Packages," below.

ORDERING PACKAGES

Packages can be ordered directly from the producers or through a local bee supply house, a local dealer, or a bee club. Advertisements by local dealers may be found in publications of state beekeeping organizations and local bee clubs, as well as in beekeeping journals, such as *American Bee Journal* and *Bee Culture*. These publications also include advertisements for many package bee dealers (see "Suppliers" in the References). Also, check websites and look at the newsletters from your local bee clubs or state beekeeping associations.

Sometimes, local bee supply dealers will drive down to pick up a load of packages, especially if they are filling a large order; you could go along to help if you want to buy a large number of packages at a reduced price. Bee clubs often order packages for their members to get better prices. Help is always appreciated, and it would be an excellent learning experience.

Place your order early. November is not too soon to place an order for package bees to arrive the next spring. A rule of thumb for selecting a date for when the bees should arrive is one month prior to the fruit bloom in your area.

If you are a new beekeeper, request **marked** queens

that have either a colored disc glued to the dorsal section of her thorax or a dab of paint applied to the same area. Marked queens are much easier to locate during a hive examination, and with a marked queen you will be able track her as she ages. If you find an unmarked queen in a colony that previously had a marked one, you can conclude that the colony swarmed or the marked queen was replaced or died. If you discover and capture a swarm in the area of your apiary, when you locate its queen, if she is marked, you will know which hive within your apiary cast the swarm. Also, if you use bait boxes to trap swarms and find a marked queen in one of your boxes, you will know if the swarm issued from one your colonies.

If the package bees are being mailed to you, visit your local post office the week before the bees are expected and leave a phone number where you can be reached (day, night, and mobile) so that the postal clerks can contact you when the bees arrive; you may need to fill out some paperwork, so check with them. When ordering, ask for guarantees about shipping (e.g., the loss of the queen, loss of bees in the package, etc.). All should be discussed before you pay for the order.

WHEN THE PACKAGES ARRIVE

When the bees finally arrive and everything seems to be in order, check for the certificate of inspection on each package. This certificate means that the bees were inspected for mites and diseases prior to their sale; see Chapters 13 and 14 for more information on bee pathogens and pests.

The bees may be buzzing loudly and wandering all over the package. Inspect each package carefully before taking it home. Some dead but dry-appearing bees at the base of the package are normal. However, a package with bees that appear to be soaking wet, along with several inches of dead bees in a similar condition, is **not** normal. The wet condition results when bees become overheated, due to high temperatures during shipment, and bees are unable to settle quietly in the package; instead they scurry about and become overheated. In an attempt to cool down, the bees regurgitate the contents of their honey stomachs; this is where the wetness comes from. In a normal colony, when the temperature exceeds 95°F (35°C), bees collect water, return to the hive, regurgi-

tate it, and by fanning their wings induce evaporative cooling. Thus, the soaking-wet bees in the package may result from the regurgitating of the contents of their honey stomachs on one another. The surviving soaked bees, even after being released, will die. Do not accept such a package from a local dealer; if the package was received by mail, notify the sender immediately. Again, inspect all packages carefully before taking them home.

Sometimes, when swarms are collected in makeshift cardboard or wooden boxes with inadequate ventilation, the same phenomenon can be seem. Such a swarm has been referred to as a "cooked swarm."

Queenless Packages

If you find the package to be queenless (where the queen has died), here are your options:

- If the queen was guaranteed in your purchase and is dead, notify the seller immediately and ask for a replacement queen. Any replacements must not be delayed, or some workers will undergo ovary maturation and begin to lay eggs. These *laying workers* can produce only drone eggs; thus, the colony would be doomed. See "Laying Workers" in Chapter 11.

- If the queenless package is installed and provided with one or two frames of both eggs and larvae under three days of age from an established colony, laying workers will not develop and the bees will rear a new queen. However, the colony will lag behind in its development. It takes 16 days from egg to adult to rear a new queen, usually 5 to 6 days before she embarks on her mating flight(s), an approximately 21-day window in which to mate, and an additional 3 days before she begins laying eggs. It will take another 21 days after she starts egg laying for the first new worker bees to emerge. Thus, a period of approximately two months has elapsed without any replacement bees; during this time the colony's population continues to dwindle. Further, if this is attempted in the spring, there may not be many sexually mature drones available; the weather may restrict queen nuptial flights as well as drone flights. Should she fail to mate during her 21-day window of time, the result will be a drone-laying queen. Now your options are very limited. You could replace her with a mated queen, but by this time the colony's population has dwindled markedly. the better option would be to remove the drone-laying queen and unite whatever bees remain in the colony with another colony.

- A short-term solution would be to add a queenless package to an existing but weak colony, or combine the package with another package into a single hive body until a new queen arrives. Then you could split that colony into two, and requeen the queenless portion.

These three solutions assume that you already have other colonies or that you, as a beginner, have ordered more than one package. New beekeepers would be wise to begin with three or more packages; this will provide you flexibility if there is a problem.

Feeding Packages

A healthy package sounds loud, with lots of buzzing bees. This is normal; the bees are not "mad" or ferocious. Feed the bees liberally by spraying sugar syrup from a spray bottle on the screen sides of the package. Do not soak the bees. Some beekeepers brush the syrup on the screen, but this can injure the bees—many of them will have their tongues and feet protruding through the screen. Spray bottles work best! As soon as possible, the package should be placed in a cool (not cold), draft-free, quiet, and darkened area, and fed heavily with sugar syrup by spraying the screen. The bees will soon become calm. Do not place the package inside the living area of your home, as temperatures may be too hot, and the bees will break their cluster around the queen cage and scatter in every direction in an attempt to escape from the hot package. Also, bits of wax from their wax glands can pass through the screens and will litter the area outside the package.

If you are spraying sugar water onto the screens, some of it will drip and litter the area surrounding the package. However, if there are no other alternatives, you can lower the house temperature, reduce the light level, and set the package on a few sheets of newspaper. The best place to store packages is a cool garage or basement, as long as they are not too cold or wet. Make notes in your hive diary about each package and how it looks.

INSTALLING PACKAGES

Before the bees arrive, prepare the syrup, which should consist of a 1:1 mixture of white granulated sugar and warm water; see "Sugar Syrup" in Chapter 7. Warm water will accelerate the process of dissolving the sugar. When feeding bees, use only pure granulated cane sugar. Sugar with impurities such as any given percentage of molasses is not a proper feed for bees; again, see "Fondant" in Chapter 7. It is also much better to be prepared in advance in the eventuality that the queen cages may not have a candy plug; therefore be sure to have mini marshmallows or some fondant candy (see "Fondant" in Chapter 7) available at the time of hiving packages.

The common practice is to install packages in the **late afternoon**. If the weather is unusually cold, wait for it to improve (but do not wait more than a few days). Continue feeding the stored packages until they can be installed. All equipment should be readied and in place well before the bees arrive. Equipment needed for installing each package should include the following: two to three deep hive bodies (the number of hive bodies is dependent on the method used to install the bees), 10 frames of drawn comb or foundation—or a combination of both, a bottom board, inner and outer covers, an entrance reducer, and syrup feeders (see "Basic Hive Parts" in Chapter 3). **Note well that entrance reducers should be used with all methods of installation.**

Two things to remember:

1. Once established, the package bee colony may not build up enough stores or population to survive through their first year. If the bee population is not sufficient, you may need to feed them in the fall, late winter, and early the next spring. On the other hand, a colony with a low population should be united to a strong one; if left alone it will not likely survive the winter.
2. To prevent mice and other small creatures from moving in and damaging the combs, each hive should have its entrance closed until the bees are installed. But remember that following installation, entrances must remain open.

For the last half hour before the packages are to be installed, feed the packages with sugar syrup. The best way to do this is to spray it onto the screens, then wait until the bees have removed most of the liquid from the screens; when they have, spray again. Well-fed bees will remain calm, are less prone to stinging, and the syrup will stimulate the development of their wax glands.

The bees in each package can be installed into hive boxes containing wax foundation, or foundation with a plastic base coated with wax, with drawn comb, or a combination of any of these. Old drawn comb should be used only if it was taken from a healthy colony; combs from colonies that died should not be reused for any purpose until the reason for the colony's demise can be learned. For example, spores from American foulbrood disease and nosema disease can remain viable for decades. A colony that died at some point during the year is referred to as a *dead-out*. If you are using frames with both drawn combs and foundation combs in the same colony, keep the drawn combs together and the foundation combs together. Do not comingle them. If you are using both, then the queen cage should be located on one of the drawn combs, not on a foundation comb. This will require excavating some of the drawn comb to make a space for the queen cage.

Whether flowers are in bloom or not, since spring weather can quickly turn from pleasant to inclement and would delay the bees from foraging, supply the packages with a protein supplement, pollen patties, or pollen substitute and fondant along with the sugar syrup. If the bees have no access to pollen, brood rearing will be delayed; see Chapter 7 for more information on feeding. Also, check out internet sites for videos on package installation; but better yet, help different beekeepers to see what methods they use. Every beekeeper has his or her own method, so helping out is a good learning experience to develop your own methods.

To install a package, suit up, light your smoker, and follow these steps. And get out your hive diary to take notes.

Indirect Queen Release, Method I

1. Take the package to the preassembled hive site.
2. With the sprayer of sugar syrup, spray the bees to coat their wings, but do not soak them.
3. Shake or jar the package so the bees drop to the bottom of the package. Two quick shakes will dislodge the bees clinging to one another, to the queen cage, the sugar syrup can, and the underside of the top of the package, so that they will drop to the bottom of the package. Some suggest the same result can be achieved by bumping the package on the ground. We like the former over the latter because bumping, which is more violent than shaking, will also throw the queen against the walls and screen of her cage.
4. Remove the lid, exposing the top of the feeder can and the slot from which the queen cage is suspended. Queen cages are suspended by various devices. In addition, the slots from which the cages are suspended may permit you to remove the cage without removing the syrup can. In other cases, the slot is so narrow that the feeder must be removed and the queen cage guided out of its slot by way of the opening provided by the removal of the feeder can. The queen cage may be attached to a metal tab, a piece of wire or screen, or in the case of a plastic queen cage, a small disc. In any case, be careful in removing the queen cage to avoid dropping it into the package.
5. Once the cage is removed, replace the package lid to keep the bees from escaping the package.
6. If the queen is alive, remove the cork or plastic cap from the end of the queen cage that contains the white candy. Poke a hole in the candy plug with a nail. Start poking with the nail slowly, because if the candy is soft, the nail will go through it quickly, and you may inadvertently skewer the queen. If the candy is exceedingly hard, it is likely the bees will be unable to remove it in order to release the queen. If this is the case, it is wise to remove two-thirds of the candy plug. The purpose of the candy plug is to delay the queen's release, allowing the bees time to accept her as their queen. Various types of queen cages are now in use; you can see the variety online.
7. If no candy is present, after removing the cork, plug the hole with a mini marshmallow or homemade fondant candy. If you don't have any on hand, you can proceed with the installation and return later to release the queen manually in a few days. Place the queen cage in your pocket, screen side away from your body, to keep the queen warm.
8. Counting from either of the lateral sides of the hive body, remove frame 4, and attach the queen cage so it sits against the comb of frame 4 with

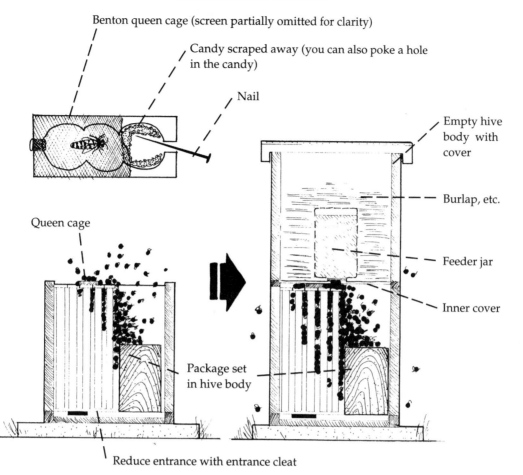

Indirect-Release Method I

Benton queen cage (screen partially omitted for clarity)

Candy scraped away (you can also poke a hole in the candy)

Nail

Empty hive body with cover

Burlap, etc.

Feeder jar

Inner cover

Queen cage

Package set in hive body

Reduce entrance with entrance cleat

the candy/marshmallow plug facing up, and the screen on the queen cage facing outward or sideways so the bees have full access to the screen. This positioning of the queen cage will orient all the bees in the package to the queen, placing a portion of those bees close to the oblong hole of the inner cover. Sitting over that hole will be the sugar feeder. This gives these bees easy access to the feeder without having to travel far from the queen cage. In cold weather, even if they have to cluster, the bees should still have access to the sugar water. The cage can be attached to the frame with a thick elastic band, string, or thin wire.

9. Make sure the feeder is not leaking syrup.

10. Place the entire package in the vacant space in the hive (where the frames have been removed), being sure that the open end of the package is face up, to allow the bees to escape (see the illustration of indirect-release method I).

11. Place pollen patties (supplement/substitute) on the top bars of the frames, then place the inner cover on the hive, rim side down, to allow extra room above the top bars to accommodate the protein supplement. Do not overfeed with the patties. **Caution:** If you are located in small hive beetle (SHB) areas, remember, they love pollen patties. Watch that the patties are eaten quickly by the bees and no beetles are seen; if you do find some, remove the patties at once.

12. Before placing the sugar syrup over the oblong hole of the inner cover, invert the syrup container outside of the hive so the initial drippings will fall into a pail and not on the ground; otherwise syrup dripped on the hive or inner cover may invite robbing bees. If the feeder leaks, get another (see "Screw-Top Jars or Feeder Pails" or change the type of feeder, Chapter 7).

13. Now place an empty hive body on the inner cover to enclose the syrup container. Block any

openings in the inner cover rims, and finally, put the outer cover over this hive body. Partially block the entrance of the hive with an entrance reducer or with grass to discourage robbing.

Advantages

- Excellent chance of queen being accepted.
- Bees disturbed less than with method II.
- Less drifting.
- Easy for beginners.
- Dead bees are left in the package and not on the hive floor.

Disadvantages

- Additional trips must be made to remove the empty queen cage and package and to reinsert the frames that were initially removed to accommodate the package.
- Egg laying is delayed until the queen is released from her cage. Although this is true if all goes well, the delay is usually less than 48 hours unless the candy plug is so hard the bees cannot evacuate the plug. Enlarge the opening to a sufficient size for the queen to walk out of her cage.
- Depending on the type of feeder used, an extra hive body may be needed.

After three days, check to see if the queen is released. Light your smoker and gently blow smoke into the entrance. Remove the empty hive body, replace or refill the old feeder if needed, and then remove the inner cover. Now you can:

- Remove the empty bee package and insert the frames that were originally removed to accommodate the package. If bees are still in the package, shake them in front of the hive so they can walk in.
- Smoke the bees down just enough to retrieve the queen cage. If the cage is empty, the queen has been released; close the hive. If the queen is still in the cage, pull off the screen or, in the case of the plastic queen cage, remove the cap that is not covering the sugar plug.
- Be aware that this caged queen may still be capable of flying off. Therefore, open these cages near the hive entrance and then quickly **push the queen cage inside**.
- Take notes in your hive diary about the condition

of each installation; please do not forget to number each colony so you can follow its progress or lack thereof.

See "After Installation," below.

Indirect Queen Release, Method II—Bees Shaken Out of Package Near the Hive's Entrance

To install a package with this method, suit up and light a smoker just in case it is needed. Again, do this in the latter part of the day when the weather is warm enough for bees to fly. **Note**: This method is not the best alternative if temperatures are cold (below 65°F [18.3°C]). This method requires shaking the bees out of the package near the hive's entrance. The bottom board of the hive sits either on the ground or on an elevated platform. If it is on an elevated platform, provide a ramp in front of the entrance. Again, have your spray bottle full of sugar syrup. Take the package to the hive, which consists of a bottom board, a deep hive body with frames, and an empty hive body that will be used to house the sugar water feeder.

The bees inside the package will be clinging to the queen cage, the sugar syrup can, the underside of the top of the package, the screens, and to one another. Give the package two quick shakes and the bees will drop to the bottom of the package.

Then proceed as follows:

1. Using the spray bottle, spray the bees through the screens to wet their wings, but do not use so much spray that the bees are soaking wet.
2. If the bees are in a wooden package, remove the lid to expose the top of the feeder can and possibly one end of the queen cage. If the bees are in a plastic package, referred to as the Bee Bus, the lid is the top of the sugar can or the top of a black plastic container with fondant as the food source. In the wooden package, the queen cage is suspended next to the feeder can by either a metal lid, a wire, a piece of metal screen, or a plastic strip. In some cases, you can remove the cage without removing the sugar can; in other cases, you first have to remove the can. If the latter is the case, you may have to give the package another shake to drop the bees to the bottom of the package. As soon as the queen cage has been withdrawn from the package, replace the lid to

Indirect-Release Method II

Close-up of Queen Cage

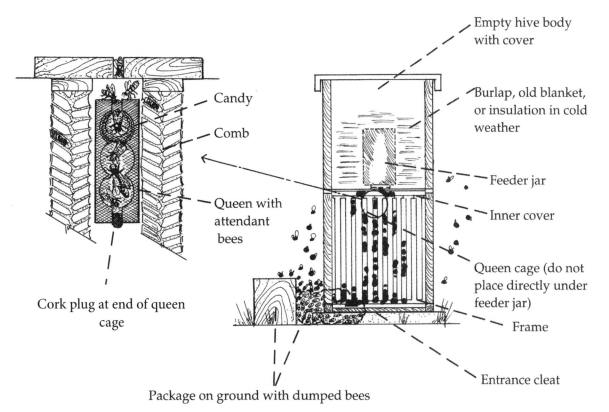

Candy

Comb

Queen with attendant bees

Cork plug at end of queen cage

Empty hive body with cover

Burlap, old blanket, or insulation in cold weather

Feeder jar

Inner cover

Queen cage (do not place directly under feeder jar)

Frame

Entrance cleat

Package on ground with dumped bees

Bees can be spread around the queen cage; dry comb or foundation can be used.

contain the bees in the package. In the case of the Bee Bus, it is first necessary to remove the sugar-water can or plastic container in order to remove the queen cage. In this type of package, the queen is in a plastic cage called the JZs BZs queen cage; this cage is attached to the top of the package by a small plastic button. Grasp this button and withdraw the cage from the package. Immediately cover the opening with a spare lid.

3. Great care must be taken when withdrawing the queen cage. Should a misstep occur and the cage drops to the floor of the package, you will have to fish it out.

4. If the queen is alive, remove the cork in front of the sugar plug of the wooden cage or the plastic cap (JZs BZs) from the end of the tube containing the candy plug. As outlined in method I, poke a hole in the candy plug with a nail, being extremely

careful not to go completely through the plug and possibly skewering the queen.

5. If no candy is present, after removing the cork plug/plastic cap, plug the opening with a mini marshmallow or fondant candy. If you do not have these with you, you can still proceed with the installation and return the next day to perform this step. Alternatively, be prepared by bringing these items with you in the event that you will need to use one of them.

6. Suspend the queen cage between the fourth and fifth frames of the hive, with the screen facing outward on a wooden cage; in the case of a plastic mesh cage, expose as much of the mesh as possible. Shake approximately one full cup of bees from the package onto the queen cage, which is positioned in the center of the comb a third of the way down from the top bar. Insert the frame back

into the hive with the side to which the queen cage is attached facing the fifth frame. By using this protective measure of placing her on the 4th frame, she will not be directly under the oblong hole; should the sugar-feeder leak, the liquid will not drip onto the cage. Sugar syrup dripping on her cage could soak the queen, which may chill her and/or mask her pheromones. Another benefit of correct placement of the queen cage is that it will position many of the bees that are clustering around her to be in easy reach of the sugar feeder; the bees will likely start drawing comb in this area, keeping them in continuous reach of a food source, which is an advantage during inclement and cool spring weather.

7. Remove the package lid and shake approximately a **cupful of bees** near the queen cage; replace the package lid.

8. Add a pollen patty on the top of the frames in the hive body containing the queen cage, add the inner cover, and then add the container of sugar syrup, locating it over the oblong hole. Surround the syrup container with an empty deep hive body and add an outer cover over this box. Winds at any season can be strong enough to blow the outer cover off the hive, so secure the cover with a heavy object such as a large stone, a cinder block, three or four bricks, or with straps.

9. Again, spray the remaining bees in the package with syrup.

10. Remove the package lid and shake a third of the bees out and onto the ground in front of the hive; make sure they start walking into the entrance. If they are not moving in the direction of the entrance, a few moderate puffs of smoke will turn them in the correct direction. If the weather is too cold, however, use the direct-release method instead.

11. The freed group of bees will soon begin to *scent* (their heads will face the entrance, abdomens raised, wings fanning), releasing the "come hither" odor or pheromone from their Nasonov glands to orient bees toward the entrance.

12. When bees begin moving toward the entrance to enter the hive, shake the remaining bees in front of the entrance. By shaking the bees outside the hive, all the dead bees that were in the package fall to the ground and not inside the hive. In many cases, the number of dead bees may be insignificant. Drifting may be a problem when the

weather is warm or it is early in the afternoon, and especially when multiple packages are being installed.

13. If any live bees remain in the package, leave it near the entrance to their new home open side up; many will make their way out of the package and into the hive. If the bees are in a wooden package, you can simply remove the screen from one side of the package and knock the package to the ground with its now-opened side facing the ground; this action will free all the bees. The plastic cage can also be opened to free the bees. If the bees are not freed from the package as in this manner, many will fly or crawl out on their own; those that fail to do so will likely perish if they remain in the package over night, especially on cold spring evenings.

14. After most of the bees have entered, partially block the entrance of the hive with an entrance reducer or with grass. Leave the entrance partially blocked for several weeks (replacing grass when needed) to discourage robbing. This will make it easier for the guard bees to protect the entrance and may also minimize drifting because bees will exit at a slower rate.

Advantages

- Excellent chance of queen being accepted.
- Syrup located in vicinity of bees and queen, so likelihood of starvation is diminished.
- Easy way to feed syrup.
- Dead bees in bottom of the package will fall on the ground and not inside the hive.

Disadvantages

- Some drifting can occur, especially if the weather is warm.
- An extra hive body is needed, depending on what type of feeder is being used.
- May take a little more time than other methods.
- Queen cage must be removed at a later date.
- Egg laying is delayed since the queen is not immediately released from her cage. However, a delay of a day or two is not critical.

Three days after installing the bees, return to the apiary with your smoker and blow a few puffs of smoke into the entrance. Remove the outer cover, remove the feeder jar/pail, puff some smoke into the oblong hole of the inner cover, remove the hive body surrounding the feeder, remove the inner cover, and apply a few light puffs of smoke over the top of the frames. Now you can:

- Remove frame 1, the frame nearest you as you work from one of the lateral sides of the hive. Now slide frames 2 and 3 toward the empty space created by the removal of frame 1. Withdraw frame 4, and if the queen has been released, remove the queen cage and return the frames in their same numerical order. Replace the hive furniture, and if needed, replenish and replace the sugar-water feeder before replacing the outer cover.
- If the queen has not been released from her cage: In the case of the wooden cage, remove the cork from the opening used to insert the queen into the cage and quickly block the exit you created by removing the cork. Some wooden queen cages have only one entrance/exit. In this situation, you must remove any obstacles that are preventing her release such as the remnants of the candy plug or a dead bee blocking the exit; then immediately block the entrance to prevent her escape. In the case of the plastic cage, remove the plastic cap, and quickly block the opening. Recall that caged queens have lost a third of their body weight and can now fly; you must prevent her escape. Now, for either type of cage, push the cage through the entrance into the hive and quickly block the entrance with a wooden entrance reducer, newspaper, or grass. By so doing, the queen will be unable to exit and will likely join the bees in the colony. After about five minutes, be sure to reopen the entrance.
- Take notes in your hive diary about the condition of each package; don't forget to number each colony so you can follow its progress.

See "After Installation," below.

Indirect-Release Method III— The Hansen Method

Follow steps 1 through 8 in the indirect queen release method II, above. Then, follow these steps:

9. Remove the queen cage from the package, and suspend it on the side of the fourth frame that faces the fifth frame, counting the frames from either of the lateral sides of hive body 2 (see illustration; this hive body will eventually sit above hive body 1, after step 13). Remember the screen side of a wooden cage always faces the fifth frame; be sure that the bees have room to access the screen. Never face the screen against the comb of frame 4; the bees will be unable to detect the queen. In the case of the plastic queen cage, if fastened in the same manner as the wooden one, most of the plastic screen will be exposed. By locating the queen cage in the center of the fourth frame, the bees that congregate around the cage will be in the vicinity of the oblong hole of the inner cover where they will have contact with the sugar-water feeder.

10. Shake a cupful of bees onto the queen cage. These bees will begin scenting, which, as the process of installation continues, will orient the remaining bees in the package to the location of the queen cage's. See illustration.

11. Remove the feeder can from the package and place the open package in hive body 1 so that the opening of the package is centered with the queen cage above it. Hive body 1 is sitting on the bottom board.

12. Now place hive body 2 above hive body 1; this hive body has 10 frames and the queen cage, plus a cupful of bees surrounding the queen cage.

13. Before placing the inner cover over hive body 2, add pollen patties onto the tops of the frames of hive body 2 and a feeder of syrup over the inner cover.

14. Place an empty third hive body (hive body 3) over hive body 2. This empty hive body will enclose the sugar syrup container. Now add the outer cover and secure it with a heavy stone or cinder block.

15. In the next day or two remove hive body 3, remove the syrup feeder, puff some smoke into the oblong hole of the inner cover, and put a lid over the oblong hole (you may use the lid that came with the package or a square piece of cardboard).

16. Set hive body 2 on top of hive body 3. Remove hive body 1, and remove the empty package that was in hive body 1. Now place hive body 2 on top of the bottom board and add hive body 3 on top of the inner cover. Remove the lid over the oblong hole, puff some smoke into the oblong hole, and replace the hive feeder over the oblong hole.

17. Add the outer cover and secure it.

Indirect-Release Method III
Hansen Method

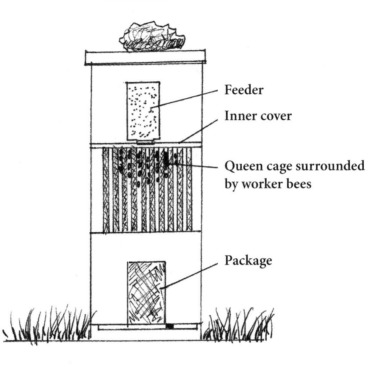

Feeder

Inner cover

Queen cage surrounded
by worker bees

Package

Advantages

- Excellent chance of queen being accepted.
- Minimal disruption of the bees in the package.
- Any dead bees remain in package.
- Initial drifting of bees is avoided; this is especially important if one is installing multiple packages in the same area on the same day.

Disadvantages

- Additional trips must be made to remove the queen cage, remove package, and rearrange hive bodies.
- Egg laying is delayed, since queen's release is delayed.
- On occasion bees may build comb in the package, not in the hive proper, and the queen may start to lay in these combs.
- Initially two extra hive bodies are needed.

Again, three days after installing the package, remove the empty queen cage or release the queen from her cage. Light your smoker and gently blow smoke into the entrance. Lift off the empty hive body and check the feeder (to see if you need to refill it), then take off the inner cover. Now you can:

- Smoke the bees down just enough to retrieve the queen cage. If the cage is empty, the queen has been released; close the hive. If the queen is still in the cage, pull off the screen or, in the case of the plastic queen cage, remove the cap that is not covering the sugar plug.
- Be aware that this caged queen may still be capable of flying off. Therefore, open these cages near the hive entrance and then quickly **push the queen cage inside**. After pushing the cage inside, block the entrance for five minutes to keep the queen from exiting, then unblock the entrance.
- Take notes in your hive diary about the condition of each package. Don't forget to number each colony so you can follow its progress.

See "After Installation," below.

Direct Queen Release Method

Follow the first procedures as outlined for the indirect queen release method I as far as removing the queen cage, and then follow this sequence:

1. If the weather is cool, place the queen cage in your pocket, screen side away from your body; if it is warm, put the cage in the shade.

2. Remove four frames from the middle of the hive.

3. Spray bees in package and remove the lid; shake all the bees onto the bottom board in the vacant space (or on ground at hive entrance).

4. Spray the bees with sugar syrup again, to reduce their flying ability, but do not soak them.

5. Dip the queen cage in sugar syrup or spray it, so the queen will not fly off when released; do not dip it if the weather is cold.

6. Carefully remove the screen from the queen cage (see illustration of the direct method).

7. Allow the queen to walk or drop gently on top of the bees.

8. As soon as the pile of bees disperses, carefully replace the frames, taking care to avoid crushing the bees (unless you released the bees on the ground).

9. Before placing the inner cover over the hive body, add pollen patties onto the tops of the frames, replace the inner cover, and then add the container of sugar syrup, locating it over the oblong hole.

10. Now place an empty hive body on the inner cover to enclose the syrup container. Block any openings in the inner cover rims, and finally, put the outer cover over this hive body. This won't be necessary if using a hive-top feeder instead of a jar-type feeder. Partially block the entrance of the hive with an entrance reducer or with grass to discourage robbing.

Advantages

- Queen released and can start to lay sooner.
- Easiest, fastest method; complete in one operation.

Disadvantages

- Extra hive body needed.
- Queen could be killed or balled.
- Bees could leave (abscond) with free-flying queen.
- Queen could fly away during installation (taking bees with her) or be otherwise lost.
- Queen could be superseded.
- Some bees and/or queen could be injured or killed when frames are replaced.
- Drifting occurs if weather is warm or it is early in the day.
- Any dead bees in the package are also shaken into the hive, making extra work for bees to remove them.
- Bees and queen could cluster some distance from the feeder and be unable to access it during cold periods.

After three days, light your smoker and gently blow smoke into the entrance. Lift off the empty hive body and feeder. Check to see if you need to refill it.

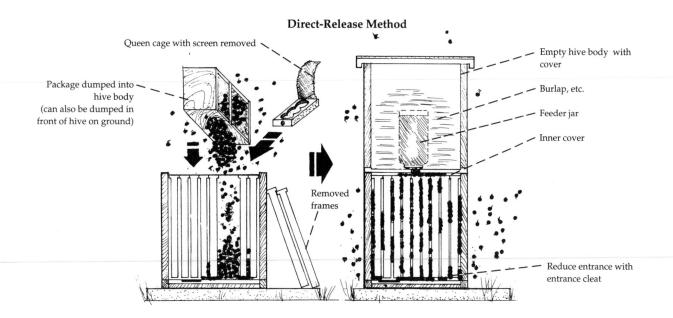

Direct-Release Method

Queen cage with screen removed

Package dumped into hive body (can also be dumped in front of hive on ground)

Removed frames

Empty hive body with cover

Burlap, etc.

Feeder jar

Inner cover

Reduce entrance with entrance cleat

AFTER INSTALLATION

Whatever method of package installation you choose, the follow-up inspections are basically the same.

On the eighth day after installation, it is time (weather permitting) to check the interior of the colony again. The purpose of this inspection is to determine if eggs and larvae are present; this will indicate there is a laying queen in the colony. If you find eggs and larvae in a compact brood pattern, things are off to a good start. Do not look for the queen, but replace any of the removed frames, placing them in the same order as when you removed them. Close the hive and leave it undisturbed for another week, but remember, always continue to replace the sugar syrup (never let the syrup feeder go empty), pollen patties, and fondant as needed. When replacing the syrup, have a lit smoker ready, and as you remove the feeder container to refill it, direct smoke into the oblong hole to prevent the bees from boiling out; return the filled container to its former position. If you have extra feeders you will not have to refill the empty one; simply swap it with a full feeder.

In all these early inspections, the key is to keep inspections as brief as possible. During cool weather (spring), you may easily chill the brood, and if the inspections take place too often, there is a likelihood the bees may abscond (abandon the hive). Bees progressing from in-house duties to outside activities first take orientation flights, which can be witnessed near the entrance to the hive. See if you can spot incoming foragers with pollen.

On the following week, in addition to eggs and larvae, you should see brood that is capped—that is, with a slightly convex wax cap over the larvae/pupae. The developing bees under these caps will soon emerge and provide the colony with its first set of new bees that will add to the colony's numbers. Also, if you used frames with foundation, the bees should be making some progress in drawing out more of these frames. Again it is not necessary to find the queen; but if the day is warm and your curiosity cannot be deterred, you can usually find the queen on a frame of open brood. In any case, minimize the time you are searching for the queen.

Continue to provide syrup and pollen patties to the colony. Once a major honeyflow is in progress, the colony may have adequate food reserves, and feeding syrup and pollen patties may no longer be required. But it will be your responsibility to determine how adequate the food reserves are, and, if inadequate, feeding must continue.

Sugar syrup will be essential to maintaining the colony's survival. A package of bees, especially if installed on foundation, must rely on the sugar syrup for its carbohydrate source.

A three-pound package contains approximately 10,500 bees, and that number will dwindle until new bees emerge. Assuming the queen begins laying eggs the day she is released, it will be 21 days before the first cadre of new bees emerge. During this period, attrition is occurring and the original 10,500 bees will have a population decline of approximately one-third of their original numbers. Remarkably, 38 days after the queen begins laying, the package colony rebounds and reaches a population of 16,000 bees, increasing the original numbers by 5,500 bees!

The temptation, especially for new beekeepers, to check the progression of their newly installed package is often more than a beginner can stand. But in this case, the **less disturbance, the better** is the rule. Although nothing is more informative as to the progress of your newly installed package than inspecting the interior of the colony, you can glean some evidence of its well-being by observing the activity at the colony's entrance. For example you may see:

- Undertaker bees removing the dead bees.
- Incoming forager bees with pollen on their legs (observe the color of pollen and try to determine from which flowers it was collected).
- Guard bees at the entrance, challenging incoming bees.
- Bees progressing from in-house duties to outside activities; these bees first take orientation flights, which can be witnessed near the entrance to the hive.
- Signs of pests, such as skunks, ants, or mice.

If time permits you to make these observations on a weekly basis, you can learn a lot about the foraging activity, colony organization and structure, and the kinds of plants blooming in your area. If you don't see these activities, there may be some problems that require your attention, so you should inspect the colony.

REASONS FOR PACKAGE FAILURES OR DELAYED DEVELOPMENT

Colonies started from packages sometimes fail or fail to develop within a normal timetable. Reasons for failure or delayed development include but are not limited to:

- Queen has been superseded due to nosema disease or other reasons delaying normal colony development.
- Queen is unmated or poorly mated.
- Queen is balled as a result of too many disturbances of the hive by the beekeeper (especially during the first ten days after installation).
- Weather has been too cold for bees to forage or obtain sugar syrup due to location or type of feeder. (**Note well**: by placing the queen cage near the oblong hole, access to the syrup feeder is enhanced during cold weather in that the bees are clustered in the vicinity of the oblong hole.)
- Bees have starved.
- Bees are infected with disease or mites and could perish.
- Bees have left the hive (absconded).
- Queen was stressed during shipment or injured while being removed from a mating box or during placement into queen cage or after her release, which may lead to diminished egg-laying capacity and/or pheromone production or a premature death.

Any of these conditions will result in substandard performance, which eventually will lead to supersedure (queen replacement by her colony); or, upon her demise, the bees will be unable to replace her because there are no eggs or young larvae that can be converted to emergency queen cells. All or any of these possibilities will delay colony population buildup, or, in the case of a queenless colony, laying workers will become established and the colony will decline and eventually die. Replacing queens whose performance is substandard may improve the overall development of the newly installed package.

No matter what method is used to install the package, there will always be a loss of bees due to drifting, as bees fail to locate their hive during and after their installation. Drifting may be more severe when multiple packages are being installed. As each package is installed, some bees will expose their Nasonov glands, releasing a homing or orientation pheromone that will attract drifting bees from other installations to their hive. Other bees will just drift away. (Methods I and III will minimize drifting.)

Installation is somewhat weather dependent. If it is a cold day, bees may cluster out of reach of the sugar syrup feeder, and if there is a succession of cold days, the cluster may starve. This situation usually can be prevented by placing the feeder close to the suspended queen cage. Bees will naturally cluster near the queen and, as a result, will be positioned near the feeder. You can position the cage close to the oval hole of the inner cover, and by the placing the sugar syrup feeder over that hole, you can ensure that the bees will be near the queen. Make sure the syrup does not leak.

Here are some other reasons packages fail to survive:

- By direct release of the queen, the cluster may form at some distance from the feeder.
- No eggs will be laid until the queen is released from her cage, and even if she is directly released (not recommended), cells need to be available for eggs to be deposited. Once egg laying is under way, the first newly emerging bees will occur 21 days later. During this hiatus, colony numbers are in decline, due primarily to natural attrition of the original adult population, death of older bees, bees lost by drifting, and foragers failing to return to their hive.
- Package bees installed on foundation have no food (honey and pollen) and thus must be provided with sugar syrup (to meet their carbohydrate requirements) and pollen or pollen substitutes (to meet their protein, fat, and vitamin requirements). In addition, since the timing for installing packages may coincide with the scarcity of flowering plants, feeding is essential.

Other things to remember:

- Feeding is also necessary to stimulate their wax glands; wax is essential for converting foundation into comb.

- Unless used equipment is available, packages will require all new equipment.
- Even though bees have been certified to be healthy, to be safe, treatments for diseases or mites may still be needed.

- Package bees may be infested with pests, such as mites and small hive beetles (see Chapter 14), so watch for these pests.
- Remember, for first-time beekeepers, this is **fun** as well as educational. Enjoy.

 **Notes**

CHAPTER 7

Feeding Bees

Recognizing when a colony of bees is in need of supplemental feeding to prevent its demise is an important step in learning bee craft and saving a colony. Especially during the winter months and in early spring, it may be necessary to supplement a colony's dwindling food supply by feeding them one or more of the following items: sugar syrup, dry sugar, sugar cake, fondant, honey, pollen, pollen substitute, and winter patties. By such intervention, you may prevent a colony's demise. Bees may be at the brink of exhausting their food stores; the colony will be hard pressed to remain alive unless provided with emergency feeding. If this situation occurs during the flowering period, the colony may continue to exist on a day-to-day basis, but its numbers will begin to decline; there will not be enough food required by nurse bees to synthesize brood food that is fed to the developing larvae. If an interval of inclement weather sets in, the entire colony will likely perish.

This chapter will offer information on when it is necessary to provide bees with food. During late summer or early fall inspections, you may discover that your bees have insufficient stores for the coming winter months, or that during the course of the winter their stores are near depletion. Some additional circumstances when bees need to be fed are: (1) occasionally during the flowering season when the incoming nectar is insufficient to sustain the colony, (2) when establishing a new colony with package bees, or (3) when replacing a queen, when rearing queens, and after hiving a swarm. In the very early fall, every effort needs to be made to evaluate the stores in each of your colonies, even those that appear healthy and well populated. If the stores are lacking or insufficient, the situation needs to be remedied immediately. During the winter months and into the early spring, in warmer southern regions, 35 pounds (16 kg) of honey would be sufficient, and in northern regions, 90 pounds (41 kg). The farther north the bees are located, the more likely additional reserves will be required.

You may also remove empty combs from the second hive body and replace them with full combs of sealed honey until sufficient stores are present. In the absences of frames of honey, begin feeding sugar syrup to the bees. A gallon of syrup (a mixture of two parts sugar and one part water, by weight or by volume) will be converted (cured) into seven pounds of stored food reserves. It will require approximately 13 gallons (49 L) of 2:1 sugar syrup to yield approximately 91 pounds (41 kg) of stored food. The number of gallons of sugar syrup needed for each colony can be determined by how many empty or nearly empty frames are found in the second deep hive body. A deep frame of sealed honey weighs between 7 and 8 pounds (3–4 kg).

Even if you think that adequate stores are in place, or you fed your bees early in the fall, it still may be necessary to provide additional food during the winter or early spring. This will likely be the case during extremely cold winters interrupted by some warm spells, when consumption of food increases in order for the bees to maintain necessary cluster temperatures required for winter survival. It may also happen when the onset of flowering plants has been delayed due to lingering cold temperatures or too dry (or too wet) conditions. If natural resources are unavailable but the colony has sufficient stores, feeding will not be necessary.

Bees should be fed under the following conditions:

- When the colony has nearly depleted its resources and will die of starvation, particularly during the winter and early spring months.
- When it is necessary to medicate with chemotherapeutic agents, using sugar syrup as the vehicle to deliver the medication.
- When installing a package or hiving a swarm.
- When bees are placed on foundation. When you install a package of bees on foundation, sugar syrup will stimulate their wax glands; feeding is absolutely necessary.
- When hiving a swarm on foundation.
- When requeening. Feed sugar syrup in order to simulate a nectar flow (bees are more likely to accept an introduced queen during a natural or simulated nectar flow).
- When rearing queens, especially when there is no strong honeyflow in progress.

At any time of the year, a colony may exhaust its food supply and thereby die of starvation. The beekeeper must always be vigilant, but particularly in the fall, winter, and early spring. Stores may be depleted as a result of:

- The beekeeper removes too much honey, particularly in the fall.
- The bees had insufficient honey reserves for the winter/spring, and this condition was not addressed.
- The bees consumed the remaining winter food by spring.
- The number of foragers (field bees) is inadequate, due to spring dwindling, failing queen, diseases,

and parasites (see "Spring Dwindling" in Chapter 8). The bees' consumption of their food reserves increases when the queen resumes egg laying in mid-January. Remember, the larvae need to be fed, and the broodnest must be maintained at temperatures required for their development.

- An anticipated honeyflow fails to materialize (i.e., specific nectar plants fail to yield nectar), or it was too wet for an extended period, preventing the bees from gathering the available nectar.
- If there were a drought during the foraging period, flowering plants would conserve water, and, as a consequence, nectar production would cease.
- The colony was robbed of its stores after the beekeeper had visited the hive and assessed that the stores were adequate.

Again, when colonies are in a condition where starvation is imminent, **the bees must be fed** to ensure their survival. The various methods of feeding bees with sugar syrup, dry sugar, sugar cakes, fondant, honey, and pollen or pollen substitutes (or supplements), and winter patties are discussed in this chapter.

SUGARS

Carbohydrates are organic compounds composed of carbon (C), hydrogen (H), and oxygen (O) atoms, with a general formula of $C_nH_{2m}O_m$; in other words, there are twice as many hydrogen atoms as oxygen atoms, and the number of carbon atoms can vary. Because carbohydrates are composed of many C–H bonds, which liberate a great deal of energy when they are broken, these molecules are an excellent candidate for energy storage. Sugars, starches, and cellulose are examples of carbohydrates.

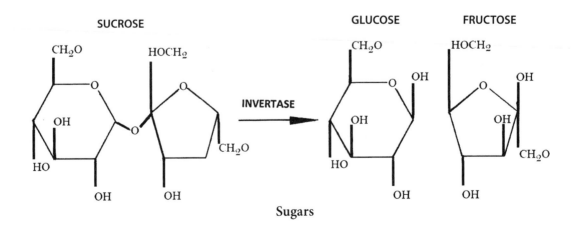

Sugars

The three different types of sugars that bees feed on, and are found in nectar, are glucose, fructose, and sucrose. The proportions of these three sugars vary; in some nectars, the principal sugar is sucrose, while in other nectars there may be in nearly equal amounts of all three sugars. While most nectar is primarily sucrose, some nectars, while rare, may have equal amounts of glucose and fructose or either one or the other as the major sugar in the nectar.

Glucose ($C_6H_{12}O_6$, also called dextrose) is found in all living cells. It is produced by photosynthesis in green plants and is the primary energy-storage unit. Glucose is a simple sugar (a monosaccharide) that comes in two forms: a straight-chain molecule and, when mixed with water, a six-sided ring hexagonal structure (see the illustration of sugars).

Fructose (levulose or fruit sugar) is another monosaccharide and has the same formula as glucose but with a different molecular structure. It is the sweetest of the simple sugars because the double-bonded oxygen is attached internally in the molecule, rather than on the end, as in glucose. Two different ring structures are possible in fructose molecules: a hexagon (six sides) and a pentagon (five sides).

Sucrose ($C_{12}H_{22}O_{12}$, or cane sugar) is a disaccharide, or two sugars, made up of glucose and the pentagon form of fructose linked together. This sugar is the end result of photosynthesis. The sucrose molecule is broken apart into the two monosaccharides by the action of the enzyme *invertase* or *sucrase* (see Chapter 12). An enzyme is a protein that serves as a catalyst in chemical reactions, building up, breaking down, or rearranging the atoms. *Invertase* is the enzyme that plants use, and *sucrase* is the enzyme animals and bees use to convert sucrose to glucose and fructose.

Sugar Syrup

A one-gallon mixture of two parts sugar to one part water (2:1 sugar to water, by weight or by volume), when stored, will increase the food reserves by approximately 7 pounds (3.2 kg). A deep Langstroth frame filled with capped honey weighs about 7.5 pounds (3.2 kg). It will require 12 gallons of 2:1 sugar syrup to yield 84 pounds (38 kg) of stored reserves. This estimate of course assumes that all the syrup will be stored. To be on the safe side, it would be wise to provide the bees with 15 gallons (57 L) of

sugar syrup. The number of gallons needed for each colony can be determined by the number of empty or partially full frames that are found in the upper hive bodies. A medium frame filled with honey will be about 6 pounds (2.7 kg), and a shallow 4–5 pounds (1.8–2.3 kg).

In general, use these proportions of sugar to water:

- 1:1 for spring feeding and spraying onto bees in packages or on a swarm.
- 2:1 for fall feeding; make up to 15 holes in the lids of gravity feeders. A gravity feeder can be made with an appropriate-sized container such as a gallon glass jar or a new gallon paint can with nail holes in its lid. When filled with sugar water and inverted over the oblong hole, the container will leak for a few seconds. But as long as the lid is secured, the pull of the pressure zone at the top of the container will counteract the pull of gravity and the liquid will remain in the container, allowing the bees to withdraw it by poking their tongues into the holes in the lid. It is important that the bees obtain as much syrup as possible in a short period of time.
- 1:2 to stimulate brood rearing. Make only two to six holes in the lids of the gravity feeders, so the bees will be able to obtain only small amounts over an extended period of time; this effect will be similar to a light nectar flow.

Use **only white, granulated** cane or beet sugar from sugar beets. **Never use brown or raw sugar, molasses, or sorghum, because these contain impurities and can cause dysentery in bees.**

Mix the desired proportions of sugar and hot water and stir adequately until all the sugar is dissolved. Hot water from the faucet is hot enough to dissolve the sugar; you can also use water that has been heated over a stove, but it must not be boiling. Never let the sugar-water solution boil over direct heat; also, syrup that is burned, or caramelized, will cause the formation of hydroxymethylfurfural, or HMF, which is toxic to bees and will lead to high bee mortality. Honey and other syrups will also form HMF if overheated. Heating the mixture over steam or in a double boiler will prevent caramelization.

To prevent fall syrup from crystallizing, some beekeepers add cream of tartar (tartaric acid) to the solution of sugar and water. Tartaric acid breaks down the sugars, but there is some concern that it is detrimental to bees; currently, its addition is not recommended. Other beekeepers add a little vinegar to keep mold from forming in the syrup; a teaspoon of apple cider vinegar added to each gallon is safe for bees.

Other things added to syrup, including salt, vitamins, and bleach, have also been advocated, but so far there is little research that these additives are advantageous. Check current journal articles for the latest information.

Bees should be fed early enough in the late summer or early fall so the sugar syrup has sufficient time to cure—that is, the bees have time to reduce the water content near to that of honey (which ideally has 18% water, and sometimes even less). If the syrup-water content is not sufficiently reduced, it could ferment or freeze; both these conditions are detrimental to overwintering bees.

High-Fructose Corn Syrup (Corn Syrup)

High-fructose corn syrup (HFCS), or corn syrup, is a sugar syrup produced by breaking down cornstarch, which is composed of two major chains of glucose (molecules linked to one another in long chains), into single units of glucose, fructose, and water. In 1969, researchers used enzymes, one that split the long chains of glucose into smaller ones, and another that reduced the smaller chains into individual glucose molecules. These enzymes are now produced by genetically engineered bacteria, replacing the process that used acid and heat. The resulting syrup contained 55 percent glucose, 42 percent fructose, and other substances.

The methods for feeding HFCS are similar to those described for other syrups. Unfortunately, HFCS may not be the best food to feed bees, and caution should be taken when feeding this to your colonies. When fed **exclusively** on HFCS, colonies may not raise enough brood to replace older foragers; therefore the colony could be too small to survive. Currently, there are products available that are mixtures of sucrose syrup and HFCS.

Corn syrup is also sold in grocery stores for making candy, but **do not** use this for bee feed. HFCS is available in bulk to commercial beekeepers, but remember, **it is very similar to honey**. If it is stored in areas that are too hot, or the storage containers are

subjected to extended periods of high temperatures, it could form HMF, which, as noted, is toxic to bees. And if this degraded syrup is mixed with water, formic and levulinic acids could form as well, which are also toxic to bees. Make sure that HFCS does **not** end up in your honey supers when extracting; you **do not** want this in the honey you sell to consumers.

When in doubt, use sugar syrup.

TYPES OF SYRUP FEEDERS

Screw-Top Jars or Feeder Pails

One of the best ways to feed bees at almost any time of the year (except winter) is with a one-gallon glass jar or a plastic pail, or a new one-gallon metal paint can. Any of these containers can be turned upside down over the top bars of the uppermost super of a hive or over the oblong hole of the inner cover.

In either case, the use of shims under the lids will provide space for the bees to access most of the perforated openings you will place in these lids. The lids in the glass jars and paint cans may be perforated with small nail holes; bees will access the sugar solution by inserting their tongues into these openings in order to withdraw the solution. You can use a shingle nail, or a three- or four-penny nail ($\frac{1}{16}$-in. [1.6 mm] diameter) to punch 10 to 15 holes in the lids. A one-inch hole can be burned out of the center of the lid of a plastic pail, and a fine metal screen can then be melted in place to cover the hole (see illustration of a plastic feeder pail); check bee supply catalogs for additional types of feeder pails.

Don't use plastic (milk) jugs, because they often collapse after being filled and inverted; they can also leak. If you use a plastic bucket, make sure it did not contain any toxic material. Although glass containers can break, they are easier to clean and inexpensive to replace, and you can readily see if they need to be refilled. At the hive, first invert the jar or pail so drippings will fall into some container or on the inner cover, rather than on the hive or the ground, so as not to encourage robbing. As soon as the dripping stops, place the feeder directly on the top bars near the cluster if the colony is weak; otherwise, place it over the oblong hole of the inner cover. Put an empty hive body rim around the feeder, and replace the outer cover. Make sure to put a weight on top of the outer cover to prevent it from blowing off.

The syrup will not leak as long as the holes are

Plastic Feeder Pail

not too large and the feeder is level; if there is empty drawn comb in the hive, the bees will remove the syrup from the feeder and store it in the comb below. Make sure the feeders do not drip on the bees—syrup leaks can kill bees or cause robbing.

Hive-Top Feeder

One type of feeder, once called the Miller feeder, or top feeder, was composed originally of either two aluminum or two plastic pans fastened together and hung within or otherwise attached to an empty shallow super. Some were made entirely of wood. In addition to feeders constructed of wood, a single-unit feeder composed of plastic with the dimensions of a hive super is now available.

An opening at one end (or in the middle) allows the bees to crawl up from below and over the sides of the feeder to get to the syrup. A control strip of wire mesh or roughened plastic allows the bees to cling to its sides without drowning in the syrup (see the illustration of the Miller or top feeder). This type of feeder is also available in nucleus sizes, which makes feeding nucs much easier. Check bee supply catalogs on the type of feeders available. The top feeder, as the name suggest, sits on top of the frames of the hive body containing the bees or above frames of foundation, if the intent is not only to draw the bees into this hive body, but to also to promote wax gland production, so as to draw out the foundation.

Check the catalogs on how to use these feeders;

some can hold up to four gallons (15 L) of syrup and can be rapidly filled or refilled. If this type of feeder is used, it must not leak and must be bee-tight on the outside, or else robbing may occur. One problem with many top feeders is that often an inordinate number of bees drown in the syrup. When the design of this type of feeder permits, add some dry pine needles to the surface of the syrup; this may help prevent the bees from drowning. The same is true for the division board feeder. There are many styles of top feeders available from dealers; check the types beekeepers use in your area.

Division Board Feeder

The division board or *in-hive feeder* is a frame-size feeder that can be inserted in place of a frame within a deep or medium super. Feeders of this type have some kind of flotation device or a strip of screen that allows the bees to reach the syrup without drowning. Other division board feeders are made with Masonite, wood, or plastic and usually have Styrofoam, screen, or wooden floats. Some plastic division boards have roughened inside walls to provide footing for bees. **Nevertheless, bees often drown in these feeders**.

Some beekeepers permanently keep one division board feeder in each brood chamber, usually on the outermost side of the hive body. On cold days, however, a weak colony will be less able to obtain syrup when this type of feeder is used, unless it is located near the bee cluster. If the cold weather continues for a long time, the bees may be unable to move to the feeder and could starve.

Conversely, if this type of feeder is left in place all the time, bees could start to construct comb in it, and the queen could lay eggs on that comb. Not only will that make it difficult to see what the comb contains, but also it will be difficult to remove, and if you fill the feeder with syrup, the brood and maybe the queen could drown if they are inside.

As with any equipment, see what is new in the bee supply catalogs and talk with other beekeepers about what works in your area.

EMPTY DRAWN COMB

Frames of empty drawn comb can be filled with syrup and placed in the hive. Slowly dip the frames

Miller or Top Feeder

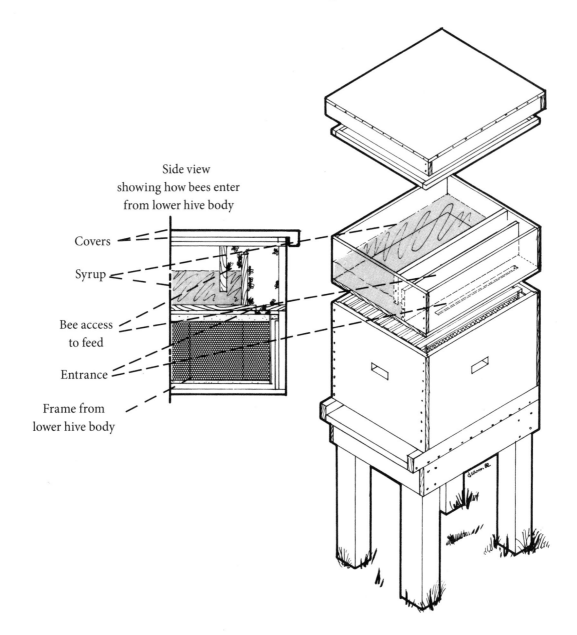

Side view
showing how bees enter
from lower hive body

Covers

Syrup

Bee access
to feed

Entrance

Frame from
lower hive body

into a tub of syrup or sprinkle them with syrup using a sprinkler can or other device. A steady stream of syrup poured directly from a container will not fill the cells because the air in the cells will act as a barrier to the liquid's entry (see the illustration on filling drawn comb). This method can be used for emergency feeding, especially if the combs are located near or adjacent to the broodnest. As with other feeding methods, you have to look into the hive (and remove these frames) to determine whether they need refilling.

This method is often used when installing package bees in a hive with drawn comb. Newly hived swarms, however, should not be fed by this method. Bees in a swarm have full honey sacs, and if they are put on drawn comb, they may regurgitate the contents of their sacs into these cells; if this honey contains disease spores and is later fed to larvae, brood diseases such as American foulbrood could result (see "Catching Swarms" in Chapter 11). It is very important to ensure that any combs used for feeding syrup have no history of brood diseases.

Filling Drawn Comb

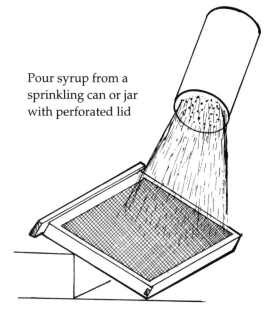

Pour syrup from a sprinkling can or jar with perforated lid

Boardman Feeder

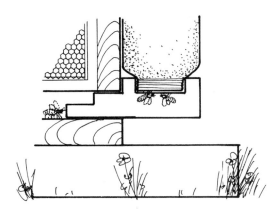

Side view
showing how the feeder fits
in the entrance

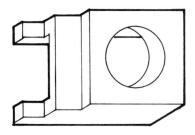

Top view

Entrance (or Boardman) Feeder

This feeder is a wooden or plastic holder for a quart-size mason jar or plastic container (see illustration of the Boardman feeder). The front portion of the holder's base is an entrance platform and is inserted into the hive entrance. The bees can obtain syrup by crawling into the holder's base to reach the jar.

This feeder is **not** recommended for feeding syrup because bees from other colonies can rob from it, which tends to encourage further robbing activity. If the weather is cold, the bees being fed will not break away from the cluster to reach the feeder and the cold syrup, and may starve. The feeder holds only a quart of liquid and would require frequent refilling when the bees are actively feeding. Furthermore, in this highly exposed condition, the liquid could freeze, or the sun may decompose chemicals in the syrup that were added to medicate the bees. If the food is in there too long, the syrup could also ferment. Unless your hives are up on stands, these feeders can be easily knocked off or taken by hungry raccoons or other clever mammals, who will scatter the jars throughout your beeyard.

The best use of the entrance feeder is to provide bees with a source of water during the summer months. If the weather is quite hot, bees can go through a quart of water in a day or two. This method of providing water also helps to keep bees from up-setting your neighbors; bees will visit nearby pools or fountains in parks or even school yards to obtain water to cool their hives (see "Good Neighbor Policy" in Chapter 4).

Plastic Bag Feeders

In an emergency, half-gallon plastic zippered bags, the kind used to store frozen food, can be used as temporary feeders if jars or pails are not available. Fill the bags one-half to three-quarters full, expel the air, and seal. Store them flat. When feeding your bees, lay one bag over the top bars or on the inner cover, and cut a slit 1 to 2 inches (2.5–5 cm) long on the upper side of the bag as it is lying on the top bars. The bees will come up and feed as the syrup oozes out. Do not fill the bags too full, as you won't be able to put the outer cover on, and too much syrup may be squeezed out, which could drip on and chill a small cluster. Remove and discard the plastic bags when empty.

NON-SYRUP FOOD

Dry Sugar

Dry, white, granulated sugar can be used as emergency food when it is too cold to feed syrup, such as in winter and very early spring. In that bees are mostly confined to their hives in the winter, they have limited access to the water that will be needed to dissolve the dry sugar. However, bees may be able to access water that has condensed in their hive; or you can spray the sugar with water. Bees also can use liquids from their salivary glands to dissolve the sugar.

The sugar should be located as close to the bees as possible. It can be spread around the oblong hole of the inner cover (rim side up), or on the back portion of the bottom board away from the entrance. You can also spread the sugar over a single sheet of newspaper, placed over the top bars of the hive body where the bees are located; the bees will chew through the newspaper to obtain the sugar.

Only strong colonies will benefit from the feeding of dry sugar; weaker colonies may not have a sufficient numbers of bees to obtain the water needed to dissolve the dry sugar.

Wet Sugar

Here is another method, called wet sugar, courtesy of the late Ann Harman, of Virginia. Mix together:

- 1 cup (8 ounces or 237 ml) of water.
- 10 pounds (4.5 kg) of white granulated sugar.

Mix well. On aluminum foil or plastic wrap, shape the mixture into blocks that will fit under an overturned inner cover (rim side down) and allow the blocks to harden overnight; they become hard as a brick. The next day remove the foil or plastic and lay the blocks over the top bars and, if possible, on top of the surface of the winter cluster or in close proximity to it; do not block the oblong hole of the inner cover. The blocks mixed with water and any available metabolic water may be sufficient for the bees to work and digest the sugar in these blocks. Any sugar bricks the bees do not eat can be rewrapped and stored until needed.

Candy Board Feeder

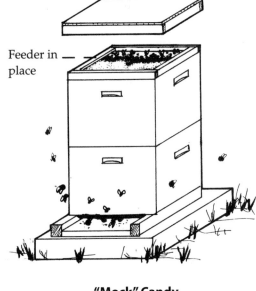

Feeder in place

"Mock" Candy

A quick candy can be made simply by mixing clean honey (not from diseased colonies or from store-bought honey) or thick sugar syrup with confectioners' sugar. But remember, confectioners' sugar can contain starch or other products to keep it from clumping up. Knead it like bread dough to form a stiff paste. Store it in the refrigerator or freezer and use it as emergency food, for queen cage plugs, or for feeding small queen-rearing nucs. As an emergency food, place a thawed piece of flattened candy on the top bars or near the inner cover hole. Store-bought honey, especially if it comes from another country, may contain chemicals or other products fed to bees; antibiotics, which are not allowed to be fed to bees in this country during a honeyflow, may be fed in other countries and could end up in the honey. Use honey of known origin; when in doubt, use a thick sugar syrup.

Fondant

Fondant candy can be made and fed to bees in small molds or placed in a rim feeder called a *candy board* (see illustration of a candy board feeder in this chapter). An inner cover can be modified to function as a candy board by adding a 1.5-inch rim to the inner cover's rim, thereby increasing the depth of the cover's flat surface. Because fondant can be difficult to make, and high temperatures are involved in its production (and the fact that in manufacturing fondant, HMF may develop, which may be poisonous to bees), some beekeepers prefer not to use it. Fondant can be purchased for winter feed for bees.

HONEY

The best food of all—when properly ripened, capped, and free of disease—is, of course, a super full of honey or several frames of honey placed next to the broodnest in colonies requiring feeding. Honey obtained from old combs and cappings, as well as crystallized honey, can be diluted and fed to the bees using any of the methods described for feeding syrup. Supers with frames that are "wet" or sticky after having been through an extractor can be placed above or below the broodnest or over the inner cover for the bees to clean.

Be careful when feeding bees with diluted honey or wet combs: the odor of honey will stimulate robbing. Therefore, feed honey or place wet combs on the hives in the early evening so that the bees will have sufficient time to remove and store it before morning. If this food is given to weak hives, reduce their entrances as a further precaution against robbers (see "Robbing" in Chapter 11). Again, make sure the equipment containing the wet comb is **bee tight** to keep out robbers. Also, supers with wet combs should not be put on colonies in the late fall or winter. The entire colony of bees might move up into them and not return to stores below, and thus they may die eventually from starvation.

Again: Store-bought honey should not be fed to bees. Not only could some honey contain antibiotics or other chemicals fed to bees; if the honey came from hives with foulbrood, there is an excellent chance that this honey contains foulbrood spores, which remain viable in the honey for up to 80 years. Using this honey could result in an outbreak of American foulbrood (AFB) in your apiary. (Fortunately for us, AFB disease spores contained in honey are not harmful to humans.) Be certain that any honey fed to the bees is free from spores that cause bee diseases.

Honey mixed with cappings, scrapings, or debris can be fed to bees if it is placed in a container or spread on a sheet of foil or on the inner cover; place an empty hive body and the outer cover over it to keep robbers out. Also, honey-filled combs that were broken, or broken combs that can be recycled, can be crushed in warm water and fed as syrup, or simply placed in a top feeder. The remaining wax can then be recovered and melted. As with any method of feeding sugar, make sure ants or other pests cannot get to the food! Honey mixed with cappings, scrapings, or debris should never be fed to bees during a dearth period, for it will attract bees from other colonies and a robbing frenzy will ensue.

POLLEN

Pollen, the male germplasm of plants, is the principal source of protein for honey bees and the building blocks of bee bread. The protein content of pollen ranges from 8 to 40 percent. An average-size colony (approximately 30,000 bees) needs 40 to 60 pounds of pollen per year (18–27 kg); other estimates are from 40 to 110 pounds (18–50 kg) a year for average-size colonies.

As forgers collect pollen, they add some honey from their honey stomachs to the pollen. This helps "glue" the pellets together as they are passed on to the corbiculae (pollen baskets). Foragers returning to the colony with their pollen loads literally back themselves into empty cells or cells partially filled with pollen, and with their first two pairs of legs, push the pollen pellets off their corbiculae and into these cells. Other bees push the pellets into cells with their mandibles, packing the pollen into a solid mass. Pollen loads are often deposited adjacent to cells containing uncapped larvae. As the pellets are being deposited and packed, house bees add nectar from their honey stomachs to the pollen. This nectar contains beneficial microbes, which start the fermentation process that will prevent the pollen grains from germinating, as well as act to preserve and enhance the nutritional content of the pollen. The processed pollen (similar to silage) will turn into *bee bread*. Pollen cells are never capped with wax, and only one-third, or up to one-half of a cell, is loaded with pollen (bee bread). Bees usually top off the surface of the bee bread with a thin film of honey, which serves as a physical barrier, while its high acidity deters spoilage and may help preserve the quality of the bee bread.

Bees on occasion "cap" a cell containing bee

bread with propolis, thus entombing the bread. The reason for entombing a pollen cell may be that it is contaminated with an excess of harmful products (e.g., pesticides). If the pollen has been contaminated with pesticides, bees could perish due to the toxic substances, or starve, as some contaminants could kill the microbes that convert the pollen into bee bread. Research also suggests that pollen containing fungicides may kill the beneficial microbes that help convert pollen into bee bread; this could result in bees starving. Make sure your bees are not in an area treated with these chemicals. Foraging bees are apparently unable to discriminate between highly nutritive pollen and that which has low nutritional value. However, by collecting pollen from diverse plant species, they achieve a complementary balance of nutrition. This balance can be disrupted if bees are forced to collect pollen from a single crop or a limited number of crops, as may occur when bee colonies are moved into locations containing a single crop, like soybeans. These areas are referred to as *monocultures* (single crop). When colonies are located in monocultures, protein supplements may be required to meet the nutritional needs of such colonies.

Bees increase their consumption of pollen and bee bread in the fall. This factor, coupled with a decrease in foraging and brood-rearing activities, seems to extend the longevity of worker bees beyond their usual summer life expectancy of six weeks. These bees are referred to as *winter bees.* Many winter workers live longer than three months, less if they are nutritionally stressed. One of the factors involved in extending the longevity of worker bees from the fall of the year into the following spring is the increased presence of *vitellogenin,* a glycolipid protein that enhances the bee's immune system, increases its life expectancy, and can be converted to brood food.

It is also necessary for nurse bees to consume bee bread, to activate their food glands (hypopharyngeal glands) so they can secrete a milky-white, protein-rich food that will feed the larvae and queen. This substance, called *royal jelly,* is fed in abundance to all larvae less than four days old and to queens during their larval stage and throughout their adult lives. Worker and drone larvae more than four days old are fed modified jellies and later a mixture of brood food, diluted honey, nectar, and pollen. Without pollen, bees could not manufacture royal jelly. About 0.0042 to 0.0051 ounce (120–145 mg) of pollen is needed to rear one bee, from egg to adult; this is around one and a half pollen pellets.

When flowers are available, bees will usually collect sufficient supplies of pollen for the hive. The demand for pollen increases in the winter when brood rearing resumes, and the remaining pollen stores can be consumed quickly. A good rule of thumb is for bees to have 500 to 600 square inches (1300–1500 cm^2) of stored pollen reserves going into northern winters. Because bees cannot forage for pollen at this time of year, there must be ample bee bread stores to enable bees to feed the larvae that will replace them in the spring. If this is used up, or not available, bees must be fed some supplemental protein.

Beekeepers feed bees pollen, pollen supplements, and pollen substitutes to initiate, sustain, and increase brood rearing. By stimulating brood rearing, beekeepers profit from the increase in populations. Such colonies can send many foragers out to harvest nectar and pollen. Beekeepers utilize strong colonies to make divisions or splits, and to supply bees for pollination or to sell in packages.

The sequence of events that initiates brood rearing may begin with an increase in the amount of sugar intake by the queen, after which she begins to lay eggs. The presence of brood activates the special glands of nurse bees, which stimulate these bees to feed on bee bread. Bee bread provides the glands (mandibular and hypopharyngeal glands) with the nutrients necessary for the production of larval food. Certain components of bee bread may activate these glands to begin the production of brood food. Our knowledge of all the factors that return a honey bee colony from a broodless condition to one with brood is incomplete. Pollen seems to be one of the factors governing the initiation and maintenance of brood rearing; however, day length, pheromones, physical and chemical stimuli, and the commencement of nectar and pollen collecting may also play an important role in the brood-rearing activities of honey bee colonies. The best protein source for bees is pollen; in second place is pollen supplement, which is better than pollen substitute (see the section on pollen supplements and substitutes below). When feeding these substances, you should provide additional stimulus by simultaneously feeding sucrose syrup. Syrup feeding also stimulates the cleaning or hygienic behavior of bees, which in turn stimulates bees to forage.

Auger-Hole or Exterior Pollen Trap

Front view showing how the trap fits in the hive

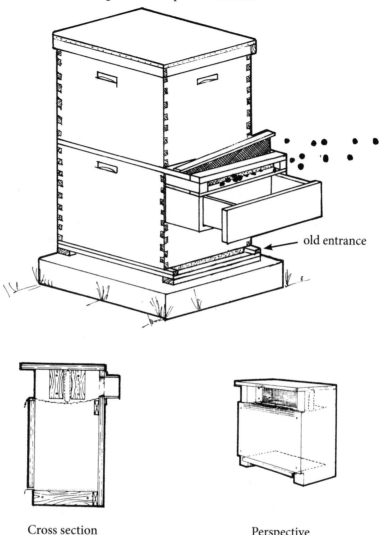

old entrance

Cross section   Perspective

Source: E.R. Harp. 1966. A simplified pollen trap for use on colonies of honey bees. Pub. 33-111. Beltsville, MD: USDA-ARS, Entomology Research Division.

Trapping Pollen

Various types of traps are available for collecting pollen, but they all have two basic elements: (1) an 8-mesh screen or perforated metal grid of approximately 8 holes per inch through which pollen-carrying bees must crawl to separate the pollen pellets from their corbiculae, and (2) a container below the grid in which to catch and store that pollen.

In any case, bees should be permitted frequent access to the free entrance of the hive so that they can replenish their own pollen stores. Check bee supply catalogs and other beekeepers for the styles of traps currently used. Bottom traps work very well, but they can collect a lot of debris, and if the pollen in the trap becomes soaked during a rainstorm it will spoil.

Beekeepers may collect pollen for a variety of purposes: to study its characteristics (amino acid or protein content), to identify the flowers bees are visiting, to add to human and pet diets, and to feed bees during periods when fresh pollen is not available. This may likely be the case during extremely cold winters and/or when strong colonies increase their rate of brood rearing. In either case, more food will be consumed under such circumstances. The need for additional feeding may also occur when cold weather in the spring delays the progression of flowering plants. An especially important reason for trapping pollen is to determine if it is contaminated with pesticides. For example, bees avidly collect sweet corn pollen; if the grower has used systemic-treated seed or sprayed a pesticide when the corn is in tassel, the pollen may be contaminated. If your bees are within range of these fields, pollen traps can be employed to collect most of the contaminated pollen, substantially reducing the amount carried into the colony. Because corn pollen is shed only in the morning hours, these traps can then be removed in the afternoon, giving bees the opportunity to collect and carry into the colony pollen that is free of pesticides (see "The Pesticide Problem" in Chapter 13).

There is always the danger that collected pollen may become contaminated with chalkbrood or American foulbrood spores, especially if the trap sits under the broodnest. Collect or purchase only pollen that has been obtained from disease-free colonies or has been sterilized by radiation. When in doubt, collect your own.

When the trap is in position on a hive, bees entering or leaving the hive are forced to pass through a mesh grid. As a bee loaded with pollen enters the grid, the pollen on the corbiculae exceeds the size of the openings in the mesh. The bee continues to struggle through the mesh in order to enter the colony. In that process, the mesh grid scrapes off the pollen load, allowing the bee to then pass feely through the grid into the hive. The pollen pellets that were removed drop into the pollen-collecting container on the trap.

Ideally, a pollen trap is placed on a colony when many plants are in flower but should only remain on the colony for short intervals in order not to jeopardize the pollen requirements of the colony. When suspending pollen collecting, only the mesh screen section needs to be removed. Some beekeepers keep the trap on throughout the summer but actually collect the pollen (set the trap) on alternate weeks or every three or four days. A colony deprived of pollen for an extended time may quickly decline. To preserve its quality, trapped pollen must be removed regularly, every day and especially if rain is imminent, as it could wash out or ruin the pollen. Even in dryer climates, remove the trapped pollen regularly

to keep ants and other pests from eating it. Clean the pollen of debris and insects by hand sorting, then quickly preserve and store the collected pollen to prevent spoilage. Fresh pollen molds rapidly, especially in hot, humid weather. Pollen loses its nutrient qualities quickly.

Storing Pollen Pellets

Drying fresh pollen pellets. Fresh pellets collected from a pollen trap can be dried in the sun for few days, in a warm oven, or with a lamp or food dryer. Heat to 120°F (49°C) for the first hour to kill yeast spores, then dry for 24 hours at 95° to 97°F (35–36°C). The pellets are ready when they do not crush when rolled between the fingers and do not stick to each other when squeezed. Store them in closed containers at room temperature.

Dry pollen may be fed directly to the bees, sprinkled into empty cells of a frame, or mixed with other dry materials. If the dry pollen is to be added to wet mixes, it should first be soaked in water or syrup for an hour.

Advantage

● Inexpensive way of preserving pollen.

Disadvantage

● Less attractive to bees if no sugar is added.

Freezing pellets. Place fresh, cleaned pollen pellets in containers and store them directly in a deep freezer at 0°F (–18°C) until ready to use; they will be moist when defrosted. Use immediately, or dry them as mentioned above.

Advantages

● Attractive to bees.
● Can be used separately or added to mixes.
● Best for human consumption, if selling pollen.

Disadvantage

● More costly to preserve.

Sugar storage. Pollen pellets can be preserved with dry sugar. Fill a container alternately with layers of pollen and granulated **white** sugar, topping it with several inches of sugar. Close the container tightly and store it in a cool place. Pollen should be mixed with twice its weight of sugar (1 part pollen to 2 parts sugar). Careful labeling of the container as to its amount of sugar and pollen will ensure that proper proportions are maintained when you are preparing mixes with other dry ingredients, to make a pollen patty. You can also freeze this pollen and sugar mixture.

Advantage

- Attractive to bees.

Disadvantage

- Difficult to separate pollen and sugar if you want to feed straight pollen.

Methods of Feeding Pollen

One method of feeding pollen is to place dried pellets (whole or finely ground) on the top bars of frames beneath which most of the bees are located. You can also pour them around the oblong hole of the inner cover, especially if it is close to the active broodnest. Dry pollen without any sugar is not always attractive to bees and is not usually recommended. A better method is to pour the pellets into frames of empty drawn comb according to the procedure described below:

1. Fill the comb on one side of a frame with pollen pellets. Insert the comb into the hive.
2. If both sides of the comb are to be loaded, to prevent the pollen pellets from spilling out of the cells of the comb in question, spray a film of thick sucrose syrup on both sides of the comb before adding the pollen.

Another method to feed pollen to bees, called *open feeding*, is to place ground pellets (grind them in a blender to break up the pellets) in a cardboard or plastic box or other container somewhere in the apiary. The container should be covered to prevent spoilage by rain or moisture. Naturally, it is necessary that the bees have access by means of a hole on the sides of the container. Because only strong colonies will benefit from open feeding, and inclement weather will prevent bees from foraging to the container, we recommend internal feeding over open feeding whenever possible. Open feeding also will attract other animals, such as raccoons and mice,

that may disrupt colonies or otherwise foul the pollen. Therefore, secure a feeder away from these pests, such as by hanging it on a clothesline.

Making Pollen Patties to Feed Bees

Pollen can be made into a dough by kneading it with clean honey or sugar syrup that has been heated before use (make sure the honey is not from diseased colonies). The dough should be stiff, not runny. Sandwich the patties between two pieces of waxed paper, not plastic wrap, to keep them moist, and roll until they are about ¼ inch (3–5 mm) thick. Although the bees will eat holes in the waxed paper to get to the patties, cut a few slits into the paper to get them started; they will then eat the pollen and discard the waste paper. An easy formula is four parts hot water to one part pollen to eight parts sugar. If you are making a lot of patties, investing in a commercial bread mixer or similar appliance will save a lot of time and effort. Make sure to store flattened, wrapped patties in the freezer.

Pollen Supplements and Substitutes

If you find it necessary to feed weak colonies or nucs, sugar syrup is preferred over honey, and trapped pollen or frames of bee bread over anything else. Only if pollen is not available or comes from a questionable source should you consider supplements or substitutes.

The terms *pollen supplement* and *pollen substitute* have been used in connection with feeding bees protein. These two terms have caused some confusion and are used here according to the following definitions: A pollen *supplement* consists of pollen and other substances of nutritional value to bees. A pollen *substitute* contains no pollen but consists of substances nutritional to bees.

Previously, soy flour made by the low-fat "expeller" or "screw press" method was a common ingredient in both, but now that the fat in soybeans is being expelled chemically, this flour is hard to obtain. Dairy products have also been used, but the milk sugars lactose and galactose are toxic to bees. Many different commercially made products are now available from bee suppliers or are advertised in the bee journals. Again, check with what others are using and experiment cautiously with new products. Brewer's yeast and soy flour are often the major ingredients in these products, with added vitamins and minerals. New products are available every year.

Pollen substitutes or other commercially purchased products available in bee supply houses may contain eggs or larvae of grain pests; these can be killed by freezing the material for several days at 0°F (–18°C); the material should then be placed in sealed containers to prevent subsequent contamination.

Feeding Pollen Supplements and Substitutes

If you find it necessary to feed colonies, it is important to start feeding in the early spring at 10-day intervals or as fast as the bees consume it. Do not stop feeding once a regimen has begun but continue to feed until the bees will no longer take it. Once natural pollen is available, bees will not take the artificial feed.

Do not be tempted to mix vegetable oil in pollen patties, as the added lipids (fats) may be detrimental to bees. Keep up with current research articles on the subject.

To make pollen patties, supplements, or substitutes:

- One pound (0.45 kg) of the dry material can be mixed with 4 cups of sugar syrup (2 parts sugar to 1 part water) to make several one-pound patties; make sure this mixture is thick enough so that it does not drip between the frames.
- Some beekeepers add anise oil, fennel oil, or chamomile oil to make the material more attractive to bees.

When you are making patties, sandwich the mixture between two pieces of waxed paper (not plastic wrap); this way they will remain moist. Tearing a few holes in the waxed paper on the underside of the patty will give the bees easy access to the food. Place the patty directly over active brood frames. Make sure it is not runny. If a patty is not consumed within a few weeks or if it starts to mold, remove it. Dry material can also be fed like pollen pellets.

Warning: Pollen or pollen substitutes should not be fed to bees until late February (in northern climates) or early in the spring when bees are able to take multiple cleansing flights and/or have resumed normal foraging activities. When bees are confined at the onset of winter, it is prudent to refrain from providing them pollen or pollen substitutes, since such food will likely result in an outbreak of dysentery. During winter confinement the metabolism of vitellogenin will serve as a substitute for pollen.

Recently, Dadant & Sons, a company that sells supplies and equipment to beekeepers, formulated a winter food for colonies light on stores; these patties are a high-carbohydrate feed.

Again: watch for pests such as the small hive beetle that love pollen patties and the like; you will be raising them as well as feeding bees. Be vigilant.

 **Notes**

CHAPTER 8

Winter/Spring Management

The survival of honey bee colonies during the winter months and early spring is largely dependent on the beekeeper's evaluation of their condition during the month of August and the first two weeks of September in northern climates. September is usually the last month when you can execute a thorough inspection of the bee colonies. A problem at this time of the year is the potential for a nectar dearth, which means if you are opening bee colonies, you may be inviting robbing. Given this potential, it is important for you to time your last inspection so it can be carried out under more ideal conditions. This is an example of where the understanding of beekeeping comes into play. In any case, it is critical to evaluate your colonies before the onset of winter.

If bees lack adequate winter stores, they will not survive. By late August or early in September, the second deep brood chamber or an equivalent number of shallows should contain nine to ten frames of sealed honey. If this is not the case, feeding must be undertaken immediately. Fortunately, if the situation is discovered early enough, you can begin feeding sugar syrup, and providing the weather remains favorable, the bees will have sufficient time to reduce the water content of the syrup. They will also be able to store the concentrated syrup in empty cells and seal these cells with wax cappings. You may delay feeding, hoping that a late summer/fall flow may materialize; if this path is chosen and the flow fails to materialize, it may then be too late to feed sugar syrup. **Timing is everything**.

In addition to an adequate supply of food, each colony should have an adequate population of bees (15,000 to 20,000), a young queen, and a low mite count. If any of these factors is not present, such colonies will unlikely survive to see the following spring.

During unfavorable weather, particularly in winter, bee colonies may be affected by any combination of adverse conditions. When inspecting a colony, if any of the following is observed, remedial action needs to be taken:

- Insufficient stores of honey.
- The absence of a queen or the presence of an old queen (over one year old).
- Laying workers.
- A small bee population (under 5000; the equivalent of two to three frames of bees).
- A high varroa mite count.
- Adult and brood diseases.
- Lack of shelter from prevailing winds.
- Hive equipment in poor condition or with inadequate ventilation.

You, the beekeeper, must take actions to correct such problems or run the risk of either losing colonies or ending up with colonies that emerge from winter with such a small number of bees and/or a high mite count that it may take a long time for them to recover into viable colonies.

This chapter is organized to provide guidance for what needs to be done if any of the adverse conditions noted above are discovered during your inspection, especially if your bees are in a location subject to the harsher months of winter and early spring. If you live in a southern climate where winter is much shorter, read the section "Wintering in Warm Climates." The next chapter, "Summer/Fall Management," is

Temperatures at Which Different Bee Activities Take Place

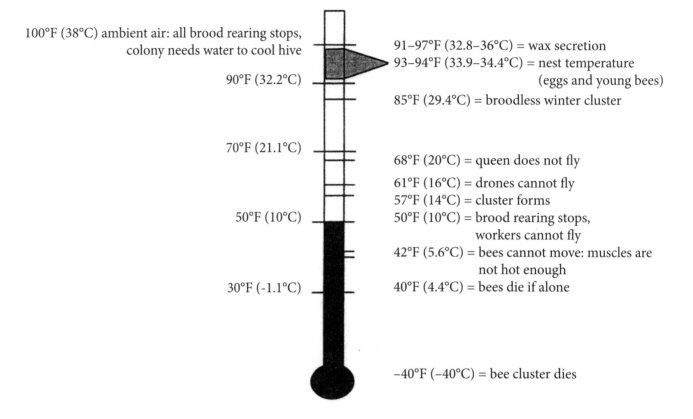

100°F (38°C) ambient air: all brood rearing stops, colony needs water to cool hive

90°F (32.2°C)

70°F (21.1°C)

50°F (10°C)

30°F (-1.1°C)

91–97°F (32.8–36°C) = wax secretion
93–94°F (33.9–34.4°C) = nest temperature (eggs and young bees)
85°F (29.4°C) = broodless winter cluster

68°F (20°C) = queen does not fly

61°F (16°C) = drones cannot fly
57°F (14°C) = cluster forms
50°F (10°C) = brood rearing stops, workers cannot fly
42°F (5.6°C) = bees cannot move: muscles are not hot enough
40°F (4.4°C) = bees die if alone

−40°F (−40°C) = bee cluster dies

designed to help you anticipate the major tasks that must be attended to during the "busy" months.

WINTERING

General Rules

Even in places with harsh winters, honey bee colonies can survive without elaborate wintering manipulations, as long as some of the conditions listed below are addressed. Following the procedures for wintering hives can make the difference for overwintering success. These are the most common wintering practices:

- Remove all bee escapes, empty supers, and queen excluders.
- Leave ample honey stores on each colony. In the lower northern latitudes, where there is significant snowfall in winter, each hive should contain 90 pounds (41 kg) of sealed honey (or cured and capped sugar syrup). Farther north and into Canada, 120 pounds (54 kg) may be required. Feed sugar syrup at a ratio of two parts sugar to

one part water, while the days are still warm, as early as August and no later than September (or in the early fall). Bees need time to cure the syrup (reduce its water content) and then cap the stored syrup. Once the bees begin to enter winter cluster, they will unlikely break cluster to retrieve the syrup from a feeder.

- Invert the inner cover, rim side down, to allow warm moist air to escape through a notch cut in the rim. You may also tilt the outer cover so that it does not block the rim opening. If you prop up the outer cover, be sure to secure it with a heavy stone or cinder block. If the inner cover does not have a rim notch, you can raise a corner of the inner cover with a small stick to allow the escape of moist air; again secure the outer cover.
- Reduce the main entrance with a wooden entrance reducer, or a piece of one-half-inch mesh hardware cloth to keep mice and other varmints from invading the hive in late summer and early fall. When to reduce the entrance is a judgment call, but when nighttime temperatures drop below 55°F (12.8°C), it is time to reduce the entrances.

Remember: these creatures do not wait until really cold weather sets in before they enter a hive.

- There are several metal reducers that make excellent guards and are available in the bee catalogs. With all the guards, it is possible that as bees die during the winter, they will drop to the bottom board and may clog the entrance. If any of the openings on a wooden reducer face the bottom board, particularly during the winter, as bees die, they will drop to the bottom board and block the entrance/exit to the hive. A better way is to position any given wooden reducer opening so that it is turned upright 90 degrees from the bottom board; in that position, the entrance will more likely remain open. In fact, one should position the entrance reducer in this position no matter what time of year it is. If the cleat is clogged with dead bees, it will prevent the bees from taking cleansing flights on warm days. **Be sure** to clear any dead bees that may be blocking the entrance.
- Provide top ventilation as mentioned, by propping up the inner cover. Some beekeepers bore an auger hole not more than 1 inch (2.5 cm) in diameter in the upper corner of the topmost hive body. This is not recommended, as it could invite robbing bees into weak or dying colonies.
- Place weights (rocks, bricks, or cinder blocks) on top of the outer cover of each colony. Winter winds can easily blow off the outer cover, exposing the winter cluster to conditions that will likely lead to their demise. Outer covers can also be secured with straps.
- Protect colonies from winter winds.

During the August (early fall) inspection, all weak but healthy colonies should be united either by combining a weak colony with a strong one, or by combining two weak ones. In the later case, the two weak colonies, when combined, should bring the combined colony numbers to over 10,000 bees. When combining a weak colony with a stronger one, the queen of the weaker colony should be eliminated, or, if time permits, she can be placed in a nucleus colony for evaluation. Colonies heavily infested with mites must be treated to reduce their mite loads.

An old beekeeping adage is "Take your winter losses in the fall."

The Winter Cluster

In the late fall and winter, bees form a *winter cluster*. Many organisms have evolved strategies that help them to survive during adverse periods—when it is too warm or too cold, when resources are unavailable, or combinations of these circumstances. Some organisms hibernate, others aestivate during conditions not conducive to their well-being. The honey bee (*Apis mellifera*) "selected" a different strategy for survival during cold periods of the year (fall, winter, spring) in northern climates.

Bees continue to consume their stores during the winter and generate heat by "shivering" (contracting and relaxing their wing muscles) and conserving the heat they generate by clustering together. Cluster formation begins when temperatures drop to 64°F (18°C). As ambient temperatures continue to fall to 57°F (14°C), the cluster takes on a more defined configuration.

Using basketballs of different sizes to represent a winter cluster, picture a basketball with two smaller balls inside it (broodless cluster) and one with three smaller balls (cluster with brood) inside the larger one. The surface of each, at different distances from one another, represents an isotherm (a curve on which every point represents the same temperature). Beginning with the surface of the largest ball, scientists found an isotherm of 44°F (7°C). The bees at the

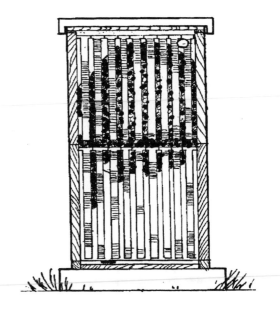

Winter Cluster

surface of this isotherm "ball" have their abdomens pointing outward and are packed tightly together. Below them may be two or more layers of bees similarly positioned. This group of bees forms the equivalent of an insulating blanket, preventing the loss of heat being generated by bees beneath them. The next isotherm is at 50°F (15°C), and when the cluster is broodless, the next isotherm is at 76°F (24°C). Once brood rearing resumes in January and/or early February, a fourth isotherm is initiated with a temperature of 92°F (33°C). Winter losses occur when there is an insufficient food reserve and/or the number of bees in the cluster is insufficient to maintain adequate temperatures within the cluster. Bees may not survive during severe interludes of low temperatures, which cause the cluster to contract in order to maintain viable temperatures. A consequence of reducing the surface area of the cluster (in order to conserve heat) is that the cluster may become separated from its food and perish. If bees exhaust their food reserves, the cluster size is unimportant.

The winter cluster expands and contracts as ambient temperatures rise and fall. Bees, with the exception of those forming the outer mantle with an isotherm of 44°F (7°C), remain active within the cluster, and the queen resumes egg laying in January (in northern climates). One of the byproducts of cellular respiration is water vapor, which rises above the cluster. If the vapor is unable to escape from the colony, it will condense into its liquid state and may precipitate back onto the cluster. Wet bees in a cold hive will likely lead to the demise of the colony. Therefore, it is imperative to have an opening above the cluster for the water vapor to escape to the outside. Ventilation during the cold months is as important as during the summer months. An opening in the inner cover or upper hive body is recommended; the use of materials that will retain the water (and thus not drip onto the cluster), such as a pile of straw, sawdust, or some other absorbing materials (e.g., fiberboard) can also be used.

Connective clusters of bees form bridges from the main cluster to reach food stores. If these connectives are cut off, or if the cold becomes severe, obligating the cluster to contract in order to conserve heat (and by so doing losing contact with the food supply), the colony will die of starvation. The cluster must be able to remain in contact with food throughout the winter. In general, bee clusters will move upward throughout the winter, which explains why bees are usually found in the upper hive body in the spring.

Bees retain their feces during the periods of confinement; this is a common situation during the winter months, even in southern states where rainy weather can keep bees confined. Periodically, air temperatures may reach over 57°F (14°C) during the winter, and on such days, bees are able to break their confinement to take cleansing flights. If the bees are confined to the hive over long periods, the hive floor and frames can become littered with fecal material; and if excessive, this is *dysentery* and can weaken the bees further. Gut residue is a result of feeding on pollen and/or pollen patties or similar food. Be careful what you feed bees in the fall.

Winter survival depends on:

- A young, vigorous queen.
- Large population of bees (20,000–30,000 or 8–10 frames, covered on both sides with bees); see the information on estimating colony strength in the sidebar.
- An adequate supply of honey or cured sugar syrup and bee bread.
- Some warm, sunny days during cold winters (to permit cleansing flights and cluster movement).
- Dry, cold winters so bees do not eat up their winter stores too fast and starve.
- Early springs with moderating temperatures.
- Colonies that are free of diseases and are entering the winter with low mite counts. Mite assays (counts) in the fall (August in northern climates) should be taken to determine whether intervention is needed to reduce the mite load. Remember: young bees being produced in August will be your winter bees, and healthy winter bees will contribute to the colony's overwintering success.
- An upper entrance for cleansing flights.
- Allowing water vapor to escape by propping up the inner cover with a small stick or shim, or reversing the inner cover so that it is rim-side down; be sure there is a notch in the rim to permit water vapor to escape.
- Protection from prevailing winds (use temporary windbreaks or snow fencing).
- Reduced front entrance.
- Maximum sunlight exposure.
- Protection from bears and other pests.
- Apiary integrity: during the winter months of

January, February, and early March, you can check your apiary to learn if any outer covers have blown off or if a fallen branch has damaged the electric bear fence.

- You can also take a quick look inside if the weather is sufficiently warm. **Warning**: it is not a good idea to disturb the winter cluster, but if you suspect that food reserves may be low, it will be necessary to be aware of this problem and to provide additional food.

Remember: An adequate supply of honey (or cured sugar syrup) and bee bread (which can be difficult to determine) is vital. In March, once bees are able to forage on a regular basis, then pollen patties can be fed. Once enough natural pollen becomes available, bees will no longer consume pollen patties, and you should then refrain from providing them. As for an adequate supply of honey or cured sugar syrup, one deep super, approximately 90 pounds (41 kg) of honey, will be required for bees in northern latitudes. If you are feeding syrup, feed an equivalent amount of sugar syrup that, when cured, will yield similar weights. Colonies located farther north and in the Canadian provinces will likely require even more. If you are using shallow or medium supers, the equivalent weights in honey need to be available.

The average loss of bee colonies over winter, before mites became established, was around 10 to 20 percent. Now with mites, small hive beetles, and new pathogens in the picture, winter losses can be much higher, and in some years can reach 60 to 80 percent. Even without these new problems, additional winter losses can be a result of:

- Wet, cool winters.
- Long, cold winters with few sunny, warm days, reducing or eliminating opportunities for cleansing flights.
- Unmated, injured, poorly mated, diseased, or drone-laying queens.
- Weak colonies in the fall, with improperly cured or poor-quality food stores.
- Queenless colony.
- Insufficient honey stores, especially if weather conditions were poor during the last honeyflow.
- Colonies placed in shaded, damp areas.
- Inadequate protection from bears.

Wintering in Extreme Climates or at High Altitudes

In regions where average temperatures during the coldest months are below 20°F (–7°C), you should leave about 90 to 120 pounds (41–54 kg) of honey on each colony (or feed an amount equal of sugar syrup). This is about one deep or two medium bodies full of honey. If sugar syrup is to be fed to bring the food stores up to these numbers, remember it must be fed to bees while the weather is still warm so it can be properly cured. Make the syrup in a 2:1 ratio of sugar to water. You may also try cellar wintering if you live in extremely cold climates where winter lasts more than five months; see the discussion in this chapter.

The essential elements for a colony to overwinter successfully, as covered above, are similar in extreme cold conditions, with the addition of providing adequate windbreaks, wrapping colonies, and insulating hives.

Windbreaks

Apiaries should be located where they will be sheltered from prevailing cold winds to reduce the amount of cold drafts in winter and spring. Situate hives where barriers such as fences, evergreen hedges, thick deciduous growth, walls, or buildings will take the brunt of the prevailing winds. When no windbreaks are present, construct temporary ones to lessen the velocity of the wind as it approaches the hives, while still permitting air drainage to take place.

A suitable windbreak would be a snow fence or slotted board fence 6 feet high (1.8 m) set up on the north side of the apiary or at other compass points as dictated by the wind direction. The boards in the fence should be about 1 inch (2.5 cm) apart to slow the wind velocity and at the same time allow air to filter through. The first row of hives should be about 5 feet (1.5 m) from the windbreak.

Wrapping or Using Hive Covers

Hives can be wrapped to protect the bees from chilling winds; also, the dark color will absorb the sun's heat. There are several covers available made of materials ranging from corrugated plastic to treated paper to builder's foam; in addition to these, there are commercial wraps such as Bee Cozy and Easy Cover. Check what other beekeepers do in your area and investigate overwintering gear from bee supply catalogs.

There are several procedures for wrapping hives, and most of them incorporate these features:

- Both top and bottom entrances.
- Top ventilation.
- Absorbent material enclosed in the cavity of a shallow super over the inner cover to absorb moisture (material such as straw, shavings, porous pads, corrugated paper, fiberglass, foam, insulating board, or other building insulation). These materials may become soaked and may need to be replaced at some point during the winter.
- Dead air space underneath the hive.
- A mouse guard at the entrance.

Advantages

- Protects colonies from piercing winds.
- May conserve a sufficient amount heat to allow the cluster to expand.
- Bees can move and re-cluster on honey if the inside temperature is warm enough.

Disadvantages

- Time consuming.
- Vapor barrier may form between hive and wrapping, resulting in excess moisture accumulation in the colony, which, if it freezes, will encase bees in an "icebox."
- The colony may warm up too much, and bees may begin premature cleansing flights before air temperatures are high enough to protect them from being chilled.
- Insulation may slow the bees' perception of suitable spring flight conditions, delaying foraging.

Insulation

Insulation will provide colonies with extra protection against cold winter temperatures. Before 1900, beekeepers would use double-walled hive bodies by adding another, wider hive body (called *chaff* hives) and filling the intervening spaces with sawdust or straw. Today, more modern insulating materials can be used, with caution. If you are experimenting, go

Types of Insulation

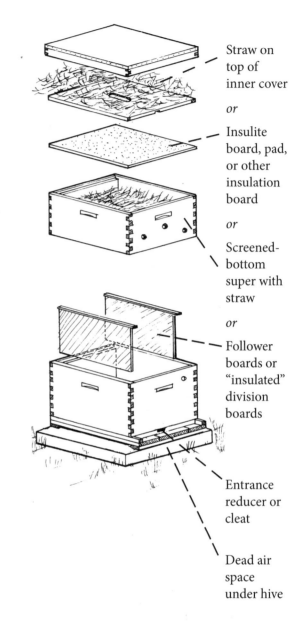

Straw on top of inner cover

or

Insulite board, pad, or other insulation board

or

Screened-bottom super with straw

or

Follower boards or "insulated" division boards

Entrance reducer or cleat

Dead air space under hive

slowly: use materials on only a few colonies at a time. Some insulating material may release toxic "gases."

Any one or a combination of these insulating materials, methods, and devices has been found to be of some aid (see the illustration on types of insulation):

- Provide dead air space underneath the hive.
- Place *follower boards* against the inside walls. A follower is a solid piece of board or other durable insulation material, the size of a deep frame, of variable thickness, that hangs like a frame. It can

be used to reduce the interior size of a hive body by substituting it for one or more frames.

● Remove each of the frames situated next to the walls on each side of a deep hive body and replace them with division board feeders filled with foam or some other insulating material; division board feeders may be used as substitutes for follower boards.

● Fashion a shallow empty super with a screen or cloth bottom and fill it with insulating material such as straw or wood shavings and place it above the second deep hive body; then cover it with the inner and outer cover.

● Place above the inner cover a 1-inch-thick foam board with a hole cut in its center matching the inner cover's oblong hole.

Some beekeepers paint hive bodies with insulating paint or put a super of dry, drawn comb on top of the inner cover (these combs can collect excess moisture and may be damaged). Supers of honey are also good insulators, so leaving extra honey on never hurts. Do **not** use any queen excluders.

Check how other beekeepers in your area winter their colonies. If you do nothing else, protect them from piercing winter winds.

PREPARING FOR THE NEXT SEASON

The following tasks should be attended to during the winter months in preparation for the coming spring:

● Order packages to arrive before fruit trees bloom in your location.

● Clean stored supers and frames of burr comb and propolis; replace old combs.

● Build new equipment for the coming year: frames, hive bodies, tops, and bottoms.

● Paint and repair equipment; replace with new equipment if necessary.

● Frames containing old, damaged, or sagging combs should be set aside and the defective combs removed and replaced with foundation. Combs more than three years old should be replaced. The reason for this is that wax absorbs insecticides and herbicides, and over time these harmful chemicals continue to accumulate in the wax.

● Sort and cut out sagging, diseased, damaged, or drone combs and replace with foundation. Old wax combs need to be replaced every three to five years.

● If you used plastic foundation coated with wax, scrape off the old drawn wax cells and paint some fresh wax over the bared foundation; the bees will quickly draw cells from the plastic base. Melt down old wax and prepare to sell or make candles.

● Order other equipment you will need for the coming year (gloves, extractors, bottles, new veil).

● Check the apiary periodically for damage from downed trees or limbs, wind, or vandals or other predators.

● Attend bee meetings, read or write for your local bee club's newsletters, and review your hive diary notes.

● Browse new catalogs, bee journals, and web pages; take a class and buy a new bee book and read it.

LATE WINTER, EARLY SPRING MAINTENANCE

When weather is suitable for hive inspections, preferably when temperatures enable bees to take cleansing flights or resume foraging activities, check on the items listed below (some of these items requiring inspection may also be revisited in the late spring and therefore may be referred to again in the next section, which will deal with spring management):

● Dead colonies should be removed from the bee-yard and placed in a shed or some other enclosed structure. If this cannot be accomplished, these hives may remain in the apiary, **but** all possible entrances to these colonies must be sealed to prevent robbing bees from gaining access to them. This addresses two main concerns: (a) to prevent a robbing frenzy, and (b) to prevent the robbing bees from carrying any disease from the dead colonies back to their colony.

● If possible, attempt to learn why the colony died. Was it from a disease (e.g., American or European foulbrood, or nosema), or was it the result of starvation due to a low bee population unable to maintain adequate cluster temperatures? Colonies that perish from American foulbrood harbor the spores of this pathogen and, therefore, combs from these colonies should not be used again unless you have access to a fumigation chamber. **Remember, never feed to other colonies honey in combs or honey that has been extracted from colonies known to have American foulbrood.**

- When inspecting a frame from a dead colony, you may observe that the dead bees are headfirst in their cells; it is likely that these bees died of starvation, even if honey is nearby. During cold weather, bees in a cluster may not break from the cluster to move to another section of the hive that may contain honey stores.

- Should you detect a foul or sour odor when holding a brood frame close to your nose, it may be caused by American foulbrood. If the colony is dead, seal it up; if it is still alive, close it down after dark. In either case, call your state bee inspector or an experienced beekeeper to assist you in diagnosing your observations. If there is an excessive amount of fecal matter (yellow/brown spots) on the frames, covers, and on the bottom boards, this may indicate that the bees have (or died from) dysentery. With a hive tool, scrape off the fecal material and dispose of the soiled equipment. If you suspect nosema is present, medicate the bees for this disease (see "Nosema Disease" in Chapter 13). If it is determined that dysentery is the problem, the honey stores may have fermented, or were not properly cured, or the bees were unable to undertake cleansing flights or were fed pollen during the winter.

- To learn what may be ailing your bees (alive or dead), freeze or place them in alcohol. Tracheal mite populations peak during the winter; see Chapter 14. To obtain a diagnosis of what is wrong with your bees, they can be shipped to the Bee Laboratory located in Beltsville, Maryland, for analysis; go to https://www.ars.usda.gov /northeast-area/beltsville-md/beltsville -agricultural-research-center/bee-research- laboratory/docs/how-to-submit-samples/.

- If temperatures begin to moderate (e.g., March and April), bees can now be fed sugar syrup and pollen to prevent them from starving and to stimulate and/or to increase brood rearing. Feed pollen, supplements, or substitutes in moist patties (see Chapter 7).

- If the air temperature gets above 75°F (24°C) and the number of adult bees in the colony is less than you anticipated, compared to other colonies you have inspected, remove a brood frame and examine the brood pattern. If the sealed brood (worker brood) is compact and covers a reasonable portion of the frame, it is likely the queen is present and in good health. Colonies with spotty brood or no brood should be of concern; in such cases the queen may be absent or failing. When examining colonies in the early spring, keep such examinations as brief as possible so as not to chill the brood.

- Prepare for the arrival of package bees and nucs by having the needed equipment to house them available in advance of their delivery dates.

- Make or order and assemble any equipment that will be needed to accommodate your plans for the new season.

If a colony is low in numbers (fewer than 3000 bees) but free of diseases, and the queen is present, the weak colony may be united to a strong one or to another weak one, thereby increasing the strength of the now combined weak colonies. When combining colonies, both of which have queens, decide which queen should be removed and which one saved (see "Spring Dwindling" in this chapter and "Uniting Weak Colonies" in Chapter 11). Other tasks can include:

- To determine if bees had a disease, collect adult bees and freeze or store them in alcohol to have them checked for disease or mites. Tracheal mite populations are highest this time of year; see Chapter 14.

- If the weather is very cold, determine the amount of stores by lifting or tilting the hive; if the hive seems light, feed the colony. Emergency food during cold weather includes fondant candy, "mock" candy, frames of honey, or dry sugar sprinkled on top of the frames or inner cover; see Chapter 7.

- If warm enough, begin to feed sugar syrup and pollen to stimulate brood rearing. Feed pollen, supplements, or substitutes in moist patties (see Chapter 7). Provide frames of sealed honey if the weather is too cold to feed syrup.

- If the air temperature gets above 75°F (24°C), check the colony for the queen's condition by examining the brood pattern: a compact pattern of worker brood indicates that a healthy queen is present. This examination should be brief; otherwise the brood can become chilled.

- If a healthy colony is near starvation, provide clean frames of honey on either side of the cluster as well as above it. Feed it depending on the present and future weeks' weather conditions.
- Prepare for the arrival of package bees or nucs by making certain you have enough equipment.
- Take steps to protect bees from insecticides, fungicides, and miticides. These compounds may have been absorbed by the wax combs (because wax is *lipophilic* or fat loving).

Even assuming you left a sufficient amount of honey stores on your colonies, this does not guarantee they will survive throughout winter and early spring. During severe cold periods, the consumption of honey increases, and strong colonies (with large populations of bees) will consume more honey. Therefore it is prudent to tilt your colonies to estimate their weight, as mentioned above; do this occasionally during the winter/spring months. If it is easy to lift up a colony, it is likely they will require more food (see Chapter 7).

If you are selecting colonies to be used for pollinating services, you will need colonies with large bee populations. Many beekeepers who rent out their colonies for pollination are often moving them to warmer southern states in the fall to help increase the bee populations. Colonies with medium to low populations will build up more rapidly when relocated in these areas where beekeepers can feed more and flowers bloom earlier. One of the early crops requiring pollination is almonds. Most of the almond orchards are located in California's Central Valley, and many beekeepers transport their hives there a month before the almonds bloom; they prepare and care for the colonies by providing sugar syrup (1 part sugar to 1 part water) and pollen (or pollen substitutes or supplements) to build up the populations. Before the colonies are moved into the orchards, beekeepers will usually equalize the colony populations so they are all at about the same strength. Orchardists pay beekeepers for their pollination services.

Young queens play an important part in growing colony numbers, as they lay more eggs and thus produce more bees than older queens. Follow the overwintering procedures discussed above, especially if the bees will remain in the North before they are shipped to other locations that require pollination services. Pollination hives are often loaded onto large trucks (since many of these trips cover long distances). Machinery such as forklifts is usually required to load and unload the colonies, which are lashed to pallets. It is imperative that the hive furniture be in excellent condition (i.e., no cracks or broken woodenware) to keep bees from being lost or the furniture damaged during transport.

SPRING MAINTENANCE

Certain apiary tasks should be started when the dandelion and fruit trees bloom in your region and when cold weather is retreating. Do the following:

- Clean the apiary of any winter debris.
- Unwrap, remove, or take down winter protection, including windbreaks and any insulating materials.
- Remove entrance reducers from strong colonies. Keep them in weaker colonies, but switch the entrance hole to the larger opening. Make note of these weaker colonies and keep track of their condition.
- Mark colonies to be requeened, to replace poor performers; see Chapter 10.
- Clean off bottom boards. Note which colonies have few dead bees and debris on the bottom board and overwintered in healthy condition; breed queens from these colonies.
- Feed hives that require additional food.

When air temperatures reach 75° to 85°F (24–29°C), inspect the hives for diseases, brood patterns, and the amount of remaining stores. In addition, you should take a mite count to ascertain mite levels in each colony. You may need to:

- Reverse the brood chambers, if the lower hive body is only being used by the bees as a passageway to reach the chamber above it. Place the upper hive body containing the queen and the vast majority of bees above the bottom board. Since bees and the queen tend to move upward in a hive, this reversal will relieve congestion and provide additional room for the bees to occupy as the colony continues to grow. During the reversal, be sure the bees have adequate stores; if not, provide them with some.

- Replace dark, old, sagging, and uneven combs, or broken frames. It is a good time to start recycling old, dark comb for newer ones, or frames of foundation. Wax has hydrophilic properties, and insecticides, herbicides, and miticides will slowly accumulate in it. Therefore, it is strongly recommended that frames over three years of age be rotated out of the colony and replaced with frames containing foundation or drawn comb less than three years old. This recommendation adds to the chores already facing the beekeeper.
- Provide additional space (hive bodies or supers) as needed.
- Make increases (in number of colonies) only when the weather is warm enough so the brood will not become chilled; see "Swarming" in Chapter 11.
- Medicate or treat colonies that show symptoms of bee diseases and/or pests, such as nosema disease or tracheal and varroa mites. Note well: before you treat your colonies with any medications for American foulbrood, you must first obtain a prescription from a veterinarian. When in doubt, consult with your state's bee inspector. Read labels carefully to ascertain the dosage requirements and when the product can be used. Certain miticides cannot be used when honey is on the colony and/or nectar is being processed. Read and follow label directions.
- Look for signs of swarming and, if necessary, initiate swarm prevention/control techniques; see "Swarming" in Chapter 11.
- Investigate clean water sources (to determine that bees are not getting into contaminated water) or provide fresh water.
- Register hives and apiaries with the state agriculture department as required.

Spring Dwindling

You may observe that some colonies fail to increase their bee population as the spring season progresses; the older bees may begin to die faster than the young bees emerge, such that the number of bees is reduced to a point at which the process cannot reverse itself and the colony dwindles to nothing. This phenomenon is often referred to as *spring dwindling*, because that is the time of year it is most frequently observed. This condition was reported as far back as the 1800s and may have several causes, such as pathogens (spiroplasmas, bacteria, or viruses), mite predation, poor queen genetics or race of bees, the condition of the colony going into the winter, poor winter stores, or unfavorable weather. It also may include the quality of the winter stores (granulated honey in the comb and/or a high water content) or weather conditions that have minimized foraging opportunities, as well as a lack of plant diversity. The reasons for dwindling can be difficult to pin down. It is noted that strong colonies entering the spring season are less likely to encounter the dwindling syndrome.

While weather is an important factor in over-wintering success of your colonies (especially when cleansing flights are limited), spring dwindling may be prevented or reversed by attending to these points:

- Make sure you have strong populations of young bees and a new, young queen in the fall.
- Winter only strong colonies with ample stores of honey and pollen, combining, requeening, or culling weaker colonies in the fall.
- Feed sugar syrup early so bees can evaporate off the excess water.
- In August, requeen colonies with young queens whose stock winters well in your location.
- Use windbreaks to protect your hives from winter and spring winds and locate colonies in areas with good air drainage.
- A weak colony (but free of disease) resulting from spring dwindling, poor queens, or other factors can be united to another weak one to increase bee numbers. The weak colony can be put above a screen over a strong colony, and, with heat being provided by the strong colony, the weak colony may be able to recover its numbers.
- Have ample colony strength in spring; if there are only three or four frames of bees, unite it with a stronger colony or destroy the weakened hive.
- Take steps to protect bees from pesticides, fungicides, and miticides that have been absorbed by the wax combs. Combs that are three years old and older should be rotated out of the colony and replaced with wax foundation. Unfortunately, now all beeswax, including foundation, contains (small amounts) of harmful chemical residues.

- Take steps to protect bees from pesticides during the previous summer. Toxic substances are found in older comb, so replace dark, old comb with new foundation in time for the bees to draw it out and store honey.
- Orient your colonies to minimize drifting. In addition, add different geometric symbols on the front of each hive to aid bees in locating their home colony. Colonies should be approximately equal in strength.
- Prevent drifting, to avoid making some colonies weaker in the fall.
- Medicate the bees against nosema disease in the fall or spring, if your colonies tested positive for nosema spores. Check local bee research on nosema, as there is now a new type of this organism that does not respond to some treatments; see Chapter 13.

OTHER WINTERING OPTIONS

The following sections, through "Building Design Requirements," were written by R.W. Currie of the Department of Entomology at the University of Manitoba at Winnipeg, with supplementary material quoted from Andony Melathopoulos of Oregon State University.

Indoor Wintering of Honey Bees

History of Indoor Wintering

In geographic regions with cold and long winters, many beekeepers choose to winter their colonies indoors. Indoor wintering can be done on a large or small scale and probably has its origins in the late 1800s, when beekeepers would place small numbers of colonies in an unheated "root cellar" in order to help them survive the winter. Wintering large numbers of colonies in specially constructed underground chambers was also attempted, but both small- and large-scale methods fell out of favor because of difficulties controlling temperature and air quality and the labor associated with carrying colonies into and out of the wintering chamber. However, some advantages of indoor wintering, including the ability to monitor and feed colonies during the cold period, and innovations that eliminated some of the disadvantages, have made this an economically viable alternative in northern climates.

Ideal Indoor Wintering

Andony Melathopoulos of Oregon State University, in "The Biology and Management of Colonies in Winter" (2007), elaborates on the technical problems involved in indoor overwintering:

> Honey bee colonies are also wintered indoors in temperature controlled buildings. Indoor wintering is only practiced in the Canadian prairies, in Quebec and in the Maritime Provinces. Bees are typically moved indoors when daytime temperatures are below freezing (typically at the end of October on the prairies). Colonies stay indoors for 5 to 6 months. Some beekeepers winter a few thousand colonies in a single building.
>
> . . . Colonies winter most efficiently at 5°C (41°F). Keeping the wintering buildings at 5°C is tricky because: 1) colonies give off heat and 2) outdoor temperatures fluctuate. In the late fall, the heat produced by the bees is greater than the amount required to keep the room temperature at 5°C (41°F). The excess heat is removed by exhausting the warm storage air and replacing it with cool outdoor air. This air is exhausted using thermostatically controlled fans. The exhausting fans also remove CO_2 and moisture, both of which are harmful to bees at high levels. Exhaust fans are typically tied into a reticulating polyethylene duct system that ensures the air moves uniformly throughout the room.
>
> When the outdoor temperature is very cold, the bees are able to generate only enough heat to keep the building at 5°C (41°F). In this situation, the exhaust fans remain closed. To prevent CO_2 and moisture from building up, a low level exhaust runs to ensure just enough fresh air is brought in. It is a task to adjust the exhausting system so that it can keep the air fresh and the temperature regulated. Consequently, I urge you to consult with your provincial apiculturalist before you begin construction. If temperatures fall lower still, backup heat is used to warm the building. In warmer climates, such as in Quebec, buildings are also equipped with refrigeration systems to cool buildings down when temperatures begin to rise in the spring.

As Melathopoulos's text makes clear, it is best to consult professionals if considering building indoor facilities for wintering bees.

Building Design Requirements

R.W. Currie continues: Whether a small- or large-scale operation, the requirements for an effective wintering room are similar. Essentially the space for hives should be well insulated, temperature controlled, completely dark, well ventilated, and with adequate volume to contain the hive and provide good air recirculation and air exchange. To create a wintering chamber, some producers modify an existing space (e.g., a hot room or garage), while others construct buildings specifically for the purpose of wintering colonies.

The shape and height of the building are not important, but it should be large enough to accommodate the appropriate number of hives and have dimensions and access points (e.g., loading dock) compatible with whatever type of equipment is going to be used to move the colonies (e.g., a handcart or forklift). Since hives are normally stacked on top of each other, the floor space required is highly dependent on how colonies will be arranged within the building. Typical recommendations for floor space vary from 2.7 to 3.2 square feet (0.25–0.3 m^2) per hive. The room volume should allow for space of 24 to 30 cubic feet (0.7–0.9 m^3) per hive. At least enough space should be left between rows of hives for the beekeeper to move and periodically sweep dead bees that fall from the hives. Higher colony densities than those recommended above are possible, but in some cases this can increase problems with air temperature regulation and air circulation.

External (white) light is highly disruptive to the bees and must be prevented. Any ventilation inlets or outlets should be shrouded with a "light trap" consisting of a black box that surrounds the fan housing and prevents the direct entry of light. The room should also be fitted with light sockets that can accommodate red light bulbs (lights are normally left off) to facilitate work within the room when necessary. Alternatively, many beekeepers carry a red-light flashlight with them when working in the wintering room. Red lights are much less disruptive to the bees.

For more information, look at websites and talk to other beekeepers.

Colony Management

Colony preparation for indoor wintering is similar to that for outdoor wintering. Colonies should have a young productive queen, be free from diseases and parasites, and have adequate food stores in preparation for the wintering period. Sucrose syrup or high-fructose corn syrup is typically the preferred type of food. Beekeepers should provide about 40 to 50 pounds (18–23 kg) of syrup for a single-hive body and 60 to 70 pounds (27–32 kg) for two bodies. Syrup is preferable to honey, as the latter can granulate during winter, resulting in increased stress to the bees. Colonies can be wintered indoors in a single-brood chamber or double-brood chamber with equal success, and small five-frame nucleus colonies are also often wintered using this method.

Colonies are typically moved into the wintering buildings in late October to mid-November, where the hives are stacked adjacent to each other in rows (usually about five colonies high), with adjacent rows separated by at least 3 feet (1 m). Hive entrance reducers are usually removed. Many producers do not feed colonies during winter storage; however, if colonies are light in weight going into the building, this is an option. Feeding of syrup (or water) can be done through entrance feeders. Pollen or pollen supplements are not given during winter, as they are believed to put additional stress on the bees (in indoor wintering, the bees do not have the opportunity to defecate until they are removed from the building).

Colonies are typically removed from the building in late March to mid-April, when the first pollen flow occurs or when temperatures are too hot to maintain the bees within the building. In order to minimize losses, bees should be removed from the building in early morning or late evening and placed in a location with no snow. Doing otherwise causes excessive disorientation of the bees and high rates of loss.

Wintering in Warm Climates

In contrast to wintering bees in the middle and higher latitudes, wintering bees in southern areas does not require the same attention or effort. Here, the daylight is longer and the nights are shorter than in northern regions. In the South, winter temperatures fluctuate between 45° and 68°F (7°–20°C). Although temperatures can fall below freezing for short periods, it is not necessary to provide the same winter protection, such as wrapping and insulation, for your colonies.

In addition, plants that yield nectar and pollen are

available almost year-round, so brood rearing is interrupted for only a brief time. It is in these climes where the package bee industry is located and where many commercial beekeepers overwinter and rear new queens for their colonies. It is also now an area where many of the pesticide-resistant mites and other resistant pathogens are passed around from apiary to apiary.

Most of the published information concerning wintering honey bees is based on apiculture in temperate climates. As a consequence, these sources frequently concentrate on wintering, because that is usually when there is the most colony loss. It is often recommended that you overwinter colonies with two full-deep chambers as a standard configuration: a brood chamber below (with a large population of bees) and a food chamber above, filled with honey and pollen stores.

In more southern regions, different conditions prevail. The cold weather is not as severe, and therefore it is unnecessary to pack colonies for protection. Brood rearing continues longer into the year; there may, in fact, be no broodless period in some southern states. The two full-depth brood chambers are often reduced to one deep brood chamber with a medium-depth honey super as the standard hive configuration.

Dr. Malcolm Sanford, retired Extension Service entomologist from Florida State University, offers some observations on wintering in southern regions. The length of daylight, or photoperiod, in warmer zones increases as you approach the equator, and warmer zones have a greater diversity of plants, which produce relatively less nectar per individual species. These two factors appear to be the most important reasons why honey production is not as pronounced or intense as in more temperate regions. In the true tropics, moisture availability, with a more pronounced summer wet and winter dry season, takes precedence over temperature.

The relatively long season for brood rearing, the warmer winters, and the sparser nectar production are the cornerstones of southern beekeeping. Thus colony management must be spread out across the year more evenly, and timing of specific tasks is radically different from that in northern areas. As examples, two management techniques extremely important in successful wintering stand out: requeening and controlling varroa mite populations.

Requeening in southern regions is done in the fall and with queen cells rather than mated queens, as is common in the northern areas. This may change, however, with incursions of Africanized honey bees infiltrating southern states. Mite treatments are scheduled differently during the longer growing season in southern regions, since varroa and tracheal mites can be present in the colony year-round. Because of the later honeyflows, a summer/fall varroa treatment must be timed carefully to keep bees free of mites but not contaminate honey. In addition, another varroa treatment may be necessary in the early spring before the colonies build up.

Here are some points to remember when wintering in southern areas:

- Winter in one brood box, with a minimum of a medium-depth (6⅝ in.) super containing honey.
- Install entrance reducers.
- Keep bee populations between 15,000 and 20,000 (5–8 deep frames covered with bees).
- Treat for diseases and mites.
- Provide an upper entrance.
- Protect colonies from cold winter winds.
- Periodically check to see how food stores are holding up; replenish as needed by supplemental feeding.
- Check water sources and provide if weather is too dry.

Timing is as critical to beekeeping as it is to most endeavors. To time your beekeeping activities properly, it is necessary to keep accurate records of past seasons, to observe carefully the flowering periods, and to recognize the needs of your colonies. Even then, there are capricious fluctuations from year to year or season to season that will make beekeeping a continual challenge. Be prepared—no two seasons are alike.

 **Notes**

CHAPTER 9

Summer/Fall Management

SUMMER MANAGEMENT

Each colony should be examined periodically during the flowering months of the year, especially before the advent in your area of major honeyflow. This period is usually marked by the flowering of a specific plant that produces copious amounts of nectar such as an annual, biennial, perennial, shrub, or tree. Check the strength of each colony to determine whether it is populous enough to take advantage of a given honeyflow. Colonies with 35,000 to 40,000 adult bees have the workforce required to take advantage of major honeyflows. One way to determine the strength (population) of a colony is by examining the deep brood frames; if 11 or more of these frames are fully covered on both sides with adult bees, such a colony has the requisite number of adult bees to exploit a honeyflow. Several researchers have provided methods to estimate the adult population of a colony based on the number of adult bees covering both sides of a deep frame; estimates range from 1400 to 2400 bees. David Cushman's "Estimation of Bees on a Frame" calculates about 1500 bees per side, for a total from both sides of about 3000 bees. Since you need 1.5 medium frames to make up a single deep frame, a medium should hold two-thirds of a deep (divide 3500 by 3 = 1167, then times 2 = 2334 bees per full medium frame). Using 130 mg per bee, 2334 bees would weigh about 0.66 pound (0.30 kg).

Honeyflows do not occur precisely on the same day or week each year; weather will play the major role. In fact, their regularity from year to year is unpredictable. Using hive scales to measure weight gains and losses, and comparing a given honeyflow in one season with a honeyflow in a subsequent season may often show a deficit in the incoming nectar. Nevertheless, by recording these fluctuations over four or five years, you will gain a general idea of the occurrences of honeyflows in the bees' locale. Keep in mind that landscapes are not stagnant; ecological succession of plants may overtake a dominant nectar plant, and with the advent of climate change, an additional variable comes into play. Record keeping is very important with respect to honeyflows; with such records beekeepers will be able to know the approximate time to anticipate honeyflows in their locale. Here is where a hive scale becomes an important tool for the beekeeper. At the commencement of a honeyflow and while one is in progress, daily dramatic increases in the colony's weight are displayed by the hive's scale. Daily weight gains from the nectar being collected of seven and more pounds are not uncommon. When hive scales are reporting such gains, a major honeyflow is under way. If you have kept careful records, you may be able to anticipate honeyflows in future years. With such information you will be able to add honey supers to your colonies in advance of the flows. During these inspections, weak colonies should be united to increase their overall populations. This may provide the untied colony with a sufficient number of bees to actively participate in a honeyflow. Even after weak colonies are united, it may be necessary, if their stores are near depletion, to feed them and/or replace both queens. The brood pattern in weak colonies may be indicative of a failing queen.

Colony Estimations

Estimating Colony Strength (all European)

- A shallow frame fully covered with bees will hold about 0.25 pound (0.1134 kg) of bees or about 875 individuals.
- A deep frame fully covered holds 0.5 pound (0.2268 kg) of bees or about 1750 individuals.
- There are about 3500 bees per pound (0.4536 kg).
- A sheet of deep wax or plastic foundation 8.5 × 16.75 inches (42.5 × 21.3 cm) equals 3350 cells per side. That is 83.75 cells linearly and 40 cells vertically; the number of cells on both sides is 6700.
- One square inch of comb has 25 cells (5 worker cells per linear inch, 4 drone cells per linear inch).
- Count the number of bees returning to the hive for 1 minute. To estimate the number of bees in the deep frames of a colony, use the following equation: number of bees returning per minute × 30 min × 0.0005. The factor 0.0005 assumes that one deep frame, both sides, contains 2000 bees, or 1 divided by 2000 = 0.0005; the time of 30 minutes assumes the amount of time needed for any one bee to make a return trip.

Estimating Weight of the Colony

- Multiply by the number of frames in the colony.
- A deep frame fully filled with honey weighs about 10 pounds (4.536 kg).
- A full medium frame (U.S.) weighs 7 pounds (3.2 kg).
- A full shallow frame (U.S.) weighs 5 pounds (1.8 kg).

Another way to gauge a colony's population, count the number of bees per minute entering and exiting the colony; if they can easily be counted, the colony is weak. On the other hand, a count of 30 to 90 bees per minute indicates a strong colony. One deep frame fully covered on both sides with adult bees equals approximately 1 pound (0.5 kg) of bees (3500 bees). For additional methods of counting population, see the sidebar "Colony Estimations" in this chapter.

Other tasks during this time should include the following:

- Requeen as necessary, such as when a colony's weakness can be attributed to its queen and when a honeyflow is underway. During honeyflows, bees readily accept the introduction of replacement queens.
- Monitor colony strength periodically during the summer, and unite two healthy but weak colonies, or unite a healthy but weak colony with a strong one.
- Check for diseases and monitor for varroa mites.
- Rear queens for late summer and very early fall requeening.
- Check the colony's food stores.
- Reverse the brood cambers if necessary. If the colony is strong and making swarm preparations, put into practice swarm prevention and/or control measures. You can also split strong colonies to reduce swarming tendencies. (A word of caution: if your goal is a honey crop, splitting strong colonies will reduce your honey yields.)
- Now is a good time to use a few colonies to raise queens to replace aging or poor performing ones.
- Add honey supers as needed; when a super is two-thirds full (six or seven full frames), add another super; see "Rules for Supering" in this chapter.
- Add frames of foundation in supers only if there is a good honeyflow; otherwise, bees will chew holes in the wax, or the foundation will warp and bend.
- Scrape burr comb and propolis off the tops of frames; collect and process. Recent research indicates that propolis's antimicrobial properties are beneficial for the bees, and as a consequence, only propolis that interferes with inspection activities should be removed. In fact, researchers encourage beekeepers to roughen the interior walls of their hive bodies to encourage bees to propolize them.

Cooling the Hives

When the temperatures are above 90°F (32°C) for an extended period, you may need to help the bees keep from overheating. Here's how:

- If the hives are in full sun, you can provide temporary emergency shade with fencing, boards, or shrubs, or break some leafy branches and anchor them over the hive cover.
- Stagger supers slightly to increase the airflow throughout the hive. To stagger the supers, the first hive body, either deep or shallow, sits flush on the entrance board's rim. The remaining supers are shifted either slightly forward or backward, creating a minor opening/gap between successive supers. Some beekeepers raise the inner cover or the front of the bottom super with a small block; others bore a ¾-inch (18.8 mm) auger hole in an upper corner of the top deep super. Plug these holes later in the year with a cork to keep robbers from entering. Do **not** bore a hole above or in the handholds! Trying to lift these supers off would be a stinging adventure you want to avoid.
- Make sure fresh water is available; this can be done with a metal or plastic cattle watering tank. To keep bees from drowning, fill the tank with sand or pebbles and fill with water to the top of the pebbles. You can also use a dripping hose or install a Boardman feeder filled with water in each hive.
- Use screened bottom boards to provide additional ventilation.
- Paint metal covers or wooden tops with white paint to reflect the maximum amount of sun.

Signs of Honeyflow

Prepare for a honeyflow in the winter or spring months; **do not** wait until the last moment to make extra frames or supers, or you could lose the honey crop! Start by repairing or making frames filled with foundation for the honey supers. (Keep fresh wax foundation sheets in plastic bags to protect them against wax moth infestation or other pests and to keep them from drying out; dry foundation becomes brittle and breaks easily.)

The honeyflow comes at the time of year when bees are able to collect surplus nectar. A honeyflow may last a few days or a few weeks. Bees are natural hoarders, and the presence of empty combs stimulates this hoarding behavior. If you provide the bees with more storage space than they need, you will be encouraging more bees to collect (and store) nectar. This behavior of storing more food than is immediately needed is a sound evolutionary trait, for the

amount of food any one colony will need or produce is unpredictable. By understanding this aspect of bee behavior, you increase the opportunity to obtain a surplus (more than the colony needs) of honey. This surplus is stored by the bees as honey in supers (located above the brood chamber). Good beekeepers harvest only this surplus and leave ample food reserves.

A honeyflow is indicated by one or a combination of the following signs:

- Fresh, white wax evident on edges of drawn comb and on top bars.
- Dramatic weight gains (as indicated by the hive scale) over several days or weeks.
- Wax foundation drawn out quickly.
- Large amounts of nectar ripening in cells.
- Bees fanning at hive entrance.
- Much foraging activity.
- Odor of nectar (ripening honey) often pervading the apiary.
- Bees are extremely docile and easy to work.

During the main honeyflow, you should avoid opening the hive to look at the brood area unless you are doing some major management operations (requeening, treating for diseases, etc.), nor should you place pollen traps on hives. Check the colonies before a major honeyflow, because if you disrupt the bees by tearing down the hive during a flow, it will upset their gathering activities and may even reduce the amount of honey being brought in for several days. You will also kill many more bees (the population is much higher at this time) and even the queen if you work bees intensively during this time. So, restrict your curiosity to smaller colonies, or set up an observation hive in your house.

If you must look at your colony, try to minimize your time inside the hive to reduce the disruption.

Apiary Tasks in Summer

In general, the tasks—and things to avoid—at the apiary just before and during a major honeyflow should include the following:

- Super the hives prior to the commencement of an anticipated honeyflow, placing honey-gathering supers above the broodnest (see next section).

- Provide adequate ventilation so bees can cool the hive and adequately cure the nectar, thereby reducing its water content.
- Reverse honey supers occasionally so the heavier one is always on top.
- Keep supers on the hive until the cells of the honey combs are capped, which will likely ensure that the final product has a low water content. Uncapped cells containing nectar that are transitioning to honey may ferment after extraction because of a high water content.
- Avoid adding too many supers at once, as bees may partially fill all of them instead of filling one completely (called the *chimney effect*). Put on only enough for the bees to fill in a few weeks; this is a judgment call, one gained by experience.
- Never super a weak colony.
- **Never feed medication** to colonies that you plan to extract honey from, unless the label indicates it is not a problem. It could be illegal, and you can contaminate the honey. Honey collected by diseased colonies that required medication must not be used for human consumption.
- **Do not feed** bees syrup during this time, because it will probably end up in the honey supers.
- Requeen weak colonies or those heavily infested with parasitic mites so as to break the brood cycle and disrupt the mites' life cycle (see Chapter 14).
- This is a perfect time to raise queens, so take advantage of a natural flow to raise high-quality queens.

Super Sizes

When a honeyflow is in progress, the bees will deposit nectar in supers placed above the brood chambers. These supers may vary in size from full-depth supers (deep hive bodies) to the shallow or the section comb supers (see the illustration of super sizes). There is no hard-and-fast rule about which super size to use; personal preference, one's physical strength, and the quantity of the expected surplus should be your guide. Some beekeepers keep all hive furniture the same size so they don't have to mix and match different sizes of supers and frames or foundation. There are pros and cons either to mixing or to keeping the super sizes the same, so chat with some beekeepers about which super sizes work best in your region.

Rules for Supering

The most important rule for supering is to keep the queen out of the honey supers. The presence of brood interspersed in honey frames will make the honey unsanitary (when extracted), because of the presence of larvae and pupae. The extracting tank will then have honey and brood mixed together, and brood would be wasted. In addition, it is difficult to cut away the cappings of brood combs that have been darkened with propolis. Set aside such frames to return to the colonies. Cull any dark and old frames and save them for other uses, such as bait for swarm traps; otherwise, melt them down for their wax.

Use one of the following methods to restrict the queen from laying in the honey supers:

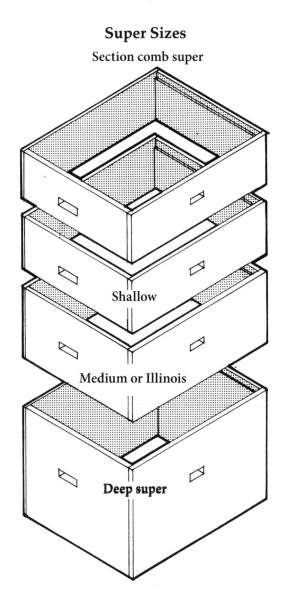

Super Sizes

Section comb super

Shallow

Medium or Illinois

Deep super

Beehive Components

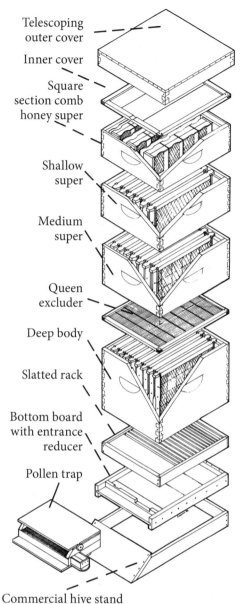

Telescoping outer cover

Inner cover

Square section comb honey super

Shallow super

Medium super

Queen excluder

Deep body

Slatted rack

Bottom board with entrance reducer

Pollen trap

Commercial hive stand

- Place a queen excluder above the broodnest (see the illustration of beehive components).
- Place a super of light-colored comb or foundation above the broodnest; as long as the queen is not crowded for space, she will prefer to lay her eggs in the darker comb.
- Keep a hive body filled with honey directly above the broodnest. Such a honey barrier often keeps the queen from moving upward. You can also place a section comb honey super above the broodnest; the queen generally will not lay in the section boxes or other types of comb honey sections.

Some general guidelines for supering during a honey-flow are listed below:

- Stagger the honey supers to hasten the ripening of honey by leaving a ¼-inch gap, especially in hot, humid areas (but do this only on very strong colonies to avoid robbing). A better option would be to use a screened bottom board.
- If you have supers that hold ten frames, instead of using all ten frames, only place nine frames of empty drawn combs in each ten-frame super, spacing them at equal distances from one another. As the bees begin to fill the cells with nectar, because of the spacing they will lengthen the cells, making it easier for the beekeeper to uncap them before extracting the honey. This type of spacing will not work with frames of foundation.
- Bait an empty honey super with a frame or two of capped or uncapped honey if the bees seem reluctant to move up; this will attract bees to move into the super.
- Some beekeepers use drone comb foundation in their honey supers; the cells are larger, and honey seems to extract readily from them. Drone foundation can be obtained from bee supply houses.
- Rearrange frames in supers periodically, so the full ones are at the outer edges and the empty ones are in the middle (bees fill the middle ones first). If you are low on supers, this exchange will give you some time to ready more equipment.

Methods of Supering

There are two basic ways to super for honey; these are *reverse supering* and *top supering* (see the diagram on supering).

Reverse or bottom supering. This method generally needs a queen excluder to keep the queen from laying in the honey supers (see the diagram on supering, and the one on the reverse supering sequence) and can also be used for comb honey production. A super with foundation or dry combs (S2) is always placed below a super at least half full of honey (S1). Because the emptier supers are on top of the broodnest, the queen excluder is necessary. As the supers are filled, they can be removed, or stored above the emptier ones. With the empty super closest to the broodnest, and the full super above the empty one, the bees will be drawn up into the empty super above

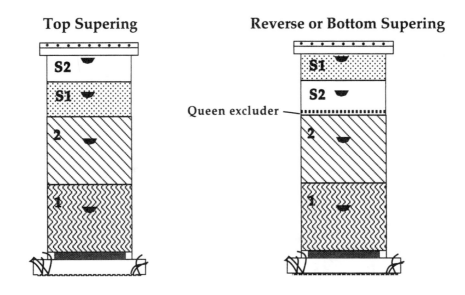

Top Supering   **Reverse or Bottom Supering**

▨ Broodnest   ◹ Dry comb   ▦ Honey   ☐ Foundation   ⋯ Queen excluder

them. **Note well:** Now that the small hive beetle is well established in many parts of North America, unless honey supers are stored in a freezer, they should be extracted as soon as possible. If you delay extracting even for several days, the small hive beetle larvae will render the honey unfit for human consumption.

Top supering. This method does not require a queen excluder because the queen rarely will go into a super full of honey. Put honey supers with dry comb or foundation (S2) **above** honey supers that are at least half filled with honey (S1); see the diagram on top supering. Keep adding supers as the bees fill the ones below, until you remove the honey.

There are many methods of supering using these two themes; talking with local beekeepers may be helpful in determining how to super in your particular area. Success of either method often depends on the type of honeyflow in your location—short in duration and intense or long but less intense. Both short and prolonged but intense honeyflows can be utilized for the production of comb honey; less intense flows should be geared to the production of extract honey. See also the diagram on the top supering sequence.

Comb Honey

Harvested honey can be left in the comb or extracted from it. Honey in the comb is referred to by various names. Normally found at fairs and honey shows, *bulk comb honey* is an entire frame of capped honey packaged without cutting. If the honey-filled comb is cut into sections and packaged, it is referred to as *cut comb honey*. Cut combs placed in a bottle and then filled with extracted honey are called *chunk comb honey*. Comb honey contained in small wooden frames or plastic rings (Ross Rounds) or plastic boxes that is **not** cut out of the frames is referred to as *section comb honey*; see Chapter 12.

Thin, unwired foundation is used to produce bulk, cut, chunk, or section comb honey. As soon as the combs are sealed, they should be removed from the hive to prevent the white cappings from becoming darkened with propolis, soiled by travel stains, or damaged by wax moths, small hive beetles, or the bee louse, which lays its eggs in honey cappings. (This last insect is rarely seen these days.)

The supers containing frames for comb honey production should be placed on only the strongest colonies, either those consisting of two brood chambers or colonies reduced to one brood chamber (as described for the production of section comb honey in the diagram). Place an excluder above the broodnest and super the hive using the same rotation method as that illustrated for reverse supering. It is a good idea not to mix the comb honey supers and the extracted honey supers in any one hive. Some colonies will fill sections quickly and with very white cappings—such colonies should be reserved for section comb honey only. The color of wax cappings (snow white or "wet") is a genetic trait.

Reverse Supering Sequence for Extracted Honey

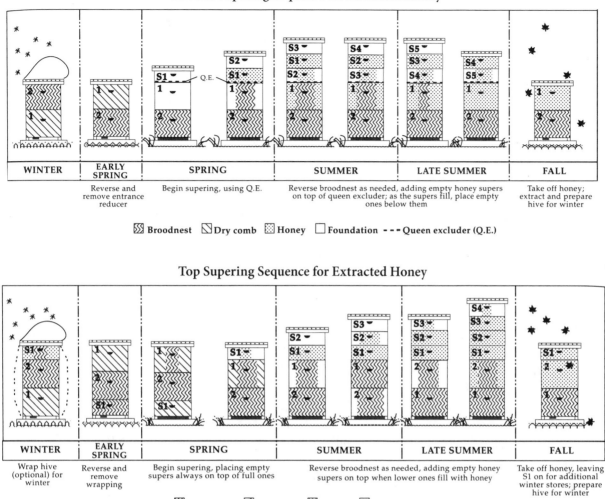

WINTER	EARLY SPRING	SPRING	SUMMER	LATE SUMMER	FALL
	Reverse and remove entrance reducer	Begin supering, using Q.E.	Reverse broodnest as needed, adding empty honey supers on top of queen excluder; as the supers fill, place empty ones below them		Take off honey; extract and prepare hive for winter

▨ Broodnest   ◹ Dry comb   ▦ Honey   ☐ Foundation   - - - Queen excluder (Q.E.)

Top Supering Sequence for Extracted Honey

WINTER	EARLY SPRING	SPRING	SUMMER	LATE SUMMER	FALL
Wrap hive (optional) for winter	Reverse and remove wrapping	Begin supering, placing empty supers always on top of full ones	Reverse broodnest as needed, adding empty honey supers on top when lower ones fill with honey		Take off honey, leaving S1 on for additional winter stores; prepare hive for winter

▨ Broodnest   ◹ Dry comb   ▦ Honey   ☐ Foundation

Supering for Section Comb Honey

Comb honey, especially section comb honey, is difficult to produce because success depends on the following: an intense honeyflow, the need to crowd a full colony of bees into one deep super with the possibility that such action may cause the bees to swarm, and the fact that bees "dislike" working in these sections. Time-consuming manipulations at the correct intervals are required. The Miller method of supering is one that is used for section comb honey; this is described below and is illustrated as method A in the diagram.

A colony used for section comb honey production is generally wintered in two deep hive bodies (1 and 2). In the spring, this colony must be built up to full strength prior to the major honeyflow, and the brood chambers should be reversed to provide ample room for the queen to lay. This may need to be done several times to maintain enough empty combs for the queen to fill with eggs.

As soon as the honeyflow begins, reduce the strong two-story colony to a single deep hive body (2). For clarification, hive body (2) is the second deep hive body, and for the purpose of consistency, it will remain with this numerical designation. But throughout this discussion, hive body (2) is situated on the bottom board. Set up this single deep hive body so it contains two empty drawn frames containing worker cells in the center of the hive body, and then fill the rest of the hive body with frames of capped brood. Hive body (2) will also contain the queen and all the bees that were once in the two-or-more story colony. Usually, during this setup, all the bees in hive body (1) or any other hive bodies that may have been above (2) are shaken off their frames in front of the entrance to the colony by the beekeeper.

Comb Honey Using Section Supers (ss) Supering Sequence

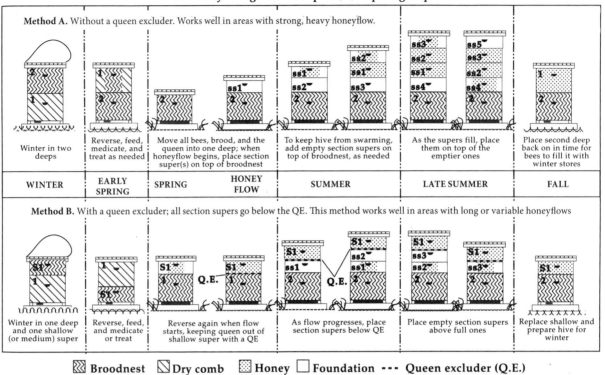

Method A. Without a queen excluder. Works well in areas with strong, heavy honeyflow.

WINTER	EARLY SPRING	SPRING	HONEY FLOW	SUMMER	LATE SUMMER	FALL
Winter in two deeps	Reverse, feed, medicate, and treat as needed	Move all bees, brood, and the queen into one deep; when honeyflow begins, place section super(s) on top of broodnest		To keep hive from swarming, add empty section supers on top of broodnest, as needed	As the supers fill, place them on top of the emptier ones	Place second deep back on in time for bees to fill it with winter stores

Method B. With a queen excluder; all section supers go below the QE. This method works well in areas with long or variable honeyflows

| Winter in one deep and one shallow (or medium) super | Reverse, feed, and medicate or treat | Reverse again when flow starts, keeping queen out of shallow super with a QE | As flow progresses, place section supers below QE | Place empty section supers above full ones | Replace shallow and prepare hive for winter |

▨ **Broodnest**   ◸ **Dry comb**   ⊡ **Honey**   ☐ **Foundation**   --- **Queen excluder (Q.E.)**

To keep section supers (ss) clean, remove them when full; this will keep the wax capping white and free of travel stains. To illustrate the positioning of comb honey supers it was necessary in the above diagram to extend the procedures from spring to late summer, but a comb honey super should be removed as soon as it is filled.

The following is the procedure for method A, without using queen excluders:

1. Reduce the colony to one deep hive body (2). Frames of honey and any remaining brood frames should be given weaker colonies, provided that these weaker colonies have sufficient bees to care for the additional brood frames.

2. On top of hive body (2), place the first section comb honey super (ss1) with thin foundations in the section boxes or rounds.

3. When ss1 is half filled with capped honey, add a second section super (ss2) beneath ss1.

4. If the honeyflow is strong, additional supers will be needed. Therefore, add a third or subsequent set of supers (ss3) above deep hive body 2. When ss2 is half full, reverse ss1 and ss2 so that the full super ss1 is directly above ss3. By placing ss1 above ss3, it serves as a barrier and forces the bees to work on drawing comb and filling ss3. If additional supers are required, place them directly over the brood nest.

5. It is best to remove filled section supers as soon as possible. One attraction and selling point for section comb honey is its clean white cappings. If these supers are left on the colony for any extended period of time, the cappings will become stained with propolis, bits of pollen, and some darkened wax. Use bee escapes or bee blowers to remove bees from these supers. **Never** use fume boards to remove the bees because the honey might be adversely flavored.

Method B is slightly different; it uses a queen excluder and a single deep brood chamber plus a shallow super. This method uses the idea to always have a full honey super below the section boxes to encourage the bees to move up and into the section boxes.

Comb honey should be marketed as soon as possible to reduce the danger of its granulating or being damaged by wax moths, the bee louse, or small hive beetles (see Chapter 14 for more information about these pests). Storing comb honey in the freezer will help eliminate these problems (see Chapter 12). After the honeyflow is over and the section comb honey production ceases, take off all section supers and unite the reduced colony with another hive or otherwise allow it to build up enough stores to overwinter in two deep hive bodies.

Harvesting the Honey

A honey bee colony, especially in northern latitudes, needs to meet two critical food requirements: it must discover adequate food sources to serve the colony's daily needs, and over the course of the spring, summer, and early fall, it must gather sufficient surplus food stored as honey to sustain the colony throughout the winter months. The distance bees are required to travel to meet these requirements will determine their chances for winter survival.

Studies by Eckert (1933) established that colony weight gains were dependent on the distance the bees traveled from their colony to the foraging areas. The net gains in colony weight were measured by the amount of incoming nectar and its conversion into honey over an 18-day period. For foragers that traveled less than a mile to food sources, their colonies gained the most weight. As foraging distances increased, weight gains decreased; at distances of five miles or more, colonies lost weight. From this information we can speculate that if it is necessary to feed your colonies sugar water annually in the fall in order for them to have adequate winter stores, your bees are likely compelled to fly great distances to locate forage. Nectar-yielding plants would include annuals and perennials as well as shrubs and flowering trees.

In some regions, two crops of surplus honey can be expected, one in the summer and another in the fall. Some beekeepers harvest the summer and fall crops separately; others harvest both at the end of the fall honeyflow. Yields vary from 25 pounds (11.3 kg) of surplus per colony to over 200 pounds (90.7 kg) or more. For hives located in temperate climates, 90 pounds (33.6 kg) or more of honey should be left as overwintering stores for each colony (see "Wintering" in Chapter 8).

For late-summer honeyflows, consider taking off the honey supers as soon as possible, unless they are to be kept on as winter stores. If varroa mite populations are high, it may be necessary to treat the bees in the fall with approved acaricides and allow enough time for the treatments to work before winter. To do this, if the treatment label instructs you to do so, you must remove the honey supers. For more information see "Varroa Mite" in Chapter 14.

Removing Bees from Honey Supers

Honey supers are often free of bees when the temperatures get very cold (in the early fall), because the bees leave the honey supers to join the warm cluster below. But remember, once you commit yourself to taking off honey, be prepared to extract or otherwise process your honey in a few days. You cannot store supers of capped honey for very long, unless you place them in a cooler or freeze them, because of the danger that wax moths, small hive beetles, or other pests will damage or destroy them. Five methods, listed below, describe ways of removing bees from honey supers.

Shaking or brushing. Remove a frame with sealed honey from the super and shake the bees off in front of the hive entrance, or gently brush off the bees with a soft, flexible bee brush or a handful of grass. Allow the bees to fall at the hive entrance. Then place the frame, free of bees, into an empty hive body and cover it with burlap or a thick, wet cotton sack (robbing cloth) to avoid initiating robbing behavior. If robbing is particularly intense, an additional cloth might be needed to cover the super you are working. The bottom of the super should also be closed off to prevent robbing bees from entering. Using the empty hive body from which the frames were previously removed, transfer frames from the next super containing frames of honey into it. If robbing becomes unmanageable, put the honey frame(s) into a vehicle and close all doors and windows; stop taking honey from colonies in that apiary for the day.

Advantages

- Able to select frames containing *capped honey* (honey covered by thin layer of wax).
- Relatively easy if bees remain calm and only a few colonies are involved.
- Inexpensive.

Disadvantages

- Timing and good judgment are critical to avoid robbing.
- Method is time consuming.
- Brushing may excite bees to sting.
- Honey could be tainted with smoky flavor, from excessive use of the smoker.

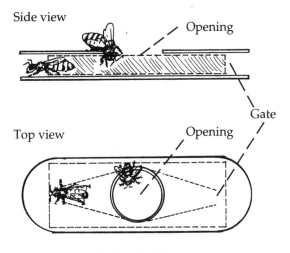

Side view

Opening

Gate

Top view

Opening

Porter Bee Escape

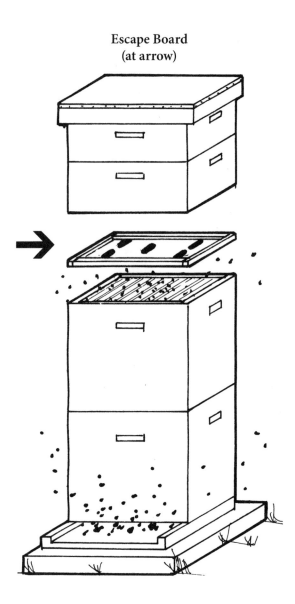

Escape Board
(at arrow)

Bee escape or escape boards. Bee escapes are used primarily to remove bees from hive bodies containing honey so that the honey can be harvested free of bees. When an inner cover or modified cover contains at least one bee escape, it is referred to as an *escape board*. The inner cover can be made into an escape board by using a small device called a *Porter Bee Escape* (see the illustration). The Porter Bee Escape is an inexpensive metal or plastic gadget that allows bees to pass through it in only one direction. The escape fits into the oblong hole of the inner cover—or any cover that has been modified to hold bee escapes—to facilitate the passage of bees.

Other styles of escape boards and bee escapes are available from bee suppliers, and all provide a means for clearing supers of bees; check bee supply catalogs and talk to other beekeepers. Some of these boards are made with cones, screens, or other escape devices. All provide passage of bees in one direction. An escape board made with several bee escapes is more efficient and will clear supers of bees faster.

Escape boards can also be used to move a hive that consists of more than two hive bodies. First, place the escape board below the honey supers. After the bees exit, the beekeeper removes the extra hive bodies, making that portion of the hive to be moved/relocated somewhat lighter (see "Moving Established Colonies" in Chapter 11).

The escape board is placed directly below the honey supers you wish to remove (see the illustration). Usually within 48 hours after the escape board is in place, the bees move down to seek the warmth of the broodnest or the bee cluster. Many of these bees

are field bees, and to resume foraging duties, they will move through the escape(s) to get to the entrance/exit. Do not leave the board on for more than 48 hours, for the bees will figure out how to reenter the supers, especially if a cone escape board is used.

There must be no cracks or holes in the honey supers placed above an escape board, because bees from the same hive, robbing bees, and other insects (such as yellow jackets) will invade and remove the honey. Tape or otherwise close off these inadvertent entrances to the unprotected honey supers. If the outer cover is warped and you are using the inner cover as an escape board, put an extra inner cover above the topmost super to close off the top and to keep out all robbers.

In extremely warm weather, install the escape

board during the late afternoon and remove the supers as soon as they are free of bees—the next day, if possible. **A word of caution**: In warm weather, the comb may melt if bees are not available to regulate the temperature in these supers.

If the supers contain a mixture of frames, some containing honey and some containing brood, or some frames containing both honey and brood, bees will be reluctant to abandon the brood and therefore remain in these supers. If this is the case, pull out the frames of brood and place them in the broodnest or into a colony that will benefit from additional brood, to allow the young bees to emerge. If the frames are of different sizes, place the brood frames into the correctly sized hive body. Once a frame is filled, you can either put it back on the same colony or give it to a weaker one. Meanwhile, you can select honey-only frames to place above the escape board.

Advantages

● Does not excite bees.
● Easy.
● Inexpensive except for the price of the escape board.
● Usually effective.

Disadvantages

● Honey could be removed by bees from the same hive or by robbers if supers are not bee tight.
● Not always effective.
● Often, as they are moving through the various escape devices, both drones and worker bees die after getting stuck in the passage ways, clogging them and preventing any additional live bees from passing out of the supers.
● Involves extra trips to apiary to insert board, remove supers.

Note: If you find brood in your honey supers, try another supering technique; it is very time consuming and wastes brood to have to sort through your honey supers. If you used a queen excluder, check it to determine how the queen got through. Whether your excluder was defective or another supering technique failed, when you place another queen excluder below the brood-filled honey supers, in 25 days the supers will be cleared of brood as the young bees emerge (providing the queen is not there). Repair, if possible, or discard defective excluders.

Fume board (fume pad). Many commercial or sideliner beekeepers drive bees out of the honey supers using fume boards with a chemical odor offensive to bees. Premade fume boards consist of a wooden frame with a metal cover and flannel lining to hold the repellent. The cloth is then saturated with a chemical that repels bees (sold as Bee Go, Honey Robber, and Fischer's Bee Quick, to name a few). Some fume boards have a black metal top that will absorb solar heat, accelerating dispersal of the repellent. These boards work best when the bees are in shallower supers. Make sure you have removed the queen excluder before using the board, so the bees will move down quickly to the brood areas. To use a fume board, pick a warm, sunny day; then:

1. Sprinkle repellent onto the pad, following the directions on the label for the repellent.
2. Remove the outer and inner covers, using smoke as needed.
3. Scrape any burr comb off the top bars.
4. Use smoke to drive the bees downward between the frames.
5. Place the board over the frames (see the illustration of the fume board on p. 142).
6. After no more than five minutes, the bees will have left the topmost super.
7. Remove the uppermost super and repeat the process for subsequent supers below.
8. Air the supers thoroughly and store them in a covered place to prevent robbing and invasion by the small hive beetle.

Use the fume board only long enough to get the bees out of the supers. Once you have removed all the supers, it is important to add at least two empty supers above the brood chambers to create room for all the bees that were driven out of the honey supers and into the two brood chambers. Do **not** use on comb honey supers, as it will impart an adverse flavor to the honey.

These chemicals work best on a hot, sunny day, **but do not overdo this**, as you can drive all the bees from the hive. Check the outside temperatures and weather reports to determine if the temperature range will be safe to use.

Federal and state laws may restrict the use of some or all of the chemicals used for bee repellents, so comply with all regulations in this regard. These chemi-

Fume Board or Fume Pad

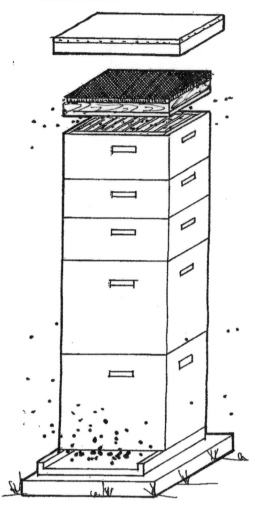

- Could adversely flavor honey.
- Rapidly crowds bees to the lower hive bodies and may force them outside.
- Need to remove queen excluder to facilitate bees exiting from honey supers.

The bee blower. A bee blower is a portable gas or electrically powered device that produces a blast of air strong enough to blow the bees off the frames and from the supers. The air blast can reach a speed of 200 mph (322 km/h). Prepare the super for blowing out the bees by setting it down on its front or back panel, with the bottom of the frames facing the blower and the top bars of the frames facing the hive's entrance. Then blow air in the spacing between the frames to flush out the bees. Some beekeepers have a special stand to hold the super. You may have to move the frames from side to side as you blow the air up through them, to clear out all the bees. Some beekeepers have a special stand that holds each super as it is blown free of bees. In this case, the stand is placed near the hive entrance, and the bees are blown down from the top. If the weather is cool, blow the bees down through a super that is still on the hive, to keep from chilling bees.

Sometimes leaf blowers can be modified to become a bee blower; they are cheap, commonly available, and adequate for most jobs. If you are too close to the parent colony, loose bees may get into the blower's intakes and clog them up; you can screen the intakes and motors. A blower is very useful if you misjudge how quickly bees will leave the supers when you are using a fume board (i.e., supers are still full of bees when you are ready to load) or if there is some brood and the bees have not left the super. Blowers are a good backup system, especially if you are harvesting more than 50 colonies. Many beekeepers use several techniques (fume boards, blowers etc.), when they have others to help.

cals are hazardous and must be treated with respect. You are responsible for following directions on the label with regard to their use, storage, and disposal. The use of chemicals is further complicated by the fact that their efficiency is governed by the air temperature; therefore, the desired result is not always certain. Never store or transport the fume boards or the supers treated with these chemicals in a closed vehicle.

Advantages

- One trip to remove honey.
- Easy.
- Relatively inexpensive.

Disadvantages

- Excites bees.
- Dependent on temperature.

Advantages

- Fast.
- Effective.

Disadvantages

- Expensive.
- During cold weather, bees blown out may be unable to return to their hive.
- Queen could be blown out and lost.

- Requires two people—one to load supers, the other to work the blower—to work efficiently.
- Requires more equipment to haul—blower, gasoline, and so on.

Abandonment or tipping supers. This is a good method, but it should be used only by experienced beekeepers because it requires considerable expertise and know-how. W. Allen Dick of Alberta, Canada, outlined his methods of tipping. To use tipping successfully, you must be able to recognize the difference between bees leaving a hive (or super) and those robbing. Over the last thirty years, black bears have expanded their territories and can be found in many areas previously devoid of them. Should this be the situation in your area, this method of removing bees from supers is not advised. Be sure you understand the conditions, both seasonal and weather related, that influence how well this method works. **Remember, conditions could change quickly.**

Here's how tipping works:

1. Choose a day when outside temperatures are warm enough for free bee flight and a good honeyflow has been on for several days. Late afternoon works well, as the bees are beginning to come home for the evening.
2. Remove full or partially filled honey supers; try to get supers that are brood free, preferably those from above a queen excluder.
3. Place each super you remove so it sits on its front or back panel, either on the ground to one side of the hive's entrance or on another empty hive body. Do not block the flight path of bees returning to their colonies.
4. In a short while, the bees in the tipped supers should finish their tasks, clean up any drips from burr comb, and fly back to the hive from which they came. This may take minutes or hours, depending on the temperature and the intensity of bee flight activity. You may even have to leave them overnight.
5. Pick up the supers when all the bees have left and take them out of the apiary. You are done.

The reason this works is that a vast majority of the foragers are goal-oriented to gathering nectar; under this circumstance, during honeyflows, the bees seem to be unaware that there are unguarded supers filled with honey. But be sure there is a good honeyflow on, or the bees could start to rob.

Advantages

- Fast and easy.
- No extra equipment or chemicals are needed.

Disadvantages

- Weather can change fast, as can the temperament of the bees. Bees that were oriented to gathering an abundance of nectar may turn to robbing honey from the exposed supers should the nectar-yielding plants shut down their output of nectar.
- Queens, if excluders are not used, may be in the honey super(s) you removed. Careful blowing, brushing, or shaking toward the correct hive is then required, as the bees may not leave the queen. Brood in the supers has the same effect; that is, bees will stay with the brood.
- Requires a second trip to the apiary to collect supers.
- Requires an experienced beekeeper to know the temper of bees in the hives and appropriate weather conditions.
- Beekeeper should live nearby; that is, use this method on the home apiary so you can monitor the bees frequently.

Getting Along with Your Back

Lifting full supers of honey might be the reward of a productive year, but it can also be a literal pain in the back. Unless you are careful in lifting these heavy boxes, you can do serious damage to your back. Proper lifting and strengthening exercises might be needed if chronic back pain is a problem. In any case, medical advice should be sought. For some general information on back care, see T. Pascoe, *Beekeeping and Back Pain*, under "Honey and Honey Products" in the References.

Extracting the Honey

Now that you have removed the honey, place the supers in your honey-processing area (a basement, garage, trailer, or special honey house); you are now ready to extract. You have already prepared this space before you took your honey supers off, right? Double-check to make sure **your extraction room is**

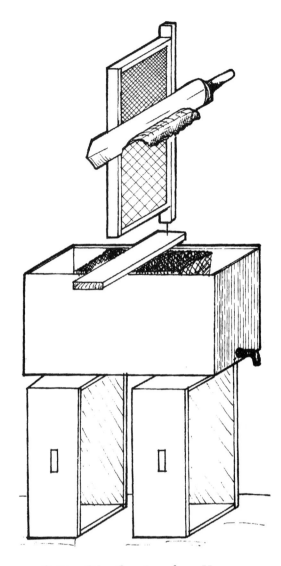

Cutting Wax Cappings from Honey

bee tight, as robber bees love to enter honey houses for free food.

Try to do all your extracting at one time to avoid cleaning up the floor, the extractor, and other equipment more than once or twice. This is a messy job and will try the patience of all involved, but the reward is worth the effort.

The usual process for getting honey out of the wax cells is to remove the cappings with a hot knife called an *uncapping knife* (see illustration on cutting wax cappings) or uncapping plane and to put the uncapped frames into a centrifuge, called an *extractor*. As the extractor spins, the honey is forced out of the cells and against the cylindrical wall of the extrac-

tor, leaving the frames of wax combs empty of most of the honey. A small gate at the bottom of the extractor can be opened to let the honey pour out into other containers. **Keep your eye on the honey** draining into these containers—overflows are costly and messy. Honey left in supers before extraction can be spoiled if it is stored in an area that has high moisture; even capped honey can absorb water vapor. One of the properties of honey is its power to absorb and retain moisture (water). This is technically referred to as *hygroscopicity*.

To prevent hygroscopicity, stack the supers in a staggered arrangement to allow for ventilation, and use either a dehumidifier or a fan to blow warm, dry air over the frames before extraction. You can also place extracting supers in a warm, well-vented room for a day to "dry down" the honey and make extracting easier. This will further reduce or maintain the existing low moisture content of the honey. Warm your supers at least one day before extraction if the frames are cold; warm honey will flow out of the cells faster and will pass through a filter in less time. If you are in a very hot, dry location, such as the southwestern United States or some parts of California, the percentage of water in honey may be below 18 percent, and this, coupled with low humidity, will impede easy extraction from the comb. You will therefore need to increase the humidity of the honey house by hosing down the floor. This is especially true for extracting mesquite (*Prosopis* spp.) honey in Arizona. Because this honey crystallizes quickly, beekeepers in the Southwest usually add humidity in their extracting rooms to help get the honey out of the frames.

Frames with honey to be extracted should be completely or almost completely capped. Uncapped cells containing nectar transitioning to honey will likely have higher moisture content. Extracting honey from too many partially capped frames will increase the moisture of the honey and invite spoilage by fermenting yeasts (see "Extracting Honey" in Chapter 12 for more information on this topic).

Remember: **Honey is a food product—someone will be eating it. Keep it as clean as possible to have a quality product**. Laws exist that require certain minimal sanitary practices, especially if you are selling your product. Honey is a quality hive product, and it is up to all beekeepers to keep it pure, uncontaminated, and clean. **Check with local or state authorities for food-processing rules and legislation**; also the National Honey Board website is a good source of information.

Note: In locations where the small hive beetle has become a problem, it is best to extract honey supers as soon as possible, otherwise the larvae of the beetles will destroy the honey combs as they burrow through them. In addition to collecting nectar from floral or extrafloral nectaries (e.g., cotton), bees collect another sugary liquid called *honeydew*. Honeydew is the sticky, sugary liquid excreted by certain plant-sucking insects that feed on plant sap. One of the principal producers of honeydew are aphids (order Hemiptera). Other insects that exude honeydew include whiteflies, gall insects, scale insects, mealybugs, and leaf hoppers; for more information see Chapter 12. In addition to collecting nectar, bees also collect honeydew in large quantities and process it in the same manner as they process nectar. Honeydew honey is usually dark in color, contains less of the two principal sugars found in nectar, and has more protein and less acidity (higher pH) than honey produced from nectar. It is considered to be a low-quality honey in this country, but is quite popular in Europe and other areas. Some years, bees collect a great deal of honeydew. If there is a demand for honeydew, to sell this product beekeepers should seek out specialty markets that deal with honeydew rather than blending it in with honey.

Harvesting Comb Honey

When you are harvesting comb honey, break apart the comb honey supers the day before harvest and scrape off any burr comb the bees may have built up between supers. Overnight, the bees will clean up any honey drippings, and the combs will be clean of leaking honey or drips.

Take off these section supers carefully, using as little smoke as possible and trying not to crack or bend the bodies or the frames, because the sections can leak if broken. Do not jar or drop such supers, and keep them covered when trucking them to the honey house to exclude dirt and debris. Once in the honey house, process them quickly, to keep small hive beetles and other pests from getting into these supers (or store them in a freezer).

Take out the sections from the frame, then cover and store them in the freezer (to kill any wax moth eggs or larvae) until they are ready for sale. Allow the comb honey sections to come up to room temperature before you put on the labels or box them; this allows the sections to dry out from water condensation (see "Selling Your Hive Products" in Chapter 12).

As with the wet extracted frames and supers, you can place any partially filled comb honey and supers back on colonies for the bees to clean up before storing them for winter; or use them as emergency food for spring feeding.

FALL MANAGEMENT

After the fall crop has been removed and the supers have been cleaned and stored, each colony should be checked and attended to as follows:

- If possible, pick a day when there still may be some opportunity for foragers to be active in the fields. After the honey has been removed, bees that have been dislocated from their boxes may be more prone to sting. Also in cool weather, most of the foragers are idle within the hive, and for some reason, under such conditions, they are also likely to sting even with a minimum of provocation by the beekeeper.
- Check for brood diseases and mites and medicate or treat as needed.
- Do not attempt to overwinter a colony found to have American foulbrood or excessive mite infestation, deformed wing virus, or other problems; destroy the colony and disinfect the equipment.
- Remove the queen excluder and any section comb honey supers.
- Remove any honey supers you don't want bees to move into over the winter.
- If requeening, check colony after seven days to see if the queen has been accepted.
- Check winter stores; about 90 pounds (40 kg) or more of surplus honey should be left on each colony in areas where winters are severe; see "Wintering in Extreme Climates" in Chapter 8.
- Feed the colonies whose stores are low; feed early enough in the fall so the syrup can be properly cured by bees (about a month).

FALL MANAGEMENT OF SOUTHERN COLONIES

Many beekeepers live in the southern states where beekeeping practices are a little different. Dean Breaux of Florida had the following suggestions for managing colonies in the South. The best way to overwinter colonies in the South is to start with young vigorous queens. Requeening in the fall is necessary because honeyflows occur so early in the year (February) that you could miss your first crop unless your bees are strong. The bees do not have a chance to utilize a queen raised in March, so begin fall management in the late fall (October) by requeening your colonies. If populations drop below five deep frames, the colony will in all likelihood produce a meager crop in the spring, or be lost in the winter. If possible, requeen with queen cells; use mated queens for the colonies that were not successfully requeened with a cell. In areas where the Africanized bees have colonized, requeening with a queen cell is not advised, because the virgin queens may mate with African drones, giving the offspring some African temperament.

The fall requeening procedure is:

1. Remove all surplus honey.
2. Treat colonies for mites and diseases.
3. Remove old queen.
4. Install queen cells (or caged queen).
5. If you have many colonies, it may be faster to not kill the old queen but to put a queen cell in the hive with a cell protector. While you don't need to find the old queen, the results are variable.
6. Return in two to three weeks to check and see that there are mated queens in the hives; if the hive is queenless, install a young mated queen in a cage.

By the middle of November, requeening should be completed. Winter preparations for these southern colonies can now proceed. Here are some tips to remember:

- Overwinter in a deep brood box with a minimum of one 6⅝-inch super of stores.
- Install entrance reducers and mouse guard.
- Provide upper entrances if you have several months of cold weather.
- Make sure the bee population is between 15,000 and 20,000 (6–7 deep frames of bees).

- Test for mite levels periodically to ensure mite- and disease-free colonies; treat if necessary.
- Protect colonies from cold, wintry winds in marginal regions.
- Periodically check to see if honey stores are holding up; feed if necessary.

Once most of the cold weather is over, prepare for the first spring honeyflows by doing the following:

- From the first of December to the end of January you can move hives from fall locations to spring yards (primary citrus groves) to prepare for the spring flows.
- Try to locate spring yards in areas that have red maple (*Acer rubrum*) trees; they bloom in late December through January and provide pollen for the bees to start building, in preparation for the early citrus flow in the beginning of March.
- Begin stimulatory feeding with light syrup the beginning of February to ensure that the queens are laying well. This feeding will ensure a strong population of young bees in March for the early honeyflows.
- Add supers to colonies in March in preparation for the flows.

BEEKEEPING IN THE SOUTHWEST

Drs. Gordon D. Waller and Gerald Loper, both now retired, worked at the USDA Carl Hayden Bee Lab in Tucson, Arizona, and gave us some insights into keeping bees in the desert Southwest. In addition, Mr. Fred Terry, of Oracle, Arizona, who keeps colonies all over the Southwest, added some advice for beekeeping in this region.

Beekeeping in the Southwest is much different from any other type of beekeeping. It presents challenges that are not seen in other regions, including sparse bee forage, little water, and extreme temperatures. Because there is such a diversity of landscapes and temperatures, from low plains and valleys to high mountain ranges, it is difficult to generalize about Southwest beekeeping techniques.

Rainfall can vary from 3 inches (7.62 cm) per year (in Yuma, Arizona) to over 30 inches (76.2 cm) in the center of the state, and comes twice a year. The slower, gentler rains come in winter, and the fast, spectacular thunderstorms occur during the summer monsoon season. Although most beekeepers locate their apiaries near irrigated agricultural areas, they often move them to take advantage of different cultivated crops and natural vegetation at varying elevations. Like most commercial beekeepers throughout the United States, many move their colonies to Southern California for almond pollination in the spring (February–March) and then move the bees out after that flow is finished.

Forty or fifty years ago, the bees in this region were mostly kept in permanent yards year-round, using only standard Langstroth deep hive bodies, with either two or three on each colony. Frames were removed and honey was extracted whenever it was available; some beekeepers had mobile extracting units set up to facilitate honey removal. Such methods have been replaced with more normal colonies, producing honey in shallow honey supers that are taken to a permanent building for extraction.

Following the spring mesquite bloom is a maintenance time, when not much is blooming. If beekeepers can find alfalfa or cotton fields, they can get another honey crop, but this is becoming more difficult, since some areas that were used for cotton have switched to other crops or are using genetically modified cotton, which some beekeepers say bees do not like as much. Bees do not normally pollinate cactus, such as prickly pear (*Opuntia* spp.) or barrel cactus (*Ferocactus wislizenii*). Mostly the colonies are maintaining until the winter rains. If bees are moved for pollination, the most commonly grown crops include melons and other cucurbits, alfalfa, onion, and carrot (grown for seed), and citrus.

Colonies overwintered in riparian areas are likely to get their first pollen from the cottonwood trees (*Populus* spp.). African sumac, planted as an urban tree (*Rhus lancia*), as well as *Eucalyptus* species, bloom in January and February. Annuals, such as London rocket (*Sisymbrium irio*), bladderpod (*Lesquerella gordoni*), filaree (*Erodium* spp.), and *Phacelia* species also bloom at this time. March is when the citrus start to bloom, and some areas produce good honey crops from this. The main desert honey, however, is from the mesquite (*Prosopis* spp.) and the catclaw acacia (*Acacia greggii*), which bloom from April into June, depending on rainfall amounts. Other plants blooming then that provide additional food include palo verde (*Cercidium* spp.), creosote bush (*Larrea tridentata*), fairy duster (*Calliandra eriophylla*), iron-

wood trees (*Olneya tesota*), and of course the iconic saguaro cactus (*Carnegiea gigantea*). In the high desert, yucca (*Yucca* sp.), sotol (*Dasylirion wheeleri*), and agave (*Agave* sp.), as well as mistletoe (*Phoradendron sp.*) and bear grass (*Nolina microcarpa*), are also important food sources.

A good desert honeyflow will depend on the winter rainfalls, which could extend or shorten the blooming times. Mesquite can bloom a second time as well but will yield no appreciable nectar if strong winds or unseasonal rains occur during bloom. Honey is usually extracted at this time, in the hottest and driest part of the year. For this reason, extracting houses are usually hosed down to increase the humidity that is necessary to allow the honey to flow more freely from the combs. Another characteristic of honey derived from mesquite, catclaw acacia, and fairy duster is that it crystallizes very quickly, so it must be removed, extracted, and bottled at once. Although the honey crystallizes very quickly, it has an exceptional consistency, with very fine, smooth crystals that stay soft and creamy—a premium honey with good flavor.

By the late summer and fall, desert areas having a heavy brush growth of salt cedar (*Tamarix pentandra*), or the related wild tree known as athel (*T. aphylla*), may bloom again and will provide bee forage throughout the summer if moisture is adequate. The plants that are good for bees include burroweed (*Haplopappus* spp.), and, at elevations of 4000 to 5000 feet, sandpaper plant (*Mortonia scabrella*) will also be blooming, as will some of the agaves. Snakeweed and buckwheat can be a primary source here. "Sugar-bush," which is a local name for buckbrush, also called mountain snowberry (*Symphoricarpos oreophilus* and *S. rotundifolius*) and Fendler's buckbrush (*Ceanothus fendleri*), *are* found at elevations over 5000 feet north of Phoenix.

Overwintering bees in the Southwest need good stores of honey, just like colonies in the northern states. Brood rearing may stop for intermittent periods when pollen becomes scarce, but otherwise brood may be found throughout the winter. If the colonies are located near large numbers of *Eucalyptus* species, trees and other ornamentals, these plants often provide sufficient pollen and nectar throughout the winter months to support continued brood rearing. An excellent reference on the bee plants in this area is Robert J. O'Neal and Gordon D. Waller's article on the pollen harvest by the honey bee (see under "Pollination" in the References).

Since there is so much information on the internet now, you can also check sources there.

Probably nowhere else in the United States are beekeepers more concerned about providing shade and water for their bees than in the desert Southwest. Summer temperatures can often exceed 110°F (43.3°C) in the shade at lower elevations, and at these temperatures, bees spend a lot of energy carrying water back to their colonies to air condition the broodnest and to keep the comb from melting. Therefore, beekeepers must either supply water or locate their apiaries near cattle watering tanks, irrigation ditches, or other places that have a continuous source of water, within a ½-mile (0.8 km) flight range.

Some beekeepers will build shade structures called "ramadas," a framework covered with shade cloth or brush. In addition, trees such as mesquite, tamarix, and cottonwood provide excellent shade. Not only is this necessary for the bees, it is also good for the beekeepers.

Other challenges to desert beekeeping include hazards such as rattlesnakes and scorpions, poisonous spiders that nest under the outer covers, and fire ants that love to invade bee colonies, not to mention the numerous plants covered with spines and thorns. Additionally, most of the feral bee colonies are now Africanized, and beekeepers have had to learn how to manage these bees. Many beekeepers are doing so successfully, however, and good honey and pollen crops can be obtained in these regions. Pests and diseases common to bees elsewhere also occur in the Southwest, such as wax moth, which can be a year-round problem, as well as the two foulbroods and chalkbrood. Varroa mites are now causing problems as well, but the Africanized bees have been seen to chew off the mite legs; these bees also swarm frequently and abscond from their nest, which reduces mite infestations. These behaviors are reported by beekeepers who manage the AHB and say that their colonies are thriving without mite treatments. AHBs have also been tested and appear to be immune to the viruses carried by the mites. For more information on Africanized honey bees see Appendix E.

 **Notes**

CHAPTER 10

Queens and Queen Rearing

Although queens may live for two or more years, the most productive queens are usually between one and two years of age. For an insect, a queen honey bee has a relatively long life. In most circumstances, she lives long enough to join swarming honey bees consisting of her daughters and sons from her colony and move with them to a new home site, where together they establish a new colony. When she departs with the swarm, she leaves behind one of her daughters, the replacement queen, in the existing colony with its nest of honey comb and reserves of honey and bee bread, as well as a substantial cadre of bees.

During the spring and early summer, mated queens can lay 1500 eggs per day—almost an egg per minute—and even up to 2000 per day for short periods of time. At either of these rates, colony numbers in the spring and early summer increase dramatically; the attrition of older bees does not adversely affect the colony's overall well being. However, these dramatic increases in colony numbers may be a contributing factor that initiates the construction of queen cups. Once these cups are constructed, the queen may deposit eggs in them. This will lead to the development of replacement queens, requiring the resident queen to depart the colony in the accompaniment of a swarm.

Occasionally, adult parasitic varroa mites are found riding on or embedded in honey bee queens; these mites may compromise the queen's health and overall longevity. After opening a wound in a bee's chitin, they feed on the lipids of both adult and developing larvae and pupae bees. In the process of feeding, the mites transmit viruses to their hosts. Older larvae about to be capped, pupae, and adult bees are

all fair prey for these mites. In order to protect bees from the varroa mite, most beekeepers world wide have come to rely on miticides as well as a number of other strategies to reduce the number of mites (mite loads) in colonies. In the case of queen bees, mites may shorten their lifespans; the use of miticides may also adversely affect their overall health. Fortunately, because queens develop from egg to adulthood in only sixteen days, female founder mites seldom select queen larvae as hosts.

The mites, the viruses they transmit, and the insertion of miticides into bee colonies are having debilitating consequences on the bees. Evidence indicates that some of the chemicals contained in some miticides cause a decrease in the viability of drone sperm and the sperm stored in the queen's spermatheca. In addition, the developing queen larvae may not be fed adequately, or with food of poor nutritional value; this may also make the resulting queens weaker, resulting in having shorter lifespans. This is one of the reasons why many beekeepers are replacing queens on an annual or bi-annual basis (see "Marking or Clipping the Queens" in this chapter). Additionally, some of these chemicals are now showing up in the beeswax, where they can persist for years.

In addition to the problems created by mites, studies have shown that a colony with an older queen is more likely to swarm than a colony with a young queen. Given this information, annual requeening may provide a threefold benefit, namely (1) a less productive old queen is replaced with a more productive young queen; (2) a younger queen will release adequate or more than adequate pheromone levels,

149

making the colony less likely to cast a swarm or supersede their queen; and (3) the process of requeening causes a break in the colony's brood cycle that will interrupt the mites' life cycle. Given the problems related to mites, their viruses, the side effects of miticides, and the need for adequate pheromone levels to keep the colony intact, annual requeening may become the norm. Naturally, at any time of the year queens whose performance is questionable should be replaced.

The effect of mites and miticides, and whether the correct time to requeen is in the spring, summer, or early fall, continue to be debated. Keep current in the bee magazines and on the internet for the latest information on this and related topics.

Bees rear new queens under three conditions: when the colony is preparing to swarm; to replace an inferior or aging queen, referred to as a supersedure; and when the queen has died, referred to as emergency. Bees preparing to swarm or supersede their queen are requeening their colony. This is natural requeening, replacing a mother with a daughter. Queen replacement resulting from swarming can produce some *afterswarms* that are accompanied by virgin queens. These are swarms that depart the colony after the primary swarm has departed from the hive. However, this is a costly way for bees to requeen their colony. Additional bees accompany these afterswarms. As a consequence, the colony's population is further reduced with each succeeding afterswarm, leaving behind an ever-decreasing workforce to carry out the colony's overall needs and diminishing any hope for harvesting surplus honey. In fact, the bees may be unable to collect sufficient food reserves for their winter needs. Although swarming is a successful strategy by honey bees to perpetuate their species, it reduces one of the major goals for most beekeepers, mainly, obtaining as much surplus honey as possible.

When a beekeeper retrieves a primary swarm, it usually contains a mated resident queen, and unless she was marked or you know otherwise, her age will be difficult to determine. Hive the swarm and monitor its development. If the queen's egg-laying rate and pheromone output are subpar, it will manifest as slow population buildup and a scattered brood pattern, and the colony's original population of adult bees will be declining. If this should be the situation, requeen as soon as possible for reasons already noted (see "Swarming" in Chapter 11). Back at the colony that

cast the swarm, the young queen that emerges is a virgin and will soon take one or more nuptial flights, during which she will mate with an average of fourteen drones. This young queen has half of her genetics from her mother and half from one of the many drones her mother mated with. As a consequence, she is unlikely to be a clone of her mother; therefore one should not expect the identical colony attributes found in her mother's colony to be manifested in this new queen's colony. The drones within her flight range may not be genetically diverse enough, and although her mother may have had many genetic attributes that are favorable to beekeepers, the new queen may not be able to sustain the same vibrant colony that her mother once headed.

If some of your colonies frequently cast swarms, requeen them with stock from several different queen breeders; otherwise, should you choose to requeen your colonies with swarm cells taken from colonies that frequently swarm, you may be promoting the tendency to swarm—swarming behavior has a strong genetic component.

By diversifying your queen stock, you are also inadvertently diversifying the drone population that congregate in *drone congregating areas* (DCAs). DCAs, where drones fly to congregate and wait for queens to fly by, can contain drones from many colonies; DCAs that have greater genetic diversity provide a powerful advantage to queens undertaking mating flights. Recent research has demonstrated that queens mated with many genetically diverse drones produce offspring that are better able to resist diseases, overwinter successfully, and gather more nectar and other resources than those in colonies with queens that have mated with a single drone. In this study by Dr. David Tarpy, queens were artificially inseminated with sperm from a single drone or with a mixture of sperm taken from a diverse group of drones. Colonies with queens with low sperm quantity and viability can decline in the fall or go into the winter with low bee populations.

Queen supersedure, on the other hand, takes place only after a colony has been declining because it is headed by a failing queen (see "Queen Supersedure" in Chapter 11). Her replacement may also be inferior, especially if colony numbers (of young nurse bees) and food stores are inadequate for rearing quality queens. She may also mate with drones whose genetic stock has poor attributes. In areas where Africanized honey bees are present, European queens may mate with Africanized drones, and within a year colonies headed by these queens may exhibit the aggressiveness found in Africanized bee colonies. Since Africanized bees rear more drones than European bees, drone congregating areas may have a higher percent of Africanized drones, providing the probability that a given percentage of the sperm in the spermatheca of a European queen is from these drones (see Appendix E on Africanized bees).

You should give serious consideration to requeening colonies that exhibit the following:

- Low bee populations for no other reason other than a failing queen.
- Bees that are prone to diseases and manifest high mite populations.
- Queen laying more drone than worker eggs; once the cells of larvae are capped, the large domed cappings over drone cells (sticking out of the cell, looking like the eraser end of a pencil) are easy to identify. If the presence of drone eggs, larvae, pupae, and adults far outnumber worker eggs, larvae, pupae, and adults, this will indicate that either the queen, if present, may not have been properly mated, the number of sperm in her spermatheca is near depletion, or the sperm are not viable.
- An unmated or injured queen laying only unfertilized eggs.
- Diseased queen, brood, or workers or a colony that never seems to expand and is always weak.
- Poor brood patterns; this may be a symptom of disease or poorly mated queens.
- Overly defensive.
- Excessive propolizing, unless you are collecting and selling propolis.
- Excessive debris on hive floor (nonhygienic).
- Poor wintering success (colony reemerges from the winter in a weakened condition with a low population).
- High honey consumption; this is measurable with a hive scale and may be difficult to discern because a large bee population may be one of many factors that result in high honey consumption.
- Poor honey production (again, if all other colonies have equal populations, only then can a true assessment be made).
- High tendency to swarm.

- Workers failing to form a retinue around queen.
- Low sperm viability; colonies peter out in the fall or go into the winter with low bee populations.

HOW TO OBTAIN QUEENS

Categories of Queens to Purchase

Queens can be obtained by purchasing them, raising your own, or obtaining queen cells from colonies preparing to swarm or supersede, or from emergency queen cells. Also, by using queen cell protectors, you can obtain uninjured queens from any one of the aforementioned queen cells.

There are five different types of adult queens available for purchasing:

1. *Virgin queens*; these will be unmated and, as a consequence, may require extra work to successfully introduce them into a colony and get them mated. Virgin queens over three weeks of age are no longer capable of mating, and as a consequence are only able to lay unfertilized eggs that will just produce drone bees.
2. *Untested mated queens* have been observed to lay eggs, but the qualities for which they were bred have not been tested; most queens sold to beekeepers today are untested.
3. *Tested queens* are not shipped until the first worker brood emerges, to determine the purity of mating. Other variables may not be observed.
4. *Select-tested queens* are tested not only for purity of mating, but for other characteristics such as tolerance to disease or mites, gentleness, and productivity.
5. *Breeder queens* are tested for one to two years, evaluated, and used to raise daughter queens to sell. Commercial beekeepers and queen breeders often buy some breeder queens to serve as the mothers of the queens in their operation. These queens are usually instrumentally inseminated with semen from selected drone stock and can be very expensive.

Whenever queens are purchased, request that they be marked and make sure they come "certified mite free" so you are not introducing more mites to your apiary. A beekeeper can choose either marked or unmarked queens. There are several advantages to either purchasing marked queens or marking your own. A marked queen has had a dab of paint or an actual disc applied to the dorsal part of her thorax. It is easier to locate a marked queen, and if your record keeping is complete, you will know its age. If she is found in a swarm, you will know which hive the swarm issued from, provided each colony's queen is marked with a different color or different numbered disc of the same color.

Further, when colonies contain marked queens, the presence of an unmarked queen in any of these colonies would indicate one of the following possibilities: in the act of grooming the queen, the bees removed the mark or the disc somehow became detached, the marked queen was superseded and subsequently died or is still present along with the unmarked one, or she departed with a swarm.

A third form of marking is for the beekeeper to clip either the left or the right forewing with scissors. This method of marking queens is not advised; often such queens are superseded.

There are some reports that commercially reared queens have been showing problems with acceptance and longevity. The causes may be related to mites, mite treatments, nutrition, or the lack of adequate and viable drones in the DCAs within the vicinity of their mating yards. If you see a colony you recently requeened with some of these characteristics, requeen it with a queen from a different source. If you are purchasing queens from a local or unknown breeder, start slowly and keep good records of how these queens performed (or didn't perform) so you can decide which breeders to use in the future.

Here are some problems to look for:

- Bees are unusually loud, as if queenless (decibel level will be above 65).
- Colony suddenly becomes queenless.
- Supersedure cells not built despite a failing queen.
- Any queen in a queen cage that does not attract workers, or the retinue does not form once a queen has been released.
- Many larval and pupal deaths are evident (be sure, however, that this is not the result of foulbrood disease); eggs and brood scattered about the broodnest instead of being in compact, concentric patterns. Brood patterns play an important role in evaluating queens.
- Pollen found deposited in cells located in honey supers or the pollen not placed in proximity to the brood areas.
- Honey stored below the broodnest or scattered throughout the colony.
- Hives are abandoned but still contain honey stores; such hives may contain toxins, viruses, or spores and should not be reused unless tested for any of these items.
- Hybrid brood cappings—more domed than worker cappings but flatter than drone caps (a cross between the two)—are seen.
- High supersedure rates.
- Bees fail to form a cluster during cold periods.
- Bees are seen scent fanning in large numbers when you open the hive.
- Bees abscond.

Purchasing and Ordering Queens

Queens can be purchased alone or in packages from reputable queen breeders throughout the United States. Purchased queens may arrive in a queen cage (with or without attendants) or in separate cages in a larger package, which includes many worker bees.

If the purpose of ordering queens is to requeen an apiary, you can order queens so they arrive during a honeyflow; bees readily accept newly introduced queens during honeyflows. Alternatively, a few days prior to introducing new queens, you can feed the colonies sugar syrup in a 1:1 mix to imitate a honeyflow. To reduce mite population, two weeks before the planned requeening date would be a good time to apply a miticide to the colonies you plan to requeen. If you plan to requeen in the fall, you can still have your queens delivered in the summer; plan to install them in nucs to get them laying in preparation for fall requeening (by uniting the nuc to the colony; see "Uniting Weak Colonies" in Chapter 11).

Order queens as you would packages, as early in the year as possible, because most breeders start making queens when snow is still on the ground in northern states. You can have your queens arrive in the spring, summer, or fall. Some breeders will ship queens with or without attendants; others will place the cages in a screened bulk shipping container (especially if you order 10 or more queens at a time) that includes loose worker bees and some kind of food. Shipping packages will vary among dealers.

Queens may arrive in separate queen cages with or

without attendants; or if you order more than several queens (10 or more), they may come in separate cages glued to the floor of a large package filled with many worker bees. The package with the multiple queen cages also contains solid food for the worker bees. After consuming this food, their hypopharyngeal and mandibular glands will produce a liquid food that they will in turn feed to the caged queens.

Handle your queens with care, and follow installation instructions below. It is prudent when installing a new queen not to apply any miticides to that particular colony for 30 days.

Many journal ads are claiming "mite-resistant" queens. Be careful and inquire as to which mites are they resistant and what is the basis for such claims. Do your homework; research the claims and talk to other beekeepers who have used such queens. Queen lines that have shown some resistance to mites include hygienic bees developed at the University of Minnesota Bee Lab (Dr. Marla Spivak) and Russian stock originating from the USDA Bee Breeding Lab in Baton Rouge, Louisiana. Queen breeders are also breeding lines of resistant queens, so check the journals and talk to other beekeepers in your area.

Designs of Queen Cages

Purchased queens are packaged in queen cages and mailed from breeders to all parts of this country and abroad. Some of the procedures for notifying the postal service of the arrival of bee packages should also be followed when queens have been ordered (e.g., call to alert the post office of your order, as suggested in Chapter 6). Again, all queens must be certified to be free of mites.

In the United States, queen bees are shipped in several kinds of cages. One style is the *Benton* (also called two- or three-hole) mailing cage; it is a small wooden block, approximately 3 × 1 × ¾ inches (7.5 × 2.5 × 1.9 cm), with wire screening stapled over the length of it to cover the two or three holes to the compartments (see the illustration of queen cages). Generally, two compartments serve as the living quarters for the queen plus several attendants, and a third compartment contains a candy plug (made of fondant).

At either end of the cage is a small, bee-size hole. The hole at one end is adjacent to the compartment that is filled with candy, while the hole at the oppo-

site end is used to place the queen in the cage; once the sugar plug is in place, it is sealed off with a cork plug, the queen is inserted in the opposite opening, and then that opening is sealed off with another cork. The candy serves two purposes: it provides food for the caged bees and delays release of the new queen from the cage into the colony. The hive bees must eat the candy plug before the new queen can be freed. This usually takes about two to three days, by which time the colony will have adjusted to the new queen's scents (pheromones), increasing the likelihood that she will be accepted.

When a cage is placed in a colony, you must be sure that the bees have contact with the screen portion of the two wooden cages and the screen portion of the plastic cage. That way bees within the colony

Queen Cages

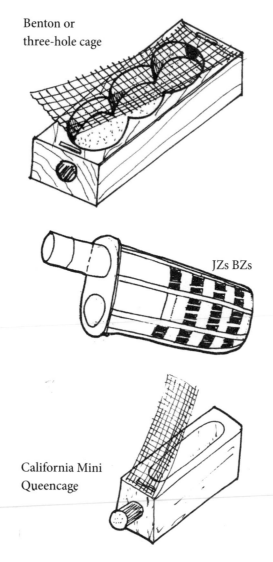

Benton or three-hole cage

JZs BZs

California Mini Queencage

can feed the queen as well as her attendants, and also become familiar with her pheromone bouquet.

One of the problems that occur when anchoring these cages between frames is that often the bee space is violated, and as a consequence, bees will construct comb in the newly created space. Once the queen is released and the queen cage removed, you need to properly space the frames again and remove the excess comb. To avoid the construction of extra comb and maintain the bee space, you can suspend the queen cage between two combs of foundation, or carve out an area of drawn comb to accommodate the cage. When following this procedure, be sure that the bees have access to both the screened portion of the cage and the candy plug.

Other cages used by bee breeders (see the illustrations) include JZs BZs plastic cage and the California Mini Queencage, developed by C.F. Koehnen and Sons. The latter is a narrow wooden block, about 2½ × ¾ × ¾ inches (6.5 × 2 × 2 cm), with one long compartment. It has a large candy-filled plastic tube inserted at one end. The smaller size of this cage allows more queens to be *banked* (or stored in queenless colonies until needed), and when it is inserted between frames, it does add extra space that bees will fill with comb. Again, queen cages can be suspended between combs without violating the bee space if some of the existing comb that will be occupied by the queen cage is carved out.

The plastic shipping cage called JZs BZs is in the shape of the lowercase letter "d." It is much smaller than the wooden cage, with its overall length, including the stem, being 2½ × 1 × ½ inches (6.5 × 2.5 × 1.5 cm). The stem of the "d" holds the candy plug.

Increasingly common is the cardboard/screened "battery box" for shipping queens. Caged queens are secured in a small screened box, and young worker bees are then added to attend all the queens in the cages. This method is preferable because the loose bees can protect the caged queens from temperature variations during shipping. Other types of cages are becoming available, so check the bee supply catalogs or talk to or visit individual queen breeders.

REMOVING ATTENDANTS

Before the queen is introduced to an established colony, many beekeepers remove the accompanying attendants (if there are any) because they are foreign bees, and the workers in the colony may become aggressive toward them and even the new queen, biting them through the wire screening of the cage or the screened openings in the JZs BZs cage. This activity may release alarm odors, alerting other bees to surround the cage, and in the process they could injure or kill the queen. In addition, these attendant workers may be infested with tracheal or varroa mites; so attendant workers should be disposed of.

Three Ways of Removing Attendants

An easy method of removing attendants from the cage is to hold the cage between your thumb and middle finger; remove the cork from the non-candied end, and place your index finger to cover the uncorked hole. Hold the cage vertically and, as attendants move up the cage, remove your index finger, permitting the bees to exit. If the queen moves toward the opening and attempts to leave, cover the hole. This method, like the others, should be done within an enclosed area. If the queen slips out, she can be retrieved—often, however, with some difficulty.

Another method is to replace the screen with a piece of plastic queen excluder that has been cut to fit on top of the cage. First, remove the staples that are keeping the screen in place while you hold the screen. Put the plastic queen excluder on top of the screen and carefully slide the screen out from underneath, taking care not to injure the queen if she is on the wire screen. Then tack the piece of excluder to the cage. Now place the cage near a light source (natural or artificial) in a darkened room, and the workers will exit through the excluder. You can also do this in a closed vehicle or room. Once the attendants have left, replace the screen, but don't forget to re-staple it.

A third way is to remove the cork opposite the candy end of the cage (or pry up one of the staples and lift up a corner of the wire screen) in a darkened room next to a closed window. Hold the cage in a vertical position, with the opening at the top, and the bees, like most insects, will climb upward. When the bees exit, they will continue to move toward the light, and you can easily recapture the queen from the window. Pick her up only by the wings or thorax and return her to the cage, and replace the corks or screen. This is a good time to mark and/or clip the queen (see "Marking or Clipping Queens," below).

Queen Cage Holder or Banking Frame

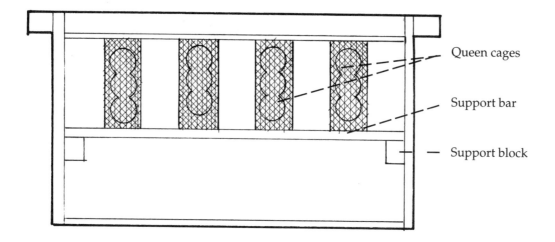

Queen cages

Support bar

Support block

CARE OF CAGED QUEENS

When caged queens arrive, they should be properly cared for and placed into a colony as soon as possible. If the cages contain candy and attendant bees, they should be kept in a warm (about 60°F [15.5°C]), dark place, free of drafts for up to a week. The attendants help keep the queen warm, but you should cover the cages with a light cloth to keep the warmth in. A very, very small drop of water should be placed on the screen twice a day. Do **not** get the candy or bees wet, as the bees could become chilled. Remember you are not providing water for horses! If you live in a very dry region, you may want to place a wet sponge in or on the queen cage to increase the humidity and prevent the bees from dehydrating.

If the queens are to be introduced by the indirect method, the attendant bees should be removed first, as we already described. If you use another method, and the attendants are needed, check the condition of the queen and accompanying bees to see if there are any dead attendants in the cage. If there are some, remove the queen and place her in a new cage with attendants of newly emerged workers from one of your colonies; older bees will in all likelihood attack the queen. If the cage becomes fouled for any reason, remove the queen and place her in a new cage, then fill one compartment with candy (made of clean honey and powdered sugar) or a marshmallow (or several miniature marshmallows).

If you are not planning to install the queen for more than a week, you can bank the queen (with no attendants) in a queenless colony for up to three weeks providing she is not a virgin. You can bank many queens this way. First, remove the attendant bees, then place the cage, screen side down, between the top bars in a queenless nuc that contains lots of young workers. Make sure there is plenty of food in the nuc, or you can feed the bees medicated sugar syrup and a pollen substitute. (You can also use a full-size queenless colony.)

You can make a special queen-cage-holding frame or banking frame (see the illustration of a queen cage holder above). One end of the cage goes against the underside of the top bar of an empty frame (without comb or foundation), and the other end rests on a bar of wood that has been nailed in to run the length of the frame. The bar is nailed to the side bars of the frame; this way 10 to 20 cages can be banked in several frames.

Insert the banking frame into a strong, queenless, or queenright colony above a queen excluder. If you surround the cages with frames of emerging bees, bee bread, and honey, the caged queens will be well cared for until they are needed. If you are storing them in a queenless colony, add some frames of capped brood once a week and feed the colony. A free queen must **not be allowed** in the queenless colony or above the excluder in a queenright colony; otherwise, the caged queens may be killed. As an added precaution, place all frames of open brood in the bottom hive body of a queenless colony with an excluder above it, in case the queenless colony rears a new queen.

Marking or Clipping Queens

It is advisable to mark the queen with a spot of paint or a color disc (with or without numbers) on her thorax, or clip her fore and hind wings on one side (see the illustration of the queen with a numbered tag). Marking or clipping the queen allows you to keep a record of the age of the queen, and the use of color will make it easier to find her, especially if she is dark in color. Clipping the wings of a queen will not control swarming, contrary to what the older literature erroneously states. In some instances, clipping may cause the bees to supersede or replace her. Therefore, our advice is not to clip the queen's wings. If you must keep the queen from leaving (e.g., if she is an expensive breeder queen), place a strip of queen excluder in the hive entrance instead. An alternative to this approach is to locate the queen in the bottom deep super with a queen excluder above the bottom board and one on top of the bottom deep super. Once this has been accomplished, drones above the first deep super will be unable to pass through the excluders. To allow any drones located in supers above the first deep super to exit the colony, make a small notch in the rim of the inner cover (one may already exist) and invert the cover rim down; this notch will provide an exit for most of the drones.

When you pick up the queen to mark her, never hold her by the abdomen; see the illustration on the proper method of holding a queen. Picking her up by the abdomen can cause injury or compromise her

**Clipped-Wing Queen
with Numbered Tag**

Proper Method of Holding Queen to Clip or Mark

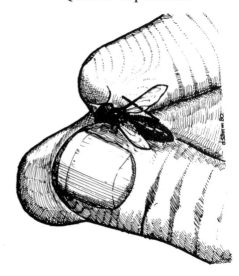

egg-laying ability. As a result of this or any injury to the queen, the colony may supersede or replace her. To avoid injuring a queen, first practice picking up a few drones. Remember, usually, mated queens do not sting unless they are mishandled; however, virgin queens are more likely to sting, especially if handled roughly.

To mark the queen:

1. Grasp the queen by both wings with one hand, and then with the other hand, grasp the sides of her thorax or a few of her legs on both sides, otherwise she may twist and break her leg(s). Once you have her in your grasp, release your hold of her wings.
2. If you are marking with paint, use a fast-drying paint, like nail polish or model paint, and mark only the thorax. Also check bee supply dealers for marking kits.
3. If you are gluing on a numbered disc, first apply adhesive on the thorax, then firmly attach the disc.
4. Allow the paint or adhesive to dry before you gently return the queen to the frame she was on, or place her on the top bar of a frame and let her walk down the comb. Sometimes the paint and adhesive odors will cause the hive bees to attack the queen; cover her with a few drops of honey to mask the odor of the paint or adhesive.

The queen can be removed from the colony without alarming the other bees; in fact, for short periods (5–10 minutes) they will not be aware she is absent.

It is becoming more common to requeen colonies every year to maintain quality stock, reduce swarming, or keep diseases and mites under control. Make sure you note down in your hive diary the date you purchased the queen, the source of the queen, and the color with which you marked her. This is becoming especially important in areas with Africanized honey bees.

You can use any color to mark the queen, but it is best to use a bright, primary color so that she will stand out. The primary colors blue, red, and yellow are ideal for marking queens. Some bee suppliers sell special queen-marking kits and colored discs, which make it easier for inexperienced beekeepers to use. If you wish, use the International Color Chart system that is on a five-year sequence for marking the queens:

White or gray—year ending 1 or 6.
Yellow—year ending 2 or 7.
Red—year ending 3 or 8.
Green—year ending 4 or 9.
Blue—year ending 5 or 0.

Advantages

- Queen can be easily found.
- Queen's age can be determined.
- The absence of a marked queen may indicate that she departed with a swarm, was superseded and then eliminated, or died. Or perhaps she is still present but the marking wore off or the disc detached. Another possibility is the colony was usurped by Africanized bees that eliminated the marked queen.

Disadvantages

- Bees might supersede "maimed" queen.
- Clipping does not prevent or control swarming.
- Queen could be injured while being handled.
- Queen that was clipped may be a virgin and thus would be unable to fly and mate.
- Virgin queen might sting when handled (but this is not likely).
- Color spot or disc may come off.

Failure to Find a Marked Queen

In the absence of finding a marked queen, you can assume that either the colony swarmed with her, she was superseded, she was accidentally lost, she died, or an Africanized or a European swarm with their queen took control of the colony. If a queen is not found, but eggs and brood are present, you can be assured that there is a queen in the hive (marked or not).

If you do not find a queen after a thorough search, and no eggs or brood are present, look for ripe queen cells, or suspect that a virgin or a new young queen is present but has not yet begun to lay eggs. Check the colony for eggs in another week; if you still find none, requeen.

A virgin queen is more difficult to locate because she tends to move more quickly on the comb than a laying queen. She might also be a little smaller than a laying queen (more the size of a worker) because she has not gained body weight after returning from her nuptial flight(s) and needs more time to fully develop before she starts to lay eggs.

The absence of eggs or brood in a colony, therefore, could mean that a virgin queen is present but has not yet mated, or a newly mated queen has not begun to lay eggs. Before you requeen such a hive, be sure that the colony is queenless, since any attempt to introduce a new queen into a hive with a virgin queen or newly mated queen is likely to fail.

If the colony is definitely queenless, you will eventually have laying workers present. If you find scattered drone brood, scattered capped drone cells, and several eggs in each cell that are attached to the cell wall instead of the cell bottom, you have laying workers (see "Laying Workers" in Chapter 11).

In areas of Africanized honey bees, your colony may have been taken over by an invasion or usurpation swarm. This is a difficult situation to change, and if you live in such an area, you may have to requeen your colonies frequently to keep your hives European.

REQUEENING

In most geographic areas, requeening can be accomplished during the spring, summer, and fall. One of the major problems with requeening a colony, other than the need to first find and remove the resident queen, is the rate of acceptance of the introduced queens. The ideal time to requeen is during periods when a honeyflow is in full progress. If you are requeening when no honeyflow is taking place, the alternative is to mimic one. A week prior to requeening, begin feeding sugar syrup to the colony (the ratio of sugar to water to mimic a honeyflow is one part sugar to two parts water). Continue feeding for an additional week after the queen has been introduced.

SPRING REQUEENING

In most geographic areas, queens may be purchased from breeders in the early spring and of course throughout the summer and fall. One of the problems with ordering queens in the spring, especially in more northern latitudes, is that bad weather during transit by mail may kill or injure queens. An alternative is to contact local package bee dealers, who usually bring back extra queens along with packages. This may be a safer way to obtain queens early in the spring.

Instead of purchasing queens, you can rear your own (see queen rearing section).

Although spring is the time of year that purchased queens are readily available from bee breeders, you may also want to raise queens from your own colonies during the spring and early summer, when bees are naturally inclined to rear queens. Remove the naturally made cells and place each in queenless nucs (you can also use the royal jelly in these cells to prime queen cups). In addition, the bee population is more manageable this time of year, so you can locate queens (or the lack thereof) easily. If you do make a mistake (queen introduction didn't work) you have all summer to manipulate the colony you are requeening into a queenright state. Before mites, beekeepers often requeened weaker colonies coming out of the winter with new queens they purchased or raised. Now, requeening takes place all season long; however, it is still easier to requeen colonies in the spring.

Advantages

- Colony is less likely to swarm if it is requeened before the swarming season has started in your area.
- A vigorous egg layer will produce large bee populations, ready for subsequent honeyflows.

- Colony will enter the winter with a large population.
- Old queen is easier to find when colony numbers are low, as is the case in the spring.
- Bees are calm and less prone to sting, run, or rob, especially during a honeyflow.
- There is less chance of swarming the following year because the queen will be only a year old if the introduction is accomplished before the advent of the two distinct swarm periods in your area. Swarming takes place earlier in a calendar year at lower latitudes and later at higher latitudes. For example, in Ithaca, New York, the primary swarm period takes place from late April to mid-July, with the most numerous swarms around the first two weeks in June. The second swarm period, markedly less intense, occurs mid-August to late September.
- There is time to assess the queen's performance and to replace her, if necessary.

Disadvantages

- If you hit a rainy or cold spell, it could be many days before you can requeen the colony.
- A queen could be superseded or killed if inclement weather sets in and bees go hungry. (This may be overcome by providing sugar water as needed.)

SUMMER/FALL REQUEENING

The major problem with summer and fall requeening is that colony populations are high, making it difficult to find the resident queen, especially an unmarked one. In addition, there may be a dearth (no incoming nectar), which compounds the problem, because as you open hives to find or replace the queen, you may initiate robbing behavior, which will compound the acceptance of a new queen.

If the colony is very populous, here is a quick tip for finding the resident queen: before you start your search, place a queen excluder between the two brood chambers and come back in a week; she will be located in the chamber containing eggs and young larvae. An exception to this approach will be when there are two queens in the colony, each in separate chambers. The plus side is you are quickly into and out of the colony, thus reducing the opportunity for robbing to be initiated.

Another alternative to introducing a mature queen

is to remove the resident queen and insert a fully developed (ripe) queen cell into the colony. The positive side of this action is there will be an interruption in the brood cycle, from the period before she emerges, after emergence, after taking her mating flight, and the additional days before she begins egg laying. This hiatus in brood production interrupts the life cycle of the varroa mites (see Chapter 14), since no larvae will be available for their reproduction.

However, there are also many risks with introducing queen cells: she has to emerge, be accepted, undergo a mating flight (or flights), and return safely. In addition, as we progress from summer to fall, the number of drones at DCAs (drone congregating areas) may be diminished, resulting in a poorly fertilized queen.

Advantages

- There is less chance of swarming the next year.
- Colony enters winter with a strong population and a young queen.
- Young queens will lay more eggs in the late winter and spring than older queens.
- Colony emerges in spring with a high bee population ready for the honeyflow.
- Timing breaks the brood cycle, thus reducing disease and pest problems.

Disadvantages

- Hive is populous, making it difficult to find the resident queen.
- If no honeyflow is on, bees are prone to sting, rob, and run when hive is opened; robbing could be serious.
- Time consuming.
- In the fall, fewer opportunities to check if queen was accepted and less time to assess queen's performance.
- Could end up with queenless colony and laying workers overwinter if the queen is not accepted.

QUEEN INTRODUCTION

Although many methods, including some ingenious ones, have been devised for introducing queens into colonies for the purpose of requeening, none can guarantee absolute success. Often, the more time-consuming ones are the most likely to succeed.

It is generally agreed that no matter what method is employed, the most opportune time to requeen is during a honeyflow. If no honeyflow is evident, feed the colony to be requeened at least one week before and during the introduction, as well as after the new queen is installed, to mimic a honeyflow. All the methods listed here, except the division-screen method, require that the colony be dequeened (queen taken out to make hive queenless) for a period up to 24 hours prior to the introduction of the new queen.

During the time you are introducing the new queen, observe how the bees react to her. If the bees in the colony tightly cluster over the queen cage and appear to be trying to bite or otherwise injure her, there already may be a queen in the colony that you are trying to requeen. On the other hand, if you observe the bees forming a loose cluster on the cage and try to feed or lick the new queen, they are probably ready to be requeened. If the bees completely ignore the queen, you may have to try another one. If they ignore a second presentation, the colony may have another queen in addition to the one removed. It may be difficult to truly discern that the bees are ignoring the queen. Check back in one week, and if the queen is present and young brood is evident, then all is well.

The methods used for requeening can be divided into two categories: the *indirect release*, in which there is a delay before the bees have direct access to the queen, and the *direct release*, in which the queen is immediately released among the bees. Some of these methods can be combined with swarm control or making increases in the apiary (see "Prevention and Control of Swarming" in Chapter 11).

After you requeen a colony, check after a week or so to see if the new queen has been accepted. Look for eggs and young larvae; if they are present, you have a laying queen. If the colony has a queen other than the one you introduced (because it has a different mark or is unmarked and the newly introduced queen was marked), you can either assess the qualities of the queen that is present for a month or so, or requeen the colony again later in the year.

Indirect Release

Push-in Cage

The push-in cage is a tried and true method of introducing a queen and is often referred to as the gold standard method for queen introduction (see the illustration of the push-in cage). This method is the

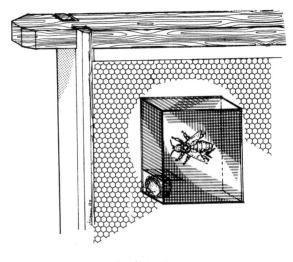

Push-in Cage

best one to use, especially if you are introducing an expensive breeder or inseminated queen.

The cage is made by folding all four edges of a square or rectangular piece of hardware cloth at right angles, to form a box with a top and sides but no bottom. The top of the formed box should be at least 3 × 4 inches (8 × 10 cm) and the sides at least ½ inch (1.3 cm) deep (see the illustration). You can also make a bigger one, 5 × 8 inches (13 × 20 cm). Pinch the corners of the box into triangles so they will fold cleanly on the edges. Use ⅛-inch (1.9 mm) hardware cloth that has been soaked in hydrogen peroxide and rinsed with water; brand-new hardware cloth has toxic substances that will kill bees and queens. Plastic push-in cages are now commercially available, and ads for these cages can be found in the bee journals and online.

Dequeen the colony at least one day (12–24 hours) before installing the new queen. Feed the colony to be requeened with sugar syrup for a few days before you introduce the new queen, and then go in and find and remove the old queen. When you are looking for the queen, use smoke sparingly, as too much smoke will cause the queen to run, making it more difficult to locate her. When you find her, remove her. What can be done with the removed queen? If she is not diseased or otherwise impaired, introduce her to a nucleus colony, and if she performs to your satisfaction, move the colony to a standard hive. If you want to dispose of her, you can dissect her to test for tracheal mites (see Chapter 14).

After the colony has been dequeened, remove a frame that has capped brood and some honey, and remove all bees from the area to be occupied by the push-in cage by shaking or brushing off the bees. Place the new queen on a patch of capped brood and honey and push the bottom of the cage into the comb over the newly obtained queen. The comb under the cage should contain capped brood, cells of open nectar/honey, and be free of adult bees. If there are no open cells of honey/nectar, uncap five cells containing honey to provide the queen access to food. As bees beneath the capped cells emerge, they will care for the queen. The frame with the push-in cage should be placed in the center of the broodnest. If the cage is not properly installed and the bees outside the cage gain access to it (in some cases, bees will chew through the wax to obtain access to the area), hive bees may eliminate or injure the queen.

Indirect Release with Queen Cages

This method of requeening employs a queen cage; here are the steps to follow:

1. Dequeen the colony 12 to 24 hours prior to replacing with new queen.
2. Select a caged queen with no attendants.
3. Remove the cork in the candy end, and if the candy is hard, use a nail to make a small hole through it to make it easier for the hive bees to free the queen by eating out the candy. The hole should not be too large; one of the purposes of the candy plug is to delay the queen's release and thus enhance her acceptance. Do not make a hole if the candy is soft. Be careful not to impale the queen with the nail! If you find the plug to be rock solid (which means the bees will likely never eat through it), it will be necessary to excavate a great deal more of the plug, or all of it, and replace it with a mini marshmallow.
4. Find a frame with uncapped larvae and cut or carve out some comb. Start about two inches below the top bar so the queen cage can be positioned vertically on the comb. Make sure the screened portion of the cage is either facing the adjacent comb or facing at a 90° angle from the adjacent comb. By so positioning the cage, the hive bees will have access to the screen, or in the case of the JZs BZs cage, the screen of the plastic cage. When installing a package on foundation, there is sufficient room to suspend the cage. If these operations are done correctly,

and the frames are coalesced so there is no violation of the bee space, bees will not construct extra comb.

5. Alternatively, place the cage horizontally between the top bars of two frames of brood, screen side down, so the bees have plenty of access to the queen and to the candy. If you place the cage between the top bars, you will be violating the bee space, and therefore you should anticipate finding bees constructing comb where it is not needed or wanted. If you remove the cage as soon as the queen is released and push the adjacent combs together again, you may avoid the construction of excess comb.

6. Another method is to find a frame or two containing uncapped larvae and cut out some of the comb to allow the queen cage to fit vertically from the top bar, screen side out. Push in the cage, candy side up, and place the frame back in the hive. Now place a frame of young brood next to the queen cage and push the frames together so the bee space is maintained. Again, if you leave a gap, the bees will fill it with comb because they do not tolerate additional space. Make sure the hive bees have access to the screened or open side of the cage and to the candy plug and queen.

7. Remove the cage after a few days to prevent excess comb from being constructed in the additional space this technique creates.

8. Observe how the bees react to the caged queen. If they are defensive, you may have a loose queen in the hive. If they cluster on the cage, trying to feed her and scent fanning, the hive will probably accept her. This observation can take place only by removing the cage after five minutes and observing the activity taking place on the screened portion. This is time consuming, and interpretation can be problematic, but it may be a good exercise for beginners.

9. Examine the colony after one week. If the queen is still in her cage because the bees have not eaten through the candy plug, enlarge the hole in the candy to let the bees finish releasing her; or release her directly by pulling out the cork in the other end, or by pulling off the screen. When you are using a plastic queen cage and the queen has not been released, poke a bigger hole in the candy plug or lift the lid that was originally used to insert the queen into the cage. Note:

these caged queens are all capable of flight, therefore these methods of releasing the queen require a great deal of caution.

10. If you are concerned that the caged queen could escape, position the cage near the hive entrance, and as soon as you have opened the cage, slide or push it into the hive. Hopefully, the queen will not turn around and fly out. You can block the hive entrance with newspaper or grass immediately after you slide the opened cage into the hive. After a few minutes, reopen the entrance.

Any delay in releasing the queen (due to a hardened candy plug) provides the bees more than sufficient time to "familiarize" themselves with the queen, unless during this period the bees began to raise a replacement queen. If no replacement is being raised, the caged queen will more than likely be accepted whether the release is further delayed or the queen is directly released.

Direct Release

Nucleus Method

This method does not require the attendant bees to be removed from the queen cage. Remove two or three frames of capped and emerging brood from a strong colony and shake or brush off all the bees back into the parent colony. Place these frames, now free of bees, in a three- to five-frame nucleus hive or *nuc*; fill the rest of this nuc with honey and pollen frames. Directly release the queen (and her attendants) onto a frame of the soon-to-be-emerging brood; any young workers will immediately accept the new queen. Use this method during the hot summer months so the brood (and the queen) will not be chilled on cool nights. On the other hand, make sure the temperatures do not get so high that they overheat the nuc. This method should be tried on a limited basis until you have a sense of its effectiveness. You can save this nuc and use it later to requeen an established colony.

To requeen an established colony with the now populous nuc, follow these steps:

1. Dequeen the colony to be requeened at least one day before.
2. Place the nuc containing the new queen next to the dequeened hive.
3. Apply a small amount of smoke into the nuc

entrance, being careful not to disturb the bees too much.

4. Remove two or three frames from one side of the dequeened hive and replace them with the frames from the nuc box containing the bees and the laying queen; the new queen should be between two of the inserted frames. Any extra frames can be given to a weak colony or to start a new nucleus colony.

5. Close the hive, put on feed, and check after one week to see if the new queen was accepted. The only way to confirm acceptance of the nucleus queen is if she is laying eggs (and is marked).

Another nuc method is similar to the one just described, except first the attendant bees are removed from the queen cage. Remove frames of capped brood from a hive and shake off any attached bees. Place the frames in a nuc and replace the cover. Remember you will be directly releasing the queen; she can fly, therefore you must immediately block the opening of the queen cage once you have removed the cork or plastic cap. Then insert the queen cage into the entrance and block the entrance with grass or a wooden entrance reducer for several minutes. If all goes well, she will move onto the frames and join the other bees. Then remove the grass or entrance reducer that is blocking the entrance. Now remove two or three frames of young, uncapped brood, making sure the resident queen is **not** on these frames, and shake these frames containing young nurse bees in **front** of the nuc so the bees can enter it. Move this new nuc to another location and check it in a week. Once the queen is accepted, you can requeen another colony later, as outlined above.

Honey Method

Dequeen the colony at least a day in advance, and then proceed as follows:

1. Open the hive and remove the nearest frame; check each frame until you find one with young larvae and honey; remove it, shaking off all the adult bees.

2. Break the wax seal over some honey and, without injuring the new queen, coat her with honey.

3. Release the queen on the frame with young larvae and then gently return the frame to the hive; replace the remaining frames and close the hive.

4. Check in one week to see if the queen was accepted.

Scent Method

To temporarily mask the odor of a newly introduced queen, beekeepers rely on the scent method. They add a few drops of peppermint, lemon, vanilla, wintergreen, onion, anise oil, or grated nutmeg per cup of sugar-water syrup, then either spray the inside of the hive with this syrup, add the syrup to the sugar-water feeder, or employ a combination of both. As the scented odor gradually diminishes, the queen acquires the odor of the hive and the bees will accept her. Dequeen the colony (or nuc) at least a day in advance and feed it with the scented syrup. Then proceed as follows:

1. Remove two or three frames containing bees and spray them with the scented syrup, but do not soak the bees. Try not to get the syrup inside uncapped larvae, as it may kill them. Do not use scented syrup during dearths, as it may attract robbing bees.

2. Spray the queen and release her onto the top bar of a sprayed frame. Guide her down between two sprayed frames.

3. After she has crawled down, close the hive and then feed the bees with more scented syrup.

4. Check after one week to see if the queen was accepted.

Smoke Method

Dequeen the colony at least a day in advance, and on the following day proceed as follows:

1. Blow four or five strong puffs of smoke into the entrance. You can also scent the smoker fuel with one of the scented oils listed above.

2. Reduce the entrance to at least 1 inch (2.5 cm) with loosely packed grass, rags, or newspaper.

3. Close the entrance for one to two minutes.

4. Open the hive entrance sufficiently to accommodate one of the three kinds of queen cages, then remove the cork opposite the candy plug. In the case of the plastic cage, remove the cap initially used to place her in the cage (**remember, caged queens can fly**), cover the cage opening with your finger, and as you begin to insert the cage, remove your finger from the opening and push the cage

all the way into the hive. Now add two quick puffs of smoke.

5. Close the entrance for three to five minutes.

6. Reopen the entrance to about one inch wide again and walk away; the bees will remove the remaining material within a few days if it is loosely packed.

7. Check after one week to see if the queen was accepted.

Caution should be used with this method if the weather is extremely warm and the colony especially populous, because the reduced entrance could make it difficult for bees to ventilate and cool the hive.

Shook Swarm Method

This method requires a nuc box that has a screened bottom, called a *swarm box*. You can make this by screening the bottom of a nuc box and attaching some legs to keep the bottom off the ground. A swarm box is normally used to collect swarms and does not contain an entrance, as it is a collecting box.

Dequeen the colony at least one day in advance, and then proceed as follows:

1. Remove four or five frames with bees and spray the bees and frames with syrup (can be scented syrup). Remember, do not use scented syrup during dearths, as it may attract robbing bees.

2. Shake these queenless bees into the swarm box or another container (the container should be large enough so bees are not overcrowded and should be screened for ventilation). After the bees have been "shook" off their frames into the screened box, remove the remaining frames with brood and clinging bees from the colony and give them to other colonies or use them in nucs for new queens as described earlier. Next, add enough frames of dry comb or foundation to equal the number of frames removed from the colony and then reassemble the colony. To feed a "shooked" swarm, place a feeder jar over the screen opening of the cover. Alternatively, if the outer cover is solid, cut a circular hole in the cover to accommodate the section of the feeder container that provides bees access to it.

3. Put the shook bees in a cool, dark place and feed them with 1:1 sugar : water syrup for at least eight hours.

4. The next morning, shake the bees (similar to dumping a swarm at the hive's entrance) out of the nuc box, directly in front of the entrance to their original hive. Before releasing the queen from her cage, spray her with sugar syrup, and then release her among the bees moving (walking) into their old hive. Alternatively, you can introduce a caged queen into the nuc box eight hours before this procedure (releasing her as instructed in the smoke method).

Snelgrove, Division Screen, or Screen Board Method

In 1934, Leonard E. Snelgrove described his double screen board, with dimensions equal to an inner cover. The board has a one-inch rim above and below it. Each rim is equipped with toggles that can be opened or closed, controlling the exiting or entering of bees into the hive body above the board or below the board or both.

Its original use was for swarm prevention. However, it has since acquired many additional functions, such as to introduce a new queen; to combine two colonies; to overwinter a weak colony; to start a two-queen system; to place two nucs above the board for winter heat; to ventilate a colony that needs to be moved during warm weather.

The board separates the queen (to be replaced) and the bees below it from the queen or queen cells and the bees above it. The small unit above the board can take advantage of the heat rising through the screens, thus keeping the smaller colony warm.

Snelgrove Screen or Division Screen

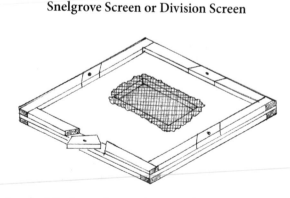

See also http://www.dave-cushman.net/bee /snelgroveboard.html.

Division Screen Method of Requeening Existing Colony without Making an Increase

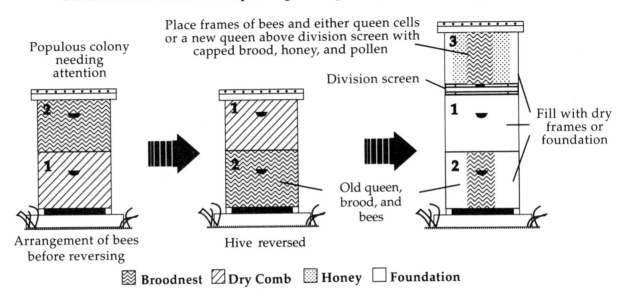

To requeen using this method, follow these steps (see the above illustration):

1. If the queen to be replaced is in the second deep brood chamber, place this chamber on the bottom board.

2. From one or both of the deep supers, remove three or four frames of capped brood (bees ready to emerge) and keep the remaining brood frames next to one another in order to keep the brood-nest intact. In each hive body (if this is the case), replace the withdrawn frames with dry comb and/or foundation.

3. Shake the removed sealed brood frames so all the adult bees drop to the main entrance.

4. Place the brood frames, now free of adult bees, into the center of an empty hive body (numbered 3 in the illustration).

5. Place two frames containing sealed honey and pollen on each side of the brood frames in (3).

6. Fill the remaining space in hive body 3 with foundation and/or dry comb. The new hive configuration should be composed of the following pieces of equipment:
   a. One deep hive body with the original queen (1 or 2).
   b. One deep hive body (3) with a queen cell or a new queen, preferable a marked one; replace the inner and outer covers.

7. As the young bees emerge from the capped brood frames, they will accept the queen or the queen that emerges from the queen cell. **Caution**: in that the queen that emerges from the queen cell must undertake one or more nuptial flights, be sure to keep the same toggle opened during that period, and preferably the one located to the rear of the main entrance; otherwise on one of her returning flights she may enter via the main entrance. If that happens, she will likely be eliminated.

8. Only one upper entrance of the division board should be opened, preferable the rear one (opposite the bottom entrance), but for the first week all entrances to the Snelgrove board should be closed, since it will take time for the bees in (3) to slowly emerge from their brood cells.

9. Check colony in one week and open the upper rear toggle.

Three weeks later, replace the division screen with a queen excluder. The hive can now be run on a two-queen system (see below) until after the honeyflow, or the hive (3) can be united (with 1 and 2) by removing the excluder just before a major honeyflow. Find and remove the older queen or just unite the colonies by removing the excluder; usually the older queen will be killed. Alternatively, you can make (3) a separate colony by placing it onto a new bottom board when the bee population is high.

Two-Queen Colony Manipulation

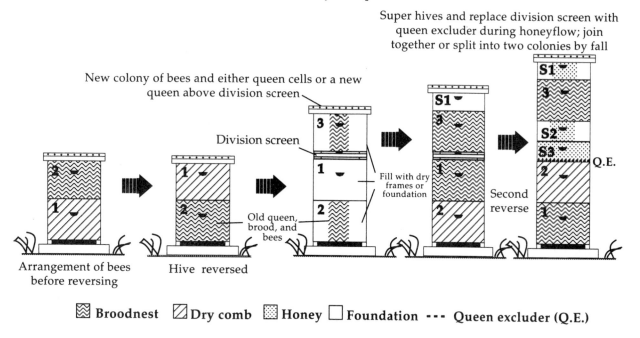

New colony of bees and either queen cells or a new queen above division screen

Division screen

Old queen, brood, and bees

Fill with dry frames or foundation

Super hives and replace division screen with queen excluder during honeyflow; join together or split into two colonies by fall

Arrangement of bees before reversing

Hive reversed

Second reverse

Q.E.

⊠ **Broodnest**   ▨ **Dry comb**   ▧ **Honey**   ☐ **Foundation**   - - - **Queen excluder (Q.E.)**

TWO-QUEEN SYSTEM

There is a general consensus that two-queen colonies produce more honey than two single colonies. In marginal areas with limited yields, management of two-queen colonies should be considered. Ideally, two-queen systems should be set up and ready to go approximately eight weeks prior to the first major honeyflow. Two-queen systems can be implemented by splitting an existing colony, using two nucleus colonies, or by using package bees.

There are two basic configurations in setting up a two-queen colony. In the *vertical two-queen colony*, the colonies are separated by a queen excluder, and, as the name implies, they are positioned vertically one above the other, with the honey supers placed above the first and second colony. This configuration poses some difficulties in manipulating the various supers.

The other configuration, the *horizontal two-queen system*, has the hive bodies containing the queens side by side, with a queen excluder covering the joined halves. The honey supers are stacked directly above the excluder. In both cases excluders without rims work best.

There are different methods of setting up and managing hives with two queens (see diagram). Here is one method:

1. Split a strong colony and requeen the upper queenless portion by the division board method, using previously described requeening methods (see illustration on two-queen colony manipulation).

2. Replace the division board with a queen excluder once the introduced queen has been accepted. Just prior to the honeyflow, or when more room is required, add supers to both units. You can also place an excluder above the brood chamber in the lower colony.

3. At the end of a given major honey flow, remove all the shallows filled with honey and, if future flows are anticipated, add additional honey supers. Before the last flow of the season, the two-queen colony can be reconfigured into a single colony or split into two separate ones, each with its own queen. If your plans are to unite the two colonies, the queens may be given the opportunity to battle it out. Or you can place the extra queen in a nuc colony while you determine what you wish to do with her.

Advantages

● Strong hive for wintering, if queen excluder is removed and colony enters the winter with one resident queen.

● Better yields, since one large colony will produce

more than two separate colonies, each having half the strength (and fewer than half the foragers) of the large one.

- More bees, which can be used later for making an increase or for strengthening weak hives.
- If one queen fails, you have a backup queen.
- Method can be combined with swarm control and requeening techniques.

Disadvantages

- Even if all manipulations are completed successfully, they will not be effective if local honeyflows and colony manipulations are not synchronized.
- Time consuming.
- Difficult to operate.
- Hard to make work successfully.

REQUEENING DEFENSIVE COLONIES

Occasionally, a colony exhibits extreme defensive behavior, such as flying at and stinging with little provocation. These colonies, at peak population, are very difficult to work, and if they release excessive amounts of alarm pheromone when worked, the defensive stance of the other colonies in the yard could be elevated as well. If this happens, you may have to quickly abandon beekeeping activities and wait for a better day.

If this colony is close to homes, schools, or high-use areas (such as parks or sports fields), it could become not only a public nuisance but also a liability. Any stinging incident results in unhappy neighbors and could lead to zoning regulations restricting beehives in residential areas (or worse).

If you experience defensive bees, first check for signs of skunk or other pest activity to see if this the cause of the heightened level of defensive behavior (see "Animal Pests" in Chapter 14). If you live in regions of the United States where the Africanized honey bee (AHB) has become established (Texas, New Mexico, Arizona, southern Utah, southern California, Oklahoma, Arkansas, and parts of the Gulf states, including parts of Florida as of this writing), extreme defensive behavior may indicate your hives are occupied by AHBs; these bees exhibit a highly defensive posture with a minimum of provocation and could be very difficult to requeen. Check the USDA website for the map of the current locations of this species. The Tucson USDA-ARS bee lab and the Department of Agriculture in Florida have the necessary equipment to differentiate Africanized bees from European races; see the References for addresses. However, there are many signs that can help you tell the two apart.

Africanized bees:

- Are very loud on the comb when you are working bees.
- Show extraordinary defensive behavior, many bees scurrying on the combs, and many bees in flight, surrounding and landing on you. They sting without provocation, such as when you are entering the vicinity of the apiary.
- Pursue you far from the apiary; see Appendix E.
- Sting animals, sometimes fatally (usually animals that are tethered or in pens).

Even European bees may have genetic traits for a heightened defensive response. In each case, the remedy requires that such colonies be requeened immediately with purchased queens of known gentle stock; or if the situation is extreme, the colonies may have to be destroyed.

Recently, it was found that AHBs do **not** get many of the viruses that plague EHBs. In the Southwest, where these bees are plentiful, beekeepers have learned how to manipulate them with great success. They do require more work but may be useful in some areas.

After you successfully requeen, the change in the behavior of the colony will be apparent as soon as the new queen's offspring populate the hive. Thus it will take almost two months for all the existing workers and larvae to be replaced; this may be unacceptable if a more immediate solution is needed. You may have to remove the colony to an isolated location until such time as the new workers replace all the defensive ones.

Regardless of which method you use for requeening, the hive first needs to be dequeened. Finding the queen of a volatile hive is not easy, especially if the colony is populous.

Here is one method of requeening, called the *non-shook swarm method*:

1. During a favorable day, such as a sunny, windless one with a honeyflow in progress and when most of the foragers (those most likely to demonstrate aggressive behavior) are in the field, move the colony (European or Africanized) to a new location.

2. If you suspect the bees are AHBs, call your county or state bee inspector, or beekeepers in your area who may help you determine whether the bees are Africanized bees or overly aggressive European bees. Once you have confirmation that they are AHBs, your bee inspector will provide you with information as how to take care of the situation. Africanized bees can be extremely volatile, and it may be best that you seek professional assistance.

3. If, on the other hand, the bees are confirmed to be European but are highly aggressive, move the hive(s) to a new location. Place a single deep hive body with all its components at the previous location, and add a frame that was removed from a gentle colony containing very young larvae, less than one day old. Another approach is to place a ripe queen cell on a frame, or a caged queen from gentle stock, filling the remaining space primarily with drawn comb. This colony can be fed sugar water to facilitate acceptance of the queen cell / queen. The foragers will exit the relocated hive and return to the original location of their colony. At around twenty-one days of age, bees become foragers. The life span of a forager is approximately three weeks; by the end of the fourth week the foragers that belonged to the aggressive hive will have perished. **Note to beekeeper:** some of the bees may still be alive. Two days after the original aggressive colony was moved, most of its foragers will have returned to their previous site. This relocated colony can now be opened, dequeened, and requeened, using any of the methods previously described. But as an added precaution, after relocating both the aggressive colony and its replacement, feed the bees sugar water to further placate them and make the requeening event less worrisome.

4. Before the queen in the aggressive hive was replaced, she was still laying eggs that will transition into larvae, then into pupae, and then into adult bees. As a consequence, it will take approximately nine weeks (three weeks from egg to adult and six weeks as adults) for the aggressive hive to be completely cleansed of aggressive bees.

The non–shook swarm method, while effective, is very time consuming, so if possible, try to requeen the problem colony at its original location if the queen can be found quickly. Or you can separate the brood chambers with one or more queen excluders. Return in four days: the queen will be in the super with eggs. If the hive is too populous, split it (separating the hive bodies and supplying each its own bottoms and tops), wait four days, then dequeen the split containing the queen. Requeen one or both splits or unite them after one has been successfully requeened.

QUEEN REARING

Natural Queen Rearing

The simplest way to rear queens is to let the bees make their own queens. Remove the resident queen and place her in a nuc box with accompanying frames containing brood honey and bees. This method has risks, as the bees may fail to rear a new queen, or the daughter queen may be inferior to her mother. A superb queen can probably be found in any apiary with five or more colonies. Obviously, if queens could be raised from the larvae of such a colony and later be introduced successfully to other colonies, the entire apiary could be upgraded, but this could also lead to inbreeding, unless you have 40 or more colonies (see "Queens" in the References).

Good queens are reared by bees in a strong colony when there is an abundance of food (honey or sugar syrup and bee bread) available to the nurse bees. Queen-rearing operations can coincide with swarm control manipulations (see "Prevention and Control of Swarming" in Chapter 11). During the swarm seasons, queen cells are robust in appearance and are likely indicative of well-developed queens within. The swarm seasons provide an additional source of queen cells.

Once the old queen is removed, provide the queenless colony with a frame of open brood with eggs and young larvae taken from a colony with desirable qualities. You can also take extra queen cells from colonies preparing to swarm; but remember, swarming is a genetic trait.

For more control, split a populous colony into two or more sections or into a series of nuc boxes. Let the bees rear their own queens. With some luck, one or both splits and the nuc boxes will rear a satisfactory queen.

Advantages
- Can be easy.
- Inexpensive.
- Will usually succeed in obtaining queens.
- Few manipulations needed.

- Can coincide with mite control, as it breaks the brood cycle.

Disadvantages

- Queen could be inferior or may mate with inferior drones.
- Cycle of the dequeened colony will be disrupted; no new brood will emerge for up to 43 days.
- Queen cells could be crushed against the bottom board and therefore no longer viable. Because swarm cells are frequently suspended near the lower section of a comb, a shim needs to be placed between the bottom board and the first hive body before placing the comb into a split or into nuc boxes.

RAISING QUEENS

Conditions Required to Rear Queens

Some beekeepers prefer to raise their own queens rather than purchase them from a commercial breeder. While educational and exciting, rearing queens can be tricky, time consuming, and often unsuccessful —but with practice it can be accomplished.

Before you begin to rear queens on a larger scale, read books on the subject, talk to beekeepers, and take a class (see "Queens" in the References); hands-on practice is essential for success. Then you should choose whether you want to (1) graft larvae into artificial queen cups (Doolittle method) or (2) cut off sections of comb so the margins of the comb contain very young larvae; the bees will then reconfigure these cells into queen cells (Miller Method). (Miller method).

Next you should ask yourself three questions: Why? What? and When? The answer to why was already covered at the beginning of this section: to replace inferior queens or to increase your colony numbers.

The next question: What is a good queen? Pick your breeder queens carefully, from stock that you can trust and that has the characteristics you seek. If you decide to instrumentally inseminate the queens, semen should be obtained from drones from your most productive colonies. When returning from DCAs, drones that are sexually mature often drift into other colonies; therefore, a drone from a colony you selected may not actually be a drone produced by the queen of that colony. However, given that you

will collect many drones from that colony, the blend of semen should mostly provide you with the genetics of that colony. If the virgin queens mate at the DCAs, they will encounter drones from many colonies, a situation that is essentially out of the control of a beekeeper with fewer than 25 hives. For commercial queen rearers, one way of controlling mating and the exchange of genetic material is to select an isolated area such as an island and saturate that DCA with preferred drones from colonies with superior queens. In his book *Practical Queen Production in the North*, Carl Jurica, PhD, of Johnstown, New York, describes another technique to guarantee that virgin queens will mate with selected drones. This is accomplished by having a superior queen's daughter remain a virgin by preventing her from mating. After three weeks, she will produce only drones. To carry the superior queen's daughter through the winter, you will be required throughout the summer and fall to continuously supply her colony with worker bees. As a result of this procedure, during the early spring months, the only drones saturating the DCA will likely be from her colony. This should ensure your virgins will mate with these drones. Then make sure your queen larvae are fed copious amounts of food, because the number of ovaries is directly proportional to the size of the abdomen—bigger queens are generally better queens.

Finally, the last question you need to ask: When do you want to rear them? Prepare and organize your bees and yourself so you can have abundant workers, food, and drones at the right time.

Conditions needed for successful queen rearing include:

- Abundance of workers.
- At least 20 pounds (7.5 kg) of honey.
- Plenty of bee bread or pollen substitute.
- Intense honeyflow, or feed syrup (1 water :2 sugar) three days before you start rearing queens.
- Sufficient numbers of sexually mature drones to collect semen from or to saturate the DCAs.
- Superior queen mothers.
- Large populations of young "nurse" bees.
- Correct-age larvae from which queens can be reared.
- DCAs with a genetically diverse drone population (extremely difficult to control).

Preselection Data Sheet for Breeder Queens

Date_____	Queen no._____	Queen origin _____	Date mated _____

Score						
	5 *Exceptional*	4 *Excellent*	3 *Average*	2 *Fair*	1 *Poor*	0 *Unacceptable*
Characteristics						
Brood viability						
Temperament						
Spring buildup						
Overwintering						
Pollen hoarding						
Cleaning/hygiene						
Honey production						
Disease	−1	−2	−3	−4	−5	Eliminated
Type						
Mites	−1	−2	−3	−4	−5	Eliminated
Type						
Color	Golden					Black
Queen[a]						
Workers						
TOTAL SCORE						

Source: Adapted from Susan Cobey, Apiarist, University of California, Davis.
[a] *Note:* Lighter queens are easier to see, but that may not be important to you.

While there are numerous methods of queen rearing (see "Queens" in the References), it falls outside the scope of this book to go into great detail on this subject; however, some methods are given below, from which you can start experimenting. You will need breeder queens. These queens should be from colonies with desirable characteristics such as surplus honey production and disease resistant, calm bees that overwinter well and swarm infrequently. Equipment needed to carry out queen rearing usually involves a starter colony that will receive the newly grafted larvae and a finisher colony where the queen cells will be transferred to, where they will complete their development. In place of a starter and finisher colony, some beekeepers rear queens using a swarm box. You will also need mating nucs and mating yards. If large-scale queen production is the goal, then colonies with desirable drones will also be needed to saturate the DCAs.

Breeder Queens

First, you must select your best queens from which to raise new ones; these are your *breeder* queens. Carefully choose these breeders by testing 3 to 20 colonies (depending on how many colonies you have) that possess the characteristics you wish to perpetuate in your bees.

To choose which queens are best, give some sort of test to potential breeder queens. The weather conditions, nectar/pollen flow, and temperature should be similar. Grade the bees, with a letter or number grade for each option, and select at least the top five colonies with the highest grades (see the sample preselection data sheet). Here are some characteristics on which to score your colonies; there may be others, so add them to your list:

- **Brood production**. A compact, solid brood pattern means good brood viability; a spotty brood pattern indicates larvae were removed because of poor viability, poor mating, disease, or an old queen, or a dearth may be in progress.
- **Disease and pest tolerance**. Hygienic behavior means bees uncap and remove dead or infested larvae in less than 24 hours. Grooming behavior means some bees remove varroa mites from each other; also important are tracheal mite resistance and chalkbrood removal.
- **Overall population**. How populous is the colony compared with others?
- **Propolis**. Does the colony collect or use excessive or low amounts?
- **Temperament**. Test without smoke; the size of the colony is not important. Grade from 1 (bees do not react) to 5 (bees sting, smoke needed).
- **Composure on comb**. How quiet or runny are the bees when you examine frames?
- **Bee bread arrangement and hoarding**. Is there a clearly defined semicircle of bee bread along the upper boundary of the brood, or is the bee bread scattered? Also, gauge the amount of bee bread stored in frames without brood, especially in the fall.
- **Honey production**. Also note that it is the natural inclination of bees to store honey above the brood nest.
- **Beeswax**. How fast are the bees drawing out foundation? Also note the color or consistency of the cappings—white or wet?
- **Swarming tendency**. Note whether the colony swarmed and, if so, how many swarms it cast.
- **Robbing tendency**. Score the number of times robbers were seen at other colonies by dusting robbers with flour and following them back to their home colony. Conversely, is the colony constantly being robbed or non-defensive?
- **Wintering ability and spring buildup**. How many frames are covered with adult bees, and of these frames, how many contain eggs, larvae, and pupae? Also, note the weekly increase in these factors during the spring months.

- **Flight time**. How late and at what temperature do the bees begin and end foraging?
- **Body color**. Are the bees light, or dark, or a mixture of colors? The evenness of the color indicates purity of mating, which is most likely achieved only when queens have been artificially inseminated.

Remember, records should be kept on **each** queen you choose to be a breeder. It would be helpful if you had similar records for your drone mothers too, as the worker offspring of your new queens will carry the genetic traits of both the breeder queen and the drones she mated with. For more information, check "Queens" in the References; you can also check the internet or your local bee club for more information.

GRAFTING METHODS
Equipment for Queen Rearing

You will need to purchase or fashion your own grafting spoon, used to lift young female larvae out of their cells. This tool is used to transfer these larvae and place them into queen cups (you can purchase commercial queen cups or mold them from your own beeswax). Note that beeswax can hold pesticides for many years, so get your wax tested (check sites online). Once the larvae are transferred into these cups, the cups are referred to as *queen cells*. As the larvae develop in these cells, the bees continue to build upon them by adding additional wax. You will also need grafting bar frames to attach the queen cups; some of these frames have bars that swivel, making it easier to transfer the larvae into the cups.

Grafting means to physically remove one- or two-day-old larvae from their cells (the larva's age is critical). Female larvae approximately twelve hours of age will develop into superior queens. These larvae are very small and, of course, difficult to see. To aid in finding and transferring such young larvae from their cells into queen cups, many queen grafters wear a magnifying shield. This procedure is tricky to accomplish and takes much practice. Nimble fingers and keen vision are needed when grafting these small larvae. Take a class on queen rearing, which will let you practice and perfect your skills; check the journals and internet for current classes.

Doolittle Frame

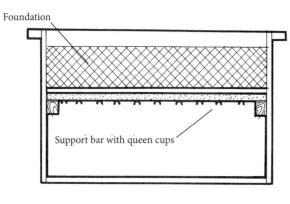

Foundation

Support bar with queen cups

Grafting Tool

This diagram is from Glenn Apiaries: http://www.glenn -apiaries.com/glenn_apiaries.html.

Doolittle Method

To graft worker larvae into artificial queen cups, you might employ the Doolittle method, named after G.M. Doolittle, a beekeeper who wrote extensively on the subject around the turn of the nineteenth century; his book was reprinted in 2008. He found that priming the queen cups with royal jelly and feeding the starter colonies were imperative to rearing good, healthy queens. Here are the major steps:

1. Fit an empty frame with two or three bars to hold the queen cups. Some frames have a 3-inch (7.6 cm) strip of foundation above the bar (see the illustration of the Doolittle frame).
2. With melted beeswax, attach the wooden bases of queen cups (or plastic cups) to the underside of the wooden bar. If you are using plastic equipment, drill holes in the bars in which the cups will fit. You can insert this frame with cups into a colony for the bees to clean and polish.
3. Two days before you transfer larvae into these cups, dequeen a strong, two-story hive and feed it with syrup.
4. The next day, shake the bees off every brood frame in the dequeened hive and remove all queen cells; these cells will provide you with royal jelly. Collect the jelly with a small, sterilized spoon, eye dropper, or siphon; you can store the royal jelly in a sterilized jar in the freezer.
5. Once you have a sufficient amount of royal jelly, remove a frame containing female larvae less than three days old from one of your breeder colonies; larvae 24 hours old or younger are ideal. You can also fabricate an enclosure from several queen excluders that will hold a deep brood frame that has been prepared for egg laying. Place the frame into the enclosure with the queen and return it to the colony. Bees of all ages will pass through the enclosure to join the queen. She will soon begin laying eggs; four days later, very young larvae will be in many cells. Remove the frame, brush the adult bees and the queen back into the colony, and take the frame to the grafting station. Larvae of the correct age are almost the size of a bee's egg. If you can easily see the larvae you intend to graft, they are too old.
6. Prime the queen cups with the stored royal jelly; place it in the queen cups, covering the bottom. Dilute the jelly, if it is too thick, with about 25 percent boiled, distilled water. Do not be tempted to lick the spoon or grafting tool.
7. With a grafting tool (see illustration of this tool) or toothpick (carve one end flat and curve it slightly upward), transfer the larvae from worker cells into the primed queen cups. Place the larvae **on top** of the royal jelly, being careful not to drown them in it, and in the same position they were found before the transfer. Because these larvae are about the size of an egg, it may help you to use a lighted magnifying glass to see them. Grafting should be done in a hot, humid room to keep the larvae from desiccating or being chilled. Cover the frames with a damp towel.
8. Insert the finished frame, now with the queen cups containing young larvae, into a dequeened starter colony or into a swarm box.

Starter Colonies

As the name implies, a starter colony is where the newly grafted queen cells are inserted, starting them on their road to becoming fully developed queens. A day before you graft, remove the queen from a robust colony, then configure the hive so it is full of young bees (essentially nurse bees) and frames of capped brood and copious amounts food (honey and bee bread). Rearrange the frames in this dequeened colony so that its lower chamber has mostly sealed brood. The upper chamber should have, in this order (assuming it is a ten-frame hive), starting at either of the longer pair of inner lateral walls of the hive: a frame of honey, two frames of older larvae, a frame of young larvae, space for a frame with queen cells, a frame of bee bread, one of older larvae, and one of honey. Feed pollen patties and syrup to make up for any shortfalls. Give up to 30 or 50 queen cells to the starter colony, keeping them in this colony for 24 to 36 hours. Because this colony is queenless, it will not have the capacity (a sufficient number of bees) to complete the development of these queen cells. The purpose of the starter colony is to jump start the development of the queen larvae by having many nurse bees available whose glands are primed with royal jelly. If you are planning on reusing the starter colony, join it to a queenright colony until you need it, or give it three or four queen cells to make their own queen. Not all the queen larvae will survive, but once the surviving queen larvae have copious amounts of royal jelly in their cells, remove the frame of developing queen cells from the starter colony and divide the cells, and then place them into several cell-builder/finisher colonies.

Although there are several different ways to set up a starter colony, we will present only one of the variations. Please note that in most other texts, discussions related to the kinds of frames that must be included in a starter colony state that two frames of pollen are needed. However, for greater accuracy, they really mean one or more frames of bee bread, which is the altered and stored form of pollen.

The starter colony should be well populated with honey bees residing in at least two deep Langstroth hive bodies, initially each containing ten frames. To reconfigure the colony into a starter colony, begin the process at least a day or two before it is to receive queen cells. Follow this procedure: First, locate the queen and place the frame she is on in a nuc box. Rearrange the frames in the bottom box so the box contains frames of sealed brood and frames of honey. Now transfer the frame with the queen from the nuc box into hive body one and immediately place a queen excluder above hive body one. Usually, the frames in either hive body one or two, which contain open brood, are covered with nurse bees. By shak-

ing these bees off their frames in front of the entrance to the reconfigured colony, you are providing a large contingent of nurse bees that will attend to and provide copious amounts of royal jelly to nourish the developing queens. After shaking the nurse bees off these frames of open brood, do not put these frames back into the starter colony; give these frames of open brood to other colonies. This puts the nurse bees in the position to direct their glandular feeding to the developing queen larvae, ensuring that these larvae are well nourished.

Above the queen excluder, add the now-empty second deep hive body and place a division board feeder against one of the longer pair of inner lateral walls of the hive. This feeder will likely take up the space of two frames. Add two frames of sealed honey and then a frame of bee bread. Leave a space that will receive the frame containing the queen cells. Follow that space with another frame of bee bread and then two additional frames of honey. After completing the arrangement in the second hive body, should there be a space for an additional frame, add a frame of dry comb. Your starter colony is now ready to receive the queen cells. A well-provisioned starter colony can attend to 100 to 120 queen cells, but of course, it is unlikely that all of the queen cells will reach maturity.

Another approach is to start your grafted cells by employing a swarm box. A swarm box can be described as a five-frame nucleus box modified to rear queen cells. The modifications include replacing the bottom board with window screen; this will keep the bees that you will be placing in this box from overheating. Adding legs onto the box so it is two or three feet off the ground will allow air to circulate through the screen. If it is very warm, locate the swarm box in a shaded area. The swarm box has no entrance or exit, and the bees are confined to the box the entire time they are attending to the queen cells—approximately two days. Add a large sponge soaked in water to this box.

Place two combs with mostly sealed honey against one of the longer inner lateral walls of the swarm box and one comb filled with bee bread in the middle of the box. Several hours before placing the queen cells in this swarm box, you must add five to six pounds of bees that are taken from a populous colony. A pound of bees equals approximately 3,500. A brood frame fully covered with adult bees on both sides will equal approximately 3,000 bees. On these frames of open brood are predominately nurse bees, bees whose hypopharyngeal and mandibular glands synthesize royal jelly, the food that plays the primary role in creating queens. Shake the bees off these frames into the swarm box. Two or three hours after the swarm box is filled with nurse bees, the grafted queen cells are added; they are removed two days later and placed into the finisher colony.

Please note that if they are to be in continuous use, starter colony and finisher colony configurations need to be constantly refreshed to reflect their original composition. Without being refreshed, they will not meet the requirements for rearing queens. Frames of open brood, frames of sealed brood, frames of bee bread and honey will not remain constant. After the queen cells are removed, the bees in the swarm box need to be returned to their original colony. If the swarm box is used again, new bees have to be introduced.

Cell-Builder/Finisher Colonies

Just as there are several versions of requirements for a starter colony, there are several versions of requirements for a finisher colony as well. It too should be very populous and reside in at least two deep Langstroth hive bodies, the lower one containing ten frames. A day before the queen cells in the starter colony are to be transferred to the finisher colony, the queen is confined to the lower hive body and restricted there by placing a queen excluder over this box. The upper box is rearranged to contain a division board feeder located against one of the longer pair of inner lateral walls of the box, followed by one frame of bee bread. Now add three frames of unsealed larvae followed by a space to be occupied by the queen cells that will be transferred from the starter colony. Then add three additional frames of open brood followed by an additional frame of bee bread. The next day, remove the frame holding the queen cells from the starter colony and add them to the second deep of the finisher colony.

Check the queen cells a few days after they have been installed in the cell-builder/finisher colony to ascertain whether the cells are receiving copious amounts of royal jelly. If not, remove those queen cells that appear to have sparse amount of royal jelly; the queen larvae should be plump and glistening white. Be sure before you begin with the cell builder that no miticides are used while the colony is rearing queen cells.

After five days from when you first installed the queen cells, the cells should be capped. Queen pupae are **extremely delicate** at this time. **Do not** bump, cool, or overheat them, or your queens will be deformed. Know the age of the queen cells (keep a diary) so that one queen will not emerge too soon and kill all the others. Do not handle the cells until they are 9 or 10 days old (after grafting). Place one ripe queen cell in a mating nuc (or you can hold several cells in an incubator if you don't have enough nucs); each cell can be placed in a small vial to keep the queens isolated.

Mating Nucs

Because the first queen can emerge 11 days after grafting, make up the mating nucs a few days before they are needed. Bees added to (or making up) the nucs should be treated beforehand so they are disease free and are taken from colonies with low mite counts; finish any treatment about 30 days before adding a new queen. Do **not** treat nucs with miticide strips, as the chemicals can deform the queens. (Handle queens as you would honey: medicate or treat 30 days before supering for honey.) If mites or small hive beetles are a problem, use alternative control measure in the mating nuc. Move the nucs to the mating yard at least 1 to 2 miles away (1.6–3.2 km) from your home apiary, if they have the following:

- No queen.
- Two or more frames of bees from a colony with a low mite count.
- Two or more frames of disease-free honey/pollen.
- An entrance that can be closed or restricted.
- Sugar syrup and pollen patty, if food is not incoming.

Transfer one or two queen cells into each mating nuc by cutting the cells from the frame bars. With a heated knife, cut the cell bases free from the wooden bars. Place cells into an insulated box (like a Styrofoam cooler) with a warm water bottle inside to transport them safely to the mating yard. After you select a nuc, wedge the cells between the top bars of two frames, making sure they get covered by bees immediately. Check food stores of each nuc, and feed if necessary. Feed your nucs with fumagillin to control nosema if it is a problem or if the weather turns wet and cold. While you should not apply miticide strips to your nucs, you can use oil patties if tracheal mites are a problem (do not use in areas of small hive beetles). Reduce the size of the entrance opening with grass or an entrance reducer to prevent robbing, and make sure the boxes will not get chilled or overheated.

Mating Yard and Drone Mother Colonies

The nucs with the queen cells should be in or near your drone mother colonies. Again, these colonies must have low varroa mite counts before the queen cells are placed in the nucs.

With the dwindling of feral bees, a result of the parasitic mites, it is important to have enough drones available to mate with your virgin queens. Because half the genetic material of your workers comes from their fathers, select drone mother colonies using the same criteria you use for your queen breeders. Choose only those colonies that show good characteristics (high-scoring colonies), but with a genetic stock different from that of your breeder queens, to avoid inbreeding.

Your goal is to produce large, healthy drones, free of mites and diseases, that were adequately fed when they were larvae. Timing is everything, so plan on having mature drones in your mating yard when the virgin queens are ready to mate; drones mature 10 to 15 days after they emerge.

Encourage drone production by placing one to three frames of drone foundation or comb into your selected drone mother colonies so the queen has ample room to lay drone eggs. You can insert frames half filled with regular worker foundation, or shallow frames in deep hive bodies, and let the bees draw out the rest of the frame. Most likely, they will draw out drone cells. Ensure there is enough food coming in, and if not, supply extra pollen and syrup. Large amounts of pollen will stimulate colonies to rear drones.

NONGRAFTING METHODS FOR REARING QUEENS

Miller Method

The Miller method of queen rearing, named after C.C. Miller, may be the easiest for the beginner and re-

quires no special equipment. Prepare an empty deep frame by fitting it with a sheet of non-wired foundation. Once this frame is installed in the frame's top bar, cut triangles on the lower half of the foundation (see the illustration of the modified Miller frame). An alternative method is to cut strips of foundation, roughly 4 inches wide and 6 inches long (10.2 × 15.2 cm) and attach them to an empty brood frame. Cut the unattached lower half of each strip of foundation to form a triangle with its apex pointing downward. Now follow this procedure:

1. Move two frames of sealed brood and queen from a breeder colony into a nuc or specially prepared breeder hive.
2. Insert the prepared frame between the two frames of the sealed brood.
3. Make sure the queen is on one of the frames and that there are a lot of young workers.
4. On either side of the brood frames, fill the hive with frames of honey and pollen (there should be no empty cells in these frames; otherwise the queen may lay in them).
5. The queen will be forced to lay eggs in the prepared frame as soon as cells are drawn (feed the colony if necessary).
6. About one week later, remove the prepared frame; trim away edges of the newly drawn pieces of foundation until you encounter cells with small larvae (preferably less than one day old, but never more than two days old).
7. Give this frame to a cell-builder colony. You may have to trim the bottom of the triangles to reach the eggs and young larvae. These larvae should be on the lower edges of the foundation. You can insert another Miller frame into the breeder colony, or return the queen to her original hive. Or use this special hive to start a new colony, requeening the parent hive.
8. Nine days after you insert the Miller frame into the cell builder, remove the sealed queen cells by cutting them from the Miller frame and attaching them to combs in queenless hives or nuc boxes, and place them in the mating yard.

Queens in these nuc hives will emerge and mate. These queens can then be left in the hives from which they mated or used for requeening after they have begun to lay eggs.

Modified Miller Frame

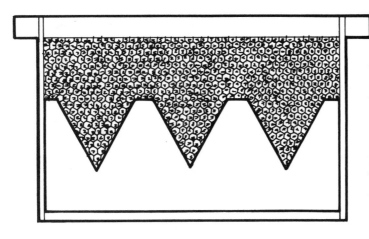

Start with unwired foundation, cut into wedges.

Place frame into your breeder colony so queen can lay eggs in the cut foundation.

After one week in a queenless colony, the frame should look like this:

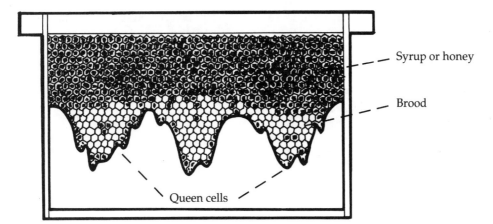

Syrup or honey

Brood

Queen cells

Alternate Miller Method

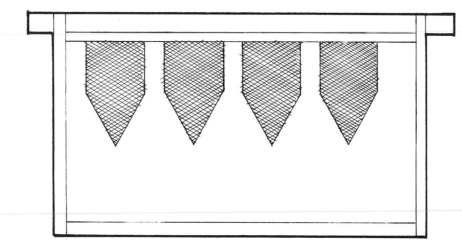

Note: Because much of the pure beeswax foundation today is contaminated with pesticides, you may need to use wax that your bees have drawn out themselves. If you have a lot of queen failures, consider having your wax foundation tested.

Queen-Rearing Kits

Bee supply companies now sell queen-rearing kits that do not require grafting and cutting comb. These come with good instructions, or you can search the internet for instructions; read through the instructions first to determine if the kit will fit into your beekeeping operation. With it you can rear many queen cells, which you must treat like grafted queen cells, giving them to cell builders, finishers, and nucs. Some of these kits are the Nicot and the EZI-Queen systems; check bee suppliers and websites.

In general, the kits are composed of a plastic container the size of a comb honey section box. On one face of the box is a removable queen excluder, and on the opposite side are evenly spaced holes into which a series of removable plastic queen cups are placed. The box, once fitted with the queen cups, is inserted into a frame and then placed into a colony for a few days so the bees can clean and polish the plastic cells.

Make sure the colony is fed heavily for a week prior to and during this procedure. Then the breeder queen is introduced into the box, where she will deposit eggs into the queen cups. After the eggs have been deposited, usually within a day or two, the queen is removed from the box and released back into her colony. The next step is to take out all the cups containing eggs and attach them to a cell bar the same way you would for hand-grafted cups; then place the bar into a cell-builder colony.

The advantage of this system is that you do not have to hunt for frames containing the right-age larvae. All you need to do is check the plastic cups to see if the confined queen has laid eggs in them. Be aware that you have the potential to rear over 100 queens at once, which could tax your facilities if you are not prepared. The downside of this system is its cost and the many little pieces of equipment you need, which are easily lost or misplaced. Ask more experienced beekeepers about their successes or failures with the system before you invest in one.

Check for availability or for new products from bee suppliers or websites, and follow the instructions that come with each kit.

 **Notes**

CHAPTER 11

Special Management Problems

This chapter covers some of the challenges in management that you may confront while examining your colonies, especially during the spring, summer, and the very early part of the fall. Bees are living creatures, and there is no single solution that can resolve all the problems you may face. However, the information contained in this chapter will provide the most accepted practices that can help to remediate some of the more common problems you will likely encounter when working bees.

WEAK COLONIES

A weak colony is one that is not thriving when compared to other colonies in an apiary. This assessment is usually made when you compare the differences between populations of each colony within your beeyard. The goal of beekeepers should be to have all their production colonies (colonies being managed for honey production) at full strength in order to take advantage of every major honeyflow. Another reason for having colonies at full strength is to provide sufficient bees for the successful pollination of various crops. Colonies can become weak (a reduction of their population) for many reasons, such as the presence of a poorly mated queen, an unmated queen, a failing queen, or the absence of a queen and the presence of laying workers. Viruses and parasites (particularly the varroa mite) can also deplete bee populations. In areas where honey bees come in contact with insecticides, herbicides, and fungicides, their colonies may be impacted dramatically, either weakened or destroyed. In addition, skunks can slowly reduce a colony's population by devouring bees during nightly visits to a beeyard. Environmental factors also can contribute to low populations in bee colonies; rain or low temperatures, for example, can prevent the bees from foraging. Also, colonies may be located in areas with inadequate flowering plants needed to sustain and promote the colony's growth. It is imperative that you first determine why a colony is in a weakened condition before you decide how to ameliorate the situation.

Before the arrival of varroa mites and the introduction of new pathogens, it was not uncommon to combine two weak colonies, as long as the combined units equaled approximately 15,000 bees. Weak colonies require sufficient stores to survive the winter months. During the spring and summer, a swarm can be united to a weak colony, or accelerate the development of a newly installed package. Combining a weak colony with a strong one near the end of August (early fall) is also done. Check the various techniques for uniting colonies below.

Obviously, it is not wise to unite a colony with high varroa mite numbers to another colony, thereby increasing the total mite numbers of the now united colonies. If, after taking an assay of the colony's mite load and discovering that the mite level is exceedingly high (so that the colony will unlikely survive the winter), you have several options. If the assay was taken in early August, there may be time to treat the colony in order to reduce the mite load. After treatment, a follow-up assay should be taken; if the second assay shows a tolerable mite level has been achieved, the colony will likely be able to produce sufficient bees to survive the winter months, assuming the colony has sufficient honey stores and/or is supplemented with food.

Also, if American foulbrood disease (AFB) is suspected, or other diseases and pathogens are detected, these problems must first be addressed. In the case of AFB, it is absolutely out of the question to consider uniting a sick colony to another (see Chapter 13 on how to recognize and deal with colonies with AFB). If American foulbrood is suspected, it is always wise to consult with your state or county bee inspector. The inspector will walk you through what needs to be done with a colony with AFB and what you can do to possibly salvage the combs and hive boxes.

Remember the cardinal rule when inspecting your hives in the late summer or early fall: **take your potential winter losses in the fall**.

Uniting Weak Colonies

Weak colonies (with small populations) can also include nucs, splits, and packages that failed to develop. Such colonies can be strengthened by adding adult bees and/or sealed brood to their units. In some cases, the queens in these units may also need to be replaced. For instance, once you have given these weak units additional bees/brood, and the queen is still unable to expand her broodnest, it is likely she should be replaced. In fact if this is the case, she may be the reason the colony failed to grow in the first place. Again, before removing adult bees from other colonies, the mite levels of both the weak and donor colonies should be checked; it will not help the weak colonies to prosper if you are adding mites along with adult bees. When adding frames of sealed brood (minus the bees covering them) be sure the colonies receiving them have a sufficient number of adult bees to maintain sustainable brood temperatures; this will enable the capped brood to complete their development.

These manipulations must begin in early August in order for these colonies to recover to full strength; that means well in advance of periods when foraging will be impossible, such as the winter months in the North. Naturally, along with population buildup, such colonies must also have adequate food stores.

If, on the other hand, some of these weak colonies fail to develop into strong units after you have performed the previously noted actions for strengthening them, the reason or reasons for their failing to recover need to be determined. It is possible that the mite loads were not sufficiently reduced, or the comb in such a colony has wax that is heavily impregnated with insecticides, fungicides, herbicides, and mite medications (some of these can last for decades in the wax). Some beekeepers make up a *hospital yard* where all their weak hives are located and where they can be easily monitored. Such a yard should not have any colonies that have, or are suspected of having, American foulbrood, nor hives with high varroa mite counts. The authors do **not** recommend such a "sick" yard.

Attempts to carry weak colonies through the fall and winter months have never been very successful. Although two weak colonies do not usually make one strong one, a weaker, healthy colony can be united to a stronger, healthy one. If you have a preference for one of the queens, eliminate the less desirable one. If both queens are productive, and it is early enough in the season, the second queen can be introduced into a queenless colony, nuc, or split. Or you can proceed with uniting the two colonies, knowing that one of the queens will be eliminated. Queenless colonies can also be united to queenright colonies.

One of the better uses you can make of a weak colony (again if it is free of diseases and has a low mite count) is to unite it to a stronger hive before a honeyflow. By increasing the bee population of the strong colony, there will likely be a greater number of foragers available to exploit a honeyflow.

If it becomes necessary to unite colonies in late summer or early in the fall of the year, be sure the joined colonies have sufficient honey stores; if not, they will require heavy feeding and/or must be provided with frames of sealed honey. Resources will become very limited late in the year.

Whenever two colonies are being united, remember that each colony is capable of distinguishing between its members and those from another colony. Hive odors will be different, and unless some precautions are taken, the bees will fight. During a honeyflow, the ability of the bees to distinguish between one another is markedly reduced; in the absence of a flow, colonies should only be united by using a sheet of newspaper, called the *newspaper method* (see below). Again, continue to check for the presence of mites or other pathogens.

Uniting Colonies Using Newspaper Method

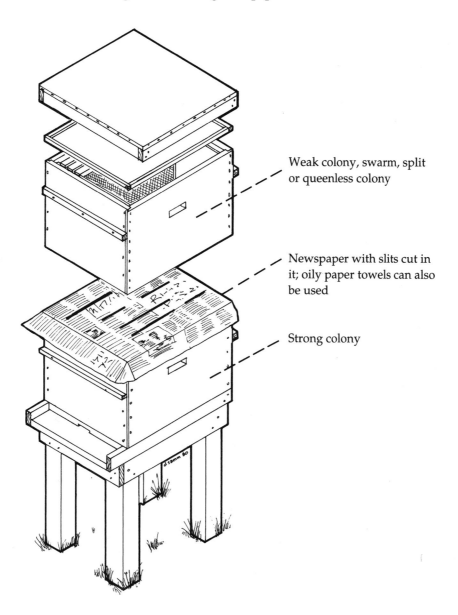

Weak colony, swarm, split or queenless colony

Newspaper with slits cut in it; oily paper towels can also be used

Strong colony

Newspaper Method

The most successful and least time-consuming method of uniting colonies is the newspaper method. This method works because the paper separates the hive odors, and as bees chew away the paper, the hive odors gradually blend. By the time a third or even less of the paper has been chewed up, the hive odors have blended enough for the bees to act as a single unit.

To unite colonies with newspaper, put a single sheet of newspaper over the top bars of the stronger colony; see illustration. Follow these steps:

1. Make sure the colonies are disease free and have manageable mite loads.
2. If the weather is warm, make a few small slits in the paper with your hive tool to improve ventilation.
3. Place the stronger hive on the bottom board; cover the top of the frames with a sheet of newspaper.
4. Set the hive body from the weak hive above the newspaper and then place the inner cover and outer covers on the top. If this manipulation is taking place in the same apiary, it may be necessary to collect the field bees that return to the

former site of the weak colony. Since this colony was weak, the number of its field bees will be limited, but they are worth being collected and then reunited with their original colony at its new site. This can be done by removing the inner and outer cover of the united colony and placing the hive box used to capture the field bees above it, and then once again replacing the inner and outer cover. This procedure may have to be repeated until the field bees die off or eventually acclimate to their new location. The bees will recognize the newspaper as a foreign object and begin to tear it up with their mandibles; the shreds of paper will be carried out from the hive and appear at or near the hive entrance.

5. If the weather is extremely hot (day temperatures over 90°F [32°C]), wait for a cooler day or unite the hives during the late afternoon.

6. If there is a dearth of nectar and pollen and the bees are unusually defensive, decrease the possibility of fighting by feeding syrup to the stronger hive for a few days before uniting. You can also sandwich a weaker colony between hive bodies; place a single sheet of newspaper below and one sheet above the weaker hive; remember to slit both sheets.

MOVING ESTABLISHED COLONIES OVER THREE MILES

Preparing the Colonies

There can be more than one reason (e.g., pollination, new apiary site, moving to another state) why it may become necessary to relocate bee colonies. If you need another location or you are buying more colonies from another beekeeper, follow the procedures for moving bees safely and according to the laws of your state and those laws and regulations for moving them to or from another state. The previous statement also applies if you are selling your colonies.

The state bee inspector should be contacted before moving bees from sites already registered with the state or county bee inspector. It is a general practice to move a hive at least 3 miles (4.8 km) from its old site; if it is moved a distance of less than that, many field bees will return to the old location or drift to other colonies.

The receiving site should be prepared in advance, and if bears are prevalent, a working electric bear fence must be in place (see "The Apiary" in Chapter 4). The best time to move hives is in the early spring, when bee populations are the lowest and the hives are light in weight. If the hive consists of more than two hive bodies (usually the brood chambers), remove any bodies above the second brood chamber (provided they are free of bees) in order to lighten the load. These extra boxes add weight to the colonies to be moved, and are also far more awkward to lift and set down. This is especially true if the colonies are in more than two deep brood chambers, and if these extra boxes contain bees and/or honey.

There are several methods to reduce the weight of the colonies and to remove bees from any extra honey supers; these include:

a) Placing a bee escape (several kinds are available) in the oblong hole of the inner cover situated above the second brood chamber.

b) Using a fume board or escape board to drive bees out of the supers.

c) Blowing the bees out, using a bee blower.

d) Shaking bees off each frame, one frame at a time.

e) Brushing the bees off the frames.

Methods c, d, and e should be done close to the hive's entrance (see "Removing Bees from Honey Supers" in Chapter 9). These manipulations should be carried out a few days in advance of moving the hives.

When reducing the colony to two deep hive bodies (or whatever equipment type you are using), be aware that any bees added to the remaining hive bodies may overcrowd the unit, and the colony may be unable to adequately control its livable temperature ranges. If this is the case, judging by how crowded the colony has become, steps must be taken to provide sufficient ventilation to prevent the colony from overheating. Avoid such manipulations during periods when extremely hot temperatures are forecast, to eliminate the dangers created by crowding bees into two brood chambers and thus "cooking" the bees.

Move a strong colony with the help of a mechanical lift or extra strong (and willing) volunteers. You can also divide it into manageable sections, leaving extra hive bodies at the original site to pick up stragglers. Alternatively, leave a weak colony at the site to collect the stray bees; move it later if necessary.

Near dusk, all the field bees will have returned to

Moving an Established Colony

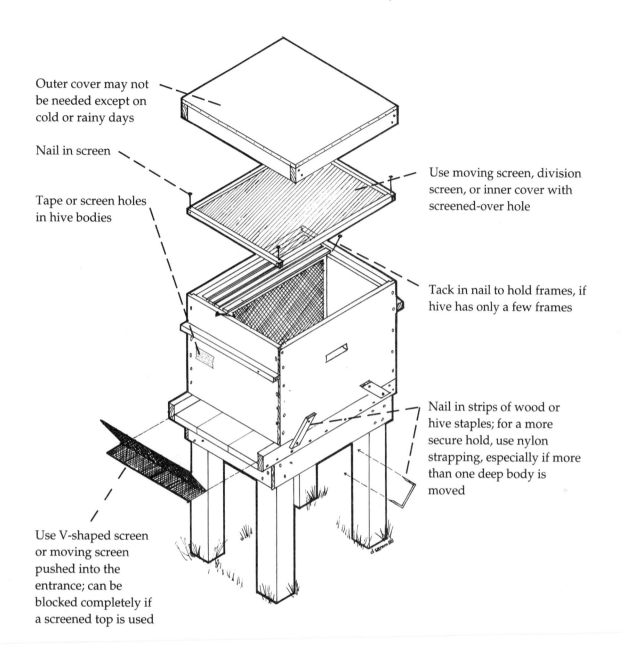

Outer cover may not be needed except on cold or rainy days

Nail in screen

Tape or screen holes in hive bodies

Use moving screen, division screen, or inner cover with screened-over hole

Tack in nail to hold frames, if hive has only a few frames

Nail in strips of wood or hive staples; for a more secure hold, use nylon strapping, especially if more than one deep body is moved

Use V-shaped screen or moving screen pushed into the entrance; can be blocked completely if a screened top is used

the colony. This is the time when the entrance(s) to the colony must be screened to keep the bees inside. The screen will keep them in and also serve to provide them with adequate ventilation. The hive can now be transported without the loss of bees; once it has arrived at its new location, the screen should be removed. **Note**: in hot weather use a screened inner cover instead of the regular one, along with an outer cover to keep robber bees away. During transportation, the bees will in all likelihood experience a somewhat bumpy ride and may be prone to aggres-

sive behavior. Therefore, before removing the screen, a few puffs of smoke from the smoker will help minimize such aggressive behavior.

Here are some steps for moving colonies:

1. A few days before the move, using hive staples, staple the hive bodies together; you can also tie them together with metal or nylon ratchet straps or bands. Use your smoker when necessary. Ratchet straps or bands are the best ways to secure hives because nails and staples will have to

be removed in order to open the hive for inspections. Furthermore, while nailing or stapling the hive bodies together, the resulting noise from the hammering will irritate the bees. And when you remove these fasteners you may find that the woodenware has been split and/or has holes in it, damage that may enlarge in time, rendering the equipment useless.

2. Duct tape or screen all holes and cracks that may be present in any exterior parts of the hive. Any leaks will provide exits for the bees you intend to confine (see illustration on moving an established colony).

3. If the hive is very populous, or if the weather is very hot, add a shallow or medium super with empty frames above the top hive body to collect the overflow of bees; otherwise bees might be hanging outside the hive when you return to move it.

4. On the evening (after dusk) of the move, or the next morning before the bees begin to fly, the hive's entrance (as well as any other openings) must be closed. Smoke the entrance to drive the bees inside, and use a piece of screen the length of the entrance and about 5 inches (13 cm) wide to close off the entrance. Slide a V-shaped piece (or a roll) of metal screen into the entrance so it will spring against the bottom board and hive body, and secure it with duct tape, nails, or staples. Entrances can be closed completely if a screened top replaces the inner and outer covers. You can move the hive that evening or early the next morning. If the weather is cool and the hive is not overpopulated (and the relocation is less than five miles), the entrance can be closed with a rolled-up piece of newspaper stuffed with grass, or a wooden entrance reducer.

5. If the weather is warm, place a screened board (like a division screen) or screened inner cover on top of the hive under the outer cover.

6. If you are moving colonies during hot weather, take off the outer cover and secure the screened top with the bands. Replace the outer cover until you are ready for the move, and then remove it when you load the hive. You can also use a screened bottom board to increase the ventilation (check with bee suppliers for this).

Loading and Unloading Colonies

Once the hive bodies have been securely fastened together, load them onto the truck. **Do not** place hives in the trunk of a car or inside a vehicle—loose bees inside a vehicle will distract the driver.

If the weather is hot, remove the outer cover while in transit so the screened top is exposed. The hives should be packed close together on the truck, with the frames parallel to the road; this will prevent the frames from sliding together if the truck stops suddenly. While you are loading and unloading the hives, keep the engine running, because the vibration of the vehicle will help keep the bees in their hives.

Once all the hives are loaded, secure them to the truck with ropes or other tie-downs. The object is to keep the hives from shifting while you drive them to their new homesite. Hives that shift off their bottom boards will permit bees to escape and will require readjusting the hive bodies and bottom boards before the hives can be unloaded. If you are moving colonies great distances, purchase a special net that can be draped over the hives in order to confine any bees that escape from their boxes.

Make sure the lids are secure as well, as they can easily blow off if they are not tied down; rocks or other weights can shift and fall off, so strap the tops or tape them down with duct tape.

Once you are at the new location, make sure the site is ready to accept the hives and you can drive close to it. An ideal site for an apiary should provide easy vehicular access. Locations for colonies should be prepared in advance (for example: hive stands in place, where the colonies are to be positioned, etc.).

Smoke the entrances just before you unload the hives, and again just before you remove the entrance screens. Unload them all, and *only then* remove the screens as you are leaving. If you moved during the day, fill the hive entrance loosely with grass to slow the bees' exit and to keep them from drifting. If they exit slowly, a few bees should be able to come out and scent at the entrance; this will help any loose bees to orient to their hive. You don't really need to do this if you move at night; just open the entrance after the colonies are all in the new location.

If this is the final site for these hives, replace the top screened board with the inner cover. Even if this is a temporary spot, replace the outer cover, to keep rain out. Inspect the hive after a few days to see if all is well.

Problems associated with moving hives are as follows:

- Hives could shift off their bottom boards, and/ or hive bodies may become displaced from one another, permitting bees to escape from confinement. If moving old, leaky equipment, cover all the hives with a traveling screen or other netting to contain loose bees.
- Bees can suffocate if weather is too hot.
- Queen could be killed, injured, or balled.
- Hive bodies and combs could break.
- If moving in winter or very early spring (when temperatures are below 50°F [10°C]), the winter cluster could lose its cohesiveness. Bees could then recluster on empty combs and starve, or existing brood could become chilled before the bees have a chance to cover the brood. The cluster may not be able to reform; many bees could die.

Moving Short Distances—I
(less than 3 miles or 4.8 km)

Follow these procedures to move an established hive less than three miles:

1. Move the hive to be relocated from its stand to its new location; this may be done at any time but preferably in the evening or early in the morning, as discussed earlier.
2. At its previous location, depending on the population of the hive you are moving, place a single hive body (for a large population) or nucleus box (for a small population) with all their components —bottom board, inner and outer covers, and empty frames of drawn comb, or a mixture of drawn and foundation (or whatever comb is available)—to collect the foragers from the relocated parent hive. These foragers from the original colony have "memorized" their hive's previous location and will return to it.
3. This step requires you to make a judgment call. If the hive you moved was exceedingly strong, fill a hive body or nuc box with drawn comb or a mixture of drawn comb and foundation, but be sure to include a frame of honey and one with

uncapped larvae at the old location, and introduce a queen or queen cell. If all goes well, you will have a new colony; or you may wish to collect the foragers at their old location and then move the collecting box three miles away and then bring it back and unite these bees to their original colony. Why would you do all this work? Well, maybe a honeyflow is about to get under way, and you want as many foragers from parent hive as you can get. Is it worth the time? That is for you to decide.
4. This new colony may be relocated as well, but must be moved at least three miles away (at least initially) and then relocated as you please.

Moving Short Distances—II

It is often recommended that when you are moving established hives very short distances, each hive be moved 1 to 4 feet (0.3–1.2 m) every few days until the hives are at the desired location (few if any bees will return to the original location). But this process is slow and not recommended unless the distance is less than 30 feet (10 m).

ROBBING

You will be amazed how quickly bees can start snooping around to see if they can get some free food. Hence, occasionally, bees will collect nectar and honey from other colonies. This type of bee behavior, referred to as *robbing*, usually occurs when bees are unable to obtain enough food from flowers (dearth periods). Whenever the weather is suitable for flight, foragers will set out in search of food (after all they are hard-wired to carry out this task). If the plants are dormant or are blooming but not yielding nectar (because it is too hot, too windy, or too dry), the foraging bees and scouts will continue to search for food. If they pick up the odor of nectar and/or honey emanating from another colony, they will make every effort to enter the target colony and remove its stores (and will even rob stored bee bread, especially in deserted drawn combs). You can recognize robbing behavior if you observe bees flying near the entrance of a colony, swaying in a zigzag fashion in front of the entrance. This back-and-forth flying pattern is different from returning forager bees that make a more direct beeline into the entrance (see "Robbing Flight" in Chapter 2). You may also witness bees fighting on the landing board, and in time, dead bees will begin to accumulate near the hive's entrance.

Nucleus hives, queen-mating boxes, and other colonies low in population, as well as hives with openings resulting from disrepair, are the most at risk of being robbed. If you observe increased activity at the entrances of these hives, you may have a robbing problem.

The robbing bees return home and communicate (by dancing) the location of the target hive from which the food was taken, and soon additional foragers, recruited to raid the unlucky colony, remove the remaining stores. The weakened colony, in fighting off the thieves, may be severely affected and may even die. Some robbing likely takes place in apiaries whenever there is a dearth. Be alert to any weakened colony when you visit the apiary. If you find one, check that it and the queen are still intact, then reduce the entrance of the hive and move it to a location with fewer strong colonies.

The best way to prevent robbing activity is to maintain colonies of equal strength; keep the entrances of weak colonies reduced in order for their occupants to adequately guard them. Other methods include avoiding adding any scents to sugar syrup, and not to break the cappings off any frames containing honey. If at all possible, avoid working colonies when there is a dearth in progress, for no matter how careful you may be, bees are equipped to quickly locate food and to communicate its location to many more bees (they can move a lot faster than any beekeeper). **Do not attempt to work bees during nectar dearths**.

If examination of some colonies is imperative, cover any exposed boxes with a robbing or manipulation cloth (or an old cloth, towel, or sheet) or place removed frames in a covered, empty hive body. (Check bee suppliers for robbing or manipulation cloths.) Quickly finish your examination to prevent robbing. Moving frames results in rupturing sealed honey in burr and brace combs; bees searching for food will detect the odor of the dripping honey, and robbing will be under way in a flash.

If you must feed during a dearth, do so with in-hive feeders (see Chapter 7), and use sugar syrups instead of honey. In the fall after you take off honey, bees are also prone to robbing, as broken combs can drip honey, attracting robbers. If newly extracted "wet" frames are to be cleaned (see "After Extracting Honey" in Chapter 12) place them on strong colonies at dusk, when all the field bees have returned for the evening. Or put wet supers on all the colonies, to keep them busy cleaning and not robbing each other.

Some beekeepers stack all their wet supers in an open field where bees will remove the remaining honey. However, this could result in the spread of bee diseases as well as attract other insects, such as yellow jackets and wasps seeking sugar (honey). A bear or other creatures may also discover these supers and the combs and will likely tear the equipment apart to get at the sweet reward (bears also will go for the brood frames). As these frames are being cleaned by bees, yellow jackets, or wasps, these robbers may turn their attention to nearby bee colonies, attacking them, and begin a robbing spree. If bears discover these supers and combs, they will likely dismantle stacked supers and break and scatter frames in all directions.

It should now be clear that you must be careful in dealing with extracted supers to avoid these potential problems.

If a robbing frenzy is under way, it is very difficult to stop it. Once started, this activity may go on for long periods (days and even weeks). Weak nucs or weak standard hives will find it difficult to stop the robbers from entering their colonies and withdrawing whatever honey is available within. It might be prudent to move these colonies three or more miles away in order to protect them. Within a single apiary, you can dust the robbing bees with flour or powdered sugar; if you go to nearby strong colonies, you can observe to which one(s) the dusted bees are returning. If by chance you are able to identify the offending colonies, move them to another location and leave the weaker colonies on site. You can also reduce the entrances of the colonies being robbed to enable the bees in these weaker colonies to better defend their entrances; of course if the weather is hot, this may make it difficult for the bees to regulate their hive's temperature (a screened inner cover helps with ventilation).

Another approach to this problem is to be proactive, and when you are aware that there is no (or very little) nectar available and you **must** examine your colonies, a day or two before doing so, place robbing screens on all the colonies. Robbing screens are designed to block off the normal entrance to the hive but have a passageway that is perpendicular to the normal horizontal entrance; check the bee supply catalogs. The bees whose colonies are fitted with these screens learn to enter and exit from this passageway; but the robbing bees are unable to determine how to gain entrance to hives fitted with these screens. It is has been reported that Carniolan and Caucasian bees are less prone to robbing, while Italian and Africanized bees are more apt to undertake such an activity.

In the fall, supers placed above the broodnest of your colonies for them to clean should be removed and stored; otherwise bees may move into these essentially empty supers and remain there, consequently to starve from a lack of food over the winter.

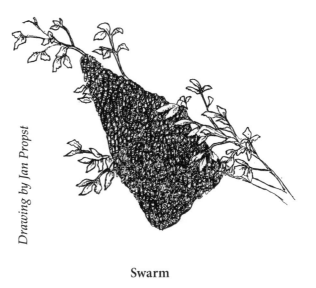

Drawing by Jan Propst

Swarm

SWARMING

From viruses to mammals, all organisms have evolved reproductive strategies to perpetuate their species (passing on their genetic code). Social insects, in this case honey bee queens, produce new individuals (workers, drones, and queens) within the colony unit, but compared to other organisms, the queens produce very few female reproductives (queens). During swarm seasons they reproduce their own replacements: virgin queens. To protect and care for their queen, worker bees feed, clean, and distribute her pheromones throughout the colony, as well as prepare in advance the cells where she will deposit her eggs. When a colony (in this case including tropical stingless bees and some ants) reproduces itself by swarming, both the colony and the swarm will end up with one reproductive (queen).

This activity of dividing the nest with new reproductives is very expensive. A colony divides, and part of it leaves for a new homesite, usually with the old queen, while the remaining members continue at the original site with a newly emerged—and later mated—queen. In this manner, a single unit becomes two. By reproducing, organisms perpetuate and protect their kind from extinction.

In the process of swarming, honey bees divide a single colony into two separate units. To do this successfully, the sole egg-laying member of the colony, the queen, must be duplicated, otherwise the division would leave one unit without a queen. This type of reproduction (fission), where the parent colony must give up approximately half its bees and its queen to the swarm, as well as provide the parent colony with a new queen, comes with its shares of risks and rewards. This method requires that after the division of the colony into two separate units, each unit will need to augment the numbers of worker bees in order to store large quantities of honey for winter survival. On top of all this, the unit that forms the swarm must also build a new nest. Nevertheless, this unusual method of cloning a single unit into two works well enough to sustain and perpetuate bee colonies.

An abundance of food, a high worker population, and the formation of many queen cells, often called *swarm cells*, indicate that swarming preparations are under way. Shortly after the swarm cells are sealed, the colony will cast a swarm. Bees will exit as a swarm on any warm, windless day, usually between 9 a.m. and 3 p.m. (earlier or later if the weather is favorable). Occasionally, bees will swarm when the weather is less than favorable.

A bee colony employs two distinct reproductive strategies. One is designed to sustain the viability of the existing colony by replacing members that die, and the other is to divide the existing colony into two separate units. If this duplication succeeds, the results are two units from one—the original colony and a new colony. An existing colony, one that is not preparing to swarm, consists usually of a single mated laying queen, the mother of all the members of the colony, and the female worker bees, which number in the tens of thousands and develop from fertilized eggs. The male bees, or drones, on the other hand, develop from unfertilized eggs; they number up to a thousand, but usually less.

When a colony is preparing to duplicate itself, it turns its energies to rearing queens; one of these newly reared queens will replace the colony's original queen. The resident queen will accompany the swarming bees as they are exiting their colony; as the bees begin to settle into a swarm cluster, usually near the vicinity of the colony from which they departed, the queen will join them. Should the queen be the first to alight at a clustering site, the swarming bees will join her. The first visible sign that a colony is preparing to divide itself is the presence of queen cups within the colony, usually at the bottom of combs. The resident queen deposits eggs in these cups; once eggs are deposited in the cups, they are referred to as *queen cells*, and the developing larvae are fed royal jelly. Within a sixteen-day period, many new queens are ready to emerge from their cells. However, before these queens emerge, approximately half the worker bees from the colony depart their nest, accompanied by the resident queen, along with a small number of drones. This cohort will eventually occupy a new cavity.

The original colony will, in a short time, end up with only one virgin queen. All the other queens will be killed, either in their cells or while fighting with one another. The surviving virgin queen will go on one or more mating flights and return to the original colony (the parent colony) as its new queen. As a result of this survival strategy, honey bee colonies "clone" themselves.

Both bee researchers and beekeepers continue to investigate the factors that set the stage for colonies to reproduce themselves by swarming. One of these factors appears to be an abundance of incoming food and a limited area to store it. The brood area of the nest has to be used to store the nectar, limiting the number of cells available for the queen to deposit her eggs, resulting in fewer larvae to feed. Nurse bees then have an excess of royal jelly and need an outlet for it. Developing queen larvae require lavish amounts of royal jelly; therefore by rearing queens, the nurse bees have an outlet for their glandular food. Another factor appears to be expanded bee populations after early spring (March), resulting in an inadequate distribution of the queen's pheromones; bees respond to this signal by rearing new queens. (This line of reasoning has its drawbacks, because the process of replacing a queen in decline, once that decline is perceived by the colony, entails the construction of only

three to five queen cells. But in swarm preparations, there may be ten or more queen cells—swarm cells—and these particular queen cells are usually located at the lower boundaries of honey combs. These cells are robust in appearance, and given their number and size, they are a more likely outlet for nurse bees to relieve themselves of an excess of royal jelly.)

As noted above, shortly after the swarm cells are sealed, the colony will cast its first swarm; this is called the prime swarm. Some colonies cast additional swarms, which are accompanied by virgin queens. A study of the time of the year when most swarming takes place was done in the Ithaca area of New York State by Cornell University researchers. Their results, shown in the graph "Swarm Emergence Dates, 1971–1976," clearly indicated that swarming is bimodal and that 80 percent of the swarms were cast from the beginning of May through mid-July, and the remaining 20 percent in August and September. Swarm populations averaged between 10,000 and 12,000 bees, about the number of bees in a three-pound bee package. Swarm periods vary in North America, beginning earlier in a calendar year at lower latitudes and later at higher latitudes. Check with your local bee clubs to learn when swarms are most likely to occur at your latitude.

Once the signal to depart the colony has been given, bees begin exiting the colony, some tumbling to the ground while others fly out and begin circling around and above the hive. Soon some bees land on a nearby object and begin fanning their wings, distributing their Nasonov pheromone complex into the air. This pheromone draws the majority of the airborne bees and the queen (if she did not alight earlier) toward the landing site, where eventually all the bees will huddle together, forming a swarm cluster. It is this cluster, readily visible to the casual observer, that is correctly called a *swarm* (see the illustration of a swarm). A small percentage of the bees that have accompanied the swarm will become scouts; these bees will begin the task of locating a new homesite for the swarm. Initially, different scouts will inform the swarm, by dancing on its surface, that they have located several potential homesites. This dancing is similar to that which tells other bees where food is located. Eventually, the scouts that have discovered the most suitable site prevail, and the scouts for alternative sites cease dancing—a true dance competition.

Research is continuing into how bees decide to

Swarm Emergence Dates, 1971–1976

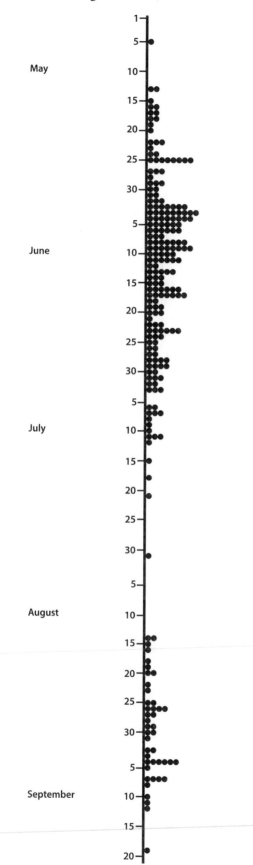

swarm, how they reach a decision on site locations, and what makes a good home. Before the swarm cluster departs from its cluster site and becomes airborne, it is necessary for the bees to raise what is referred to as their flight temperature. Once the cluster is airborne, the scout bees who have discovered the new home site guide the swarming bees to it.

Bees in a swarm are usually very gentle and not prone to sting. Two reasons have been given for this behavior. For a period of over ten days before the swarm issues, the bees engorge themselves with honey. These food reserves serve three purposes: they nourish the swarm for a few days while the scouts are searching for a home site, they stimulate the bees' wax glands which is essential for comb construction at the new home site, and they provide food for the bees at the new home site should inclement weather prevent them from commencing foraging activities from their new home. Other researchers have speculated that with the honey sac loaded with honey, the bees may have difficulty bending their abdomens to insert their stingers. Another reason for the absence of an offensive posture may be the fact that they have no colony to defend at this time. Lastly, the normal division of tasks is suspended once the bees have departed from their colony and are now concentrating on a single goal, namely getting to a new homesite, where the normal division of labor, including guarding, will resume.

Reproductive Swarming

Reproductive swarming is when a healthy colony divides itself into essentially two separate colonies by casting a swarm. Swarming usually occurs early enough in the season for the original swarm to build sufficient comb, augment its population, store sufficient reserves for the winter, and likely replace its queen with a new one. Although researchers and beekeepers have yet to tease out all the factors that lead up to a colony casting a swarm, observations of

The dates of swarming in the area of Ithaca, New York (data from 1971 through 1976). Each dot represents a swarm that emerged on the date shown.
Source: T.D. Seeley, R.A. Morse, and R. Nowogrodzki, Bait hives for honey bees, Cornell Cooperative Extension Publication, Information Bulletin 187:3.

colonies in swarm mode have revealed the following internal and external factors that may initiate swarming behavior:

- The cavity the bees reside in becomes congested, leaving insufficient room for the ever-increasing number of bees. A period of inclement weather can also contribute to congestion.
- An imbalance in the number of bees that carry out specific age-related tasks.
- The presence of queen cups along the bottom edges of brood frames. This is the first visible sign that the bees are making swarm preparations.
- The clogging of the broodnest (the area were the queen can lay her eggs) with nectar.
- With less area to lay eggs, fewer larvae are present, and the numerous nurse bees have no outlet for their glandular food, in particular royal jelly. An outlet for royal jelly would be the presence of multiple queen cells.
- With the rapid expansion of the adult population, the distribution of the queen's pheromones may be compromised to the point that they cannot prevent the bees from constructing queen cups and rearing queens.
- Age of the queen: a colony with queens older than one year are more likely to swarm than colonies with younger queens (less than a year old).
- The race of European bee: Italian honey bees are more prone to cast reproductive swarms than other races.
- An ever-increasing day length.
- An abundance of resources (nectar and pollen).

Non-reproductive Swarming, or Other Reasons Why Bees Leave

Under certain conditions an entire colony may depart its abode; this is referred to as *absconding* and may be caused by:

- Disease or mites, particularly the varroa mite.
- Starvation.
- Wax moth or other pest infestation (e.g., small hive beetle infestation).
- Fumes from newly painted or otherwise treated equipment.
- Poor ventilation.
- Excessive disturbance of the colony by the beekeeper or vandals.
- Excessive disturbance by animals pests such as skunks and bears.

Observable Signs That a Colony Is Preparing to Swarm

Signs that a colony is in some stage of swarm preparation are clearly visible during routine hive inspections. The list below presents a rough chronology of the various signs you might see in a colony that may ultimately swarm:

1. The presence of numerous queen cups (10 to 40) along the bottom bars of frames.
2. The presence of many queen cells, both opened and sealed, along the bottom bars of frames.
3. The presence of many drones.
4. Queen egg laying tapers off; this can be noted by a decrease in the number of young larvae in the brood area.
5. Queen is being chased by worker bees rather than slowly traversing the combs.
6. Field bees are less active and can be seen congregating at the hive's entrance during the swarm season. Bees can be seen congregating at the hive's entrance when it is too hot; this type of assembly is called *bearding*.
7. The broodnest (area where eggs, larvae, and pupae are located) cannot be expanded because the combs there are fully occupied with brood, nectar, and/or bee bread.
8. The wax has been removed from the tips of the queen cells, exposing the fibers of the cocoon (referred to as the bald spot).
9. Few bees are foraging (minimum flight activity near the entrance) compared to other hives with an equal number of bees.

Swarms May Issue from a Colony without the Accompaniment of a Queen

On occasion, bees will commence exiting the colony as a swarm, but the queen may fail to become airborne because she may have had her wing(s) clipped. As a consequence, she could be lying on the ground outside the entrance or, given the chaotic event inside the colony as the signal is given to exit, the queen fails to exit from the colony. The swarming bees will remain airborne, waiting for the queen to accompany them, or form a cluster on a nearby branch; but when the queen fails to join them, the airborne bees or the cluster will soon become aware of the fact that their queen is absent and return to their hive. Once the bees have returned to the colony and have temporarily reengaged in more routine activities, find the queen and cage her. Given that the colony has demonstrated its intentions to swarm, it will now be necessary to employ various methods of swarm control. Here are some steps to follow:

1. Find and cage the queen either before or after the swarm returns.
2. Move the parent hive from its stand and replace it with a new hive of foundation or dry drawn comb.
3. When the swarm returns, let the queen walk in with them. If the swarm has already returned to the parent hive at the old location, shake half of the bees in front of the new hive; the bees will enter this hive. Release the queen so she can walk in the hive entrance with the bees.
4. Check after 10 days.
5. Requeen the colony with new stock; the queen might have swarming instincts and is probably old. Any virgin queens emerging from the original colony will likely have this swarming instinct.
6. The parent hive will have several frames with capped queen cells as well as some unsealed queen cells. The colony will swarm prior to the emergence of a virgin queen. To prevent swarming from occurring, perform the following actions: divide the frames with queen cells and clinging bees into several nuc boxes. In most cases, the queens that emerge will undertake mating flights, resulting in many new colonies for your apiary. When placing frames with queen cells in nuc boxes or larger hive bodies, be sure that you first place a shim between a hive body and bottom board—otherwise, you will likely injure the queen cells when the lower section of the queen cell comes in contact with the bottom board.

Another method would be to let the swarm return to the original colony after you remove the queen cells. Check after 10 days to remove any additional queen cells, or demaree the hives (see "Demaree Method" below). Requeen the colony later with non-swarming stock.

Prevention and Control of Swarming

Swarm prevention involves measures a beekeeper takes to prevent the construction of queen cups and/or to prevent the queen from laying eggs in existing cups. Swarm control is put into practice when the beekeeper discovers there are larvae/pupae in the queen cells. This is clear evidence that the colony is preparing to swarm. The manipulations required for swarm prevention and swarm control may be similar but usually are employed under different circumstances. In **prevention**, the beekeeper is observing very early indications, such as congestion, that the bees may have the inclination to swarm. In **controlling** swarms, queen cup construction and queen cells are already present, as well as a clogged brood area and a very populous colony; it is clear in this case that swarm preparations are in an advanced state.

Swarm Prevention

To prevent a swarm from issuing, you need to relieve the crowded conditions and congestion of adult bees by adding more brood boxes. You can also:

● Relieve congestion in the brood area by reversing the brood boxes (by the end of winter the bees are usually all located in the top brood box).
● Prevent the bees from clogging the brood area with nectar by adding honey supers above the brood chambers before the dandelions bloom in your area (early spring).
● Remove all queen cups and all queen cells.
● Consider requeening the colony (resident queen's pheromones are inadequate, prompting the bees to replace her, in this instance by swarming).

Reversing

Reversing the brood chambers at regular intervals, or as needed, beginning in the spring, is one method used to relieve congestion in the hive. Through the winter, the colony and its queen move upward through the hive bodies (see "The Winter Cluster" in Chapter 8). We will assume that in the fall each colony was reduced to brood chambers with the second chamber filled with 9 to 10 frames of sealed honey. During the fall, the location of the majority of bees in a colony can vary depending on the number of boxes that make up the colony, the location of the brood, and the variation in day/night tempera-

Reversing Hive Bodies

Reversing with Two Deeps and a Shallow

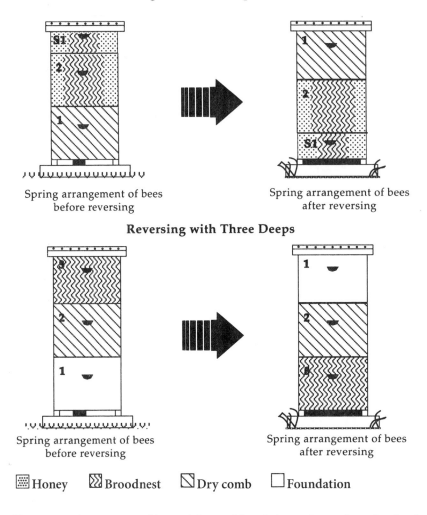

Spring arrangement of bees
before reversing

Spring arrangement of bees
after reversing

Reversing with Three Deeps

Spring arrangement of bees
before reversing

Spring arrangement of bees
after reversing

⊞ Honey   ⊠ Broodnest   ◩ Dry comb   ☐ Foundation

S1 super can also go on top of deeps. A frame of foundation or dry comb can be placed
in top deep (#1) to encourage bees to move up.

tures. Assuming the colony has been reduced to two deep-hive bodies and the second deep is filled with frames of sealed honey that will serve as the bees' winter food and that there is still some brood in the lower hive body, you may expect to find many bees still occupying the lower hive body. However, as the days grow shorter and brood rearing markedly declines, the bees will slowly move up into the second deep hive body where they will eventually form their winter cluster. Insects, including bees, tend to move upward and are reluctant to move downward (in this case into the lower brood chamber), except to traverse through it to exit from the colony.

Furthermore, by late winter (January, February) the bees, brood, queen, and the food will in all likelihood be located solely in the top chamber. However,

if this situation is not remedied early in the spring, the colony will become congested and likely begin preparations to swarm. There are two kinds of congestion: too many adult bees in a limited space, or a limited space for brood rearing; a strong colony in the early spring may manifest both. Given this situation, the beekeeper needs to reverse the hive bodies, so the top brood chamber containing most of the bees, the queen, and the brood is placed on the bottom board, and the brood chamber that was on the bottom board is placed above it (the brood boxes are reversed).

The queen and a contingent of bees may now move upward into the empty chamber to expand the brood area and to provide more room for the bees. A similar procedure can be used if the bees are housed in three deep brood chambers or two deep brood chambers and one shallow, or if all the hive bodies are the same size. The equipment containing the brood is placed on the bottom board and the empty hive bodies above it.

A word of caution: at times the queen may expand the brood rearing area into the upper third area of the lower brood chamber. If you reverse boxes with this condition you will in effect split the continuity of the brood nest, and the bees may be unable to maintain adequate brood temperatures in both boxes. In such cases, you may have to wait for the queen to continue laying in the lower box; in the meantime, add another brood box above the second one. This will at least relieve some of the congestion, and the queen may move into this box. Remember, do **not** split the brood area during reversing.

While carrying out these activities, be sure to ascertain the amount of food that remains in the colony; if the food reserves are low, feeding may be required.

Here is a quick outline for reversing hive bodies (two deep and one shallow are illustrated):

1. Take a clean bottom board to the beeyard.
2. Move the colony housed in the brood chambers off its stand.
3. Place the clean bottom board on the stand.
4. Remove the outer cover and inner cover from the top brood chamber (the one containing the queen, the brood, and most of the bees).
5. Place the brood chamber filled with bees/brood on top of the clean bottom board.
6. Place the empty brood boxes above it and replace the inner and outer covers.
7. Repeat this process with the remaining colonies, as needed.
8. Repeat this process every two weeks until the colony presents no evidence that it has intentions to swarm.

Other Ways to Relieve Congestion and/or a Congested Brood Area

Colonies may be congested (crowded with bees) due to a lack of sufficient space in the hive bodies, or the unwillingness of the bees to move downward into the empty brood box. The brood areas may also be restricted due to old, broken combs and/or brood combs filled with nectar/honey, or the unwillingness of the queen to move downward.

Listed below are some techniques for ameliorating these conditions:

- Congestion: reverse hive bodies and/or add additional hive bodies.
- Restricted brood area: remove combs clogged with honey and replace with empty drawn combs, or foundation.
- Separate most of the brood and the queen:
  1. Place the queen, with unsealed brood, eggs, and bees in the lowest body.
  2. Above this, place a hive body with foundation.
  3. Above this, place a body filled with capped brood and the rest of the bees. Note: this in effect splits the brood nest and should be done only if there are enough bees to care for the separated brood.

Decrease the number of bees or brood in the hive by splitting the hives to make additional ones, called *increases* or *splits* (see the illustration on feeding a weak colony). This means, of course, that you are making additional colonies and expanding the size of your apiary. If you do not want to increase the number of your colonies, you can always sell your extra hives or donate them to your local bee club to help get someone else started. Alternatively, splits made to control swarming can be reunited later in the season to consolidate colony numbers. To make a split, follow these steps:

1. Move frames of capped brood, honey, and bees from the congested hive into a new box (e.g., one hive body). You should try to leave the frame with the queen in the original hive, the one you are splitting. You can also put these frames into nuc boxes and requeen to have spare nucs available for requeening later (see Chapter 10), to augment the number of your colonies or to sell to others.
2. If you are combining frames of capped brood and bees from different colonies, spray each frame of bees with syrup to reduce any fighting among the bees.
3. Give the new hive a frame of open brood (either a frame of eggs or one of newly hatched larvae from your best hive), so they can rear their own queen. Requeen the split with a new queen, or provide some queen cells (usually swarm cells). Whatever is provided, place it in the middle of frames of emerging brood.

Feeding a Weak Colony (Split, Nuc, or Swarm)

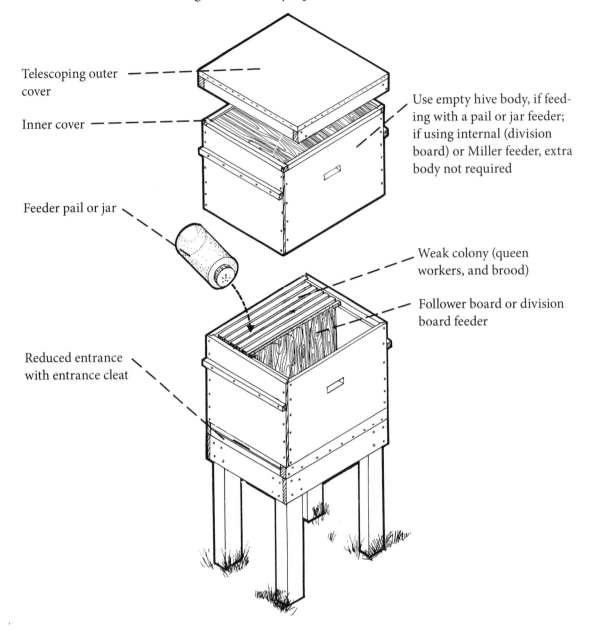

Telescoping outer cover

Inner cover

Use empty hive body, if feeding with a pail or jar feeder; if using internal (division board) or Miller feeder, extra body not required

Feeder pail or jar

Weak colony (queen workers, and brood)

Follower board or division board feeder

Reduced entrance with entrance cleat

4. If you made an increase using a deep hive body, it should have a frame of open brood; frames of capped, emerging brood; frames of foundation; and frames of honey and bee bread, or empty drawn comb filled with syrup. These frames should be arranged as follows: the frame of open brood is placed in the middle of the box, frames of capped brood are placed on either side of the frame with open brood, and the two frames of emerging brood on each side of the frames with sealed brood. Add a frame of honey and a frame containing bee bread or a frame with a combination of both honey and bee bread, and fill the

remaining spaces with drawn empty comb or foundation. Be sure there are sufficient adult bees in this hive body to attend to the open brood, the sealed brood, and the emerging brood.

5. Reduce the entrance to discourage robbing; check after one week.

6. If you are using nuc boxes, a good rule of thumb is to have each contain two frames of bees and brood, two of food, and one of foundation.

Demaree Method

The Demaree method, described first by George Demaree in 1884, makes it possible to retain the com-

Modified Demaree Method of Swarm Control

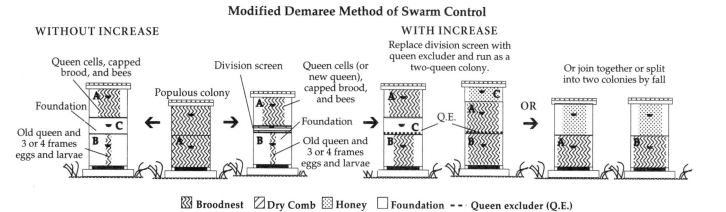

WITHOUT INCREASE

WITH INCREASE

Replace division screen with queen excluder and run as a two-queen colony.

Queen cells, capped brood, and bees

Foundation

Old queen and 3 or 4 frames eggs and larvae

Populous colony

Division screen

Queen cells (or new queen), capped brood, and bees

Foundation

Old queen and 3 or 4 frames eggs and larvae

Q.E.

Or join together or split into two colonies by fall

OR

▨ **Broodnest** ▧ **Dry Comb** ▨ **Honey** ☐ **Foundation** -- **Queen excluder (Q.E.)**

plete population of a colony while practicing swarm prevention and control. Basically, it separates the brood from the queen and decreases the congestion. Here is one way to "demaree" (see the illustration of the modified Demaree method):

1. Select a populous colony (A). If you are not making an increase, follow the left side of the diagram (without increase). If you are increasing, follow the right side of the diagram (and go to step 9).
2. Have ready a hive body (B) filled with frames of dry drawn comb (from which brood has already emerged) and a hive body (C) with foundation. If no honeyflow is on, use less foundation, because the bees will chew it; if you have only foundation, feed bees with syrup so foundation will be drawn out.
3. Place the empty hive bodies beside the hive to be demareed (A). Find the queen and place her on a frame containing very young larvae and eggs. Make sure there are no queen cups or cells on the frame with the queen; if they are present, remove them or replace the frame.
4. Remove two frames of foundation from the middle of B and place the frame with the queen and clinging bees there.
5. Remove A from the bottom board for the moment and set B in its place.
6. Add two or three frames of honey and bee bread from A to B and fill the remaining spaces with dry comb or foundation.
7. Place a queen excluder above B and then place hive body C (full of foundation or dry comb) above the excluder.
8. If you do not want to make another colony (an increase), add three or four frames of eggs and larvae plus two or three frames of honey in A and

place on top of C. Fill in the rest of the space with frames of foundation.

9. If you are making an increase or split, separate the old queen in B below (fill up the space as described in step 8) with a division screen. Above the division screen, place A, filled with queen cells (or a new queen), capped brood, and bees. If you are requeening A with a purchased queen, remove all queen cells from the brood frames. Any extra frames can be given to other colonies.
10. Once you have two established queens in A and B, separate them with a honey super (C) placed above a queen excluder, and run the colony as a two-queen unit (remove division screen). You have created a two-queen colony. Add additional honey supers as needed.

At some point in the season you may wish to return the two-queen colony into a single queen colony or into two separate colonies. The latter can be achieved by placing the hive body (or bodies) above the queen excluder on its own bottom board and adding any necessary equipment to it (i.e., inner and outer cover and any additional super). In the process of assembling the new colony, be sure to remove the queen excluder, unless you wish to keep the bees in the colony below from entering any honey supers you plan to add to the lower colony. By separating the two-queen colony into two separate units, each with its own queen, you have added another colony to your apiary.

On the other hand, you may not wish to increase your colony numbers; in this case remove the queen excluder, and the two queens will eventually meet, and if all goes as planned, one queen will kill the other, thereby returning the colony to a single-queen system. In this scenario, you have united the two-queen colony and returned it to a single-queen colony. Either procedure is usually accomplished after harvesting the honey from the original two-queen colony. Regardless of which procedure you choose, be sure the bees have sufficient supplies of honey by the early fall.

Advantage

- The population of bees in a two-queen system outnumbers that of a single-queen system; two-queen systems produce more honey than single-queen colonies.

Disadvantages

- Before adding the second queen, you must find the original queen and place her below the queen excluder.
- There is no guarantee the second queen, introduced above the queen excluder, will be accepted.
- Many manipulations are required.

Interchanging Hives

In an apiary where the hives are in long rows (never a good idea), the bees tend to drift toward the row ends. As a result, the colonies in the middle may be weaker than the colonies on the ends; see "Hive Orientation" in Chapter 4. The stronger colonies may become more congested and begin swarm preparations. In addition, you may misinterpret the abilities of the queens in the weaker colonies. These queens may be superior queens, but because of drifting, you may think the queens need to be replaced, when in fact it is the long, straight rows that have led to a misinterpretation of the observation.

If a hive is very populous and seems likely to swarm at some point but has not yet made preparations to do so, interchange it with a weaker hive. Most of the foragers will return to the stronger hive's previous location, and as a consequence of the switch, they will actually be entering the weaker hive, augmenting its population. Conversely, the strong hive will have a sudden decrease of incoming field bees, and any idle bees that might have normally initiated swarm preparations will begin foraging. However, should you opt to take this approach, you will likely sacrifice the potential honey crop from the strong colony.

Foreign bees entering the switched hives should not fight if there is a honeyflow in progress. To decrease chances of fighting, wait for a good honeyflow

before interchanging the hives. Very weak colonies or nucs should not have their numbers increased by substituting the location of a strong colony with a weak, unless the queens in the weak colonies are caged for a few days and then released, or caged in a queen cage with a candy plug to delay their release while the incoming foragers become acquainted with (what is essentially to them) a foreign queen. Again, this is something you may wish to try on a few colonies, but be advised that if you opt to do this during a honeyflow, you are sacrificing some or all of your honey crop to strengthen a weak colony. Very weak colonies or nucs should not be strengthened in this manner, as the incoming strange foragers can overwhelm and kill the queen. Recent research claims to have found that often, when such an exchange is made, the weaker colony becomes stronger and remains so, while the former stronger colony takes on the characteristics of the weak colony. These observations appear counter intuitive.

Other Factors

The following factors may also be of importance in helping to decrease swarming in some hives:

- Having queens less than a year old: established colonies headed by young queens are less likely to swarm than colonies with older queens (i.e., more than a year old). However, even colonies with young queens will swarm if they are not provided with adequate room for expansion of their broodnest and if there is congestion of adult bees in the colony. If available, queens from non-swarm stock or hybrid queens with fewer tendencies to swarm may reduce swarming tendencies. However, swarming is natural colony reproduction, and it is likely that all colonies, no matter the type of queen, will eventually swarm. By providing colonies with young queens and sufficient space for the growing population of adult bees, swarming may not be eliminated but at least controlled to some extent.
- Ventilation to increase airflow within a hive:
  - Hive bodies can be staggered so they do not sit directly on top of one another. The first box above the first deep super is moved slightly forward or backward allowing a space for air to escape. Continue to stagger any additional boxes; however, these openings created by stag-

gering the boxes may create an opportunity for robbing.

- Inner or outer covers can be propped up.
- Screened bottom boards can replace solid bottom boards; this gives bees more ventilation and can be kept on the hive year-round. To avoid all these manipulations, we strongly suggest requeening your colonies each year.

The first two ventilation techniques might encourage robbing when the honeyflow is over; thus, only strong colonies should be manipulated in the ways described. To avoid all this extra work, requeening any colony that has an old queen may be the best solution.

CATCHING SWARMS

Bait Hives for Capturing Swarms

It is generally not possible to check your home apiary on an hourly basis, and to give such attention to your outyards throughout the swarm season is nearly impossible. Despite good management procedures for swarm prevention or control, a given number of colonies will cast swarms. The loss of any swarms from your beeyard is costly; colonies that cast swarms take time to rebuild their population and may produce less surplus honey. In addition, often colonies that cast swarms are likely to be healthy and prosperous ones.

Given these circumstances, beekeepers have turned to using bait hives to recover swarms before they take up an abode elsewhere. Bees captured in bait hives may come from swarms emanating from your apiary or from someone else's yard.

Bait hives can be defined as containers whose cavity size is sufficiently attractive to scout bees for them to convince the swarm to move to it. In addition to providing the proper cavity size, the addition of combs (when appropriate), lures, and the physical position of the bait hive (in regard to both its height and the compass direction of the entrance) will play a role in enhancing its attractiveness to scout bees.

Research on bait hives has revealed that scout bees do not randomly select any available cavity. Scout bees show a marked preference for bait hives that are situated 15 feet (about 5 m) above the ground, with a small opening near the bottom of the cavity and with the opening facing south or southwest. Bait hives located at lower heights also work but should not be positioned where they are in full sun for most of the day. Areas that provide partial shade are preferred over those in full sun. Bait hives can be placed in the vicinity of your apiary or in other locations that can be monitored by the bee "trapper."

When using a standard ten-frame Langstroth brood chamber as a bait hive (the ideal cavity size preferred by scout bees), include inside one or two frames of empty drawn comb. These frames should be located next to the entrance, and then the remaining cavity is filled with frames of foundation. The reason for filling the bait hive with drawn combs and foundation is twofold: (1) when transferring the frames and bees from the bait hive into a permanent hive, you can simply remove the frames from the bait hive and relocate them into the permanent hive; and (2) the drawn comb with its associated odors and propolis deposits helps to lure the searching scout bees to the bait hive.

Check beekeeping supply catalogs for what is available as a bait hive. Note: wax, propolis, and other odors will also attract mice and wax moths. To prevent birds, mice, etc., from entering bait hives, cover their entrances with hardware cloth, which will permit bees to enter; remove and properly store all bait boxes at the end of bimodal swarm seasons.

Once the bait hives containing movable frames are occupied by bees, the frames can easily be transferred to another hive body. Today, various lures (pheromones), readily available from most beekeeping catalogs, make the bait box more attractive. These lures can be applied inside and outside these hives.

The nonstandard swarm boxes that do not lend themselves to the inclusion of drawn comb can create a problem when you have to transfer the bees and their comb into movable frame boxes. Cutting the comb and wiring it to empty frames is cumbersome and messy, but effective. **Note:** The use of bait hives and other swarm traps in areas inhabited by Africanized bees is not recommended.

When can you be certain that your swarm trap has been occupied by a honey bee swarm? When you observe bees entering the bait hive with loads of pollen, then and only then can you be sure the trap is now being occupied by a bee colony. Remember, other insects or animals may also inhabit the trap, so be alert.

Types of swarm traps include:

1. Bait hives constructed from wood and containing drawn comb or foundation (or a combination of the two). Many beekeepers leave their swarm traps in place year round, refreshing them with lures at the beginning of each new swarm season. Others remove these traps at the end of the swarm season. Remember that the swarm season is bimodal: the most intense swarm period begins in early spring (May) and extends to the middle of summer (July); but a second, less intense swarm period occurs in late summer (August) and extends into the early fall (September). **Note:** the swarm season in the Gulf States begins in February, and at the end of March into April in the lower Eastern states. Check with your local beekeepers for the approximate dates of the bimodal swarm periods in your location.

2. Many bee trappers use lures such as Nasonov pheromones and lemon grass. These lures can be placed both inside and near the exterior entrance to these traps. Some of these lures are available in containers that slowly release their contents over an extended period. In most cases, however, these lures need to be reapplied after a few weeks, since they are volatile chemicals.

3. In addition to bait boxes assembled from plywood, or standard hive bodies, wood fiber or peat pots are also available for use as swarm traps. Originally used to trap Africanized bees, these pots have the disadvantage that once they are occupied, you must cut and remove the combs and then place/attach them to movable frames.

Again, helping local beekeepers collecting swarms will show you different techniques that could help you in your swarm capturing.

Advantages to Swarm Traps

- Relatively easy to assemble, especially a ten-frame Langstroth hive body that has been modified to serve as a trap (it has the ideal cavity size preferred by a high percentage of swarms).
- Package bees can be expensive; capturing a swarm provides free bees.
- In almost all cases, the colony that cast the swarm is likely to be a healthy one with good survival attributes, and with some reservation we will add that it may be more locally adaptive.

Disadvantages

- Swarms that occupy containers that do not include movable frames will attach their combs to the top and sides of the container; transferring these combs from the container into movable frame hives is tedious, time consuming, and may trigger robbing behavior, as the honey from these combs begins to leak all over the place. In addition, in the process of transferring the combs, brood may be killed, and the queen may be lost.
- Bait hives need to be securely fastened to trees, poles, or buildings. This involves climbing, using ladders and other equipment to both secure and remove the swarm traps. Inexperienced individuals should not risk life and limb to accomplish these tasks.
- Bait hives may attract other stinging insects to take up residency in them, particularly Africanized bees. Therefore you should approach an occupied bait hive with caution, especially if Africanized bees are in your area.
- Bait hives require weekly monitoring during the swarm periods.

Swarm-Collecting Containers

As mentioned, there are many ways to collect swarms that have clustered on a tree limb or trunk, on the side of a building, on a fencepost, on the ground, or a variety of other surfaces. How you collect a swarm and what equipment you will need to do so will depend on the location of the swarm and the swarm-collecting equipment available to you. Examples of swarm-collecting equipment include:

- A cloth sack with a wide opening will allow you to surround a cluster situated in the middle of a branch. By shaking or jarring the branch, you can cause the swarm to drop into the bag, after which the opening is quickly closed. If on the other hand the swarm is located on the tip end of a branch, envelop the swarm with the bag, tie the end, and cut the branch. So don't forget to bring pruning or lopping shears.
- A basket with a secure cover can also be used to capture a swarm.
- A cardboard box that can be closed quickly will also work in a pinch.

- A large wooden box.
- A spray bottle with sugar water. By spraying the cluster, you will accomplish several objectives that will make retrieving the swarm easier. If it is a swarm that has been there more than three or four days it may have exhausted its food reserves, and providing it with sugar solution will help to keep the bees from becoming defensive; wet bees soaked with sugar water are less able to fly, allowing you to retrieve most of them. When you arrive at the swarm site, the swarm might be at the point of taking flight because the signal to break cluster and take flight has been received by the clustering bees. But if you immediately spray them with sugar water, it will likely disrupt their signals to become airborne, thus giving you time to retrieve the swarm.

As mentioned, you can also use a deep hive body with a bottom board and inner and outer covers and containing one drawn empty comb, with the remaining space filled with foundation. Colonies that cast swarms are usually healthy colonies. However, some beekeepers refrain from using empty drawn comb, fearing that it may contain American foulbrood (AFB) spores, or that bees from the swarm may have spores in their honey stomachs, and when they deposit the content of their honey stomachs into the drawn cells, it will include the foulbrood spores. However, the use of at least one drawn comb will encourage the swarms to move into the hive and will provide cells ready for the queen to resume laying, as well as cells to deposit the honey stored in their honey stomachs. As in all things, there are times when you must make a decision as which path to take. After retrieving the swarm from its clustered site, shake the bees in front of the entrance of the deep hive body.

With the exception of the cloth sack, all the other collecting containers should have a portion of their surface covered with window screen. Swarms in containers often "cook" themselves (die) for lack of ventilation. Along with these collecting containers, it is always a good idea to include some queen cages, in the event you spot a queen and are able to catch her. You can then place her into the queen cage, and unless there is more than one queen, once she is in the cage the swarm will enter any container that contains the caged queen.

Swarms, especially large ones, need plenty of ventilation and must be kept out of direct sunlight. Often bees are "cooked" or smothered when collected in an inappropriate container (one that is too small or too airtight). Bees in a cooked swarm will look wet and will be crawling on the bottom (similar to a "cooked" package of bees); these bees will not recover. Therefore, hive all swarms in standard hive equipment as soon as possible. If you cannot hive the swarm within a reasonable time, the container holding the swarm should be placed in a cool, dark place. If at all possible, hive the swarm the same day or as early as possible the following day. Bees in a swarm are geared to build comb, and as long as they are retained within a container, the construction of comb will get under way. This will lead to a waste of wax and any honey stored in their cells.

The Swarm Call, or How Beekeepers Can Get Word Out That They Are Available for Retrieving Swarms

Yearly, beekeepers can notify their local Humane Society, surrounding police and fire departments, as well the state's bee inspector by letter, phone, or email that they wish to be put on the list of beekeepers willing to retrieve swarms. Most beekeepers appreciate being notified that there is a swarm to retrieve and usually do not request a fee for their service.

In today's age of lawsuits, both by homeowners (for any property damages caused by the beekeeper while retrieving the swarm), or by the beekeeper (who may be injured while capturing the swarm), it may be wise for the beekeeper to obtain insurance that provides protection from any form of liability or injury. Some beekeepers become professional bee retrievers, especially in areas occupied by both European and Africanized honey bees.

Before responding to a swarm call, try to determine from the caller if in fact the swarm is a honey bee swarm. This will save you time and prevent your going on a wild goose chase. Here are some questions to ask:

- Ask the caller to describe just what he or she sees. Is the caller sure these are honey bees? If there is a gray, paperlike nest, then the insects are hornets, and a professional exterminator should be contacted.

- How high off the ground is the swarm? If the swarm is higher than 10 feet (3 m), forget it, unless you are prepared and equipped for tree climbing or are skilled at using a ladder.
- Can you drive to the swarm's location? Is there vehicle access to the swarm, or can one easily walk to the swarm's location?
- What is the swarm clustered on (e.g., a tree branch or trunk, or on a post?). If it is on the terminal end of a branch, will it be permissible to cut the limb?
- How long has the swarm been there? If it has been there more than a day or two, either it will be very hungry and therefore volatile, or it may take flight before you arrive. In this case, don't forget to bring a spray bottle with sugar water, to spray the cluster with.
- How large a swarm is it? The size of a basketball? A softball? Smaller swarms may contain a virgin queen or be queenless, or Africanized, and may not remain at the cluster site by the time you arrive.
- Are these bees the caller's to give away? Is there a beekeeper next door or in the immediate area? If so, as a courtesy, that person should be contacted first.

Once these questions have been answered to your satisfaction, get directions on where to find the swarm, and with swarm-capturing equipment, off you go. It is wise to obtain the caller's phone number and address and to have a map showing the location and/or bring along your GPS unit. Do not rush to the site. If the swarm departs before you arrive, you will live to retrieve many more over the years. Your life and limb and those of others are far more important than obtaining a single swarm.

Bees in a swarm are usually well engorged with honey and have no permanent home to defend, and it is believed that for these reasons swarms usually show no defensive behavior and are easy to handle. However, on occasion, you may encounter a defensive swarm; possibly the bees have depleted the contents of their honey sacs and/or have been attacked by some predator, or someone has thrown rocks at the cluster, causing the bees to become defensive. Given that a swarm may not be gentle, it is always a wise decision to approach all swarms with the same protective gear you use in your own apiary while working bees. In the case where Africanized bees are present, wearing protective gear is essential.

These are the basic steps for collecting and hiving a swarm:

1. Spray the cluster gently with sugar syrup from your spray bottle. If the swarm is clustered on a limb, with the owner's permission, cut away excess branches, leaves, or flowers. Avoid shaking or jarring the cluster.
2. If the swarm is jarred and the bees begin to break the cluster, spray the bees and wait for the cluster to re-form.
3. While you steady the limb or branch with one hand, saw the limb or clip it free from the tree.
4. Shake the swarm into a hive or collecting container prepared for the bees or, if possible, put the entire cut limb into the collecting container.
5. If the swarm is on a post or flat surface, brush or smoke the bees into a hive or container, directing them gently with puffs of smoke. Look for and try to capture the queen and put her in a queen cage; this will make collecting the swarm much easier. After you put the caged queen in a hive or in a container, all the other bees will quickly enter.
6. If you don't have a soft brush, you can use a piece of cardboard or a feather duster to scrape bees gently into the container or in front of the hive entrance.
7. If you have collected most of the swarm and its queen (there can be more than one queen in a swarm), secure the container and leave the site. Inform the homeowners that a few bees may linger, likely scouts that were away searching for a homesite who have returned to find the cluster gone. These bees will eventually dissipate. If in the process of collecting the swarm you fail to capture a significant proportion of it, then it will be necessary to leave a collecting apparatus on the ground, hoping that the queen will enter it. If this is the case, and she remains in the container, the remaining bees will all enter the container prior to nightfall. This situation will require you to make a second trip to retrieve the container. At times, and especially if the queen is not in the container, all the bees within the container will fly out and re-cluster at the site where the queen settled. In retrieving swarms, it is fair to say that sometimes things do not proceed as smoothly as you hope.

Collecting a Swarm

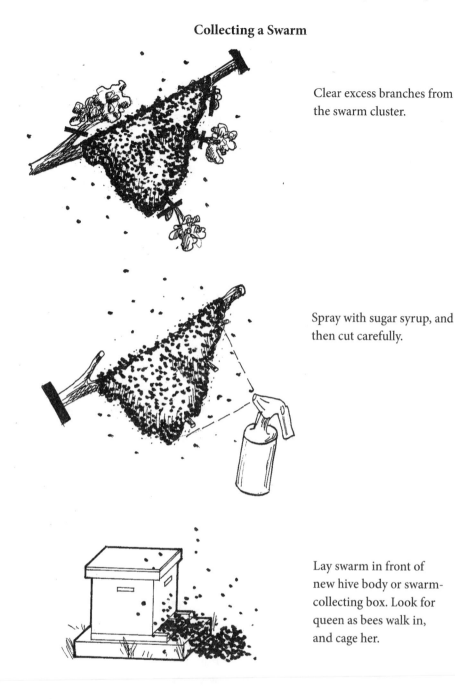

Clear excess branches from the swarm cluster.

Spray with sugar syrup, and then cut carefully.

Lay swarm in front of new hive body or swarm-collecting box. Look for queen as bees walk in, and cage her.

Back at the apiary, if you collected the swarm in a regular hive body, place it in a desired location in the vicinity of your other colonies. If you collected the swarm in a cloth sack or a cardboard box, you need to transfer the bees into regular hive equipment. A Langstroth hive with ten frames of foundation is one approach (or whatever equipment you are using); that way the bees, since they have engorged themselves with honey/nectar and their wax glands are geared up to produce wax, will draw the comb nicely.

This will provide the quickest way to get your foundation drawn into combs. Not all swarms stay put once they are properly hived. To prevent the swarm from absconding once hived, many beekeepers add one frame of open brood so the bees are more likely to remain.

Another technique to keep the bees from leaving is to temporarily put a queen excluder between the bottom board and the first hive body. Even if the bees begin to exit the hive, without their queen they will be obliged to return. Swarms can also be used to strengthen a weak colony; but why place a swarm with full wax glands in a weak colony with drawn combs? A better approach would be to unite a small swarm to a weak colony with the hope that the addition of bees to such a colony will allow it to grow and prosper. You can also start a nuc colony with a small swarm (less than 2 lbs., or 1 kg).

A month after the swarm is hived, it will be time to evaluate the queen's performance. Usually a primary swarm contains the queen of the parent hive, usually in her second or third year of life and likely to be superseded before the winter. Here the beekeeper comes to a dilemma: upon inspection, the queen is laying well, and there are solid patterns of sealed brood. The beekeeper wonders: shall I keep her or replace her with a young queen? We strongly suggest she should be replaced (see "Breeder Queens" in Chapter 10); but after being replaced, she can be transferred to a nuc colony, and, if she continues to demonstrate her value, she is a keeper; move the frames from the nuc into a standard hive body and prepare this new colony with adequate food needed for the winter. **Note**: remember, swarming instinct can be genetic.

As in many management opportunities with honey bees, there are several options available to the beekeeper who has had the good fortune of capturing one or more swarms. Here are some options:

- The swarm can be housed in its own Langstroth hive (or in other equipment, such as a top bar hive, a Warrè hive, a Long hive, etc.).
- If you were able to capture the swarm's queen and cage her, you can unite her bees to a weak colony to strengthen it. However, a word of caution: it is possible the bees in the swarm may be carrying American foulbrood spores in their honey stomachs. A way around this potential problem is to place the swarm in a separate hive body containing only foundation, with its own bottom board and inner and outer covers. Then situate the swarm box above the weak colony, with its entrance facing in the opposite direction of that of the weak colony. After the passage of one week, these two units can be united with a sheet of newspaper. Instead of using a bottom board for the top colony, a Snelgrove board will also work; in this case the odors of each hive will blend, and

the board can be removed in one week, uniting the two colonies (See "Uniting Weak Colonies" earlier in this chapter)

- The swarm queen can be introduced to a queenless nuc.
- If you know which colony swarmed, it should have numerous capped queen cells. This can be an opportunity to divide the frames with the queen cells and clinging bees into a number of nucs. One precaution, however: often the queen cells extend downward from the bottom bars of the frames on which they are located, and as a consequence they can be damaged if they come in contact with the bottom board. To avoid damaging these queen cells, it may be necessary to add shims to the top (or bottom) rims of nuc boxes so that the queen cells will not come in contact with the bottom board.

Precautions

Before we discuss precautions, let us make it clear that most colonies that cast reproductive swarms (not absconding swarms) are in all probability reasonably healthy and free of American foulbrood. But as an added precaution, hiving the swarm on foundation will serve as insurance should foulbrood spores be present in the bees' honey stomachs. These bees, however, will be carrying with them any number of adult varroa mites. A day after the swarm is hived, sprinkle the bees with powdered sugar, and if there is a sticky board beneath the screened bottom board, you may be able to determine the swarm's mite load. The powdered sugar can also be applied while the swarm is in a container with a screened bottom. By following either of these procedures, you will start the newly housed swarm with fewer mites—a good way to begin a new colony.

Destroying a Swarm

If a swarm must be destroyed—for example, it is in a schoolyard or is Africanized, or it is a dry swarm and its location makes it difficult to get control of it—a thorough spraying of soapy water (½–1 cup liquid detergent or soap per gallon of water) will kill the bees quickly. Make sure to wet all the bees thoroughly.

QUEEN SUPERSEDURE

Supersedure is how the colony replaces an old or inferior queen with a young queen. The workers in the colony build a few queen cells, and when a new queen emerges, she destroys the other queen cells and may destroy the old queen (sometimes mother and daughter queens coexist for a short time). Swarming does not usually take place when a queen is superseded. Some of the reasons for the supersedure are listed below:

● Queen is deficient in egg laying.
● Queen produces inadequate amounts of queen substance (pheromone) due to age, injury, lack of nourishment when a larva, or other physiological problems.
● Queen is injured as a result of clipping, fighting among virgins, or temporarily balled by workers when released.
● Queen was injured when removed from or placed into a queen cage.
● Queen is defective, not raised under ideal conditions, or poorly mated.
● Colony and queen have nosema disease or tracheal mites.
● Weather has been inclement for extended periods (other than winter).
● After installation of a package, when numbers of adult bees decline and no new ones emerge for 21 days, the remaining older workers may undertake supersedure activities.
● Queen's pheromone levels are low.

Queen Cells

Young queen larvae can begin their development in queen cells or in worker cells. Queen cells developed from worker cells are either supersedure or emergency queen cells. The sudden loss of a queen usually forces the bees to modify worker cells into emergency queen cells, and as the larval queens develop, the cells' edges are slowly enlarged by the added wax. These cells will eventually take the shape of a peanut shell and be suspended vertically from the comb. The outer peanut-shell shape is characteristic of all queen cells. In addition, supersedure and emergency queen cells are fewer in number (two to five) than swarm cells and are usually located on the surface of combs. Another way to distinguish emergency/supersedure cells from swarm cells is that the former two are dark in color and not as robust as swarm cells, which are usually light tan in color (constructed from fresh wax).

Queen cells that begin as queen cups, usually located along the bottom edges of frames, are referred to as swarm cells (cells containing queens that will replace the resident queen after she departs with a swarm). These queen cells are numerous, some containing larvae, some pupae, all of different ages. Again, they are commonly found suspended along the bottom bars of combs (see the illustration of swarm and emergency queen cells on this page). By contrast, supersedure/emergency queen cells are few in number, contain larvae or pupae of the same age, and are similar in size. In addition, they can be found near the central region of the brood comb. A queen issuing from an emergency queen cell replaces one that is no longer present! In the case of the queen emerging from a supersedure cell, she replaces a queen who may still be present!

Swarm cells

Supersedure or emergency queen cells

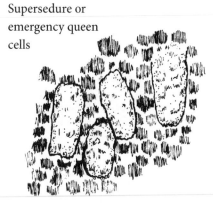

Swarm Cells vs. Supersedure (or Emergency) Queen Cells

The state of the colony (e.g., low bee population or bubbling over with bees), the time of the year, and the location of the queen cells should provide enough information for you to know exactly what the colony's intentions are.

In Supersedure

- Few (between one and five) cells are made.
- The colony is usually not very populous.
- The brood pattern is scattered or almost nonexistent because the queen is injured, diseased, or failing.
- Queen cells converted from worker cells are located near the center of the comb.
- The age of queen larvae is the same, so all the queen cells mature nearly at the same time.
- Drone brood often appears in worker cells.
- If queen cells are present outside of the two swarm periods, they are not abundant, and they are not located in the areas one finds swarm cells.
- Wax to make cells is darker in color.

In Swarming

- Numerous (between 10 and 40) queen cells are present on the lower edge of combs.
- The colony is populous.
- Numerous frames of capped brood are present, and there is a diminished number of cells with uncapped brood.
- The ages of queen larvae are varied, which may result in multiple swarms as different queens emerge over time.
- There are two separate swarm periods in northern latitudes, the more intense period that runs from late April into the middle of July, and the less intense period that takes place between August and September.
- The queen cells are light yellow in color, robust, numerous, and located near the lower areas of combs.

In Emergency Replacement

- Few (between one and three) cells and will be roughly the same age.
- Queen cells are usually constructed on the face of the comb amid capped worker brood.

Attempts to make queens from older worker larvae may result in intermediate forms of queens on maturation. **Note:** emergency or supersedure queens rarely look as robust as queens developed during the swarm season.

LAYING WORKERS

When a colony loses its queen and is unable to rear a replacement as a result of an absence of fertilized eggs or young larvae (less than three days old), some workers will eventually begin to lay unfertilized eggs (approximately after 21 days without a queen). The limited reproductive capabilities are no longer suppressed by, among other things, the presence of a

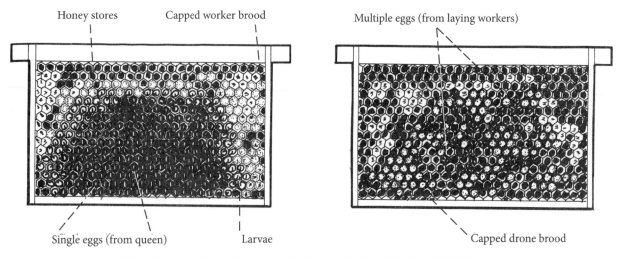

Brood Frame from Queenright (L) and Laying Worker (R) Colonies

living queen and her complement of pheromones. Worker bees, although females, are not programmed to undertake nuptial flights and lack some of the reproductive anatomy found in queen bees. The ovaries of these females mature, and, after they are fed royal jelly, the eggs in the ovaries will mature. Since workers are incapable of mating, the eggs they lay will be unfertilized; therefore, only drones will develop from them. As a consequence, the worker population within a colony with laying workers will slowly decline because the rearing of new workers stops with the loss of the queen.

Laying worker colonies are easily identified by the number of eggs seen in each cell; when a queen is present, there is only one egg per cell, and it is laid at the bottom of the cell. If more than one egg is present in a cell, and they are glued to sides of the cell walls instead of to the cell's bottom, laying workers should be suspected (see the illustration of the brood frame of a laying worker). If you come upon such a colony six weeks into this activity, the number of mature drones and capped drone cells will be somewhat striking.

One approach to returning the colony to a queenright state is to shake all the bees from their frames at a distance of 100 yards (91 m) or more from the parent colony just before the introduction of a queen, queen cells, or young larvae. This action supposedly prevents the laying workers, who are too heavy to fly, from returning to their colony, and, at least in theory, it is specifically the laying workers who prevent the acceptance of a new queen or a queen cell and prohibit female larvae from becoming a queen. In addition, it would also preclude all the bees that have yet to take orientation flights from returning to the colony, leaving the colony with only forager bees. Under these circumstances, you may introduce a mated queen, a virgin queen, a queen cell, or a frame of young larvae. As you can see from the previous information, this method leaves the colony with very few bees and in general does not work.

Unfortunately, attempts at rescuing a colony of laying workers from inevitable doom have never worked to anyone's satisfaction. Frequently the colony will reject the introduced queen or queen cells, or it will rear workers rather than queens from the introduced larvae. You also lose valuable honey-making time. So the best solution is to unite a colony of laying workers with a queenright hive. Experimenting with the other methods or inventing your own, however, may be worth the experience.

Notes

CHAPTER 12

Products of the Hive

HONEY

Honey is a bee product. The word *honey* will always be connected to honey bees and flowers. Honey is the substance produced when nectar, which is the sweet secretions from nectar glands of flowers, is gathered, modified, and stored in honeycomb by bees. Bees forage for and collect sugary secretions from flowers (called *nectar*). Nectar, primarily a mixture of sugars and water, also contains minerals, vitamins, pigments, enzymes, and aromatic compounds. Honey bees concentrate and alter the nature of some these ingredients in the process of converting them into honey. Bees also collect sugary secretions called *honeydew* from aphids and scale insects, which feed on the sap of plants.

Bees will collect other sugary solutions from sources such as discarded soda cans, candy, or discarded candy wrappers. Such solutions, even when processed and stored by bees, are not, by definition, honey. We will take the position that honey is a product whose principal raw materials are the molecules and compounds derived from the nectaries of flowers and from *extra-floral* nectaries (nectar glands on other parts of a plant), which some plants also have.

Nectar is produced by flowering plants to attract pollinators. In addition to honey bees, organisms such as, but not limited to, birds, bats, beetles, flies, butterflies, moths, wasps, and bumble bees also seek floral nectar. The close association of pollinators and flowers is found in the fossil record, and the development of nectaries and their specific visitors coevolved to the benefit of both (see Chapter 14); hence the different colors, scents, and shapes of flowers (e.g., bird-

pollinated flowers are often red, as bees cannot see into the red range). Additionally, flowers pollinated at night are often white in color and are very fragrant (e.g., night-blooming cereus).

Nectar
Physical Changes

The percentage of water versus sugars in nectar can be large: some nectars contain up to 80 percent water. In order to concentrate the sugars contained in the nectar, a drastic reduction in the water content of nectar is necessary. Under ideal conditions, the final product honey should have a water content around 18 percent. Bees visiting flowers for their nectar remain loyal on each foraging trip to the same flowering plant species; this behavior is referred to as *flower fidelity*. One may speculate that this repetitive behavior makes the gathering of nectar more efficient from the bee's standpoint. In turn, by visiting the identical flowering plant, the bee, while collecting nectar, inadvertently transfers the flower's pollen (sperm) to the female part of the plant. This transfer results in the fertilization of the flower's eggs, and thus the formation of seeds. Bees make possible plant reproduction (seeds), and the plant rewards the bee with nectar (and pollen).

The honey bee is equipped with a special organ, called the *nectar crop* or *honey stomach*, which is posterior to the esophagus. A valve at the posterior end of the honey stomach prevents the nectar from continuing into the remaining parts of the bee's digestive tract. On returning to the hive, the forager with its load of nectar seeks a receiving bee to unload this bounty. By unloading the contents of its honey stomach (or "honey sac") to a receiving bee, the forager is free again to exit the hive and continue its pursuit of collecting nectar. As one receiver bee passes nectar to another, and so on, they are also adding the enzyme *sucrase* (formerly known as *invertase*) to the nectar. This enzyme splits the sucrose sugar molecule into two simple sugars, glucose and fructose; see "Sugars" in Chapter 7.

Since the water content of nectar can be as high as 80 percent, it has to be reduced in order to increase the concentration of its sugar component. The two simple sugars glucose and fructose are the principal sugars found in honey, with fructose at a slightly higher percentage over glucose. Additional sugars include small amounts of sucrose, and some complex sugar molecules are also present, depending on the nectar source (see the table on properties of honey). The honey stomach also has many kinds of lactic acid bacteria (first discovered by Swedish scientists) that were found to aid in bee health.

In addition to sucrase, another enzyme in honey is *glucose oxidase*, which converts glucose to *gluconic acid* and *hydrogen peroxide*. These substances endow ripening honey with antibacterial properties and high acidity (low pH) but are very delicate and heat sensitive. Honey's high sugar content (about 80%) and low pH make it a hostile environment for bacteria. The enzymatic activity in nectar represents its chemical alteration into honey.

When honey is reduced to ash, trace amounts of minerals are found: these are calcium, chlorine, copper, iron, magnesium, manganese, phosphorous, potassium, silica, sodium, and sulfur. Other components of honey are acids, proteins, amino acids, and vitamins—all in trace amounts (see chart on chemical properties of honey on p. 206).

As the receiving bees in the colony pass nectar from one receiving bee to another, the nectar's water content is being evaporated, and the sugars become concentrated. Further evaporation of the water takes place as receiver bees regurgitate small portions onto the base of their tongues. By stretching out their proboscises, the nectar comes in contact with the air flow within the colony, evaporating more water from the nectar.

Eventually, the nectar is stored in honey cells, where the process of evaporating excess water continues. When the cells are filled and the water content has been reduced to around 18 percent, the bees cap these cells with a wax cap. If the wax caps are in direct contact with the honey, they appear wet (yellowish to brown in color). If the cappings, on the other hand, are not in contact with the honey beneath them, the cappings are referred to as *dry cappings* and appear to be white in color.

The evaporation of water from the nectar serves two purposes: by reducing the volume of water, the bees are able to store a more concentrated sugar solution in less space, and secondly, high concentrations of sugars make the product uninhabitable for fermenting organisms, and thus it will store for a long time. During strong honeyflows, some beekeepers stagger their honey supers by a half inch (1.3 cm) to increase the evaporation of water and thereby hastening the "ripening" process of the honey. Although this may accelerate reducing the water content of the nectar, if not monitored closely when the nectar flow ends, robbing bees will have easy access to such colonies.

We believe that honey bees, if given a preference, will visit flowers for their carbohydrate needs rather than other sources (e.g., open soda containers). However, during dearths, bees are attracted to other sources of sugar. If nonfloral sources are abundant, their water content will be evaporated, and the sugar solution will be stored in the comb. Two recent incidents of bees collecting nonfloral sugary solutions are worthy of note. Several beekeepers in the New York City area noticed bees storing a red liquid in their honeycombs. A close examination showed this liquid contained Red Dye No. 40, used in maraschino cherry juice. Bees were collecting this juice at Dell's Maraschino Cherry Company on Dikeman Street, in Brooklyn. In another case, beekeepers in northeastern France noticed that their bees were producing honey in shades of blue and green. In this case it was discovered the bees were collecting a sugary solution from the discarded remnants of colored M&M candy shells. Other beekeepers have noticed similar strange "honey."

During a dearth, bees will find other sources of sugar, including partially empty soda containers; this liquid can produce "honey" with off color or flavors. If you notice odd colors in your nectar, you may want to explore the area for any uncommon sugar sources (e.g., recycling centers, factories, amusement or other parks, dump sites). Use the internet to locate your

Chemical Properties of Honey

Principal components	Percentage
Water	17.2
Fructose or fruit sugar	38.2
Glucose or grape sugar	31.3
Sucrose or cane or beet sugar	1.3
Maltose and other reducing disaccharides	7.3
Trisaccharides and other carbohydrates	4.2
Total sugars	99.5
Vitamins, minerals, amino acids	0.5

Enzymes

1. Invertase (inverts sucrose into glucose and fructose, from bees).
2. Glucose oxidase (oxidizes glucose to gluconic acid and hydrogen peroxide plus water; from bees).
3. Amylase (diastase; breaks down starch to dextrins and/or sugars; may aid in bee's digestion of pollen).
4. Catalase (converts peroxide to water and oxygen) and acid phosphatase (removes inorganic phosphate from organic phosphates; from plants).

Aroma constituents

Alcohol, ketones, aldehydes, and esters.

Sources: J.W. White, Jr., M.L. Riethof, M.H. Subers, and J. Kushnir. 1962. Composition of American honeys. Washington, DC: USDA Bulletin 126. E. Crane. 1990. Bees and beekeeping: Science, practice, and world resources. Ithaca, NY: Comstock. National Honey Board, http://www.honey1.com and Bee Product Science, http://www.bee-hexagon.net.

At specific times of the year, there can be a very heavy nectar flow in your area from a single source, such as basswood, black locust, citrus, sourwood, or tupelo trees, or goldenrod and buckwheat plants. If you have empty drawn combs in your honey supers and place them on your colonies just as one of these specific nectar flows begin, you will have the opportunity to harvest *varietal honey*—that is, honey from a single source. Varietal honeys, when extracted, will provide a wide range of colors, from water white (clear) to black. These honeys command premium prices and provide a direct connection between your bees and the specific flowering tree or plant. In areas where strong honeyflows are uncommon or infrequent, most honey will be a mixture from several to many floral sources. For a complete description of the Standards for Extracted Honey, refer to the Pure Food and Drug Laws on the USDA website (see "Honey and Honey Products" in the References section).

If you wish to verify that a certain honey is from a mono-floral source, send samples to independent laboratories, and for a price they will extract the pollen from the honey and identify the flower that produced the pollen and hence the nectar that the bees converted to honey. A list of labs can be found on the National Honey Board website (www.honey.com). **Note:** The Honey Board is going to be changed to the "Honey Packers and Importers Research, Promotion, Consumer Education and Industry Information," or PIB, for Packer-Import Board (check the website).

A definition of honey from the National Honey Board is "the substance made when the nectar and sweet deposits from plants are gathered, modified and stored in the honeycomb by honey bees . . . without additions of any other substance." Other, international definitions are available to protect against the sale of substances that look like honey or are packaged like honey; such imitations must be clearly labeled.

colonies on an aerial map (such as Google Maps) and determine what the bees could be visiting within their two or so mile flight range. If you discover that your bees may be visiting a questionable location, you may consider moving your apiary. However, some beekeepers lease their colonies near recycling centers for the bees to "clean" used soda cans.

Forms of Honey

Honey is packaged and sold in several forms—as all liquid, a combination of liquid and comb, as all comb, or in a granulated or crystallized form. *Extracted honey*, the liquid form, is removed from the comb and packaged in bottles or jars. Honey is classified into colors ranging from water white (clear) through amber (gold) to dark (black). Light-colored honey tends to be mild in flavor, whereas darker honey has a more pronounced, stronger flavor.

Comb honey is sold when its honey is in the liquid state beneath the sealed cappings. Beneath the cappings, the honey may in time begin to form crystals (granulate), either rapidly (e.g., canola honey) or slowly (e.g., buckwheat honey). Consumers purchasing comb honey expect the honey to be in a liquid state. Honeys that tend to granulate rapidly are extracted as soon as possible. If your bees are located in areas were the major nectar source is, for example, canola or rapeseed, you must be prepared to extract the honey as soon as it has been capped.

Most honey will eventually granulate, or crystallize (becoming semisolid), if not treated (see "Granulation" below). Granulated comb honey cannot be extracted except with heat, which will melt the wax and honey until it is warm enough to separate. To anticipate if your comb honey will granulate, record the blooming dates of your major honey plants so you can identify which flower nectars granulate quickly.

The basic types of comb honey are:

- *Section comb*, consisting of individual round sections, called Ross Rounds (but these sections were originally designed by the late Dr. W.S. Zbikowski), or of individual square wooden boxes.
- *Bulk comb*, consisting of the entire comb on a frame.
- *Cut comb*, in which sections of sealed honey are cut out of their frames, and after the cells that were opened as a consequence of being cut have drained off, the sections are packaged in plastic containers.

Comb honey is the purest form of honey; the delicate aromas and flavors of its floral sources remain sealed beneath the cappings until the cappings are removed, usually as a result of being chewed or cut into smaller pieces. Unfortunately, many honey customers do not know how to eat comb honey or what to do with the wax as they chew it. A sales technique used by beekeepers to improve sales of comb honey is to provide samples for customers at farmer's markets and other venues. Once customers have been introduced to comb honey, they usually become permanent fans of this form of honey. In addition, providing recipes for the many uses of all forms of honey also helps promote honey sales (see "Selling Your Hive Products" below and "Honey and Honey Products" in the References section). Remember, most honeys, even comb honey, may eventually granulate or crystallize (become semisolid). Proper storage of most forms of honey will delay this process.

Granulated or *finely crystallized honey* is a unique form of honey. The crystallizing of honey into very fine crystals was perfected by Elton Dyce while a graduate student at Cornell University and is still referred to as the Dyce method for crystallizing honey.

Liquid honey or extracted honey is bottled in liquid form and is the one most familiar to consumers of honey.

Honeydew

If we adhere to the strict definition of honey, that it is produced by honey bees from nectar gathered by them from floral and extra-floral nectaries, then *honeydew honey* is a separate hive product. In fact, we may infer that is the reason that honey from floral sources is always labeled with its floral source (e.g., "Clover Honey"). In the case of honeydew honey, the source of the sugars is not indicated on the label; it may be assumed that packers, by labeling the product honeydew honey, are proclaiming it is not floral honey. Furthermore, the word "honeydew" does not convey to the typical consumer the meaning of honeydew. Nevertheless, it is certainly of great interest to learn that in addition to the sugary nectar produced by flowering plants, bees can gather another liquid sugary solution that they can convert to a product containing some of the sugars found in floral honey. We rest our case in this matter by noting that one of the foremost authorities on honey, the late Dr. Jonathan W. White Jr. of the US Department of Agricultural Research Service, referred to this nonfloral source of sugars as "honeydew," not "honeydew honey" (see Chapter 10 in J.M. Graham, *The Hive and the Honey Bee*, 1992).

Honeydew refers to the sugary excretions produced by plant-sucking insects, such as aphids (Homoptera: Aphididae). These insects are found feeding on the tender shoots of conifers such as pines (*Pinus* spp.), fir (*Abies* spp.), spruce (*Picea* spp.), and larch (*Larix* spp.), as well as from the leaf phloem (sap) of some deciduous plants such as oaks (*Quercus* spp.), maples (*Acer* spp.), willows (*Salix* spp.), and some others.

The aphids are sucking the sap from these plants to extract the amino acids and other nutrients, and excreting the excess water and sugars onto the leaves and twigs. If the secretions are profuse, they are eagerly collected by honey bees, because they are rich in sugars normally found in honey and contain some other complex sugars. "Pine honey" or "forest honey" (*Waldhönig*) is popular in some parts of the world, especially Europe. This "honey" comes from honeydew found on some evergreens in European forests.

Honeydew honey comes in varying flavors and colors, depending on the plants on which the aphids have fed. One characteristic of some honeydew honey is its rapid crystallization. It is possible that many honeys contain some honeydew, especially if bees are collecting nectar in areas next to plants with heavy aphid populations.

Microscopic components in honey can be used to identify its source; these include pollen for nectar honey and algae and sooty mold particles from honeydew honey. Dark honeydew honey can be somewhat bitter in flavor, because of its higher mineral and ash content. Proponents of this honey claim it has higher antibacterial activity.

THE HONEY HOUSE

A sanitary honey house for extracting, bottling, or otherwise handling honey is very important, and in some states it must meet certain health and food safety requirements. It is imperative that it be kept clean—running water and a washable floor are necessities—and be as insect-proof as possible. Bees and other pests attracted to the smell of honey or wax can get into everything, including newly strained or uncapped bottles of honey. Fecal material from insects and animals is also a problem in any shed that is accessible to insects and small mammals.

If you have only one or two hives, you can extract in the kitchen or a clean basement. Put down some newspaper to keep the floor from getting sticky. Cut the cappings off a frame or two of honey, let the honey drip into a pan; if the room is warm, the honey will drip out faster. One of the properties of honey is that it is *hydroscopic*, which means it has the ability to absorb moisture from the air. When the water content of honey is increased, the product will begin to ferment (spoil). Damp basements therefore are not a good place to extract honey.

For those who have 10 or more colonies, you may have to commit to one location designated for extracting, and all the other necessary associated tasks involved in honey bottling, rendering wax, etc. If you decide to move in this direction, before building or renovating a room or rooms as a place to extract and bottle honey, show your plan to your local and state health officials and ask for their suggestions and recommendations. An alternative is to rent or get permission to do your extracting in a facility approved by the local and/or state health departments.

Use these guidelines in planning a good honey house; include the following necessities:

- Electricity.
- Hot and cold running water.
- Restroom facilities.
- Washable floor (concrete or ceramic tile) with center drain.
- Washable ceiling and walls, coated with "food approved" paint.
- Hose and water source to wash floors.
- Easy loading/unloading area for moving in heavy honey supers.
- An uncapping/extracting area where sticky, wet frames are handled and honey is strained.
- A hot room, to warm unextracted honey supers and dry down those not fully cured.
- Dehumidifier, if area is excessively damp.
- Storage space for empty, dry honey supers in an unheated portion of a building.
- Dust-free space for bottling, labeling, and storing honey.

Remember: Before building or renovating a place to extract and bottle honey, show your plan to your local and state health officials and ask for their suggestions and recommendations.

Make sure when using metal equipment for honey processing and storage to use only stainless steel, because the acid nature of honey will react to all other metals. Plastic is safe to use. Though not essential, the following equipment may also be stored in a honey house, if the structure is large enough:

- Uncapping knives.
- Extractor(s).

- Cappings tank or tray.
- Honey pumps.
- Nylon mesh straining cloths.
- Capping baskets.
- Screen to drain cut comb honey.
- Bottles and labels.
- Holding tanks for bottling and barrels for bulk storage.

If the area has sufficient additional space, you may be able to use a section of the space for assembling frames and repairing and storing equipment. You can store your smokers, hive tools, gloves, suits, and veils here as well. You could also think about putting in an extra washing machine and a refrigerator or freezer devoted to your beekeeping operation.

Some excellent floor plans have been published in various books; check out the "Bees, Beekeeping, and Bee Management" and "Honey and Honey Products" sections in the References. The 2015 edition of Graham and Langstroth's *The Hive and the Honey Bee* (Dadant & Sons) has some basic floor plans, as does *The ABC and XYZ of Bee Culture* (A.I. Root, 2018). Talk with other beekeepers, go online and look at various bee sites, and visit honey houses in your area, large and small, to see how other beekeepers handle their hive products. Some bee suppliers also sell floor plans for extracting rooms.

Beekeepers with fewer than 10 to 20 hives can easily manage by using any sanitary space (garage, basement) and an extractor, uncapping knife, and some of the equipment listed above. Or consider getting together with other beekeepers in your area, or members of a bee club, to buy or share equipment and space. Many beekeeping organizations have expensive equipment that members can borrow at a minimal cost. For instance, the Connecticut Beekeeper's Association has a honey extractor program; every county in Connecticut has one of these extractors that beekeepers can rent. An arrangement whereby beekeepers help one another pull and extract honey supers, dividing the finished product proportionally, should be made ahead of time, to avoid misunderstandings.

If you have 100 or more hives, congratulations, you are no longer a hobby beekeeper. However, more recently, beekeepers with a small number (up to 20) of colonies are now considered as *small*, *part-time*, or *sideliner* beekeepers rather than hobby beekeep-

ers (compared to full-time or commercial beekeepers). **Note:** Dr. N. Ostiguy from Penn State University (https://ento.psu.edu/directory/nxo3) reported that this change in wording from hobby to part-time is a result of comments by congressional members when approached to support research on honey bee diseases and the current decline in honey bee colonies. Congress doesn't provide money for research on hobbies, but it does provide money for research irrespective of the size of operations.

Consider going into this as a business and purchasing more professional, commercial equipment. If you are heading in this direction, work with some bigger or commercial beekeepers to see how they run their operations before investing a lot of your hard-earned money.

Precaution: Now that small hive beetles are here, stored honey supers should be monitored, because this pest can invade and destroy honey still in the comb (see "Small Hive Beetle" in Chapter 14). You can alleviate this problem by circulating the air, via fans, on your supers or putting the supers into a cooler; the best solution is to extract your honey within a few days.

Extracting Honey

The best time to extract honey is when a frame of honey is 100 percent sealed with wax cappings. Usually frames of honey that are not completely sealed are safe to extract without increasing the water content of the honey to any appreciable degree. **Note well**: in humid climates, only frames that are completely capped or at least 80 to 90 percent capped should be selected for extracting.

Bee equipment with frames of honey ready to be extracted should be placed in a bee-tight room or honey house. If the room temperature is between 80° and 90°F (26.6–32.2°C) the honey can be extracted with ease. If you are removing honey supers in cold weather, place these in a warm room until the honey is room temperature and easily flows, to facilitate extraction. To prevent the granulation of honey, avoid storing honey supers in temperatures below 57°F (13.9°C); or you can freeze them if you have the room, as this will slow the granulation of the honey.

The wax cappings that seal the honey in the cells are commonly cut away with a heated or an electric uncapping knife or rotating chains or blades. Small-

Radial Honey Extractor

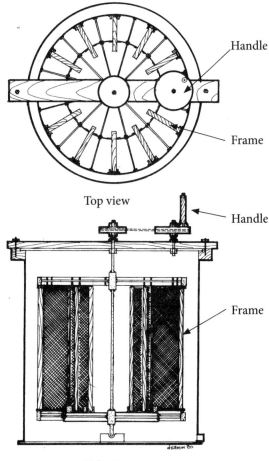

Top view

Side view

Basket Extractor

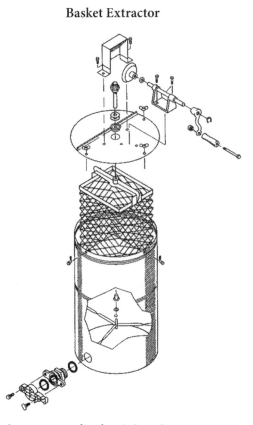

Used with permission of Dadant & Sons, Inc.

scale beekeepers generally use an uncapping knife, whereas the commercial outfits have more mechanized equipment. Cut the cappings off both sides of the frame, letting them drop into a screened basket or into some other container that will permit the honey to drain off. These cappings are later melted down to separate them from any remaining honey (see "Wax Cappings" below).

Place the frames into an extractor (radial or basket type) such that frames of equal weight are opposite each other (see illustration of honey extractors). If the weight is distributed unequally, the extractor will wobble and vibrate wildly, due to the unbalanced load.

If using a *basket-type* extractor (frames are parallel to the sides), start with a slow spin and gradually increase the speed. Spin the frames on one side for three minutes, then reverse them and spin on the other side for three minutes. Continue to reverse the

frames until most of the honey has been extracted from the combs. Because of the viscosity of honey, it is impossible to remove all the honey from the frames.

If using a *radial* extractor (frames are perpendicular to the sides), there is no need to reverse frames—both sides of the frame are extracted simultaneously because of their perpendicular placement. Start with a slow spin of about 150 revolutions per minute (rpm) and gradually increase speed to 300 rpm. Spin at the maximum rate for about five minutes. Honey can be draining from the extractor, into a strainer on a bucket, while it is spinning, but keep your eye on the bucket, as it will quickly fill up and overflow. On occasion, a queen may enter a honey super and lay eggs in the frames. In such cases, some of the frames in a honey super may contain a mixture of honey, eggs, and open and sealed brood. Depending on the time you removed the super and when you started to extract, the brood may no longer be viable. There are options available to you in this case: you can return the super to a colony and wait for all the brood to emerge, or you can allow the bees to remove the dead brood and, depending on the season, the bees may then fill the vacated brood cells with honey.

It is very important **not to feed antibiotics** to colonies that you expect to remove honey supers from, and not to apply mite treatments, other than those that permit their use when honey is on a colony. **No medications should be applied to bee colonies that have honey supers**. There is **no tolerance** for these materials in honey for public consumption.

After Extracting Honey

Extracted honey should be strained immediately, to remove wax, bees, and other debris, and may require a second or a third straining through finer mesh. The strainer can be made of nylon or metal screen—any material that is easy to wash and will not become easily clogged. Stay away from cloth, such as cheesecloth, for straining honey, because it can leave lint and other foreign matter in your honey. Remember, warm honey strains much faster than cold honey.

After extracting:

1. Honey can be stored in a variety of containers ranging from large storage containers or holding tanks, to 55-gallon barrels, 60-pound tins, or smaller 5-gallon containers. Honey in a storage tank should sit overnight, covered, in a warm room before being bottled. During this time, fine air bubbles and wax particles rise to the top. Skimming off this layer before bottling results in a better, cleaner finished product. The honey can be sold in 55-gallon barrels, 60-pound tins, or 5-gallon containers, or bottled into 1-, 2-, or 5-pound jars.

2. Remove the empty, wet frames from the extractor and place them in empty supers; return these to hives at dusk to allow bees to clean the wet frames. To avoid robbing problems from yellow jackets and other bees, duct tape over any holes or cracks in the hive bodies to block such entrances. If no other honeyflows are anticipated, remove the cleaned supers after a week or so and store them. Wax moths will destroy combs that are not properly stored—that is, if not stored at freezing temperatures or otherwise protected (see "Wax Moth" in Chapter 14).

3. Extracted combs should not be stored wet, because any remaining honey will crystallize in the cells, providing the "seeds" that will hasten the granulation of next year's crop. Wet combs can also ferment and mold, in which case you should melt them down and install fresh foundation. Now is a good time to cull combs that are distorted, broken, or otherwise damaged.

4. Remove as much honey as possible from the cut wax cappings and then melt the wax cappings down (see "Wax Cappings," below).

Physical Properties of Honey

Dr. Kirsten Traynor, in Graham's *The Hive and the Honey Bee*, lists many of the physical properties of honey, which include refractive index, viscosity, optical rotation, density and specific gravity, hygroscopicity, thermal properties, crystallization, bacteria, yeast, fermentation, color, and hydroxymethylfurfurnal (HMF). Let's look at some of these properties in detail.

Hygroscopicity is the ability of a substance, in this case especially fructose, to absorb moisture from the air. Honey, which contains different amounts of fructose (levulose), will absorb moisture from the air. Honey exposed to moist air (such as uncapped honey in the comb, or honey in uncapped bottles or in other uncovered equipment, like holding tanks) will absorb moisture. In areas of low humidity (desert regions), honey will "dry out," returning moisture to the air. Any exposed surface of honey will begin to accumulate moisture. It is critical to keep honey in sealed containers, bottles, vats, or storage tanks. Leaving extracted honey even overnight in an open container is not a good idea. When honey incorporates too much water, sugar-tolerant yeasts begin to multiply and will break down the fructose in honey into alcohol and carbon dioxide. **Mead**, also called honey wine, may have been accidentally produced when an open container of honey absorbed water from the air, thus creating an environment amenable to yeast, which could lead to fermentation.

The good side of this hygroscopic property of honey is that it benefits the baking industry by keeping certain baked products moist. Honey as an ingredient in cosmetics serves the same purpose.

Fermentation. Sugar-tolerant or *osmophilic* yeast spores (genus *Zygosaccharomyces*), under high moisture conditions, are able to germinate in honey and metabolize its sugars. These are "wild" yeasts, not the ones used in bread or beer. As the sugars of honey are metabolized, these yeasts produce alcohol and carbon dioxide as by-products, which will spoil honey. Originally, these yeasts may have been used to produce early alcoholic drinks; today "cultured" yeasts give more predictable results, and are used to make mead (honey wine) and honey beer. Probably all honeys contain osmophilic yeasts in the form of dormant spores; given the right conditions, these spores will germinate and multiply.

Conditions that will influence fermentation are:

- Water content of the honey.
- Temperature at which the honey is stored.
- Number of yeast spores in the honey.
- Granulation of the honey, which results in an increase in the water content of the remaining liquid portion.

Fermentation of honey can be prevented if its moisture content is less than 17 percent, if it is stored at temperatures below 50°F (10°C), or if it is heated to 145°F (63°C) for 30 minutes, which kills the yeast. Heat the honey rapidly (it must not reach a temperature of more than 180°F, or 82°C), to prevent it from burning and darkening; stir carefully while heating. Cool honey quickly to retain the delicate flavors. Once honey has fermented, it cannot be saved and should not be fed to bees because the alcohol content may poison them.

Granulation. Most honey, which all begins in the liquid state, will eventually granulate, returning to a more solid state. Some honeys, such as canola (rapeseed), may even granulate in the comb. Other honeys will granulate in just a few days after being extracted (pure canola, aster, mesquite, and goldenrod), whereas other varietal honeys remain in the liquid state for weeks, months, and even decades (alfalfa, sourwood, and tupelo). The granulation of honey results when the sugar glucose loses water and becomes glucose monohydrate. It takes on the form of a crystal; in time, these crystals form a lattice that immobilizes other components of honey, creating a semisolid state.

There are two kinds of crystals that glucose forms: one is a simple sugar crystal, and the other is a glucose hydrate crystal. The latter contains some water, and because of this, when honey crystallizes, the chances increase that it will also ferment. Honey that granulates naturally can also produce undesirably large crystals (as does canola honey) or very fine crystals (as does mesquite honey).

In order to keep stored extracted honey in a liquid state, it has to be heated to 145°F (63°C) for about thirty minutes. To keep out any "seed" or particulate matter, air bubbles, wax particles, or pollen, commercial packers force "flash-heated" honey through special filters to strain out any foreign particles. The cleaned honey is then "flash cooled" and stored or bottled.

Partially granulated but unfermented honey, often considered "spoiled" by the uneducated consumer, is in fact perfectly good to eat. Larger honey producers make a form of controlled granulated honey, or "cremed" honey, by a method called the Dyce process (see below in this chapter) and sell it as a spread. (Note the spelling of "cremed"—it is missing the "a" of cream, so as not to suggest it contains any dairy ingredients.) This honey has very fine crystals and does not ferment. The ideal temperature for honey to granulate is 57°F (14°C). Unless the object is to produce this kind of honey, store honey above this temperature. To prevent granulation during long-term storage, honey may be kept in a freezer.

To liquefy small quantities of granulated honey, place the container (glass jar only; plastic will soften) in a pan of warm water until the crystals are melted. Do not let the honey overheat, as many of the flavors and aromas of honey are volatile and are destroyed by heat. Five-gallon buckets or barrels of honey are generally liquefied in a "hot room." Another method is to put an uncapped container of honey into the microwave and heat it for one minute and then in additional 30-second increments, until it is liquefied. Never leave honey containers unattended when heating.

Thixotrophy. A few rare honeys have thixotropic characteristics. In the comb, the honey appears to be solid and cannot be extracted because of its thick, viscous nature. However, if the honey is subjected to vibration with a special type of extractor, or as the honey is being spread on bread, it will liquefy. As soon as the vibration stops, the honey reverts to a thick, gel-like solid. A single, particular plant protein imparts the honey with this unique property. The most famous thixotropic honey is from ling heather (*Calluna vulgaris*), commonly found growing on the moors of Europe. Another is pure grapefruit honey

(*Citrus paradisi*) and Manuka honey (*Eucalyptus* spp.) from New Zealand.

Formation of HMF. Hydroxymethalfurfural (HMF) is produced when fructose breaks down in the presence of high temperatures and some metals. HMF is measured if honey is sold on the world market; if levels are too high, the honey may be rejected. It is best to properly dispose of honey containing a high percentage of HMF.

Factors that favor HMF formation are:

- Fructose (starting material); HMF will not form in sucrose syrup.
- Heat.
- Time.
- Low pH (acidic conditions).
- **Metal ions** (such as iron, stabilize HMF); for example, if the honey (or high-fructose corn syrup) is stored in an untreated metal container.

If HMF levels are very high, such as when honey or high-fructose corn syrup have been stored in the sun, do not feed this to bees. If you rehydrated the syrup by adding water, HMF breaks down and forms levulinic and formic acids. Both these acids are toxic to bees and make the honey or syrup unpalatable. It is best to throw away this honey or syrup.

Medicinal Properties of Honey

Honey is known for many healing properties. Some have been scientifically studied, and some have not. Various groups maintain that honey is helpful in:

- Retaining calcium in the body.
- Counteracting the effects of alcohol in the blood.
- Providing quick energy.
- Deterring bacterial growth, especially on burns.

Medicinal applications of honey (such as for burns and wound healing) have been reported, showing that honey:

- Contains an anti-inflammatory agent.
- Stimulates new tissue growth.
- Has hygroscopic properties that promote healing and prevent scarring.
- Provides a protective thick layer.
- Has antimicrobial properties (due to the low pH and low water content and the presence of hydrogen peroxide).
- Contains antioxidants; darker honey contains more antioxidants than lighter honey.

Check the website for National Honey Board (www.honey.com), or the Packer-Importer Board or PIB; they should have the latest information on the medicinal properties currently studied in honey. Also, very good information is in *The ABC and XYZ of Bee Culture* (A.I. Root, 2018).

Bottling Extracted Honey

After all your honey is extracted, you should begin bottling it within a month or sooner, because the honey in a large holding tank could crystallize, which will make it more difficult, and require more heating, to make it free-flowing again. Before starting to bottle, have everything ready in a dust-free and clean room. If not done already, heat the honey to 140° to 160°F (60–71°C) to kill yeast spores, and begin to fill the bottles and then cool the honey quickly. Bottling dispensers or gates at the bottom of tanks or buckets make filling small bottles easy; for larger operations, a nondrip and accurate bottler is worth the investment; check bee supply companies or visit a honey packing plant to see how they operate and what equipment they use.

Jars used for honey are usually glass, called "queenline" jars, with a distinctive shape (see illustration of various types of honey containers). They are the best because they can be recycled and heated easily. Some plastic jars are also good but may collapse when heated. Honey bears now come in many styles and are a popular and recognizable container. Containers can be ordered from suppliers advertised in bee journals or online.

Fancy jars for liquid honey are also available; the Muth or square-shaped jar embossed with a skep is particularly distinctive; such containers are eye-catching and can be used to promote a varietal honey or given as gifts to special clientele.

Jars must be washed before being filled, even newly purchased jars straight out of the box, to eliminate dust and other contaminants. Wash them in a dishwasher and dry upside down so that all the water will drain out. Make sure your caps are also clean and dust free, and use some kind of gasket or sealer

Various Types of Honey Containers

1. Plastic squeeze skep
2. Glass hexagonal jar
3. Plastic squeeze honey bear
4. Plastic queenline jar
5. Glass queenline jar
6. Muth jar

1. 2. 3. 4. 5. 6.

to ensure honey-tight lids. Different styles of tamper-proof lids and seals are also available, and if you are selling honey to commercial stores, you need to investigate these kinds of seals. Labels can be printed from your home computer or purchased preprinted for a professional look.

A typical mid- to large-size bottling operation should include:

- Jars (stored in cardboard boxes).
- Dispenser or filling machine with automatic shutoff.
- Automatic capper.
- Labeling area and labels.
- Packing area, where filled jars are placed into the boxes.
- Shipping and storage area for boxes of jars.

The shelf life of bottled honey is around six months, depending on the kind of honey (some crystallize faster than others) and storage temperatures. If your bottles are stored longer and the honey begins to crystallize, you can put them in hot water baths to reliquefy; then label and package. For more information on selling your product, go to the end of this chapter.

Cooking with Honey

Honey is a natural sweetener, often used to replace sugar in cooking. Because honey is a combination of sugars that are broken down by the bees into the simple sugars (fructose and glucose), it is very digestible. Honey is also able to keep baked goods moist because of its hygroscopic property discussed above.

When substituting honey for sugar in any basic recipe, the following should be noted or observed:

- Measure honey with a greased utensil.
- Honey has 1½ times the sweetening power of sugar.
- Reduce the liquid in the recipe by ¼ cup for each cup of honey used to replace sugar.
- Use a mild-flavored honey, unless the flavor of the honey is a necessary part of the product.
- Some people add ¼ teaspoon of baking soda per cup of honey to counter honey's acidity.
- Reduce the cooking temperature of the final product by 25°F.

If honey is stored in a dry place (not the refrigerator) or if it is frozen, it will granulate (crystallize) much more slowly and will not ferment. Honey

Other Measurements of Honey

- Specific gravity of 15% moisture @ 20°C: 1.4225.
- Acidity: average pH for honey = 3.9; contributes to honey's antibacterial nature (pH for vinegar, Coca Cola, beer = 3.0; for tomatoes = 4.0).
- Osmotic pressure: more than 2000 milliosmols/kilogram (makes honey a hyperosmotic solution); also contributes to antibacterial activity.
- Refractive index: 1.55 if water content is 13%, 1.49 if water content is 18%.
- 1 gallon (3.785 liters) weighs 11 pounds 13.2 ounces (5.357 kg).
- 1 pound (0.453 kg) has a volume of 10.78 fluid ounces (3.189 milliliters).
- Caloric value: 1380 calories/pound.
- 1 tablespoon = 60 calories.
- 100 gram = 303 calories.

Thermal Characteristics

- Specific heat: 0.54–0.60 cal/g/°C at 20°C (68°F).
- Conductivity of honey at 21% moisture at 21°C: $12.7 \times 10-4$ cal/cm^2/sec/°C.
- Conductivity of honey at 17% moisture at 21°C: $12.8 \times 10-4$ cal/cm^2/sec/°C.
- Freezing point at 15% moisture: from –1.43 to 1.53°C (32.4°F).

Sweetening Power

- 1 volume of honey = 1.67 volumes of granulated sugar.
- 1 pound = 430 grams (0.95 lb.) of sugar.
- 1 gallon (3.785 liters) = 9.375 pounds (4.25 kg) of total sugars.

Color

Color	Pfund scale (mm)	Optical density
Water white	< 8	0.0945
Extra white	9–17	0.189
White	18–34	0.378
Extra light amber	35–50	0.595
Light amber	51–85	1.389
Amber	86–114	3.008
Dark amber (molasses color)	> 114	—

Sources: Honey Technical Glossary, http://www.arkadiko-meli.gr/default.asp?contentID=568; Airborne Honey Company, http://www.airborne.co.nz/manufacturing.shtml; E. Crane, ed. 1975. Honey, a comprehensive survey. London: Heinemann Ltd. H. Shimanuki, K. Flottum, and A. Harmon (eds.). 1992. The ABC and XYZ of Bee Culture. Medina, OH: A.I. Root. Also check References under "Honey and Honey Products."

stored this way could last for years. For more information see "Honey Cookbooks" in the References or search the internet for honey recipes.

Cautionary note: It is not advisable to feed honey to infants less than one year old. Spores of the bacterium *Clostridium botulinum* exist everywhere and are able to survive in honey. In the digestive tracts of infants under twelve months of age, the spores may progress to the vegetative stage of their life cycle and produce toxins that could prove fatal or injurious to infants. This disease is known as infant botulism. Again, check the Honey Board websites for more information.

Preparing Comb Honey for Market

If you are cutting comb honey from a frame (cut comb honey), use a warm, sharp, thin-bladed knife and cut the comb on some kind of screen to allow the honey to drip off. (Wire fencing material over a large flat pan works well). Use a template when you cut the squares of honey to fit into your containers; that way all the sections will be uniform. Allow the honey to drip out overnight, and carefully package each section in a clear plastic cut comb box or other clear container. Make sure no boxes leak when they are displayed or sold. Leak-proof containers are available from a variety of sources; check the internet, bee journals, and beekeeping suppliers. Special labels are also available for these boxes. These cut comb sections can also go into liquid honey jars to make chunk honey.

Section comb honey, whether in plastic, in wooden boxes, or in plastic cassettes (or other comb honey

---

Dyce Process

This process for making crystallized honey was developed and patented by E.J. Dyce in 1935. If made properly, this cremed honey is a granulated honey product with very fine and smooth crystals that spreads like peanut butter. To make cremed honey, follow this procedure.

Have on hand a "seed" or starter honey. The starter should be from store-bought crystallized (cremed) honey that has a very fine-grained consistency. You can also reuse your own granulated honey, as long as the crystals are ground very fine. Heat honey that has between 17.5 and 18.0 percent, water, until it reaches 150°F (66°C), stirring constantly. This pasteurizes the honey to kill any yeast spores that would otherwise cause the product to ferment. Strain through two to three layers of very fine nylon mesh to collect all pollen grains, sugar crystals, wax particles, and other impurities. This straining is very important; otherwise the honey may "seed" on these particles instead of on the seed honey you add. Cool honey rapidly to 70° to 80°F (21–27°C), stirring slowly. Do not add any air bubbles into the honey; do this by stirring slowly without breaking the surface of the honey.

When the temperature is between 70° and 80°F (21–27°C), add 10 percent by volume the starter seed (i.e., if you have 10 lbs. honey, add 1 lb. starter). Incorporate this into the honey without the addition of air bubbles, as bubbles will create a frothy top on the finished product. Let this settle for a few hours, skimming froth off the top (air bubbles will produce a frothy appearance) and then pour the mixture into the final, marketable containers. Now store this honey at temperatures between 45° and 57°F (7–14°C) for a week, until completely crystallized. Label and sell. Do not allow these containers to overheat, because they will become liquid again. Store crystallized honey in a cool, dry place until ready to sell.

Source: The ABC and XYZ of Bee Culture: An Encyclopedia of Beekeeping (A.I. Root, 2007).

---

section boxes), is relatively simple to prepare for the consumer. It takes no special equipment except labels to fit onto the boxes. One problem you will have is what to do with stained, damaged, or half-filled sections. Half-filled ones can be used the next season to bait new section supers. Stained or damaged sections can be sold as seconds, put in a jar and filled with liquid honey (to make chunk honey), or crushed to extract the honey.

All the sections need to be as clean as possible, free from debris, propolis, and other foreign matter. A sharp knife is used to scrape off the propolis that accumulates on the rim of the sections. This job is both fast and easy, but make sure you do not nick the comb or allow debris to fall inside the finished product. Nicking the comb or otherwise damaging it, or causing it to detach from the section box or holder, will cause the honey to leak into the final package. Leaky section boxes are not only unattractive but also do not sell well and discourage future customers.

Store your comb honey sections in the deep freezer (either boxed or in the section super) to kill any ants or other hitchhikers and to keep the comb from granulating. Be careful not to bump or drop these delicate honeycomb sections; the slightest bump will cause a crack in the cappings that will surely leak when the honey warms up.

If you are using the old-fashioned wooden sections, as an extra precaution place each section in a plastic sandwich bag and then store them in the freezer in a box. Once they are back at room temperature, slip each section inside the cardboard display box (that you have purchased). Do not freeze the wooden sections in the cardboard boxes because the boxes may get wet and fall apart when defrosted.

If you have frozen your other comb honey sections, the same rule applies: attach the labels after the boxes have come to room temperature, or the labels will wrinkle on the wet containers. Print the price and weight on each section and stamp it with your name, address, and zip code. Whichever section comb system you buy, make sure you purchase the correct labels and other accoutrements. Be sure to use indelible ink to keep it from washing off. In general, comb honey has a shelf life of two to three months.

Honey Products

Most extracted honey is sold for table use or for use in baked goods. However, there are many other products that contain honey that can be sold as unique or value-added products. You can also experiment with your own unique brand of manufactured goods. Some suggestions listed here have been on the market for a long time, and sales continue to grow. Check online for other ideas for honey products.

You can make honey beer, honey wine (mead), and honey vinegar. Other products include honey butter and candy made with or filled with honey, honey ice cream, honey drinks, and dried fruit and nuts stored in or mixed with honey. Some of these products are very perishable and will need to be refrigerated. If you add other food to honey, it then becomes a "food product"; check and comply with all federal and state food laws before selling these products.

BEESWAX

The domestic wax industry can obtain only two-thirds of the beeswax it needs from US beekeepers; the rest must be imported. Beeswax is used for cosmetics, candles, and foundation. Beekeepers probably are the most common users of beeswax, recycling their wax back to bee supply companies to make into wax foundation for their frames. Minor uses include pharmaceuticals and dentistry concerns, followed by foundries and companies that make floor polishes, automobile wax, and furniture polish. Beeswax is a minor ingredient in some adhesives, crayons, chewing gum, inks, specialized waxes (grafting, ski, ironing, and archer's wax) and woodworking supplies, and for arts and craft projects.

Wax combs are lipophilic, having an affinity to combine with fats and lipids. Nearly all agrochemicals used today are lipophilic and readily dissolve in wax; these chemicals have been reported in wax as old as 20 years. Recent studies have detected a host of chemicals in wax combs, including insecticides, fungicides, and a high level of miticides, which are used by beekeepers to control the varroa mite. Today, as a consequence of the lipophilic property of beeswax, all commercial comb foundation rendered from beeswax contains insecticide, fungicide, and miticide residues. It is recommended that brood combs that are three years old be rotated out and replaced with foundation. Some beekeepers now practice rotating out two frames from each brood chamber each year, resulting in a complete replacement of brood combs every five years. Others place a small starter strip of foundation in a brood frame and let the bees drawn their own comb from the strip, rather than placing an entire piece of chemically contaminated foundation into a frame. By so doing, the newly drawn combs, with the exception of the strip, will be temporarily free of lipophilic chemicals (see "Beeswax" in the References for a list of labs that analyze samples).

Because of the ability of beeswax to become contaminated with such wax-soluble pesticides, some "organic" beekeepers are letting the bees make their own comb instead of using commercial foundation. They are worried that commercial foundation may contain a low level of chemical compounds that bees come in contact with while foraging. Again, beeswax can hold many such impurities for years, and frames should be changed regularly.

Wax foundation is expensive, and beeswax is a valuable hive product; you should make every effort to save all cappings, old combs, and bits and pieces of extra wax scraped from frames and other hive parts. Melt these pieces in the solar wax melter, and trade or sell wax blocks to dealers. Separate out old dark combs from lighter cappings or new wax, because the lighter shades can be worth more when selling wax products.

Cappings, old combs, and wax scrapings should be kept in airtight containers or frozen until you are ready to process them. Wax moths can quickly infest such wax if not protected, or their eggs may be already present, waiting to hatch. Melt cappings separately from the old combs, since the latter contain non-wax substances that would impregnate and reduce the value of the almost pure wax cappings. Use extreme caution when melting wax—it ignites easily, and wax fires are difficult to put out (see the information on beeswax in the sidebars).

If you have used chemical miticides in your colonies, you can use your wax to make nonedible products such as candles, ornaments, or furniture wax or polishes, or sell bars for craft or sewing projects. Nevertheless, beekeepers need to be more judicious when making wax products because of the contaminants now known to be found in beeswax. Wax products should be made from the cleanest wax available, such as the fresh capping removed from honeycombs.

Characteristics of Beeswax

- Melting point: 142–151°F (61–66°C); wax from cappings: 146.7°F (63.7°C). Paraffin melts at 90–150°F (32–66°C).
- Solidification point (liquid wax becomes solid): 140–146°F (60–63.5°C).
- Flash point (wax vapor ignites): 490–525°F (254–274°C), depending on purity of the wax.
- Temperature of plasticity: 89.6°F (32°C).
- Relative density at 68°F (20°C): 0.963 (water is 1.0, so wax floats).
- Saponification value: 95.35.
- Acid value: 16.8–24.0.
- Ester value: 66–80; ratio of ester to acid is 3:4.3.
- Refractive index (light bending property) at 176°F (80°C): 1.4402.
- Electrical resistivity: $5–20 \times 1012$ ohm m.
- Shrinkage: If melted to 200°F (93.3°C), shrinks 10 percent (by volume) when cooled to room temperature.

Solubility: Insoluble in water; slightly soluble in cold alcohol; soluble in benzene, ether, chloroform, and fixed or volatile oils. Beeswax is stable for thousands of years.

Beeswax Components

Formula: $C_{15}H_{51}COOC_{30}H_{61}$

Component	Percentage of total
Monoesters	30–35
Diesters	10–14
Hydrocarbons	10.5–14
Free acids	8–12
Hydroxy polyesters	8
Hydroxy monoesters	4–5
Triesters	3
Acid polyesters	2
Acid esters	< 1
Free acids	12
Free alcohols	< 1
Unidentified	6

You can send in samples of your wax for testing (at a price); but then you can sell it as "tested" and pesticide free; check with your state or local bee organizations.

Melting Beeswax

Use extreme care when processing beeswax, as it is highly flammable and can cause serious burns.

Follow these simple precautions when melting beeswax:

- Extended exposure to high heat over 185°F (85°C) will darken the wax.
- **Never** use an open flame; use a double boiler over electric heat.
- Use only aluminum, nickel, tin, or stainless steel containers to melt wax, as other metals will discolor wax.
- Never use direct steam to melt wax, as it permanently changes the composition of wax.
- Separate dark, old wax from light, new wax, and never add propolis to wax.
- Don't put plastic frames in the wax melter: they can warp.
- Remember: many pesticides are lipophilic and will bind with wax. Never store such chemicals near beeswax or combs. This includes miticide strips and other pesticides.

Separate white, new wax and cappings from the older, darker wax from old combs and scrapings. Lighter wax is more valuable; both colors of wax can be melted with one of these devices:

- Cloth bag set in boiling water.
- Electric wax melter.
- Solar wax melter.
- Double boiler made of aluminum or stainless steel; iron or copper will darken the wax.
- Steam chests or hot boxes.
- Wax press that is steam heated.

Older, dry comb that may have wax moth silk or ample amounts of bee bread or fermented or crystallized honey can be soaked in soft rainwater. Place the old comb and scraps in a cloth bag (or a bag fashioned out of cheesecloth or an old pillowcase) and submerge it in a tub or barrel of rainwater or acidic water (**never** alkali water, as the wax will become

More Information on Beeswax

- Bees need to eat between 7 and 9 pounds of honey in order to produce 1 pound of beeswax.
- 1 pound of beeswax will make 35,000 wax cells.
- 1 pound of beeswax stores 22 pounds of honey.
- There are 500,000 wax scales per pound (one scale measures ⅛ inch [3.175 mm] in diameter).
- Under optimal conditions, 10,000 bees produce 1 pound of beeswax in three days.
- Wax cappings yield 90 to 97 percent beeswax, depending on the extraction method; these numbers reflect conventional, not commercial, methods.
- Culled combs yield 37 to 83 percent wax; brood combs yield 15 to 16 percent wax.

spongy and not be fit for other uses). Place stones or bricks in or on the bag to help keep it submerged. Heat the water between 150°F and 180°F (66–82°C) for several hours, occasionally poking the bag with a stick to allow the wax to move through the fabric to the surface of the water. After the wax has melted, remove from heat and allow the water to cool. The wax will solidify on the surface of the water, and most of the debris will fall to the bottom of the tub. Render (melt) and filter the wax cake a second time.

None of the methods will be sufficient to render all the wax found in old combs; the remaining mixture of wax and debris should not be discarded, but saved and taken to a dealer who has the special equipment needed to render it.

Solar Wax Melter

The solar wax melter is essentially a box painted black outside and white inside and covered with a piece of glass, Plexiglas, or plastic and made airtight. It is put in a sunny location and tilted at a right angle to the sun's rays. The sun heats the interior of the box, as it would a greenhouse, melting the wax inside, which collects into a pan. For greater heating efficiency, use two pieces of glass or plastic, separated by a ¼-inch

(6.4 mm) gap. The inside of the box contains a tray, fashioned from sheet metal, onto which the wax comb and scraps are placed (see the illustration of the solar wax melter).

The melter will render cappings, new burr comb, and old comb, but it will not melt frames of old comb completely. After old, dark comb has been in the melter for a few days, collect the black, gummy remains (sometimes called *slumgum*) and either compost it (it is rather acidic) and use as a soil amenity or discard. If you have a lot of it, and it is feasible, take the slumgum to a bee supply dealer who has equipment to extract the remaining wax chemically. The amount of wax one can derive from broken pieces of foundation, cappings, and old or burr comb is significant and well worth the effort.

Wax Cappings

When extracting honey, you will generate buckets of wet wax cappings and honey. If you have only a few buckets, you can place them into a hot room (where it heats up to about 143–151°F [62–66°C]). Some beekeepers make an insulated box heated with a single lightbulb; this will melt all the wax in 24 hours. The remaining honey, if not overheated, can be eaten or bottled or fed back to bees (if it contains no foulbrood spores). If it is overheated and dark, discard this honey, which will be unacceptable to both people and bees, or use it to feed wax moths, if you are rearing them (see Appendix F). For larger amounts of cappings, you may want to invest in a cappings spinner (an extractor to separate out most of the honey) and some sort of water-jacketed stainless steel melter. The best way to determine what you need is to visit several beekeeping operations and see how they handle their wax.

White/yellow cappings, when rendered, will yield the best wax and is worth the effort to recover, as whiter wax is preferred for candles and other wax products. After you separate out the honey, melt the wax and pour it into plastic tubs or loaf pans (first coated lightly with talcum powder or cornstarch). Weigh the finished block, so you can estimate how many pounds of wax you have. Again, if you have not used any chemical compounds for mite controls, you can sell the wax in the blocks, or render it into salable products, such as lotion bars; otherwise, make candles or ornaments.

Solar Wax Melter

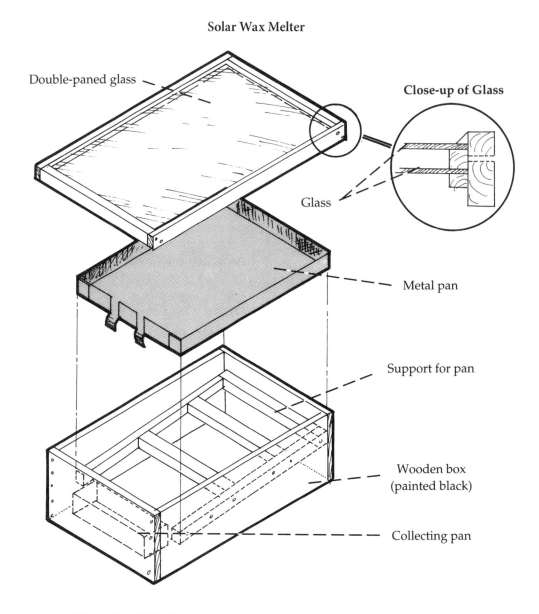

Rendering Wax: Final Filtering

Once you have collected the chunks or blocks of wax from your wax melter, you must further refine it to remove the fine particles of dirt, honey, and other impurities. Use a large double boiler or other wax melter (check bee supply catalogs) and filter through sweatshirt material (fuzzy side up), heavy cotton, or paper towels. If you are using this wax for value-added products (such as candles) you may need to filter it again, using coffee filters or paper towels.

When melting small amounts of wax to pour into molds, or the like, use an electric frying pan, an old electric coffee pot, or double-boiler arrangement (in which you place the wax in a small can or pan within a larger pan of hot water). Buy used metal pans at a thrift store and use them only for wax. When you do a final filter, pour the hot wax into a filter-covered can and, to keep the wax liquid, put the can into a warm oven until all the wax has drained. This final, fine-filtered wax can then be used for candles, in batik, or for whatever you decide (see "Beeswax" in the References).

BEE BROOD

Generally an unexploited product of the hive, bee brood is rich in proteins, fats, and other substances required in our daily diet. Bee brood consists of over 15 percent protein, 4 percent fat, and mostly water (about 77%), compared with beef, which is 23 percent protein, 3 percent fat, and 74 percent water. It

Honey Bee Soufflé

½ cup butter
½ cup sifted flour
1 ½ teaspoons salt
½ teaspoon paprika
Dash of hot pepper sauce
2 cups milk
½ pound sharp cheese, grated
8 eggs, separated
½ cup marinated bees (larvae and young pupae
    in 1 cup soy sauce, ¼ cup sake or sherry,
    1 garlic clove, 1 dried hot pepper, 2 table-
    spoons ginger root, grated)

Melt butter in double boiler over boiling water and add next four ingredients; mix well and gradually add milk, stirring until sauce thickens. Add cheese until it melts; remove from heat. Separate egg whites from the yolks. Beat egg yolks until lemony and stir slowly into cheese sauce. Beat egg whites until stiff and fold sauce into whites. Layer bottom of 1-quart greased soufflé dish with bees and cover with sauce. Bake at 475°F for 10 minutes; reduce oven temperature to 400°F and bake 25 minutes longer.

Source: R.L. Taylor and B.J. Carter. 1992. Entertaining with insects; or, The original guide to insect cookery. Yorba Linda, CA: Salutek.

has over 100 times the vitamin A present in milk and over 60,000 times as much vitamin D.

The value of this hive product does not yet compensate the cost of removing brood from comb and the reduction of the adult colony population that ensues if too much brood is removed. Honey bee brood is currently used, on a small scale, as food for birds, reptiles, and fish. Drone larvae are often used for fish bait; and if you cut out drone brood for varroa control, you may be able to sell it in these markets. Bee larvae also make a good quiche or soufflé (see sidebar).

BEE VENOM

Bees require pollen in their diets in order to synthesize some of the components of venom. The synthesized venom is stored in the poison sac of worker and queen bees (see Appendix C). Venom contains a complex array of chemical substances, such as water, amines, amino acids, enzymes, proteins, and peptides, as well as some sugars and phospholipids. Venom also contains histamine, which reacts adversely with the body chemistry of some individuals (see the table on the analysis of bee venom in this chapter).

To collect substantial amounts of venom, either for medical or for other uses, an electrical grid is placed inside a hive. This special grid produces a mild shock, and bees that land on it react by stinging a sheet of nylon taffeta below this grid. The venom is deposited

Honey Bee Brood Components

Component	Larvae	Pupae	Beef	Milk	Egg yolk
Water (%)	77	70.2	74.1	87.0	49.4
Ash (%)	3	2.2	1.1	3.5	
Protein (%)	15.4	18.2	22.6	3.9	16.3
Fat (%)	3.7	2.4	2.8	< 5	31.9
Glycogen (%)	0.41	0.75	0.1– 0.7	—	—
Vitamin A (IU) —	89–119	49.3	1.6	32.1	
Vitamin D (IU) —	6130–7430	5260	0.41	2.6	

Source: E. Crane. 1990. Bees and beekeeping: Science, practice, and world resources. Ithaca, NY: Comstock.
Notes: % = Percentage of fresh weight. IU = International units/gram fresh weight.

Analysis of Bee Venom

Class of molecule	Component	Percentage of dry weight (%)	Effects
Enzymes	Phospholipase A$_2$	10–12	Bursts cells by degrading cell walls. Decreases blood pressure and inhibits blood coagulation. Causes pain; activates arachidonic acid to form prostaglandins (which regulate inflammatory response). Wasp toxin contains phospholipase A$_1$.
	Hyaluronidase	1–3	Dilates the capillaries, causing the spread of inflammation. Hydrolyzes connective tissue. Helps spread other components.
	Acid phosphomonoesterase	1.0	
	Lysophospholipase	1.0	
	Alpha-glucosidase	0.6	
Other proteins and peptides	Melittin	40–50	Releases histamine and serotonin from mast cells. Bursts blood and mast cells. Depresses blood pressure and respiration. Induces production of cortisol in the body.
	Apamine	1–3	Increases cortisol production in the adrenal gland. Is a mild neurotoxin.
	Mast cell degranulating peptide (MCD)	1–2	
	Secapin	0.5–2.0	
	Procamine	1–2	
	Adolapin	1.0	Is anti-inflammatory and analgesic because it blocks cyclooxygenase.
	Protease inhibitor	0.8	
	Tertiapin	0.1	
	Small peptides (with less than 5 amino acids)	13–15	
Physiologically active	Histamine	0.5–2.0	Causes allergic response: itching, pain.
Amines	Dopamine	0.2–1.0	Increases pulse rate.
	Noradrenaline	0.1–0.5	Increases pulse rate.
Amino acids	Gamma (γ)-aminobutyric acid	0.5	
	Alpha-amino acids	1.0	
Sugars	Glucose and fructose	2	
Phospholipids		5	
Volatile compounds		4–8	

Sources: E.M. Dotimas, K.R. Hamid, R.C. Hider, and U. Ragnarsson. 1987. Isolation and structure analysis of bee venom mast cell degranulating peptide. Biochimica et Biophysica Acta, 911:285–293. R.A. Shipolini. 1984. Biochemistry of bee venom. In: A.T. Tu (ed.), Handbook of natural toxin, vol. 2, pp. 49–85. New York: Marcel Dekker. E. Crane. 1990. Bees and Beekeeping. Ithaca, NY: Comstock. See "Venom" in the References.

on and collected from a glass plate located below the nylon portion of the device.

Research is still in progress concerning the benefits obtained from honey bee venom for persons with arthritis and other diseases. Bee venom therapy, called *apitherapy*, is becoming more popular today. In addition, more recent research indicates that some of the components of venom are much more effective than other serums in desensitizing persons who are allergic to bee venom (see Appendix C and "Venom" in the references, or search online).

ROYAL JELLY

Royal jelly consists, roughly, of 66 percent water, between 4.5 to 17 percent ash, 11 to 13 percent carbohydrates, 12 percent protein, 5 percent fat, and 3 percent vitamins, ether extracts, enzymes, and coenzymes. Royal jelly, the sole food of queen larvae, is manufactured by young nurse bees from two glandular secretions: a white jelly from the hypopharyngeal and mandibular glands, and a clear jelly, which is a hypopharyngeal secretion mixed with honey.

Royal jelly has long been collected and used in Asian countries for medical purposes. Modern uses include cosmetics, lotions, and dietary supplements. None of the curative properties of royal jelly have been extensively studied in the United States, nor should any be claimed on product labels. In addition, some people can be allergic to this product, and an appropriate warning should be applied to any label.

Queen breeders often collect and freeze royal jelly to use during queen rearing operations. If you are considering buying royal jelly from outside the United States, **be warned** that it can contain viruses that may be detrimental to bees; this discovery was reported by researchers studying the causes of colony collapse disorder.

PROPOLIS

Propolis is a resinous mixture used by bees to seal cracks in the hive. The word is derived from the Greek words *pro* (before) and *polis* (city), and refers to its use by bees to reduce the hive entrance ("in front of" or "before" the "city") for winter protection and defense against ants.

The original semiliquid material comes from the sticky exudations of trees and buds—such as the alders, poplars, mesquite, and some conifers—and is collected by foragers, either in its natural state or from abandoned equipment (such as mite- or disease-killed colonies). Foragers transport the material back to the hive on their pollen-collecting structures. Inside the hive, other bees assist the resin collector in removing the resin from the collector's corbiculae, and it is mixed with 40 to 60 percent beeswax and a third, unknown, substance. It is this mixture that is called propolis.

Propolis is used by bees to:

- Coat the inside of the hive walls.
- Seal the inner cover, bottom board, and outer cover to the hive body.
- Strengthen comb.
- Glue frames together.
- Seal hive bodies together and caulk any small crack or crevices.
- Embalm foreign objects too difficult to remove (e.g., dead mice, snakes, or large insects).
- Fill small, tight spaces and cracks that are smaller than the bee space.

To beekeepers, propolis is a sticky, gummy mess in the hive and makes separating the hive bodies difficult, even when using a hive tool.

Studies have documented the antimicrobial activity of propolis, which may help bees keep the hive clean and kill foreign microbes that would otherwise live in small cracks. Dead animals coated with propolis are virtually mummified and will not decompose.

Some of the components identified in propolis include flavonoids, benzoic acid, cinnamyl alcohols and acids, alcohols, ketone, phenols, terpenes, sesquiterpene alcohols, minerals, sterols, volatile oils, beeswax, sugars, and amino acids. The compounds in propolis that have some pharmacological activities (antiviral, antibacterial, antioxidant, antifungal, antiinflammatory, and antihistamine) are quercertin, pinocembin, caffeic acid, caffeic acid phenethyl ester, acacetin, and pinostrobin.

Beekeepers who collect propolis lay a special plastic screen (check bee supply stores) on the top bars of the uppermost honey super and allow the bees to fill it with propolis. Once filled, the screen can be frozen, then the propolis knocked off and cleaned. Propolis is even appearing in capsule form in health food stores, as a health supplement. None of its purported curative properties have been extensively studied in the United States, nor should any be claimed on the label. As with other bee products, some people can be allergic to propolis, thus an appropriate warning should be applied when selling this product.

Because propolis is not water soluble, use rubbing alcohol to remove it from hands and clothing. Other people find that a mechanic's hand cleanser works well to remove propolis. You can scrape it or burn it off hive tools.

Check with other beekeepers or online bee sites for more current information.

POLLEN

Pollen, the protein-rich powder produced by the male parts of flowers, is collected and sold by beekeepers to health food stores, to pollination businesses, to bee dealers (for bee food), and to allergy patients (as a desensitizing agent). Pollen traps are put on hives to collect pollen pellets from foraging bees. Collected pellets should be stored properly; see "Pollen" in Chapter 7 and "Products of the Hive" in the References. Be careful that any pollen collected comes from flowers that have not been sprayed with pesticides, because they can contaminate it.

Pollen pellets, as a vitamin supplement, are sold in many health food stores. Pollen contains a number of nutrients, such as carbohydrates, proteins, and minerals. The composition varies according to the plant species, and bees are not able to distinguish the more nutritious ones. Pollen constitutes 10 to 35 percent of the protein in the diet of bees. Remember, pollen that bees convert into bee bread supplies the essential amino acids needed to manufacture their own proteins and enzymes and contains such vitamins as A, C, D, E, B_1, B_2, B_6, and B_{12}, as well as minerals, such as sulfur, nitrogen, phosphorus, and many minor elements; see the description of pollen components in the sidebar. Some substances found in pollen have yet to be identified.

Pollen pellets are also sold for human consumption. Some claim that pollen is a superior food with curative powers and provides extra energy. Authoritative reports as to the value of pollen as human food are contradictory. To survive the rigors of sun and rain and preserve its precious cargo, pollen grains are protected by a wall of intine and exine. Some pollen grains have been excavated intact (good enough

Pollen Components	
Component	**Percentage of total (%)**
Water	7.0–16.2
Crude protein	7.0–29.9
Ether extracts	0.9–14.4
Carbohydrates	
Reducing sugars	18.8–41.2
Nonreducing	
sugars	0–9.0
Starch	0–10.6
Ash	0.9–5.5
Unknown	21.7–35.9

Other components include lipids, organic acids, free amino acids, nucleic acids (plant), enzymes, sulfur, nitrogen, flavonoids, and growth regulators (plant).

to identify plant species) from acid bogs over 10,000 years old.

Bees have special stomachs that humans lack to grind off some of the exine layer and digest the protein content of the pollen grain. Recent studies indicate that the protective layers of some pollen can be ruptured and proteins released into the human digestive systems.

Whatever the claims, there is a market for pollen, and it may be lucrative for beekeepers to collect and sell it, since the importation of pollen pellets from overseas is not permitted (because they could contain chalkbrood, chemical toxins, and other pathogens).

Before selling or consuming pollen, be certain that it was not contaminated by pesticides (including fungicides, which will kill beneficial fungi in the bee colony). Beekeepers trap pollen using equipment called pollen traps, sold by bee supply companies, while some crops are being sprayed with insecticides, to reduce the amount of contaminated pollen from entering the colony. Such pollen should never be sold for human (or bee) consumption. Know the source of pollen before buying or selling it. There are labs, books, and websites that will identify the plant species of your pollen, which is a fascinating and informative activity. Other labs will test pollen (and other hive products) for contaminants, which may be worth the cost if your business depends on selling a pure product. The National Honey Board has a list of labs on its website.

Some companies will buy bee-collected pollen, so check the bee magazines for advertisements. But remember, none of its curative properties have been extensively studied in the United States, and some people are hypersensitive to this product and could suffer a severe allergic reaction. Add a warning when selling pollen.

Any bee products can cause an allergic reaction in some people, and you can face potential lawsuits. Look into getting product liability insurance if you will be selling bee products. **Do not rely on your homeowners insurance**. Also, do not make any health claims.

WAX MOTH LARVAE

Rearing wax moths for the fish-bait industry or as pet food can be a lucrative and successful side business to beekeepers. Rear the moths in an area away from your bee operations so you don't add more pests to infest your bee equipment. For information on rearing wax moths see Appendix F.

SELLING YOUR HIVE PRODUCTS

There are many avenues open to you to sell your hive products: the office, local retail outlets, roadside stands or orchards, flea or farmer's markets, fairs, your local church, or other organizations to which you or your friends belong. Keep it simple at first; large grocery chains are expensive and difficult to buy into and require a year-round supply.

Here are some rules, researched by the National Honey Board, which you should follow to be successful in keeping your customers happy. In addition, check out local agriculture extension offices, other beekeepers, and the internet for other ideas.

- Have a reliable supply of quality product.
- Local honey or hive products are best. Use slogans such as "produced by bees in Your County" for regional appeal, especially in souvenir shops or highway markets.
- Keep your jars clean; people do not like sticky jars.
- Do not heat your honey higher than 120°F (49°C), or you will ruin the delicate, floral flavors.
- Make sure your bee products are clean and free from debris.
- Use eye-catching, attractive labels; in general the public does **not** like to see bees or beehives on the label.
- Use consumer-friendly words, such as "pure," "natural," or "raw," on natural hive products. Honey can be labeled "raw" if it has been minimally filtered or heated.
- Conduct taste tests, give out free recipes, dress honey bears in holiday outfits, make gift baskets, or use other "value-added" techniques to help sell your products.
- Use "organic" only if you comply with the USDA organic food standards.
- Try making "cremed" honey using the Dyce process (in this chapter).
- Use different kinds of honey jars (see the illustration of various honey containers).
- If you wish to mix honey with dried fruits, nuts, or flavorings, you must label these carefully, as these are considered "food products" and must meet federal and state standards. Instead, you can give small jars of this away as gifts.

Advertising is also an important part of selling your products. Take advantage of promotions available from the National Honey Board or other organizations, many of which are free. Many beekeepers simply sell honey from their home; but before expanding your operation to a point at which you need an outside sign, check with your homeowners insurance policy and any local zoning codes: forewarned is forearmed. A small classified ad in local papers may be all you need to have a steady supply of customers.

Honey Labels

Jars of honey have to conform to state and federal pure food laws. The guidelines below will meet most requirements, but check with your state agriculture department for more information.

The following information must appear on each jar or section comb of honey you sell:

- Name, address, city, state, zip code of manufacturer, packer, or distributor.

- Name of product (e.g., Orange Blossom Honey).
- Net weight in pounds (oz.) and metric (grams) in the lower 30 percent of the display label (e.g., Net Wt. 8 oz. [227 g]).

You may also want to add additional labels or enhancers (Local Honey, Award Winning, Made by US Bees) or nectar source labels (Pure Fireweed, Oregon Wild Flower Honey, or Sourwood Honey, only if it that is verified by analysis).

Labels can be purchased from most bee supply dealers, or label companies that advertise in the journals. You can have a printer make up your own label preprinted with your name, address, etc., but the initial cost is rather high. Instead, you can buy blanks and have a stamp made with your name and other information that you can later imprint with indelible ink. Labels for round or square-section comb honey are also sold by bee supply companies or label companies that advertise in the bee journals.

You can also print your own labels on your home computer using special label paper. Make sure you use label paper and indelible ink. In this way you can change the label to fit the season, and personalize it, with pictures of your bees or even for special customers.

Nutritional labels are required, especially if you claim a nutrient content or make a health-related statement. Again, check with the National Honey Board or your state food and labeling regulatory agency for particulars. Some beekeepers put on additional labels giving instructions on how to reliquefy honey that is crystallized. Unfortunately, any instructions you supply could make you liable for any injury to a person following those instructions. Carefully label your products, but you may want to check with a lawyer about the wording.

 **Notes**

CHAPTER 13

Diseases of Honey Bees

Honey bees, like all organisms, are subject to a variety of diseases and maladies, and are preyed upon by other organisms. Some of these may affect individual bees, and some may lead to the demise of an entire colony. When organisms are crowded together, diseases can spread rapidly. Unfortunately, the honey bees' lifestyle is one where many bees (10,000–40,000) live together in a somewhat restricted space. In addition, they exchange food with one another, groom each other, and often bees from one colony will enter another colony. Such close contact and tactile activity helps to spread pathogens and parasites from one adult bee to another, from adult bees to larval bees, and from one colony to another. The greater the number of colonies in a given apiary, the greater the likelihood for the transmission of diseases, pests, and viruses from colony to colony.

Beekeepers tend to be concerned with maintaining a healthy environment and prefer to keep their bees healthy and free of diseases. When you introduce chemicals into the colonies to keep your bees healthy, you must first ascertain if the treatment is permitted by law. Most treatments (e.g., miticides) come with very specific directions as to when they can be introduced into a colony. If you have questions concerning what chemicals can be introduced into a colony, check first with the state or county bee inspector or seek advice from your state's beekeeping organization.

Commercial beekeepers are especially vulnerable to having their colonies at risk, because they keep thousands of colonies, moving them many miles to pollinate our fruits and vegetable crops, which ex-pose them to other beekeepers' colonies that may carry diseases and pests, thus spreading them.

COLONY COLLAPSE DISORDER

Bee losses have been recorded as early as 1896 and called by various names, such as May disease, spring dwindling, disappearing disease, or autumn collapse. Since the 1970s, there has been a further decline in the number of "domesticated" as well as feral bee colonies. Factors attributed to this decline include urbanization (loss of forage), pesticide use, and introduced parasitic mites (tracheal and varroa). In the spring of 2006, large numbers of overwintering bee colonies in North America were found with virtually no adult bees. While such disappearances have occurred throughout the history of apiculture, the term "colony collapse disorder" (CCD) was first applied to these bee disappearances—an all-encompassing catch phrase used to cover a host of bee problems for which it was difficult to determine the cause. Other countries have not been immune, and that same year, European beekeepers began to observe similar symptoms. The causes are still not fully understood, and no specific cause has yet to be determined, but factors such as diseases, including the new *Nosema ceranae*, and varroa mites have been found in dying colonies. Other possibilities include environmental stresses, poor nutrition, and pesticides, as well as the rigors bees go through in migratory beekeeping operations. Other unproved possibilities (and some wild speculations) implicate cell phone radiation and genetically modified (GM) crops. But weather is also a factor; too much or too little rain will influence the amount of nectar flowers secrete and, ultimately, the amount of honey stored and thus the ability of bees to overwinter successfully. Other possibilities are nutrition, stress, and the exposure of bees to the pesticide treatments beekeepers themselves use to kill mites.

A survey by the U.S. Department of Agriculture's Agricultural Research Services (USDA-ARS) and the Apiary Inspectors of America (winter 2008) revealed that in the United States, 36 percent of 2.4 million hives were lost to CCD. The survey was obtained from 20 percent of the 1500 commercial beekeepers and showed an increase of 11 percent over the 2007 losses and 40 percent over the losses of 2006. Reports of large losses in other countries have also been published. However, beekeepers should keep current with research in the bee journals, as this phenomenon needs historical information to put it in perspective. Most of the colonies that have been affected by CCD were in large-scale commercial operations, not in small-scale beekeeper colonies.

Bee colonies that succumb to CCD generally have all of these conditions occurring simultaneously:

- Absence of adult bees, with little or no accumulation of dead bees in or around the hive.
- In the abandoned colonies there are food stores, both honey and pollen, but:
  - The stores are not immediately robbed by other bees.
  - If hive pests are present (such as wax moth and small hive beetle), damage is slow to develop.
- Occurrence of some capped brood (bees will not normally abandon a colony until the capped brood have all emerged).
- The queen is evident, but the workers appear to be all the same age (young), and they are slow to feed on supplemented food (syrup and pollen patties).

Here are some recommendations:

- Do not combine collapsing colonies with strong colonies.
- When a collapsed colony is found, store the equipment where bees will not have access to it. **Equipment from a collapsed colony should not be used again; burn it, combs and all, unless you can find a gamma irradiation chamber.**
- If you have determined that a colony is in the process of collapsing—that there is a marked decline in the bee population or there are signs of an infection such as European foulbrood— contact your state or county bee inspector for guidance. Certain chemical treatments to ameliorate infections now require a prescription from a veterinarian.

Again, check the web pages and especially bee research labs for updated information.

DISEASES OF ADULT BEES

Nosema Disease

Nosema, the most common adult bee disease, is caused by a microscopic protozoan or microsporidium (now considered to be a fungus). Nosema disease is common in insects and has been used as a biological control for locusts and grasshoppers (*N. locustae*) and wax moths (*N. galleriae*). Other species of *Nosema* also affect silkworm moths, bumble bees, and leafcutter bees. *Nosema* species are specific to the insect they attack; therefore grasshopper *Nosema* will not infect honey bees.

Currently two *Nosema* species are found to infect honey bees. One is *Nosema apis* (Zander 1909), which is prevalent in the four races of European honey bees. A second, more virulent species, *N. ceranae*, was first described in China in 1994. In 2004, *N. ceranae* was discovered in Spain, where it was responsible for large colony losses. Originally described from Asian honey bees, *N. ceranae* is thought to be one of the contributing factors in CCD. More information is being released on this new disease, so keep current in the bee journals.

Both *Nosema* species are spore-forming microorganisms that invade the digestive cell layer of the midgut of bees. Up to 30 million spores can be found in a single bee one and a half weeks after initial infection. The spores of both of these species are very difficult to tell apart, requiring molecular techniques. If you want a positive identification, send samples of bees to the USDA-ARS Bee Lab in Beltsville, Maryland (log on to its website for information on how to send samples).

Nosema disease is now distributed worldwide, wherever bees have been introduced; it is a serious problem in temperate climates. *Nosema ceranae* is now more prevalent than *N. apis*. Normally, *N. apis* is prevalent in the spring, especially after winter weather has confined bees to their hive. If it does not kill the colony over winter, *Nosema* greatly reduces the life span of all castes of adult bees, reduces honey yields, and is a factor in the supersedure of package bee queens, further delaying the growth of a hive in which packages are installed. Check websites, especially those from government and university bee labs, to stay current with the spread of this disease.

Life Cycle of *Nosema apis*

Both *Nosema* species have similar life cycles. The only way for bees to become infected is from picking up the spores from other bees or from contaminated food or water. It takes only a few spores to infect bees. When the spore enters the ventriculus (midgut), a filament forms that attaches to and then protrudes into the cells lining the bee's gut. Once inside, the filament increases in size and multiplies. This vegetative state is not infective but produces more spores within 5 to 10 days, which can burst the cell walls, releasing the spores into the bee's midgut. These new spores either reinvade other cells within the bee or pass out of the host bee with the feces. If the spores reinfect the same bee, digestive function of the bee can be repressed within two to three weeks.

The spores are viable for up to one year in the fecal material and can be found outside the colony, especially at common drinking areas, within the colony on the comb, and in bee bread and honey. The disease is spread by drifting and robbing bees coming in contact with fecal material on frames and combs.

Nosema can be a serious disease if not checked, especially now that *N. ceranae* is here, which can kill bees within a week, and weakens the foragers to where they don't return to the colony, leaving a skeleton population behind. Because nosema disease is often confused with symptoms of other diseases (such as pesticide poisoning, or amoeba or mite predation), diagnosis is important to treat this disease properly. Some effects of severe *Nosema* infection are:

- Reduced longevity of workers (by 50 percent).
- Reduced honey yield (by 40 percent).

- Queen supersedure; queen's egg-laying capacity in decline.
- Reduced function of hypopharyngeal (food) glands, resulting in poor brood rearing ability, because the nurse bees are unable to produce enough brood food.
- Disruption in hormonal development, causing bees to age faster and forage earlier in life than normal.
- Disruption in the secretion of digestive enzymes, causing bees to starve to death.
- Bees will not consume sugar syrup.

Although bees with this disease display no specific symptoms, listed below are some signs to watch for. These symptoms can also be associated with pesticide poisoning, mite injury, or other pathogens; the only way to obtain a positive diagnosis for nosema disease is to dissect a bee. Nevertheless, if most of these symptoms are observed in the spring of the year after winter confinement, nosema disease should be suspected. While these symptoms could be apparent at any time of the year, they are most noticeable in spring and fall:

- Bees unable to fly or able to fly only short distances.
- Bees seen trembling and quivering, colony restless.
- Feces on combs, top bars, bottom boards, and outside walls of hive; also correlated with dysentery or diarrhea (see "Dysentery" below).
- Bees are seen crawling aimlessly on bottom board, near entrance, or on ground; some drag along as if their legs were paralyzed.
- Wings positioned at various angles from body (also called *K-winged*)—that is, wings are not folded in normal position over abdomen but with the hindwing held in front of the forewing.
- Abdomen distended (swollen).
- Bees not eating when fed syrup.
- Bees abandoning the colony, leaving the queen and a few emerging workers.

Again, the only way to diagnose positively for nosema disease is to dissect the bee. A field test, which is not very reliable but good in a pinch, is to pull apart a bee until the viscera are visible. If the midgut (ven-

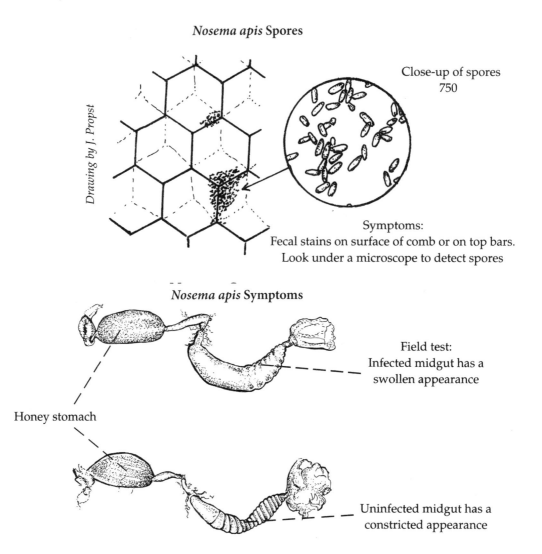

Nosema apis Spores

Drawing by J. Propst

Close-up of spores
750

Symptoms:
Fecal stains on surface of comb or on top bars.
Look under a microscope to detect spores

Nosema apis Symptoms

Honey stomach

Field test:
Infected midgut has a
swollen appearance

Uninfected midgut has a
constricted appearance

triculus) is swollen and a dull grayish white in color, and the circular constrictions of the gut (similar to constrictions on an earthworm's body) are no longer evident, then *Nosema* is the culprit (see the illustration). The normal gut color is brownish red or yellowish, with many circular constrictions.

A better way to get a positive diagnosis is to use a hemacytometer; this is a special slide with a grid etched onto it. Here's how:

● In that older bees provide the best evidence of the existence of this disease, collect foragers departing from or returning to the colony. Place them in a container of alcohol or in a plastic bag; store these containers in a freezer.
● Collect 30 to 60 bees from a suspected colony.
● Cut off the abdomen of the bees and grind in a mortar and pestle.
● Add 25 ml water per 50 bees.

● Grind a second time, then remove a sample with an eye dropper, place the sample on a microscope slide, and examine the contents under a microscope.

For complete instructions, go to bee lab sites (e.g., https://www.beelab.umn.edu/).

If you place the sample on a plain microscope slide, *Nosema* spores, if present, will be seen clearly at about 40x. They are small, smooth, ovoid bodies, much smaller than pollen grains, which will also be present in the gut. It is best to look at the spores under higher magnification, over 100x; they measure 4 to 6 micrometers long and 2 to 4 micrometers wide. If you have any questions, and for a positive identification, send samples in to the USDA Bee Lab in Beltsville (go to their website, or your state's university bee lab, for complete instructions).

As we have already mentioned, *N. ceranae* is the

A Bee's Encounter with Pathogens and Parasites

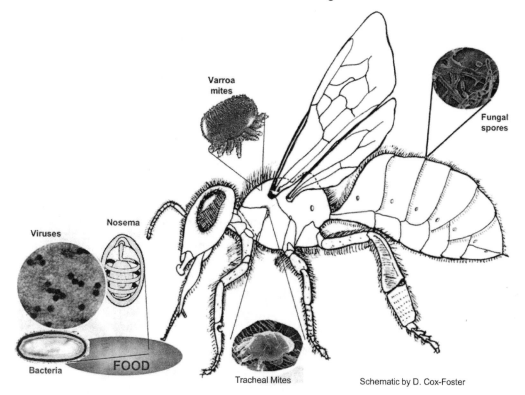

Schematic by D. Cox-Foster

more serious disease, especially if your apiary is near a commercial operation, or if you have purchased queens or package bees. We need to stress that the symptoms are not always clear and are often confused with other diseases, pesticide poisoning, or mite predation; therefore diagnosis is important (see "Diseases and Pests" in the References).

Differences between *Nosema apis* and *N. ceranae*

It is now evident that *Nosema ceranae* is more prevalent than *N. apis*, and there is no drug treatment available. The only way to eliminate the spores from a colony that has succumbed to *N. ceranae* is to have the combs and hive furniture irradiated. If this is not an option, **do not reuse** the equipment of the affected colony. Check out USDA bee lab and other sites online as well as local state apiary sites for current information.

Treatment for Nosema

Good management practices, including apiary sites that protect bees in winter from cold winds and shady conditions, can help control the disease and ensure healthy colonies. To control *Nosema* and prevent it from spreading:

Nosema Spores

N. ceranae   *N. apis*

Bar = 5 micrometers

- Provide fresh, clean water; individually feed water via Boardman feeders to each colony, if you have a small apiary.
- Provide a young queen.
- Locate hives at sunny sites, sheltered from piercing cold winds but with good air drainage.
- Maintain adequate stores of bee bread, honey, or cured sugar syrup; if stores are short, bees should be fed a heavy (2:1 sugar : water) syrup in early fall and protein supplements.
- Keep only clean combs; sterilize or dispose of those that are soiled with fecal material.
- Provide upper hive entrance during winter to improve hive ventilation.
- Keep bees on sunny slopes, with the entrances facing south.
- Reduce stress to bees, such as moving bees often for pollination. Also, make sure the bees have adequate honey and bee bread stores during inclement weather.
- If bees are not feeding on the syrup, spray the bees (if the weather is above 60°F) so they can clean it off each other and thus ingest it.

Combs with *Nosema* spores must be sterilized, especially if you are going to reuse frames from dead colonies. If you do not disinfect the frames, you are just spreading the disease. Many methods to fumigate combs have been described; however, before you start, check with your state apiary inspector or Department of Agriculture and comply with state and federal regulations, especially if you have hundreds of supers. In addition, some states have fumigation or gamma chambers for disinfecting diseased combs. Irradiation of hive furniture contaminated with disease spores, including *Nosema*, is an effective way of reducing other diseases in older combs. Be sure to check this out.

You can kill spores by heating the supers with combs in a hot room or autoclave to 120°F (49°C) for 24 hours; combs should be free of honey and bee bread (temperature should not get above 120°F, or the wax will melt). Some compounds that have been found to kill *Nosema* spores are 10 percent bleach and 60 percent acetic acid.

If it is not possible to fumigate combs, you can replace old, fecal-stained and dark combs with new foundation. A good rule of thumb is to replace frames when you can no longer see through them when you hold them up to the sun. This will not apply when using comb with a black-colored plastic base. Set up a rotation cycle by putting a unique mark (such as the year) on the top bars every time you install new foundation, so you can easily determine the age of each frame. In other words, as you go through your hives in the spring, pull out the older combs and replace them with new foundation. Put the marking system in your field notes.

Nosema Chemotherapy

The drug used to control both *Nosema* species was bicyclohexylammonium fumagillin (isolated from the fungus *Apergillus fumigatus*), or fumagilin, and was sold as Fumigilin-B (and other names). **This drug may no longer be available from beekeeping supply companies. Again, check with your state's bee inspector.**

Amoeba Disease

Another protozoan is the amoeba in the kingdom Protista (which includes mostly single-celled eukaryotes such as the slime molds, algae, and the Sarcodina or amoebae). The amoeba that affects honey bees is called *Malpighamoeba mellificae* Prell, which forms resistant spores called *cysts*.

A single bee can have a half-million cysts within three weeks of the initial infection. All adult bee castes and drones can be infected. Cysts are ingested by bees from infested food, water, or elsewhere in the hive, and then migrate to the Malpighian tubules (the kidneys) of bees. The cysts germinate in the intestines and make their way into tubules from the hindgut within 24 hours; what the amoebae are feeding on is not known. However, some researchers report that once the infection is under way, the abdomen becomes distended and the tubules cease functioning, taking on a glossy appearance when filled with the cysts. These spherical cysts measure 6 to 8 micrometers, big compared to the smaller *Nosema* spores, which are elliptical (and not found in the tubules). Once the cysts migrate down the tubules into the hindgut, they are voided with feces.

Amoebae are found mostly in bees infected with nosema disease, and the cysts are observed in fecal material along with *Nosema* spores. Although mostly present in workers, queens can be affected as well.

There are no clear symptoms other than dwindling populations, as bees die away from the colony.

Heavy infestation results in reduced honey yield and impaired functioning of the tubules. Development time of the cysts is slowed in cool weather (68°F [20°C]) but increases as the temperature in the broodnest reaches 86°F (30°C). Therefore, spring is the time when amoebae infections are most severe, peaking in May in the Northern Hemisphere.

The only control is hygienic conditions in the apiary and at the water source, and decontaminating frames (as in *Nosema*) or replacing equipment. Some reports indicate requeening has been successful in saving an infested colony (as long as the queen is not herself infected).

Minor Adult Bee Diseases

Septicemia

Septicemia is caused by several different bacteria found in the hemolymph (blood) of bees, the most common being *Pseudomonas aeruginosa* (=*P. aspiseptica*). Although septicemia rarely if ever debilitates bee colonies, it destroys the connective tissues of the thorax (legs, wings) and antennae. The disease can be recognized by these symptoms:

- Dying bees are sluggish, and the hemolymph turns white (instead of clear).
- Dead bees decay rapidly.
- Dead bees become dismembered when touched as the muscle tissue degenerates.
- Dead bees have putrid odor.

Bees come into contact with the bacteria in soil, water, and infected bees by way of their breathing tubes (tracheae). It is still not clearly understood how the disease is transmitted nor how to treat it, but some success has been found by requeening colonies and placing hives in locations that are sunny and dry and have good air drainage.

Spiroplasma

Spiroplasma is a coiled, motile prokaryote without a cell wall. It was first described in the nectar of some honey plants, but infection with it is rarely diagnosed.

There are other minor diseases not mentioned here, and which many beekeepers will never see. For a complete look at all the diseases of bees see "Diseases and Pests" in the References.

DYSENTERY

Dysentery is not caused by a microorganism and is not a disease at all, but primarily the result of poor food and long periods of confinement. This is a condition generally seen in northern states in later winter or in overwintering cellars when bees are placed in controlled environments. Even though workers can retain 30 to 40 percent of their weight in fecal material, if fed improperly or unable to take cleansing flights, bees will defecate inside the colony, resulting in spotty combs and fecal-streaked entrances. In general, dysentery is caused by:

- Fermented stores.
- Syrup with impurities such as those found in "raw" or brown sugar.
- Dampness near hive or in apiary due to poor drainage.
- Long periods of confinement.
- Too much moisture in the hive.
- Honeydew in stores.

Symptoms

The symptoms of dysentery are similar to other adult bee diseases:

- Sluggish bees.
- Swollen abdomens.
- Staining on the hive furniture with yellow to brownish fecal material.

Because this is not a disease but a condition, the only way to treat dysentery is to:

- Provide a winter exit, so bees can take cleansing flights on warm winter days instead of defecating inside the hive.
- Ensure winter stores of low water concentrations (properly cured honey and sugar syrup); fermentation of stores with high water content may be a factor contributing to dysentery.
- Feed thick syrup (2:1 sugar : water) in fall if bees need more stores going into winter.
- Clean dirty combs and hive bodies by power washing and rotating combs with new foundation every three to five years.

THE PESTICIDE PROBLEM

Farmers, beekeepers, growers, orchardists, and homeowners apply pesticides—substances used to control pests in order protect their crops and therefore their investments. Pesticides include herbicides, insecticides, fungicides, and miticides, to name a few.

Most are labeled safe for bees and presumably cause few problems. But bees are insects and can be killed by a variety of pesticides. Honey bee / pesticide problems arise when insect pests or fungal (or other diseases) threaten crops that bees are working (when the crops sprayed are blooming), and the growers use chemicals to protect their crop. Recent investigations into the cause of CCD have led researchers to analyze pesticide residues in beeswax, and unfortunately, low levels of pesticides (mostly fungicides and acaricides used for mite control) are becoming more prevalent contaminants, even in new wax foundation. If you look at the breakdown list of pesticides (see the table "Families of Pesticides"), you can see that we are flush with chemicals in our environment, used for lawns, in Christmas tree farms, as rat bait, to kill bugs, and on and on. Some products break down fast when exposed to sun and water, but others (e.g., DDT) can remain in the environment for years. In addition, some of the breakdown products, or a combination of pesticides, may be **more** toxic to bees; much more research needs to be done here.

Bees come in contact with many chemicals while foraging. The best way to minimize the interaction of bees and pesticides is to know what dangers bees may encounter within the two-mile (3.2 km) foraging distance from your apiary. Get a map of your location and scout problem areas (see illustration on forage areas for honey bees in Chapter 2). Google Maps or other internet sites makes this task much easier nowadays; also drive around to get an idea where your bees are foraging and to see if there are any likely areas of pesticide applications for mosquito control or other possible pesticide applications.

If you are located near orchards, tree farms, nurseries, or other areas such as parks that may be sprayed, bees may be exposed to pesticides. Bees are also exposed to chemicals in their environment, and certain crops are sprayed during bloom (with fungicides). Again, being vigilant on what is sprayed on which crops in your area will keep you better informed. Here are some situations indicating bees could or have come into contact with poisons:

- A sudden reduction in number of bees (usually thousands) in all colonies whenever a specific plant or crop is being sprayed when flowering. In some cases, apple trees are sprayed before the apple blossoms open, and below these trees, dandelions are in bloom.
- Contact with recently applied compounds on target crop. Depending on the formulation, bees may die in the field or return and die in the hive.
- Consumption of contaminated water, nectar, or pollen. Field bees, hive bees, and larvae will die inside the hive.
- Misapplication of material, including direct application to non-target plants (drift from treated areas, the use of inappropriate chemicals or application methods). This will kill field bees, hive bees, and larvae.
- Chemicals (e.g., fungicides) that are sprayed directly on blooming plants or are systemic in the plant tissues (neo-nicotinoids and others). This may kill or disrupt foraging bees, plus bees may be bringing back contaminated pollen and/or nectar. Many applicators mix chemicals together in one tank to reduce the amount of time needed for spraying. Such mixtures can be a problem and are usually illegal.

It is important to know the telltale signs displayed by bees exposed to pesticides, as some of these may be confused with symptoms caused by bee diseases. In general, you will notice:

- A sudden reduction in the numbers (thousands of bees) in what was a strong colony.
- Excessive numbers of dying and dead bees, within 24 hours, in front of the hive, on the bottom boards or on top bars.
- Dying larvae crawling out of cells.
- A break in the brood-rearing cycle, disorganization of hive routine.
- Queen supersedure when circumstances prior to the discovery of supersedure cells would not indicate the need for supersedure: for example, you recently introduced a new queen to the colony whose observed performance was not deficient in any way.

Families of Pesticides

Type	Action
Algicides	Control algae in lakes, canals, swimming pools, water tanks, and other sites
Antifouling agents	Kill or repel organisms that attach to underwater surfaces, such as boat bottoms
Antimicrobials	Kill microorganisms (such as bacteria and viruses)
Attractants	Attract pests (for example, to lure an insect or rodent to a trap). (However, food is not considered a pesticide when used as an attractant.)
Biocides	Kill microorganisms
Biopesticides	Biopesticides are certain types of pesticides derived from such natural materials as animals, plants, bacteria, and certain minerals
Disinfectants and sanitizers	Kill or inactivate disease-producing microorganisms on inanimate objects
Fumigants	Produce gas or vapor intended to destroy pests in buildings or soil
Fungicides	Kill fungi (including blights, mildews, molds, and rusts)
Herbicides	Kill weeds and other plants that grow where they are not wanted
Insecticides	Kill insects and other arthropods
Microbial pesticides	Microorganisms that kill, inhibit, or outcompete pests, including insects or other microorganisms
Miticides	Kill mites that feed on plants and animals
Molluscicides	Kill snails and slugs
Nematicides	Kill nematodes (microscopic, wormlike organisms that feed on plant roots)
Ovicides	Kill eggs of insects and mites
Pheromones	Biochemicals used to disrupt the mating behavior of insects
Repellents	Repel pests, including insects (such as mosquitoes) and birds
Rodenticides	Control mice and other rodents

Sources: https://ipm-info.org/pesticides/chemical-families-pesticides/.
https://www.epa.gov/sites/production/files/2015-01/documents/pesticide-types.pdf.
US Environmental Protection Agency Office of Pesticide Programs Index to Pesticide Types and Families, and Part 180 Tolerance Information of Pesticide Chemicals in Food and Feed Commodities.
Notes: Pesticide tolerance information is updated in the Code of Federal Regulations on a weekly basis. EPA plans to update these indexes biannually. These indexes are current as of the date indicated in the pdf file.

Bees are also exposed to insecticides, and each kind of insecticide affects bees in a different way. Some of the kinds of insecticides in general use today include:

- Nicotinoids.
- Chlorinated hydrocarbons.
- Carbamates.
- Dinitrophenyl.
- Botanicals.
- Pathogenic.

Pathogenic insecticides, unless specific for hymenopterous (beelike) insects, are not toxic to bees. Some, such as *Bacillus thuringiensis* (Dipel, Biotrol, Thuricide), are used to control many lepidopteran insects (like the wax moth or gypsy moth) or polyhedrosis virus.

See the table for a more complete list of families of pesticides. The symptoms for bee poisoning can be found in recent editions of J.M. Graham's *The Hive and the Honey Bee* (Dadant) and Shimanuki, Flottum, and Harma's *The ABC and XYZ of Bee Culture* (A.I. Root); check the References section as well, and investigate the *Compendium of Pesticides Common Names* and other sites online.

What Beekeepers Can Do

First, register your beehive locations with the state bee inspector's office; that office may alert you if pesticide application in the area of your apiary is imminent. If you aren't registered, they can't notify you. If your state doesn't have a bee inspector, check with local beekeepers in your area.

To lessen the chances of hive exposure to pesticides, beekeepers can take the following steps:

1. Careful selection of apiary location:
   - Locate and meet farmers, landowners, or land renters within a two-mile (3.2 km) radius of the apiary.
   - Contact beekeepers in the area to learn about past problems.
   - Check plat (legal description), county, or air photo maps or online maps, even soil type maps, to assess apiary locations in relation to areas that may be sprayed (parks, orchards, tree farms, or residences).
   - Become familiar with the crops grown, production methods, rotation practices, and past pest or disease problems.
   - Be aware of planting, blooming, and harvest dates of target crops.
   - Become knowledgeable with spraying schedules for that orchard, crop, or location.

2. Assess chosen apiary site; weigh these chemical danger potentials:
   - Spray drift from nearby treated areas.
   - Frequency of sprays during the season.
   - Cyclic or unexpected outbreaks of insect pests (such as gypsy moths or mosquitoes).
   - Need for sprays during crop blooming period (fungicides).
   - Application methods used (air or ground, low volume, ultra-low volume, standard, or electrostatic equipment).
   - Herbicide or fungicide applications.
   - Parks or roadside spraying for mosquito or other pests.

3. Become familiar with:
   - The identification of crop pests in your area.
   - When other chemicals will be applied (e.g., after a rain).
   - Pest population levels that require spray treatments (economic threshold).
   - The types of pesticides used locally, their common names and formulations.

   - Registration procedures for apiary sites, so applicator can locate your hives.

4. Know formulations of pesticides: Formulations with the designation WP (wettable powder), EC (emulsifiable concentrate), MC (microencapsulated), and D (dust) will kill on contact and may be picked up by bee feet or body hairs. Pesticides in these forms may also settle on pollen or drift to water puddles near sprayed areas. The addition of stickers or spreaders to the sprays may significantly reduce problems caused by these formulations, making the chemicals less accessible to bees by sticking to plants instead. However, recent studies suggest that some stickers themselves are toxic to bees, so keep current with the research.

   Systemic pesticides, those that are absorbed into the plant tissues and kill insects that feed on the plants, are now common. Such pesticides are used as seed treatments or are placed in or on the soil; others are painted on tree trunks. The danger in systemic chemicals is that because they are flowing in plant sap (to kill insects such as aphids or other sap eaters), some of these compounds or their breakdown products are being found in both pollen and nectar as well as in beeswax. Keep current with the new research in this area.

5. Determine:
   - Local weed/wildflower blooming periods; learn to identify local honey plants.
   - Where bees will forage next (sequence of blooming plants).
   - What time of day bees are on particular target crops.

6. Anticipate:
   - Changes in cropping practices (has the spray schedule changed? Or are new pesticides being used?).
   - Scheduled and unscheduled sprays.
   - Crop blooming periods and sequences.

When you observe dead bees near the hive entrance, the bee inspector should be contacted. The inspector will provide instructions on your next move. Have handy the phone number and address of the local apiary inspector, if such exist in your state, or your state or county cooperative extension or agriculture office. Find out what legal recourse you have and how and where to send dead or dying bees for analysis as a result of a suspected poisoning. In most cases, **if your bees are not registered with the state agriculture department**, you will have no recourse.

Remember, many times exposure to pesticides may lie *not* with the farmer or orchardist spraying but with a neighborhood homeowner killing those "pesky" insects on the backyard rosebush or "weedy" dandelions in the lawn.

Protecting Bees from Being Poisoned by Pesticides

If you have time and it is practical, the best protection method of all is to move your hives at least two miles from their previous location and target area. This is the most expensive but most successful protective method. If you can't move your hives, here are some general protective measures you can take before spraying occurs; use one or a combination of several methods:

- Check your county extension office or other authority for information on protection programs in your area.
- Make routine contacts with landowners, renters, applicators, and county agents for updates on pest problems.
- Post your name, address, and phone numbers (or a neighbor's number so you can be immediately contacted) in a conspicuous place near apiaries.
- Paint hive tops with a light color for easy aerial identification.

In some states, applicators must notify a governmental agency that they will be applying a pesticide at a specific address, and the agency will inform the applicator if there are apiaries within the area to be sprayed. The applicator in such circumstances is obligated to provide the beekeeper with a heads-up as to the date the spraying will take place, the plants being sprayed, and the pesticides that will be applied.

When spraying is imminent, here are some quick methods to protect your bees:

- Prevent the bees from foraging for a day or so, but in doing so, screen off the entire entrance and provide additional ventilation to keep the hive from overheating. Do this by adding shade and more space within the hive(s) by adding su-

pers, if practical (a few colonies only). Close the entrances with 8-mesh hardware cloth or screen, to confine bees, and place a screened cover on top, with a third of the screen covered with a wet cloth or blanket. Keep this cover wet, especially if weather is hot, with a sprinkler or watering can, for *at least* 24 hours. This is a dangerous step to take, because even with wet cloth, hives could overheat and die very quickly.
- Saturate the hive with sugar syrup, by pouring it directly on top of the frames (bees will stop foraging to help clean it up); pour in about a quart twice a day for one day prior to a spray, and once a day for two or three days following a spray.
- If weather is cool, screen entrances with 8-mesh hardware cloth, and screen the top. Once bees are permitted to resume foraging activities, place pollen traps at the entrances to collect any pollen that has been contaminated (this pollen should be destroyed after it has been collected and identified).
- Feed colonies with syrup, water, and clean pollen patties during **and** after the spray period.

Once a kill has been experienced, you must immediately help those colonies that have been affected. After all, if it is early enough, they may still yield a surplus of honey, or at least store enough for winter. Here are some things to do:

- Combine weakened colonies to increase populations.
- In the event the queen is killed, requeen the colony with a mated queen rather than delay the rebuilding of the colony by allowing the bees to rear their own queen.
- Destroy contaminated stores, combs, or equipment; supply new equipment and clean combs or foundation.
- Feed syrup, bee bread, or pollen substitutes to support the colony's recovery, aid in queen acceptance (if requeened), and stimulate brood rearing.

Beekeepers should strive to cooperate with neighboring growers for the benefit of all. But the ultimate responsibility for a colony's protection rests with you, the beekeeper, not the farmer, landowner, or applicator. You need to make your site selections as safe as possible and be alert to the expected problems, while anticipating the unexpected. Use the numerous resources available to help you and your bees. Such sources of information, if still available, are:

- State and country extension offices (extension entomologists, agronomists, horticulturists) if they exist in your state, and their publications on crops, insect identification, insecticide lists, and formulations. Check websites as well.
- State apiary inspector, to register your apiary locations.
- Regional, state, and local beekeeping organizations.
- Libraries or city/state agencies, and their internet sites, for maps and other references.

BROOD DISEASES

Brood diseases can be devastating to both novice and commercial beekeepers alike. Recognition of healthy and diseased brood is an important part of colony management, and awareness of how these diseases are transmitted and spread may prevent a serious outbreak. Become familiar with recognizing diseases by taking workshops and seminars.

Diseases are transmitted by:

- Beekeepers, moving between diseased and clean colonies and not cleaning hive tools, gloves, or other bee wear and equipment.
- Interchanging brood frames within or between apiaries.
- Buying old, diseased equipment or combs and interchanging or mixing them with clean, healthy colonies.
- Honey fed to bees, especially store-bought honey or honey obtained from a diseased colony by robber bees.
- Pollen, honey, or royal jelly sold commercially.
- Package bees or queens.
- Swarms.
- Foraging honey bees inadvertently spreading viruses and other pathogens during flower visits.

Bee diseases are not mutually exclusive. A colony could have both nosema and foulbrood diseases at the same time. Some conditions, such as chilled brood, might be mistaken for some brood diseases. Careful attention to the symptoms, the condition, history, and mite levels of the colony is necessary in order to distinguish between chilled brood and diseased brood.

Bacterial Diseases: American Foulbrood

American foulbrood disease (AFB) is caused by the bacterium *Paenibacillus larvae* subsp. *larvae* (formerly *Bacillus larvae*) and exists in both a spore and a vegetative stage. The disease is transmitted by the spore, and the infected brood is killed by the vegetative stage, when the spore germinates in the larva's gut. This is the most destructive of the brood diseases and is why apiary inspection laws were first enacted.

Once the vegetative stages appear in a colony, the disease is spread rapidly, and the colony weakens; in most cases, the hive will eventually die unless it is resistant to AFB. Spores can remain viable in hive products (honey, wax, and propolis) **for up to 80 years**! It takes about 10 spores fed to a one-day-old larva for it to become infected. Older larvae require a higher number of spores in order to be infected. Death of the developing larva occurs after the cell is capped over and therefore is not immediately visible (unless you remove the wax capping). Once AFB has killed the larvae, the larvae will have turned black, often with a gluelike consistency; rarely you may see (if the cells are uncapped) the so-called pupal tongue sticking out at a 90° angle from the base of the cell.

Symptoms

The symptoms of AFB are varied and sometimes are confused with other diseases or even with mite infestations. AFB is transmitted from bee colony to bee colony in a variety of ways.

Here are some things to look for:

- Brood pattern (open or sealed brood) is irregular rather than compact.
- Larvae die in an elongated position, not twisted on the bottom of the cells like larvae that die from European foulbrood.
- Healthy prepupae and early pupal stages are pearly white; diseased ones turn from pearly white to light brown to dark brown. AFB is a disease that primarily impacts larvae, rarely the pupal stage. However, on occasion one uncaps a cell and observes a dead pupa with its tongue protruding from the cell at a right angle to the

American Foulbrood Disease

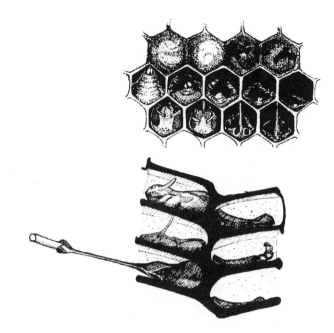

Symptoms:
Cappings are sunken, brood appears dark brown and "melted" down, and pupal tongue sticks up

Test:
See if the contents of the cells are sticky and will extend at least an inch when drawn out

pupa. This phenomenon is referred to as the pupal tongue and is rarely seen.

- Since the death of larvae and pupae often occurs after their cells are capped, the cappings become concave, and some will be punctured by bees attempting to remove the dead brood (see illustration). These puncture marks are very prevalent.
- Prepupae of bees that have been dead for a considerable time have a gluelike consistency, making it difficult for cleaning bees to completely remove these corpses from their cells.
- Eventually the dead larvae desiccate, and their remains form scales that adhere to the bottom, back, or side walls of their cells; these scales are extremely difficult to remove.
- An unpleasant, rank odor is often present as well, which has been described as the smell of long-unchanged socks.

Cells that contain newly emerged larvae may contain AFB spores, which can be ingested by the larvae as they consume larval food. Make sure to keep diseased equipment from coming in contact with any other colonies.

Any bad-smelling hive, especially if winter-killed, should be suspect, and if you look at the brood frames in good light, you should be able to see scales in the cell bottoms, or perhaps a few protruding pupal tongues.

AFB is transmitted in these ways:

- By beekeepers, with diseased equipment, tools, or bee suits.
- Cells in which larvae hatch may contain spores.
- Spores, present in honey and/or bee bread, are passed on to the larvae by nurse bees feeding them.
- Cleaning bees, spreading spores throughout hive when attempting to remove dead brood or scales.
- By robbers from diseased colonies, entering an uninfected hive, or scout bees robbing a diseased colony.
- By bees drifting from diseased to clean colonies.
- By swarms that have AFB.

Important note: American foulbrood disease needs to be attended to immediately. Listed below are the steps needed to take if AFB is either detected or suspected. Delay is not an option; professional advice and assistance is required. Remedies are limited. Seek the assistance of a bee inspector or seasoned beekeepers.

If a colony is suspected of being infected with AFB, follow these steps as soon as possible:

1. Reduce entrance to minimize robbing.
2. Distinguish an infected colony from the rest by

Characteristics of Some Diseases and Pests

Common characteristics	Possible problem
Adult bees are absent with little or no accumulation of dead bees in or around the hive. 1. Any food stores are not immediately robbed by other bees. 2. Damage by other pests is slow to develop. 3. Some capped brood is present in colonies (bees will not normally abandon a colony until all the capped brood have hatched). 4. Queen is evident, but the workers appear to be all the same age (young). 5. Bees are slow to feed on supplemented food. 6. Large areas of brood are not covered by bees.	Colony collapse disorder (CCD)
1. Bees are unable to fly or able to fly only short distances. 2. Bees are seen trembling and quivering; colony is restless. 3. Feces is found on combs, top bars, bottom boards, and outside walls of hive. 4. Bees are crawling aimlessly on bottom board, near entrance, or on ground; some drag along as if their legs were paralyzed. 5. Wings are positioned at various angles from body (also called *K-winged*); that is, wings are not folded in normal position over abdomen but with the hind-wing held in front of the forewing. 6. Abdomen is distended (swollen). 7. Bees are not eating when fed syrup. 8. Bees are abandoning the colony, leaving the queen and a few emerging workers.	Nosema, virus, pesticides
1. There is a sudden reduction in numbers (thousands of bees) in a previously strong colony in the middle of the summer season. 2. Excessive numbers of bees are dying or dead, within 24 hours, in front of the hive, on the bottom boards, or on top bars. 3. Dying larvae are crawling out of cells. 4. There is a break in the brood-rearing cycle, and disorganization of hive routine. 5. Inappropriate queen supersedure occurs.	Pesticides
1. Brood pattern is irregular rather than compact. 2. Larvae turn from light brown to dark brown and die upright, not twisted, in cells. 3. Cappings are sunken and punctured. 4. Surface of cappings is moist or wet rather than dry. 5. Larvae long-dead develop the consistency of glue and are difficult for bees to remove. 6. The dried-out brood or scales adhere to the bottom, back, and side walls of the cell and are difficult to remove. 7. Dead pupae have their tongues protruding at a right angle to their scale or straight up. 8. There is an unpleasant putrid "foul brood" odor, which can permeate apiary if many colonies died over winter.	American foulbrood (AFB) disease
1. Larvae die coiled, twisted, or in irregular positions in their cells. 2. Discolored larvae are clearly seen. 3. Dry scales are easily removed from their cells, unlike with AFB disease. 4. Some larvae die in capped cells, scattered, discolored, concave, and punctured. 5. A sour odor may be present. 6. Dead larvae are rubbery and do not adhere to cell walls. 7. Drone and queen larvae are also affected.	European foulbrood disease
1. White, mummified larvae are found. 2. Infected larvae are usually removed from their cells. 3. Dried mummies turn dark gray to black. 4. Mummies are found in brood frames and on the bottom board.	Chalkbrood

Characteristics of Some Diseases and Pests

1. Young workers (and drones) emerge twisted, deformed, and wrinkled. 2. Bees are smaller, underweight, and discolored. 3. Varroa mites are present.	Deformed wing virus
1. Larvae are darkened from white to yellow; eventually they will turn dark brown. 2. Older larvae develop leathery skin and dark head regions. 3. Black-headed larvae are bent toward cell center. 4. Larvae fail to pupate and die with heads stretched out. 5. Diseased larvae are easily removed in liquid-filled sacs (the larval skin). 6. Scales are dry, brittle, and easily removed.	Sacbrood virus
1. Populations of bees are dwindling. 2. Weak bees are crawling on ground with K-wings. 3. Hives with plenty of honey stores are abandoned in the spring. 4. Hive bodies are spotted with fecal matter.	Tracheal mites, virus, nosema
1. Capped drone or worker brood is infested; cappings can be punctured, as in foulbrood. 2. Adult bees are disfigured, stunted, with deformed wings, legs, or both; there can also be crawling bees on ground. 3. Bees are discarding infested or deformed larvae and pupae. 4. Pale or dark reddish brown spots are found on otherwise white pupae. 5. Spotty brood pattern is seen and diseases are present. 6. Dead colonies are found in the early fall. 7. Queens are superseded more than normal. 8. Foulbroods and sacbrood symptoms are present. 9. AFB symptoms exist, but no ropiness, odor, or brittle scales.	Varroa mites, virus
1. Tunnels are observed in combs. 2. Silk trails are seen, crisscrossing one another. 3. Small dark specs are found on bottom board or in the silk trails in a hive. 4. Silk cocoons are attached to wooden parts. 5. Capsule-sized depressions are carved in wooden hive parts. 6. Piles of debris and larvae are on bottom board. 7. Moths may be present.	Wax moth
1. Whitish to tan larvae are present on cells or on bottom board. 2. No silk tunnels are present. 3. Honey is slimy and has particular orange odor. 4. Bees do not rob deserted colonies. 5. Dark, small beetles are seen on comb or hiding on bottom board.	Small hive beetle
1. Few queen cells are present in the center of the comb. 2. The age of queen larvae in the cells is the same. 3. Scattered brood pattern is observed. 4. Drone brood often appears in worker cells.	Failing queen

marking or painting it a different color, to reduce drifting of bees and for visual identification by the beekeeper.

3. Either destroy diseased colonies (see below) or begin a medication (chemotherapeutic) program immediately; see "Foulbrood Disease Chemotherapy" below. If you do not want to use medications, you can also apply integrated pest management (IPM) practices outlined in the next chapter.

Call your state bee inspector or other officials, or your local beekeeping organization for advice and to confirm diagnosis. If official help is unavailable, first isolate the suspect colony so other bees won't rob it (move it or close the entrance to its smallest opening), and then collect and send a sample of the suspected brood comb. To do this, select a suspect frame and cut a sample of comb, about 4 × 5 inches square and free of honey, and containing the diseased brood.

Wrap it in **newspaper** so it will not get moldy; do not use any other kind of wrapping. On a separate piece of paper, write your name and address and place it and the sample in a sturdy cardboard box and mail to your state bee lab or to the USDA Beltsville Bee Research Lab. Send a letter, or call or email, stating that you are sending samples under separate cover. Describe the problem you are having with the hive, and include the following information:

● Name and address of beekeeper.
● Name of address of sender (if different).
● Location of samples and source.
● Number of samples sent (each labeled and numbered in a different package); indicate if the samples are from the same or different apiaries.

Testing for AFB

Use the "ropy test," described below, on larvae that have been dead for about three weeks. Since it is difficult to determine how long a larva has been dead, randomly test between 5 and 10 cells containing dead larvae from several frames. An accurate way of determining how long a larva has been dead is by checking the presence or absence of its body segments or constrictions (like earthworm constrictions). If they are absent, the larvae have been dead for at least three weeks.

Insert a match, toothpick, stem, or twig into a cell, stir the dead material, then slowly withdraw the testing stick (see illustration on p. 239). If a portion of the decaying larva clings to the twig and can be drawn out about 1 inch (2.5 cm) or more while adhering to the stick, its death was probably due to AFB. Be sure to burn the test stick. Scrub your smoker and hive tools with a soapy steel wool pad and wash your hands, gloves, or bee suit thoroughly in hot soapy water. Bleach does not kill AFB spores.

Treatment for AFB

It must be stressed that AFB is a very serious and contagious bee disease and is the reason why the bee inspection program was started. Unfortunately, many states no longer have apiary inspectors, and it is up to **you**, the beekeeper, to be diligent and keep your bees healthy and disease free. Your choice of action, should you find a colony afflicted with AFB, is somewhat limited and will depend on both national and state regulations. Many states no longer recommend that you medicate colonies with AFB, leaving you but a single choice: burn the colony, while strictly following the procedures required.

Before you burn any equipment, be sure to check with your local fire department or town hall as to whether a burn permit is required.

Some countries do not allow bees to be treated with antibiotics and only permit burning of all the infected equipment. This is a good option if you purchased old, diseased equipment; and it was the only option before the availability of chemotherapy and decontamination chambers. For chemotherapeutic treatments see "Foulbrood Disease Chemotherapy" in this chapter. **Do not** use chemotherapy drugs as a prophylactic—**treat only if you have a disease outbreak**. Now that drug-resistant strains of foulbrood are present, the only way to ensure healthy bees in the future is to use stock that is from queen lines resistant to diseases (and mites). So your choice is to either destroy the affected colonies or begin medication, if permitted.

Important note: A veterinary feed directive (VFD) form is now needed for obtaining medications for honey bees! This means you need a prescription from a licensed veterinarian. Check with your state bee inspector or your local bee groups for more information.

Burning hive equipment. If you do not wish to medicate the colony, and/or you are not permitted to use antibiotics, an integrated pest management (IPM) approach is to burn the colony.

To do this you can (1) kill all the bees in the evening after all the foragers have returned from the field and then burn the dead bees and the equipment; or (2) save the adult bees by installing them on new equipment containing only comb foundation (be sure your state or province permits you to take this action). Kill all adult bees by spraying at night all frames using a fast-acting insecticide; you can also spray bees on frames using a 3 to 4 percent soapy water solution (1 cup liquid detergent per gallon of water). If possible, contain all bees and equipment in a tarp or plastic bags. Here's how:

● Remove entire colony with dead bees inside, to a field in which a pit has been dug at least 18 inches (45 cm) deep and contains a hot burning fire.
● Burn all the brood frames, whether they contain brood or not. It is permissible to extract the

honey, which can be consumed by humans without any ill effects, but such honey must not be fed to bees, for often the honey may contain foulbrood spores. After extracting the honey, these frames should also be burned.

- You can support the larger hive bodies with tree limbs across the pit. Make sure all is consumed and turns to ash. You can save the wax (to be sent to a rendering plant), as long as it is securely closed against robbing bees.
- When done, cover ashes with dirt and refill the pit with the soil you excavated.

Another option is to save the bees. Cage the queen, then shake the adult bees off all frames into a new hive body filled with foundation. The reason for placing the adult bees from an AFB colony on foundation is to prevent them from regurgitating the contents of their stomachs, which likely contain foulbrood spores. By the time cells become available, the spores will have been passed along to other parts of their digestive system and no longer be viable. This practice is not advisable without the support of your apiary inspector, because some of these adult bees could drift to other colonies containing drawn combs and deposit the contents of their honey stomachs, along with the foulbrood spores, into existing cells. Check with your bee inspector to learn if this practice is permitted in your state or province. We do not consider this a sound practice.

If, after you try this approach, the colony develops AFB, destroy the colony by burning.

If there is a medical or veterinary incinerator in town, you could see if it will burn your (intact) equipment. Make sure to extract any honey from these diseased colonies and bottle it. Honey that contains AFB spores is perfectly safe for human consumption, but **do not** feed honey from a colony with AFB back to other bees. **Remember: AFB spores can last up to 80 years!** There are many other tests for ascertaining whether AFB is present. One easy test is the Holt Milk Test. Check the internet for information as to how to perform this one.

You can save newer hive bodies, if they are not too coated with wax and propolis (which also may contain AFB spores). Invert hive bodies, so rim edges and handholds are upside down, and stack them three or four high. Scrape the rims and the interior of the hive bodies so they are as free as possible of wax

and propolis. Be sure to contain the scraped-off wax and propolis, and burn these items as well.

- Fill the inside of the stack with newspaper and ignite it; when the insides of hive bodies are scorched, extinguish the fire.
- You can also "paint" the insides of the equipment with kerosene and then ignite the kerosene by putting a match to newspapers stuffed inside the hive equipment.

Here is a more thorough way to sterilize equipment:

- A propane torch can also be used for the tops and bottom boards as well as hive bodies; wood should be lightly browned, including all edges; give special attention to the seams.

These methods, while satisfying, may not completely sterilize hive equipment. Before burning, again check with your state's bee inspection program to be sure the procedure is legal. Gamma radiation chambers or other fumigation facilities are the most effective method to decontaminate bee hive equipment, but many have been closed. A few states may have some chambers available, so check it out online or with your state's agriculture department.

Sterilization. Another way to sterilize equipment is to boil the woodenware in a lye or paraffin bath (we strongly urge beekeepers **not** to use this method). Both these approaches will require extreme skill in order to avoid injury.

You can go online to learn if gamma radiation or fumigation facilities are available in your area.

Bacteria Diseases: European Foulbrood

European foulbrood (EFB) is caused by the spore-forming bacterium now called *Melissococcus plutonius* (formerly *Streptococcus* or *Bacillus pluton*). Other bacteria may also infect larvae at the same time, producing similar symptoms. EFB is commonly found in weak colonies, or in those stressed due to poor foraging resources, or when colonies are trucked across the country to carry out pollination activities.

The disease is usually prevalent in the spring, slowing the growth of the colony, but may disappear with the onset of a good honeyflow. Larvae

European Foulbrood Disease

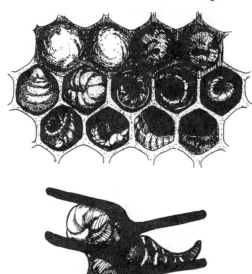

Symptoms:
Bacteria infect younger, uncapped larvae,
turning them brown

Test:
Larvae are twisted in the cells, no pupal
tongue is evident, no "foul" smell is present,
and the ropiness test does not apply

more than 48 hours old are at greatest risk, and thus those that die are usually **not** capped but are visible in the bottom of the cells. The bacterium is found in feces, wax debris, and is on the cell walls containing infected larvae. The dead larvae dry in their cells to form soft scales that are easily removed by bees. Not as serious as American foulbrood, EFB can be treated with antibiotics, such as Terramycin (if permitted), by requeening, or by strengthening the colony with additional adult bees.

Symptoms

The symptoms of a colony infected with EFB are different from those for AFB. Learn to recognize these signs:

- Larvae die in a coiled, twisted, or irregular positions in their cells (see illustration).
- In EFB, unlike AFB, the larvae usually die well before their cells are capped over. This makes it easy to view the dead larvae and also informs you that the disease is likely EFB and not AFB.
- The color of the dead larvae may change from pearly white to a light cream, then to brown and then a gray brown, darkening further as the dead

larvae desiccate; healthy larvae remain pearly white throughout their larval stages.
- Dry scales—the remainder of the larvae—are easily removed from their cells, unlike AFB scales, which are difficult to remove.
- Some larvae die in capped cells, scattered over the brood comb; cappings may be discolored, concave, and punctured.
- A sour odor may be present.
- Dead larvae are normally not ropy, as in American foulbrood.
- Drone and queen larvae are also affected.

EFB is transmitted from colony to colony in the following ways:

- Robber bees or bees that have drifted from an infected colony may spread the disease when they enter a disease-free colony.
- Contaminated equipment.

If you have any questions about what is wrong with your colonies, send a sample of brood to the USDA Beltsville Bee Research Lab. Use the method previously described in the section of AFB testing. Read about varroa mites, in this chapter.

Control

These methods have been used to control or eliminate EFB from living colonies:

- Requeen to break the brood cycle; this allows the bees to clean out dead and infected larvae.
- Use chemotherapeutic agents (if permitted) to treat the disease (see "Foulbrood Disease Chemotherapy," below).
- Feed with sugar syrup and pollen supplement/ substitute; make sure pollen you purchased is irradiated, as it could contain EFB spores or other disease organisms.
- Restrict drifting between colonies by relocating or redistributing hives (see the illustration on "Hive Orientation" in Chapter 4).To strengthen colonies or to make splits, brood is often transferred from one colony to another or from a colony into a split. **Be sure the brood combs being transferred are disease free**.

Foulbrood Disease Chemotherapy

Drugs can be given to bees for both AFB and EFB once the disease has been diagnosed. **As already mentioned, you will now need a prescription from a licensed veterinarian to obtain these drugs**.

Since the 1940s, beekeepers have used the antibiotic oxytetracycline (sold as Terramycin) to control AFB. Unfortunately, since the 1990s, this drug was found to be no longer effective for AFB. It is, however, used to treat EFB. In the case of AFB, drugs will kill the bacteria but not its spores; therefore after the drug is withdrawn, the disease may manifest itself again. Before using any drugs, you must first ascertain whether a given drug is permitted for use with honey bees.

Currently, the new antibiotic for AFB is used **only to treat** the disease, not to prevent it. This drug is tylosin tartrate, sold as Tylan soluble powder. Remember, the antibiotic drugs **do not cure** the disease.

Never use any chemotherapeutic drug during a honeyflow. If drugs are used during a honeyflow, the honey must not be used for human consumption. **Follow the label directions before using any drugs on bees**. Otherwise, you may not be giving enough medication to treat the disease and are contaminating your honey.

Some beekeepers are deciding **not** to treat their bees with **any** drugs. If you decide to do this, check with your state bee inspector and assess where other beekeepers are within the flight limit of your colonies, to be sure that you are not picking up pathogens or parasites from other bee colonies, or transmitting them to those colonies. A better alternative would be to practice an integrated pest management (IPM) scheme, outlined in the next chapter.

Fumigation Chambers

Special fumigation chambers or gamma radiation can decontaminate empty combs and equipment. This method kills the disease spores and allows the equipment to be reused. Most states no longer have these chambers. Check with the local bee inspectors or state agriculture extension offices for the licensed operators in your area.

Fungus Diseases: Chalkbrood

Although common in Europe for decades, chalkbrood was first reported in the United States in 1968 on leafcutting bees (Megachilidae) and on honey bees in 1972; it is now spread throughout the country. The causative agent is the fungus *Ascophaera apis* (Maassen ex. Clausen), and it may reduce honey production but usually will not kill a colony. Currently, chalkbrood disease has been reported on the rise, and in some instances it can kill colonies weakened by mites or other diseases, or by stress (moving bees frequently for pollination) or exposure to pesticides (including fungicides). Some genetic lines, especially inbred bees, can be more susceptible than others, so one control may be to requeen a diseased colony with a resistant strain of bees.

There are about twenty other *Ascophaera* species, and some of them affect other pollinating insects. Infections are seen in the late larval stages, and drone brood can be especially vulnerable, because they are located along the edges of the brood frame.

Evidence that chalkbrood is present in a colony includes white mummified larvae in brood cells, on the surface of the bottom board, and/or on the ground in the vicinity of the entrance. These mummified larva resemble broken pieces of chalk, hence the term chalkbrood (see illustration).

The most susceptible larvae are four days old. The spores of the fungus are resistant to degradation and can be viable for 15 years; spores are transmitted from bee to bee during food exchange, to or from queen bees, and by drifting bees. Contaminated combs and tools will also carry the disease.

Chalkbrood Disease

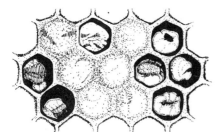

Symptoms:
Fungus infects young larvae, which appear chalky white in the cells, but turn gray and eventually black

Mummified larvae may resemble pollen packed in cells, but bees pull out these mummies, which will be clearly visible on the bottom board

This fungus is transmitted throughout an apiary by:

- Wind.
- Soil.
- Nectar, pollen, and water.
- Drifting bees or robber bees.
- An infected queen.
- Equipment.

While chalkbrood may not be considered a serious disease to a hobby beekeeper, in severe cases a colony's population can be compromised, with a subsequent reduction in honey production.

In some commercial operations, chalkbrood can be a serious problem. Pesticide exposure may be responsible for lowering bee immunity or killing beneficial fungi, but investigations in this area are still too premature. Staying current with bee research will help you become a better-informed beekeeper.

There is no chemical registered for use against this disease, but maintaining strong, healthy colonies keeps many diseases under control. Addition methods that have worked are:

- Move hives to sunny location, with plenty of ventilation and dry conditions.
- Remove infected combs and burn them; replace with new foundation. Adding bees to colonies with low population that are infected with chalkbrood helps weak colonies overcome this problem.
- Requeen if disease is severe, especially with hygienic queen lines.
- Make sure the bottom board is dry throughout the year.
- Feed sugar syrup and protein supplements during the spring, especially when prolonged inclement weather (three or more days) curtails foraging activities; colonies with large populations tend not to come down with chalkbrood (see Chapter 7 and "Diseases and Pests" in the References section).

While there are other diseases caused by fungi, such as stonebrood (*Aspergillus flavus* and *A. fumigatus*), they are rarely seen, or they go undiagnosed. If you have any questions about a disease, send samples to the Beltsville or other bee labs.

VIRAL DISEASES

Viruses differ from bacteria and other organisms mainly because they lack organelles and are unable to self-replicate without invading the cells of other organisms. Viruses contain fragments of DNA or RNA (nucleic acids), which are enclosed in a protein "envelope." They are capable of replicating themselves by attaching to the cell membranes of other organisms, and once attached, they inject their DNA into the host cells. Then the viruses use the cells' nucleic and amino acids to manufacture more of their own DNA and protein coats. In the process of replicating themselves, viruses consume the amino acids and nucleotides of their host, killing the cell they invaded in order to replicate themselves and often inflicting great harm to the host organism, even leading to its demise. Because of their microscopic size (over 100 times smaller than bacteria), they cannot be viewed with conventional microscopes but can only be seen and identified with electron microscopes.

Viruses are difficult to kill because of their mode of reproduction, unlike bacteria, for which antibiotics often exist. If an organism is injected with a particular virus that has been disabled, the host will produce antibodies to combat any identical virus that may invade it. Bees, however, have a very limited antibody defense system and therefore are more vulnerable to viral infections, which often disable or kill them. Although bee viruses have existed in the past, the problem has become more acute with the presence of the varroa mite in honey bee colonies. While the mite is feeding on the bees' fat bodies, it is transmitting viruses to the bee it is feeding on. The mites not only wound the bees to obtain nourishment, but their mode of feeding also increases the transmission of viruses from the mite to the bee. The only way to reduce this transmission is to continually work to keep varroa mite levels as low as possible.

Following is a current list of the more important viruses affecting honey bees. The list was supplied by Dr. Humberto Boncristiani, honey bee husbandry researcher at the Honey Bee Research and Extension Laboratory, Entomology and Nematology Department, the University of Florida, Gainesville (www .ufhoneybee.com). He starred the viruses most important to date (2018):

Deformed wing virus (DWV)*
Black queen cell virus (BQCV)*
Israeli acute paralysis virus (IAPV)*
Kakugo virus
Chronic bee paralysis virus (CBPV)*
Acute bee paralysis virus*
Sacbrood virus*
Kashmir virus*
Slow paralysis virus

Lake Sinai virus*
Apis iridescent virus
Berkeley bee virus
Arkansas bee virus
Big Sioux virus
Varroa destructor Macula-like virus
Aphid lethal paralysis virus
Cloudy wing virus
Filamentous virus
Bee virus X
Bee virus Y

Viruses are spread by mating of drones to queens (venereal transmission), through the egg to larvae and adults (vertical transmission), and between bees, via food and fecal matter (horizontal transmission). Some viruses are also found in bee bread and wax combs. The prevalence of bee virus is still not known, but with intensified CCD research, better surveys and testing will help give us a clearer picture of the spread and extent of these pathogens.

Prior to the introduction of parasitic mites, the only virus most beekeepers came in contact with was sacbrood virus. Now in the post-mite era, **up to 48 viruses have been identified** (to date) that cause bee diseases; the seven outlined below are the most commonly encountered (in order of most seen to least). For more information see "Diseases and Pests" in the References.

Deformed Wing Virus (DWV)

This disease was first identified in 1991 and only became noticeable after the discovery of varroa mites, with which it is associated. It is now distributed worldwide. Symptoms include young workers emerging with wings that are twisted, deformed, and wrinkled. Bees with deformed wing virus are underweight, discolored, and usually die within 72 hours after emerging from their cells as adults. DWV bees are usually evicted from the colony and can be seen crawling on the ground in the apiary; their presence is an indication that the colony they come from is likely on the verge of collapse. This virus is vectored (transmitted) by the varroa mite, and there is no known treatment, other than keeping mite numbers in colonies as low as possible. Some reports suggest requeening with an uninfected queen may help save the colony, if done early enough in the season; how-

Deformed Wing Virus

ever, to date, there is no way to test if queens are clean and virus free.

Black Queen Cell Virus (BQCV)

First isolated in 1977, this virus kills capped queen larvae and prepupae. The diseased larvae inside the cell turn yellow and have a sacbrood-like covering. It is commonly found in commercial queen-rearing operations in the spring and early summer, but is also found in worker brood and adult bees. Infected queen larvae or prepupae sealed in the cells turn black, and the cell walls turn brown-black. Virus particles are transferred by nurse bees from their food glands (hypopharyngeal and mandibular), and bees with *Nosema* can also be infected with BQCV. Infected queens may have a poor brood pattern, so requeening would certainly be advised. Further studies need to be done, and perhaps controlling *Nosema* could reduce the incidence of this virus.

Israel acute paralysis virus (IAPV)

This virus, discovered in 2004, was associated with dying colonies in the 2007 outbreak of CCD. Trans-

mitted by varroa mites, it causes paralysis in bees, which then die outside the hive.

Acute Bee Paralysis Virus (ABPV)

This virus was first identified when bees were experimentally inoculated with chronic bee paralysis virus (CBPV) in 1963; the bees were seen trembling and crawling on the ground with dislocated wings. ABPV is found in brood and adult stages and is often seen during the summer when varroa populations are high. It appears that the mite also activates the virus in infected bees, but its replication in bees or mites is still not clearly understood

Sacbrood Virus (SBV)

Sacbrood is the most widely distributed virus, first identified in the United States in 1913. Larvae and adults can both get the disease, but SBV is most easily seen in larvae that are two days old or older. Nurse bees become infected when cleaning out diseased larvae, and the viral particles are found in the food or hypopharyngeal glands. Thus the virus is spread throughout the colony, including foragers, who can infect the pollen loads when they regurgitate nectar onto their pollen baskets. Young bees can then become infected by feeding on virus-laden bee bread. It is easy to diagnose when seen in the larvae, but the adults with the disease may not live long. If you see a scattered brood pattern, examine the brood carefully; larval symptoms are clear and obvious:

- Larvae are darkened from white to yellow; eventually they will turn dark brown.
- Older larvae develop leathery skin and dark head regions.
- Black-headed larvae are bent toward cell center.
- Larvae fail to pupate and die with heads stretched out.
- Diseased larvae are easily removed in liquid-filled sacs (the larval skin).
- Scales are dry, brittle, and easily removed.
- Disease is often seen in the spring and early summer months.

Sometimes bees remove these diseased larvae quickly, but if there is any question as to the identification, send a sample to the USDA bee lab (see method in "American Foulbrood Disease" above). Sacbrood can also be found in the spring after many foragers have been killed by pesticides. A new strain of SBV from the Asian honey bee *A. cerana* has been identified in India, called Thai SBV.

Strengthening a colony with clean food and more bees may help in its recovery. Because sacbrood is a viral disease, medication is ineffective; requeening the colony may remedy the situation.

Kashmir Bee Virus (KBV)

First isolated from Asian honey bees in 1977, KBV was subsequently found in Australia and New Zealand and later in North America. This virus attacks all stages of bees and brood with no clear symptoms; it is closely related to acute bee paralysis virus (ABPV). While the virus could be present as an inapparent infection (no symptoms), KBV has been shown to be activated to a lethal level when varroa mites are present in high numbers; the mites can transfer the virus.

Chronic Bee Paralysis Virus (CBPV)

CBPV was first identified while researchers were studying the tracheal mite outbreak in the UK. Later, virus particles were extracted from dying bees. This virus is now found worldwide, except in South America. It has two forms: One form is represented by trembling bees crawling on the ground with abnormal wing posture and distended abdomens. The other form is called hairless black syndrome, because the bees lose their hair and appear shiny black or greasy; such bees are not allowed back into the hive and therefore are found on the ground. Both forms can be present in the same colony. CBPV is less virulent than ABPV, which takes only one day to kill bees. It is found in overcrowded colonies where the hairs of bees can break, exposing the cuticle, so it is not transmitted by varroa. It is not usually seen anymore.

Again: There are now many internet sites that can be helpful in detecting and illustrating these viruses and other pathogens of honey bees, so check them out.

 **Notes**

CHAPTER 14

Pests of Honey Bees

MITES

Mites are arthropods belonging to the superorder Acari, which also contains ticks; all have eight legs (see illustration). They differ from spiders because spiders breathe (exchanging oxygen and carbon dioxide) using "book" lungs, while most mites have a respiratory system consisting of an array of tubes referred to as a tracheal system.

This is an amazingly diverse and rich order of animals. Many of them are microscopic and so are not often encountered; they range in size from less than 100 micrometers to 2 mm (0.8 inch). In adults, there are two body regions, the *gnathosoma* (head) and the *idiosoma* (abdomen and legs), the region that bears four pairs of walking legs; larvae usually have six legs.

Over 50,000 named species of Acari have been identified, but estimates of over one million species have been suggested. Their habitats are even more diverse than those of insects, and include in and on soil, water (fresh and salt), plants, arthropods, vertebrates, and invertebrates. Some more interesting habitats are sea snake nostrils, monkey lungs, sea urchin guts, and snail shells (the mites live in special pockets on the shell and eat snail slime). Many, like the sarcoptic mange mites (family Sarcoptidae), are parasitic. Demodex mites (family Demodicidae) live in the hair follicles of mammals, including humans. House dust mites (family Pyroglyphidae) originally inhabited rodent nests but moved into our homes—a much bigger nest. Other mites live in and on food products, such as cheese, flour, seeds; others live on fungi or in the soil. Because they have no wings, mites are dispersed by wind, water, insects, plant seeds, molds, fungal spores, and birds, to name a few. The scientific study of mites is called acarology.

Dr. G.C. Eickwort, in chapter 40 of *Africanized Honey Bees and Bee Mites* (J. Wiley & Sons, 1988), listed around 40 mites found inside a beehive. To date, the only two parasitic mites inhabiting honey bees found in the United States are listed below. The third one, *Tropilaelaps* sp., may still make its way out of Asia and inhabit new areas.

Tracheal Mites or Acarine Disease

Tracheal mites, or *Acarapis woodi* (Rennie), are the causative agent of Acarine disease, which was originally called Isle of Wight disease (where it was first reported in 1919). When these mites were first discovered, all importation of honey bees into the United States was halted, and the Honey Bee Act of 1922 was passed by Congress. Tracheal mites were not found in the Western Hemisphere until 1984, when they were reported in Mexico. From there they quickly spread into the United States and Canada primarily by migratory beekeepers and by way of bee packages.

Many bee colonies were initially lost (due to tracheal mite infestation) before the varroa mite appeared. Distribution of the tracheal mite is now worldwide except in some Pacific islands and perhaps in some African countries. While it was first identified in Europe, its true origin is not known.

This small mite lives inside the thoracic tracheae (breathing organs) of adult bees, where it feeds, mates, and lays eggs. After mating in the tracheal tube, the females emerge from the tube via its spiracle opening and crawl onto the hairy body of their

Questing Female Tracheal Mite

Stages of Tracheal Mite

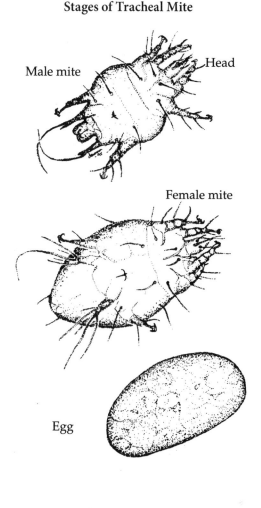

Male mite

Head

Female mite

Egg

Actual size ·

host. From there they quest in order to find and attach to a young, newly emerged or *callow* bee. Once on the new host, they enter the bee's spiracle, which allows them access to the tracheal system (see illustrations). There the mite can mate and lay about 1 egg a day, for 8 to 12 days (see chart on parasitic mites on p. 261). After the eggs hatch, the immature mites, or larvae, live as parasites inside the tracheae of adult bees, feeding on bee hemolymph by piecing the walls of the tracheal tubes. Young male mites exit the tracheae of their host after 11 to 12 days, while young females emerge from their host in 12 to 15 days. After exiting, they seek to infest and enter the tracheae of young bees, and this cycle is repeated over and over.

The bees at risk are young, newly emerged bees (workers, queens, and drones) up to three days old, which are distinguished and selected by female mites over older bees (see illustration on p. 252). During the summer months, the number of tracheal mites can be quite low because of the short life of the workers. However, during winter, when bees are clustered, mites can infest and reinfest bees of all ages. Since there are few young bees in the winter, the mite populations inside each bee will reproduce and reinfest all bees inside the colony. Eventually, the mite population can become high enough to overwhelm

the colony, which will result in its demise. Usually, the first signs of the tracheal mite are dead colonies in the early spring containing ample stores of honey. Should you find such conditions, you should suspect the colony's demise might be due to a tracheal mite infestation.

These mites can cause severe losses in temperate climates but cause fewer problems in warmer climates. However, they may carry virus diseases or cause queen supersedure; purchased queens can have tracheal mites. Because of their small size, tracheal mites are often overlooked, and their impact is now overshadowed by the larger varroa mite. But they have **not** disappeared. If your colonies are not thriving, if your varroa counts are low, and you have checked for all other problems, you should look for tracheal mites.

Life Cycle Chart

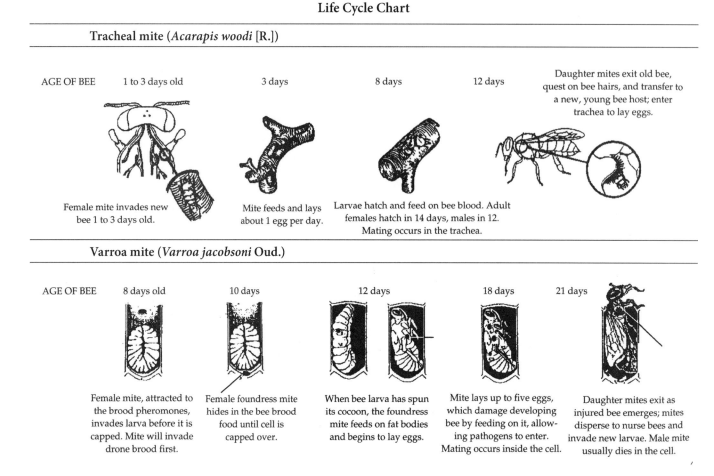

Tracheal mite (*Acarapis woodi* [R.])

AGE OF BEE   1 to 3 days old   3 days   8 days   12 days   Daughter mites exit old bee, quest on bee hairs, and transfer to a new, young bee host; enter trachea to lay eggs.

Female mite invades new bee 1 to 3 days old.

Mite feeds and lays about 1 egg per day.

Larvae hatch and feed on bee blood. Adult females hatch in 14 days, males in 12. Mating occurs in the trachea.

Varroa mite (*Varroa jacobsoni* Oud.)

AGE OF BEE   8 days old   10 days   12 days   18 days   21 days

Female mite, attracted to the brood pheromones, invades larva before it is capped. Mite will invade drone brood first.

Female foundress mite hides in the bee brood food until cell is capped over.

When bee larva has spun its cocoon, the foundress mite feeds on fat bodies and begins to lay eggs.

Mite lays up to five eggs, which damage developing bee by feeding on it, allowing pathogens to enter. Mating occurs inside the cell.

Daughter mites exit as injured bee emerges; mites disperse to nurse bees and invade new larvae. Male mite usually dies in the cell.

Collecting Bees to Test for Tracheal Mites

External signs of tracheal mites are unreliable but include dwindling populations of bees, weak bees crawling on the ground with K-wings, and abandoned hives in the spring with plenty of honey stores. A positive diagnosis of the tracheal mite by gross examination of the colony, or by bees walking around on the ground, cannot be done. Bees must be dissected for a positive identification, as some of the visible symptoms are not always reliable and are not necessarily due to this mite. In order to determine if you have these mites, you **must** dissect bees. You can also send a sample of your bees to your local state bee inspector's office or the USDA Beltsville Bee Research Lab, as you would send brood samples for disease identification; follow their shipping directions. If you suspect that your apiary is infested with this mite and you want to collect the specimens yourself, here is how to do so:

1. Sample at least 50 percent of the colonies in any one apiary.

2. Collect only "old" worker bees; they are most likely to have a mite infestation in their tracheae and are the easiest to diagnose (see below), thus facilitating identification. Old bees can be found on the inner cover, at the entrance, or as they return from foraging, not near the broodnest.

3. The best times of year to collect bees for tracheal mites are early spring or late fall.

4. Collect drones or old queens (that you are replacing), as they may have mites as well.

5. Place the collected bees in a 70 percent ethanol (alcohol) or isopropyl (rubbing) alcohol or freeze them in a glass or plastic jar or bag.

6. Send or deliver these specimens to the state bee inspector, state entomologist, or USDA lab, with the following information:
   - Your name and address.
   - Location of apiary tested (state, county, township).
   - Number of colonies in the apiary.
   - Source of bees (i.e., packages from dealer X).

Mites in Tracheal Tube

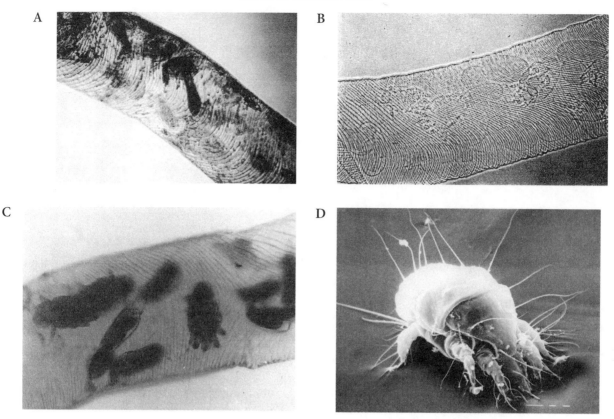

(A) Damage mites do to tracheal tube; (B) tracheal tube freshly invaded by mites—note no damage, but many mites inside; (C) mites (dyed with colorant) now visible inside the tube; (D) tracheal mite magnified 600 times. All photos by D. Sammataro. For video on how to look for tracheal mites go to https://www.ars.usda.gov/ARSUserFiles/31186/sammataro-vp(32-4)-november%204.pdf.

If sending them through the mail, use bees stored in alcohol, or place the frozen bees in a very small amount of alcohol. Pour off as much alcohol as possible to reduce the weight, if shipping by post. Check the USDA Beltsville lab website for mailing instructions.

Dissecting Bees

If you want to dissect bees yourself, or for a science class project, follow the procedure outlined below. Patience and practice are the most important requirements for a successful dissection. Practice on drones first, especially those collected in the later summer or early fall; they are easy to hold, and their tracheal tubes are larger. If requeening a colony, check your old queen, for she may have infested the entire colony. Finally, collect some old summer or early spring bees and dissect them. If you have used chemicals to control varroa mites, you may not find

a lot of tracheal mites, so collect and examine bees before you treat a colony.

A dissecting microscope (at about 40–60×) and a pair of fine jeweler's forceps will be needed. You can find older microscope models online or at school or university surplus departments. Now follow these steps:

1. Soften a frozen bee by holding it in your hand a few seconds. If the bee was stored in ethanol, it is already soft enough, but if it has been in alcohol for a few months, the tissues will be darkened, and it may be difficult to see mites.
2. Place the bee on its back and pin it (on a piece of corkboard or wax) through the thorax, between the second and third pairs of legs. You can make "petri" dishes by filling a jar lid with wax, letting it harden, and using that to pin bees.
3. While looking through the microscope, remove

Dissecting Bees for Tracheal Mites

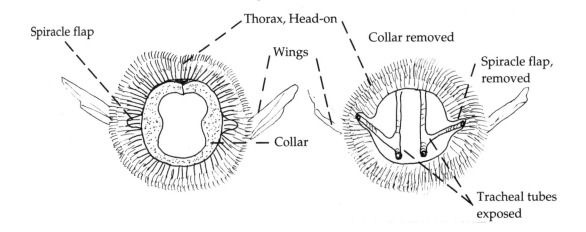

the head and pull off the collar surrounding the thoracic opening with the forceps (see illustration on dissecting bees). The thoracic trachea will be exposed once the collar is removed. In a healthy bee, the tracheae are pearly white, similar in color to a new, white dryer hose.

4. If it is not clear and the tubes contain mites, you can remove the tubes and increase the magnification of the microscope to examine them more closely.

If mites are present, the trachea will have shadows or be spotted; the different-colored spots are mites in various states of development. In severe infestations, the tracheal tubes can be completely brown or black in color. Darkened tracheae will be visible to the naked eye, while healthy tracheae will be white and shiny. You can use this method to detect heavy infestations (spring and fall) but not light infestations, such as in the summer. You should examine at least 25 bees per sample, as light infestations can be missed by sampling too few bees.

This is one of several ways to look for tracheal mites. Check websites or your local beekeeping organization for more information. Search online for a video of tracheal mite dissection.

Controlling Tracheal Mites

Chemical acaricides. Menthol, from the plant *Mentha arvensis*, is sold in crystal form (98% active ingredient) at many bee supply companies; each two-story colony takes 1.8 ounces (50 g) of the menthol, or one packet. The problem with menthol is that it is temperature dependent. Menthol vapors will sometimes cause the bees to leave the hive if the ambient temperatures are high. Conversely, the crystals will be ineffective if the temperature is low, because not enough vapors will be released. The crystals should remain in the colony at least two weeks. Remove all menthol at least one month before the surplus honeyflow, to keep honey from becoming contaminated. As for all treatments, follow the label directions.

Other chemical control methods include formic acid and a commercial product (ApiLife Var), both of which are approved in the United States for use to control tracheal mites. These compounds are also used for varroa mites. Check with state agriculture or apiary inspectors, bee suppliers, and bee journals for the current status of other chemical controls.

Oil patties. An alternate method is to use oil patties; a vegetable shortening (**not lard**) and sugar patty, kept in the colony all the time, has been shown to protect bees against these mites. If it is difficult to put these patties in all the time, make sure to place them in overwintering colonies, which are the most at risk.

To make the patties, use a 1:2 ratio of vegetable shortening to white sugar, or enough to have the patty hold its shape. Each colony should get about a quarter-pound (93.3 g) patty, about the size of your hand, on the top bars of the broodnest in late fall. This will help protect emerging bees during the winter months. The patty should last about a month; after that, replace it with another one. Some colonies will remove the patty much more quickly; they may be displaying hygienic behavior, a good trait. Because young bees are continually emerging, it is important to have the patty present in the colony for an extended time before winter.

The best time to treat is when mite levels are increasing—which is usually in the fall and spring (see chart). Check for the presence of small hive beetles, as they relish pollen patties and may also be attracted to oil patties.

Varroa Mite

The varroa mite (*Varroa destructor* Anderson and Trueman 2000) and the tracheal mite both carry out their reproductive activities within the confines of a honey bee colony. *Varroa destructor* was at first mistakenly identified as *V. jacobsoni* (Oudemans 1906) and was exclusively a parasite of the Asian honey bee (*Apis cerana*). When European honey bees were introduced into Southeast Asia, the mites "discovered" a new host that had no defenses to mount against them. These mites quickly spread with their new host throughout most of Asia and then from Russian into Europe and South America. In 1986, the first varroa mite was discovered in a honey bee colony in Florida. There is credible evidence that the varroa mite was introduced into the United States by a beekeeper who returned from Europe with queen bees that had mites attached to them. In the following year, varroa mites were discovered in 23 additional states. Today this mite is located on every continent where bees are kept, with the exception of Australia.

This mite is the major factor contributing to death of entire bee colonies and now the major focus of most researchers working with honey bees. Researchers worldwide are working diligently to learn more about this mite's life cycle, how to raise it artificially, and how to maintain healthy bee colonies.

The varroa mite has presented both beekeepers and researchers with a most formidable problem. Recent published work by Dr. S. Ramsey et al. (2018) indicates that the varroa mite feeds on the fat bodies of honey bees; adult mites feed on both larvae and adult bees. In addition, these mites employ adult bees (preferable nurse bees) to transport them to open brood cells, where they "dismount" and enter the brood cells to parasitize larval bees. Here they can rear their own young, or, while attached to another adult bee, end up in a neighboring colony.

Until recently, when a mite was attached to an adult bee, it was assumed that it was using the bee solely for transportation; this was referred to as *phorsey*, an association between two species in which one transports the other. However, it is now known that, often, the adult varroa is feeding on the adult bee, using the bee for both transportation and feeding, so this association can no longer be referred to as phorsey. It has been suggested that this phase now be referred to as an ectoparasite or dispersal phase.

The movement of commercial pollinators and package bees has accelerated the mite's spread throughout the United States. Because of the serious harm this mite inflicts on European honey bees, it is aptly named *Varroa destructor* and has spread across most of the beekeeping world.

Life Cycle of Varroa

This is a large mite; adult females measure 1 mm long by 1.5 mm wide (0.04 by 0.06 in.); it is easily seen with the naked eye and is about the size of a large pinhead (see illustration). Their life cycle is closely tied to the life cycle of bees. These mites have no visual organs and no observable antennae. However, they are able to navigate within the dark cavity of the bees' nest by touch and by smell. Only mature female mites are found on adult bees, feeding on their fat bodies after piercing their soft tissues, particularly between their abdominal segments, and/or when they are "using" bees to be transported. Male mites die within sealed brood cells.

Adult female mites are able to home in on drone and worker bee larvae by way of the bees' brood pheromones. Mites clearly demonstrate a preference for drone brood, which require 24 days from egg to adulthood.

The foundress mites (those seeking to lay eggs) enter open cells containing bee larvae (seven or eight days from the egg stage). There they hide at the bottom of the cells beneath the larvae. These cells also contain liquid food for the developing bee larvae, a place where the mites could potentially drown. However, the mite is equipped with two lateral *peritremenes* or breathing tubes, which are extended above the liquid food, ensuring the mite's survival. By hiding beneath the developing bee larvae, the foundress mites are able to avoid detection by the cleaning bees that would normally remove any foreign objects from the cells. Soon the bees will cap the larval cells with wax, and the larvae will straighten out along the length of their cells and spin their cocoons. Once completed, they will advance from the larval stage to prepupae, then on to the pupal stage; finally they emerge as adult honey bees.

Before the larvae bees begin spinning their cocoons, whose silk threads can entangle the mite, the mite crawls onto the metamorphosing bee and clings to it until the spinning process is complete. After the cocoon is formed, the mite pierces a hole through the pupal bee's epidermis and begins feeding off the bee's fat bodies. Approximately thirty hours later, the foundress mite will lay an unfertilized egg, which will produce a male mite; about every 30 hours or so she will lay fertilized eggs, which will give rise to female mites. In between her egg laying, the foundress mite will keep the wound open so her offspring can have a continuous source of sustenance.

When sexually mature (in five to six days), the male mite, who is small compared to his sisters and whose chitinous shell never fully hardens, will mate with his sisters (common in mites), unless more than one foundress mite has entered the same cell. Then a male mite may mate with several female mites that are not his sisters. The male mite dies in his natal cell. When the wounded bee emerges as an adult from its cell, the foundress mite and her daughters also exit the cell. The foundress, along with her daughters, begins the reproductive cycle again, which increases the mite load (numbers) of the colony over time.

The number of daughter mites produced in each cycle depends on whether the foundress mite invades a cell containing a worker or drone larva. Worker bees take 20 to 21 days to reach adulthood (from egg to adult), and it takes 18 days for young female mites to reach adulthood. As a consequence, usually only one and sometimes two daughter mites will fully mature in a worker cells. On the other hand, drones take 24 days to develop, allowing more time for immature female mites to fully mature. As a consequence of this longer period for a drone to develop, up to three mature female mites, along with their mother, may emerge from drone cells (see illustration). The foundress mite and her female offspring will live for a time on other adult bees until they invade cells containing larvae, and the cycle is repeated again.

The young, wounded bee that emerges from a mite-infested cell usually emerges alive but is not in a wholesome state. The injured adult bees exhibit the following characteristics:

- Reduced flight activity of foraging bees.
- Weight loss (6–25%).
- Shortened life span for most adult bees (by 34–68%).
- External damage (chewed wings, legs, stunted growth) if more than five mites in one cell.

Clearly, detection and treatment are imperative to keep your colonies from perishing.

Varroa Mite

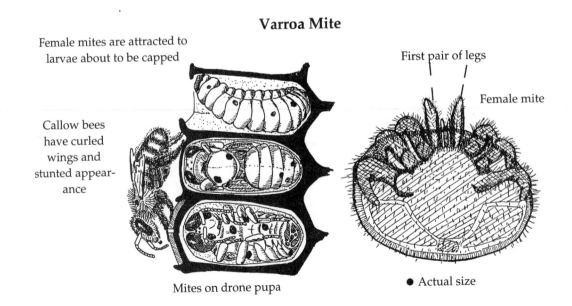

Female mites are attracted to larvae about to be capped

Callow bees have curled wings and stunted appearance

Mites on drone pupa

First pair of legs

Female mite

● Actual size

LIFE CYCLE OF VARROA MITES

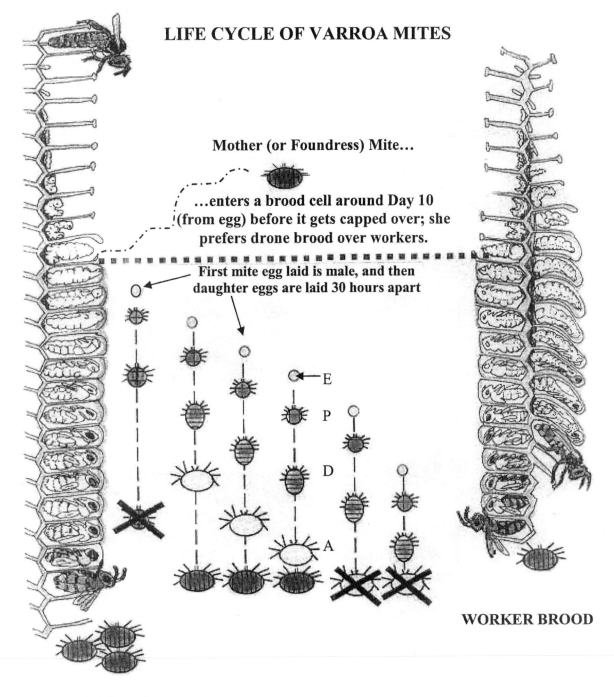

Mother (or Foundress) Mite...

...enters a brood cell around Day 10 (from egg) before it gets capped over; she prefers drone brood over workers.

First mite egg laid is male, and then daughter eggs are laid 30 hours apart

E
P
D
A

WORKER BROOD

DRONE BROOD

Up to three daughter mites can develop in drone brood because of the 24-day time frame it takes for drone adults to emerge. If the foundress mite infests worker brood, on average only one daughter has time to reach maturity; if in drone brood, 1 to 3 daughters can emerge. The foundress mother mite can re-infest at least one other brood cell. Males mate with their sisters in the capped cell and never leave the cell; they are killed by nurse bees cleaning the cell. Mites *very rarely* enter queen cells. E = egg, P = protonymph, D = Deutonymph, A = Adult. Schematic by D. Sammataro; see also http://www.ars.usda.gov /Services/docs.htm?docid=2744&page=14.

Symptoms of Varroa Infestation

This condition, originally called *bee parasitic mite syndrome*, now called parasitic mite syndrome, or PMS, was first coined by researchers to explain why colonies with both mites (varroa and tracheal) were not thriving. PMS suggested that bee mites vectored viruses, making bees susceptible to other pathogens. If your colonies have both mites, and are not thriving, here are some common signs to look for:

- Infested capped drone or worker brood; cappings can be punctured, as in foulbrood.
- Disfigured, stunted adult bees with deformed wings, legs, or both; you may also witness bees crawling on the ground.
- Bees discarding infested or deformed larvae and pupae.
- No predominant bacterial disease evident.
- Pale or dark reddish-brown spots on otherwise white pupae.
- Spotty brood pattern and the presence of diseases.
- Dead colonies in the early fall, right after honey has been harvested.

Other symptoms can be present any time of the year and include:

- Queens superseded more than normal.
- Foulbrood and sacbrood symptoms present.
- AFB symptoms may exist, but no ropiness, odor, or brittle scales present.
- No predominant bacterial disease present.

Detecting Varroa Mites

There are four basic techniques you can use to detect varroa mites. An essential part of beekeeping now requires that you, as a beekeeper, periodically test your colonies to determine the mite levels (numbers). Once the levels of this mite exceeds the threshold (in your area), you, as a beekeeper, are obligated to make every effort to reduce the affected colony's mite levels to tolerable numbers; otherwise the colony will eventually succumb to the mites and will spread them to other colonies.

When using the various approaches for monitoring mites, we are only sampling the number of mites found on adult bees, not the total number of mites in a given colony. Many, possibly the majority of mites, lie beneath the capped brood. In the absence of brood, **all** the mites will be found on the adult bees. By obtaining a sample of the number of mites found on the adult bees, we are obtaining the number of mites per 100 adult bees. If the number of mites obtained in a given sample exceeds the threshold level, intervention is required in order to reduce the mite count to "acceptable" levels.

Threshold levels vary in different sections of the United States and Canada, depending on the month of the year and when a colony reaches its peak population. In Connecticut, for example, this is usually around the middle of June. Many methods for detecting and treating varroa mites have been developed. In many cases, it is prudent to first ascertain the varroa mite levels in a colony before applying a treatment that may help to keep the colony alive.

Because mites are now resistant to acaricides, you must choose treatments that will control mites without contaminating hive products. Check the mite levels in at least 10 to 20 percent of the colonies in each apiary to determine whether treatment is required.

A word of extreme caution: Many of the chemicals applied to obtain the number of mites in a colony, **and** many of the chemicals used in miticides, are also **extremely toxic to humans**. Some miticides, if mishandled, can cause irreparable eye damage and injury to your lungs. It is extremely imperative that you **follow all the instructions** on miticide labels.

Following are some of the various methods used by beekeepers to detect mites in a given colony.

Ether Roll

Although this is one of the methods used to estimate the number of mites in a honey bee colony, the "alcohol wash" (see below) has been determined to be the most accurate.

Steps needed to execute an ether roll include:

1. Remove a brood frame from a colony that contains sealed brood and open brood (fifth instar larvae) about to be capped. Be absolutely certain the **queen is not** on the frame selected. If she is there, place her on another frame or put the frame in a nuc box and select another brood frame. Make sure you **never include the queen** in any of the methods described.
2. There are two ways to collect the 300 bees (which should equal half a measuring cup) required for

obtaining a mite count. Shake all the bees off the brood frame into a basin, then scoop up half a cup of bees (dedicate a cup for this use only).

3. You can also scoop them into the cup directly from the frame: start at the top of the frame, press the cup (open side up) against the comb and run the cup down the comb; the bees will drop into the cup. Continue until the cup is filled with bees.

4. Empty the half cup of bees into a wide-mouth mason or pickle jar and quickly replace the lid.

5. Tap the jar so most of the bees drop to the bottom of the jar.

6. Now quickly remove the lid and squirt two or three bursts from a can of carburetor starter fluid (ether) into the jar, and then quickly replace the lid. **Although this is a small amount of ether, constant or inappropriate use of this product or other ether products may result in respiratory paralysis, intoxication, or dermatological problems; and it is highly flammable. This will also kill the bees.**

7. Now agitate or roll the jar for 10 seconds; this action will help to dislodge the mites from the bees. The mites will stick to the walls of the jar (see illustration).

8. To obtain an optimum mite count, remove the lid, add water to the jar, and then strain the liquid through a coarse hardware cloth into a collecting basin, where the mites can more accurately be counted.

The number of mites may vary depending on the time of year the sampling is performed. The current research suggests that if the sample contains 3 or more mites per 100 bees, the bees need to be treated to lower mite levels.

Powdered Sugar Roll

This method was developed at the University of Minnesota and will not kill the bees being sampled for mites. Researchers often use this technique to obtain live mites for their research work. Beekeepers who are reluctant to kill bees while assaying the mite loads in their colonies prefer the powdered-sugar roll approach. Here's how:

1. Collect bees in a jar, as above for the ether roll. This time, use a canning jar (such as a mason jar with a removable lid that has been replaced with

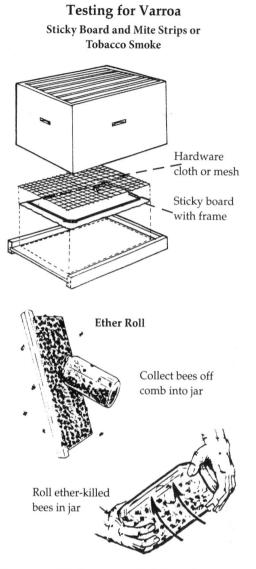

Testing for Varroa
Sticky Board and Mite Strips or Tobacco Smoke

Hardware cloth or mesh

Sticky board with frame

Ether Roll

Collect bees off comb into jar

Roll ether-killed bees in jar

Courtesy of M. Burgett and A.I. Root Company.

a piece of 8-mesh hardware cloth, cut to fit inside the metal rim.

2. When you have about 300 bees, add about 2 to 4 teaspoons of powdered (confectioner's) sugar. You may need more if there are more bees. Shake the jar vigorously and set aside for one minute. The bees should be thoroughly coated with the powdered sugar.

3. Next, invert the jar over a piece of white paper coated with petroleum jelly, or over a basin (use a white plastic basin) filled with water and some dish soap. Shake the jar; the mites will pass through the 8-mesh hardware cloth into the basin or onto the coated white paper.

Alcohol Wash

This will achieve the most accurate count. The jar used in the sugar roll can be used in the alcohol wash; again, use the same, dedicated jar for collecting bees. After the bees are in the jar, instead of using powdered sugar, add enough rubbing alcohol to cover the bees. You can also purchase a container called Varroa Easy Check produced by Veto-pharma (for additional information, see http:///vimeo.com/174841012).

The alcohol will kill the bees, and while this is repugnant to many beekeepers, it is a small percentage of the total number of bees in the colony; the purpose of obtaining an accurate mite count is to save the entire colony from perishing.

One standard suggests that if you get 3 mites per 100 bees from an alcohol wash, mite treatment should be initiated. Recent information (2018) informs us that we may have to reevaluate all the procedures used to obtain varroa mite counts; please be alert to new research as to how you can get an accurate assessment of a colony's varroa mite population.

Cappings Scratcher

You can look for mites inside capped bee cells by using a cappings scratcher (with forklike tines), which is used for scraping off the wax cappings from the honey frames. Use the capping scratcher to pull up capped drone pupae. Here's how:

1. Pick a frame of drone brood or find a large patch of drone brood on several frames.
2. Hold the forked tines parallel to the comb and insert the tines into the top third of the cappings, into the pupae below.
3. Pull the capped drone pupae straight up, or lever the handle end of the fork, leaving the tines on the comb, until the drone pupae are pulled out of their cells.
4. Examine the pupae carefully. A heavy infestation is at least 2 mites per cell; a moderate level is 5 mites per 100 pupae. The mites are clearly visible: females are reddish brown and look ticklike on the white pupae. Immature female mites are white or light brown.

Sticky Board

A sticky board is a treated sheet of cardboard or corrugated plastic; the latter can be purchased from a bee supply house. You can purchase ready-made boards from companies that make traps to monitor insect pest populations (see "Diseases and Pests" in the References section), or you can make a board using card stock or other stiff paper. The sheet is coated with petroleum jelly or an oil-based pan spray. If you live in hot climates, use Tanglefoot to keep the oil/jelly from melting.

Now place the sticky board under the screen of a screened bottom board. When bees are grooming one another or themselves, they may dislodge mites, and/or on occasion mites may drop off a bee or comb and fall through the screened bottom board and stick to the coated sticky board. The board is removed after three days to obtain a count of the number of mites that have dropped.

This technique has many variables and is not as reliable as a sugar roll or an alcohol wash. However, it can provide information that there are mites in the colony, or that a particular treatment is working (use boards for a before-and-after-treatment count).

Place these boards into the hives in your apiary. These boards can be placed on the interior surface of a bottom board or beneath the screen of a screened bottom board. When a sticky board is placed on the inner surface of a bottom board, the sticky board needs to be covered with a sheet of 8-inch hardware cloth with its edges turned under, to prevent the bees from getting stuck on the sticky board. You may also staple the mesh to ¼-inch-thick (6.35 mm) wooden lath strips for the same purpose.

Remember, there are many factors that influence the results when any test is administrated, such as time of year, the amount of brood in the colony, if the colony is queenless or not, the size of the bee population, and the race of bees. High numbers of mites (more than 50 a day) may indicate the colony requires intervention.

These tests will provide you with a **general indication** as to the number of mites in a given colony. Remain current with the literature as to what threshold is recommended for intervention (treatment). However, if you wish to be reasonably sure of this, retest with a sugar or alcohol roll or with the ether test.

Many bee supply companies sell screened bottom boards that have grooves beneath the screen for sliding in a sticky board. To make counting the mites easier, a company in Michigan (Great Lakes IPM) produces a board with a third of the grid blackened; see https://hal.archives-ouvertes.fr/hal-00891743 for a tip on counting mites.

Parasitic Bee Mites and Their Honey Bee Hosts

Bee Host	Mite Species									
	Varroa destructor	*V. jacobsoni*	*V. underwoodi*	*V. rindereri*	*Euvarroa sinhai*	*E. wongsirii*	*Tropilaelaps clareae*	*T. koengerum*	*T. mercedesae* n.sp.	*T. thaii* n.sp.
Apis florea			X Nepal, S. Korea		X					
A. andreniformis						X				
A. cerana	X	X	X							
A. koschevnikovi				X Sumatra						
A. nuluensis Borneo		X	X?							
A. nigrocincta Sulawasi		X?	X							
A. dorsata dorsata Asia, Indonesia, Palawan	## Korea							X Sri Lanka	X Palawan, Sri Lanka	
A. d. breviligula							X Philippines (not Palawan)			
A. d. binghami Sulawesi									X**	
A. laborisoa Nepal								X	X Vietnam	X Vietnam
A. mellifera	X Japan and Korean haplotypes	X Papua N.G, Irian jaya	##				X Philippines		X	
A. m. scutellata Africa	X									

Notes: X = Positive identification; ## = Incidental visitor; ** = Currently unresolved; ** = Currently unresolved. *Mesostigmatic mites parasitizing honey bees, arranged according to host bee species. Sources:* Compiled by D. Sammataro and D.L. Anderson (from Anderson and Morgan 2007 and Navajas et al. 2010), published in *Honey Bee Colony Health: Challenges and Sustainable Solutions,* edited by D. Sammataro and J. A. Yoder (Taylor and Francis, 2012).

Sequence of Suggested Treatment Times for Bee Mites

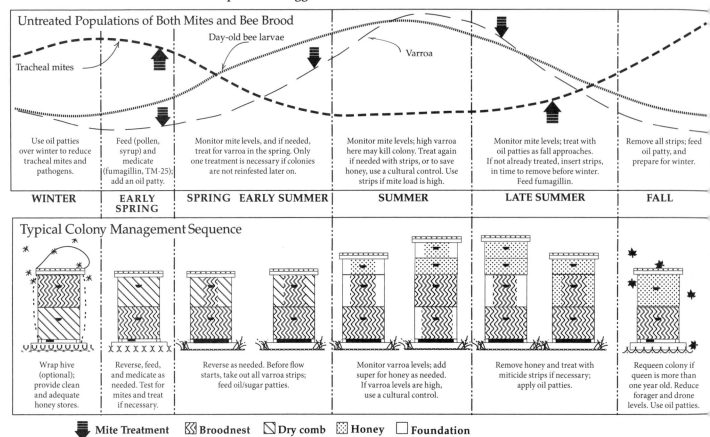

WINTER	EARLY SPRING	SPRING EARLY SUMMER	SUMMER	LATE SUMMER	FALL

Untreated Populations of Both Mites and Bee Brood

Tracheal mites — Day-old bee larvae — Varroa

| Use oil patties over winter to reduce tracheal mites and pathogens. | Feed (pollen, syrup) and medicate (fumagillin, TM-25); add an oil patty. | Monitor mite levels, and if needed, treat for varroa in the spring. Only one treatment is necessary if colonies are not reinfested later on. | Monitor mite levels; high varroa here may kill colony. Treat again if needed with strips, or to save honey, use a cultural control. Use strips if mite load is high. | Monitor mite levels; treat with oil patties as fall approaches. If not already treated, insert strips, in time to remove before winter. Feed fumagillin. | Remove all strips; feed oil patty, and prepare for winter. |

Typical Colony Management Sequence

| Wrap hive (optional); provide clean and adequate honey stores. | Reverse, feed, and medicate as needed. Test for mites and treat if necessary. | Reverse as needed. Before flow starts, take out all varroa strips; feed oil/sugar patties. | Monitor varroa levels; add super for honey as needed. If varroa levels are high, use a cultural control. | Remove honey and treat with miticide strips if necessary; apply oil patties. | Requeen colony if queen is more than one year old. Reduce forager and drone levels. Use oil patties. |

▼ **Mite Treatment**   ▧ **Broodnest**   ◇ **Dry comb**   ▦ **Honey**   ☐ **Foundation**

Treatments for Varroa

Russian Bees and Hygienic Bees

Efforts to find genetic lines of bees that can keep the mite population at tolerable levels are ongoing. Bees from eastern Russian have demonstrated some resistance to the varroa mite, as have some hygienic bee stocks. However, both these stocks require monitoring of their varroa levels and may require treatment should mite numbers exceed tolerable levels (see procedures for monitoring mite levels).

If treatment is required during a honeyflow and/or when honey is on the colony, it is the law that you use only miticides that are authorized (current examples are Hopguard or Mite Away Quick Strips). Other miticides will contaminate both the honey and the wax.

Chemical Acaricides

To date there are several chemical controls registered for varroa. Check online for registered products. For treatment options go to university bee lab sites, e.g., BeeLab.umn.edu.

All registered chemicals are available from bee supply companies, and it is important to follow the label directions carefully. **Do not be tempted to use other, homemade chemical cocktails, as some of these are now being detected in honey and especially in wax, making these hive products unfit for human consumption, use, or sale.** Timing of treatment is very important.

Formic acid is registered for use. It is effective, but in its liquid form it is very dangerous to use. It acts as a fumigant, killing both types of parasitic mites, but it can be toxic to bees too if not applied correctly. It is **extremely caustic to humans, so respirators must be worn**.

Another method of chemical control is the use of oxalic acid. This is a powerful acid and must be used with extreme caution! Oxalic acid in liquid form is used by carpenters and others to bleach wood. It can be either applied as a liquid dribble on top of bee frames to kill varroa mites or heated to where it sublimates to a gas and applied with a wand to reduce the mite load in the fall. **Use of either the liquid or the** gas requires a professional respirator and gloves. If applied incorrectly, oxalic acid may cause great harm to the beekeeper. Take this advice seriously.

If you don't want to use chemical controls, an integrated pest management approach (or IPM) offers multiple options, the last of which is chemicals. See more about IPM in this chapter.

Dusting with Powdered Sugar

Another method of reducing mite levels in bee colonies is to dust the bees with powdered (confectioner's) sugar. When the bees become coated, they begin to groom themselves in order to remove the powdered sugar coating their bodies. In the process of grooming they also dislodge mites attached to them. A sticky board beneath a screened bottom board is required to capture the falling mites. Research indicates this method, although effective in removing some adult mites from bees, does not significantly reduce mite loads to the extent of precluding the need for additional treatment protocols. It is generally used to determine if a particular mite treatment is working, using before-and-after boards to see if the treatment dropped fewer mites after the treatment. This method can be used in nucs or other smaller colonies without sacrificing bees.

Drone Trapping

Drone trapping is another approach that can be used to reduce the number of varroa mites in bee colonies. If you elect to use this method for reducing the mite loads, be advised that sampling for mite levels must be continued, and the use of miticides may still be required. Remember, mites prefer drone brood over worker brood, if given a choice between the two.

As already stated, drones require 24 days to reach adulthood and therefore take an additional three days to mature, compared to workers. Mites produced in drone cells have those extra three days to reach adulthood, and as a consequence, approximately 2.8 female mites plus the foundress mite will emerge from a drone cell, compared to "only" 1.8 mites from a worker cell. Researchers have proposed that beekeepers use frames containing only drone cells. A queen will lay unfertilized eggs in these drone combs, and reproductive mites will readily select the developing drone larvae in these cells over worker brood. This can be done during the summertime, when the mites are most active.

Other Parasitic Bee Mites

There are two other parasitic mites, but they are not yet found in the United States: *Euvarroa* and *Tropilaelaps*. *Euvarroa* was first identified from *Apis florea*, the dwarf honey bee from India, in 1974 and is reported to parasitize only drone brood. These mites are smaller than *Varroa*, and currently two species have been identified: *Euvarroa sinhai* on *A. florea* throughout its natural range, and *E. wong-sirii* on *A. adreniformis* from Malaysia and Thailand (see table on p. 261). *Euvarroa* generally have long setae or hairs on the posterior edge of the pear-shaped body shield and so far are not a threat to our bees.

The *Tropilaelaps* mite is the newest threat to global apiculture but for now is confined to its home range in Asia, where four species have been identified (see the "Diseases and Pests" section of References and the table on p. 261). This mite is an important pest of the European honey bee *A. mellifera* wherever it has been introduced throughout Asia. Of importance is that the adult mites, both male and female, are fast, active, and will invade bee brood before it is capped. They are reported to be phoretic for only a few days, however, and may not survive for long in a broodless colony. One method of control is to cage the queen for a week or so to keep the colony broodless. These mites carry some of the same bee viruses as do *Varroa*. Be alert when monitoring mite levels in your colonies to watch for this new mite (see illustration comparing *Varroa* and *Tropilaelaps*). If you do find it, contact your local apiary inspector or the National Bee Lab in Beltsville.

Bar is 1 mm

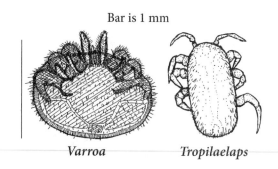

Varroa   Tropilaelaps

Comparing *Varroa* Mite and *Tropilaelaps*

Keeping accurate records will be required, and if you plan to use drone frames to reduce the colony's mite numbers, you need to remove the drone combs on the 22nd day after the queen has deposited eggs in them. If you are just beginning this practice, the bees will first have to draw out the drone comb foundation before the queen can begin laying eggs. If these combs are not removed in a timely manner, you will greatly expand the mite loads in your colony. Instead of waiting until the 22nd day to remove drone comb, remove it just as soon as most of the drone cells have been capped, saving the bees the need to spend time keeping these cells incubated. You will also be able to insert another drone comb or two, hastening the capture of more foundress mites.

A comb with only drone cells is the Pierce Drone Comb, which is colored green. When the drone comb is removed, place it in a freezer for 48 hours; this will kill the drone pupae and all the mites. The drone comb can then be reinserted into the hive, and the bees will uncap the cells and remove the dead drones and mites. You can also put in frames empty of comb and allow the bees to draw out comb, usually drone comb, which you can then freeze before varroa emerges. Keep alert for all the drone combs showing up in your colonies.

An alternative method for killing the drones and mites was invented by Dr. Zachary Huang of Michigan State University; it consists of a special frame of drone comb, which is connected to terminals; by hooking up the terminals to battery posts for two to three minutes, resistant heat elements in the comb will heat and kill both the mites and the drone pupae without the necessity of removing the frame from the hive. This drone frame is trademarked as the Mite-Zapper. The dead brood can be fed to chickens or frozen and used as pet food; see below.

MAJOR INSECT ENEMIES

Wax Moth

First reported in the United States in 1806, the wax moth was probably introduced with imported bees. The female greater wax moth (*Galleria melonella* L.) is about ½ to ¾ inch (1.3–1.9 cm) long and is gray-brown (color varies somewhat). This moth holds its wings tentlike over the body instead of outstretched or upright, like a butterfly. The wax moth is thought to have evolved with honey bees from Asia and commonly inhabits nests of all honey bee species.

This moth deposits eggs in cracks between hive parts or in any other suitable place inside the hive. After hatching, the larvae are quite active, moving up to 10 feet (3 m) to infest other hives, where they tunnel into the wax combs, hiding at the mid-rib to keep from being discovered by house bees. The dark wax of brood combs contains the shed exoskeletons of bee larvae and some bee bread, both of which are highly attractive to wax moth larvae. The larvae can grow to 1 inch long (2.5 cm) in 18 days to 3 months, depending on the temperature. As these larvae tunnel along, silk strands mark their trails through the combs (see illustration). Before pupating, the larvae fasten themselves to the comb face, on the wooden frames, or inside the walls, inner covers, or bottom boards of the hive and spin a large silk cocoon. The moth larvae can damage the hive furniture by chewing into the wooden parts. Left untended, wax moths can destroy weak hives within one season.

Symptoms of wax moth damage are:

- Tunnels in combs.
- Silk trails, crisscrossing over combs.
- Small dark specs (excrement of wax moth larvae; also called *frass*) on bottom board or in the silk trails in a hive.
- Silk cocoons attached to wooden parts.
- Destroyed comb, piles of debris, and moth larvae on bottom board.

Control

To control wax moths, use these methods:

- Maintain strong colonies (the best defense against wax moths).
- Inspection: Wax moth are active when outside temperatures are over 60°F (15.5°C), so inspect weak colonies (nucs) once a month to prevent problems.
- Freeze empty combs or comb honey at 20°F (–7°C) for five hours; at 10°F (–12°C) for three hours; or at 5°F (–15°C) for two hours. If treating a lot of comb honey, it is best to freeze for at least 24 hours to kill any eggs.
- Store empty combs in cold places; cold temperatures like 40°F (4.4°C) will slow down the rate

Wax Moth

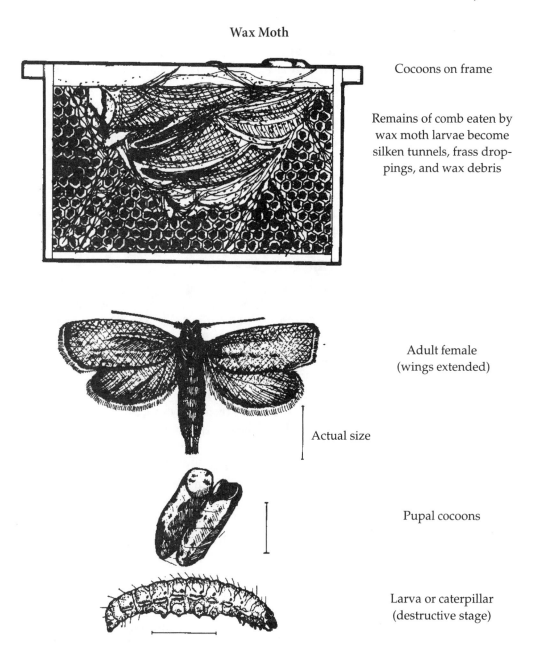

Cocoons on frame

Remains of comb eaten by wax moth larvae become silken tunnels, frass droppings, and wax debris

Adult female (wings extended)

Actual size

Pupal cocoons

Larva or caterpillar (destructive stage)

of growth and deter the adult moths from laying their eggs.

- Brood combs can be stored if they are exposed to light 24 hours a day; female wax moths cannot lay eggs if light is present.
- Heat can also kill all stages of the wax moth. Allow empty combs to come to 115°F (46°C) and "bake" them for 80 minutes. You can also raise the temperature to 120°F (49°C) for 40 minutes, but any higher than that will soften the wax, and it will sag in the frames.

- Fumigate dry combs with 74 percent CO_2 and 21 percent N (nitrogen), at 50 percent RH (relative humidity) at 100°F (38°C) for four hours, 115°F (46°C) for 80 minutes, or 120°F (49°C) for 40 minutes. Be careful, as beeswax melts at 143° to 147°F (62–64°C). Check bee suppliers for new fumigants.

Some other chemicals can be used to fumigate combs, but their permitted use varies from state to state. The state bee inspector or extension entomologist should be consulted before using chemicals.

A cultural practice is to put combs in the sun; in areas that have fire ants (*Solenopsis invicta* Bunen) you can leave moth-infested combs near an active ant nest—the ants will kill wax moth larvae effectively. But use caution, because these ants also kill bee colonies and can bother beekeepers.

Wax moths are naturally beneficial because they destroy diseased wax combs of feral or varroa-killed colonies, thus eliminating diseased comb and providing a new clean nesting space for bees. In addition, they are a valuable commodity, sold as fish bait and pet food for reptiles and other exotic animals. They can be reared off beeswax, using baby cereal, glycerin, and honey, and are often a secondary business to many beekeepers; see Appendix F, "Rearing Wax Moths."

The lesser wax moth (*Achroia grisella* Fabricius) does similar damage to wax comb, but unless the infestation is great, the damage is minor compared to that of the greater wax moth. But if left unchecked, this moth can get into dry goods (flour), seeds, and grains and other stored food products.

Small Hive Beetle (SHB)

The small hive beetle *Aethina tumida* (Coleoptera: Nitidulidae), our newest bee pest, was first identified in a Florida apiary in the spring of 1998. However, unidentified at the time, specimens of this beetle were collected in the United States as early as 1996. Before it was discovered in Florida, this beetle was endemic in the sub-Saharan regions of Africa. The rapid distribution of this beetle in North America may have been a result of the movement of colonies around the continent, and beetles inadvertently shaken into bee packages along with the bees. We have yet to learn how this beetle arrived in Florida.

In the sub-Sahara, it is not considered a pest of honey bee colonies; however, in North America and on other continents, it is a major pest that can cause honey bee colonies to decline and eventually perish. These beetles are in the Nitidulidae family, which includes picnic beetles, known for their attraction to fermenting fruit. The SHB has been found in traps containing fermenting pollen and bee bread, cantaloupe, pineapple, and bananas. Adult hive beetles may consume fruit as a food source if bee colonies, with their stores of honey, pollen / bee bread, and bee larvae and pupae are unavailable.

The beetle is now found in all the states and parts of Canada, but it is especially a problem in the southeastern states, where the winter weather is moderate. In most northern areas of the United States and Canada, the small hive beetle can now build up to sizable numbers during the summer months to cause colonies to collapse. They can inflict major damage to frames of honey and often cause colonies to abscond.

As of 2002, the beetle was also found in Australia; seehttp://entnemdept.ufl.edu/creatures/misc/bees/small _hive_beetle.htm for a map and more information.

Description

The adult beetle is about one-third the size of a bee, around ¼ inch (5.5 to 5.9 mm) long, ⅛ inch (3.1 to 3.3 mm) wide, reddish brown or black, and covered with very fine hair (see illustration on p. 267). The larvae are cream colored and similar in appearance to young wax moth larvae. You can differentiate the beetle larvae from wax moth larvae by examining their legs. Beetle larvae have three sets of legs just behind the head. Wax moth larvae, like all moth and butterfly larvae, have three sets of legs behind the head, and in addition have a series of paired *prolegs*, which run the length of the body. Prolegs are absent in beetle larvae. Small hive beetle larvae have two rows of short spines along their backs; the two at the posterior end are pronounced.

Here is a quick checklist to identify the SHB:

Larvae

- Color: white to tan.
- Spines along the dorsal side (back).
- Size of mature larva, ready to pupate: ¾ inch (10 mm).
- Three pairs of legs, behind the head.
- No webbing or tunnels with black droppings (wax moth frass).

Adult

- Size: around ¼ inch (5–7 mm); see the illustration comparing sizes.
- Antennae: clubbed.
- Color: reddish brown to black.
- Very short hairs covering the body.
- Scurries away from illumination and seeks concealment in dark areas, such as hive crevices or the rear of the bottom board.

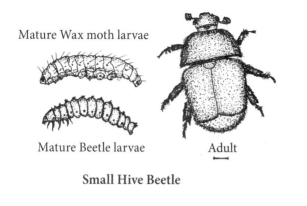

Mature Wax moth larvae

Mature Beetle larvae   Adult

Small Hive Beetle

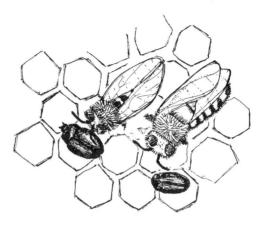

Size Comparison of Small Hive Beetle with Worker Bees

Life Cycle

Adult females are strong fliers and can fly up to two miles (3.22 km) or more in order to reach and invade bee colonies. They reach sexual maturity a week after emerging, mate, and then lay their egg masses in cracks and crevices in the beehives, just beyond the reach of cleaning bees. Eggs laid under brood cappings and in empty cells have also been reported. A single female beetle is capable of laying 1000 eggs during her lifetime of approximately six weeks. The eggs hatch in one to six days, yielding a large number of larvae, which consume pollen (bee bread), honey, bee brood, and wax. While they resemble wax moth larvae, the legs are much larger and are close to the head; also they do not spin webs or cocoons (see illustration). They complete their larval stage in 10 to 16 days and then drop to the bottom board and crawl out of the hive's entrance. When infestations are heavy, beetle larvae by the thousands have been seen crawling out of the colony entrance; even in strong colonies, queens will stop laying eggs, and the bees may abscond.

Once outside, they fall to the ground and can walk up to a distance of 100 feet (30 m) from the colony, and then burrow themselves into the soil, where they will transition from larvae to pupae and then to adulthood. They seem to have greater success in light or sandy soils. Adults emerge from the pupal stage in approximately 3–4 weeks and may return to the colony where they were born, or go to other colonies in the same apiary or to colonies located elsewhere.

Adult beetles live 30 to 60 days, depending on food resources, temperatures, and moisture or soil type; they are able to thrive in both hot humid and temperate climates. During the winter months in temperate climates, adult beetles will be found huddled within the winter cluster inside bee colonies. Should they depart from the cluster, or the cluster perishes over winter, so too will the beetles. Their reproductive activities are suspended during the winter in cold climates.

Destruction Caused by These Beetles

In areas of infestation, the SHB can cause significant economic harm to bee colonies. These beetles and their larvae are able to infest and may often overwhelm even strong colonies with little resistance from the bees. In addition to consuming a colony's resources, both the adult and larvae beetles defecate on the honey. Since their feces contain yeasts, this causes the honey to ferment. Often beekeepers who move honey supers into their honey houses for extraction but fail to extract them in a timely fashion may find the supers are infested with both adult and larvae SHB. This can render the honey unfit for human consumption and unpalatable to bees, to the extent that they will not eat it. The fermenting honey smells similar to the odor given off by rotting oranges.

Full honey supers stored in the honey house or on hives above bee escapes, as well as weak colonies with honey but few bees, including nucs and small queen mating colonies, are vulnerable to infestation by the SHB.

Detection

All hive inspections should be done with an eye open for this pest. When a hive containing beetles is opened, they can be seen running across the combs to find hiding places. Adults may also be detected under top covers or on bottom boards. If an infestation is heavy, both adults and masses of larvae may be seen on the combs and bottom board. Combs can be full of holes, but these larvae do not produce silken tunnels, webbing, or cocoons in the hive (as wax moth larvae do); yet wax moth could also be present. It is important to be able to identify the beetles; check out the SHB traps, which can help control light to moderate infestations.

If beetles are suspected in a honey super, firmly shake it over an upturned outer cover; dislodged beetles can be seen running to find a hiding place. Fermented honey exuding from full supers in storage waiting to be extracted, or on active colonies, is a sign that hive beetles may be present; the "decaying orange" odor will be detectable.

Another place to check is pollen patties and even syrup feeders; beetles can be raised on artificial pollen patties. Corrugated cardboard with the paper removed from one side and placed on the bottom board at the rear of the hive has been successfully used in detecting adult beetles. Plastic corrugated "cardboard" is preferable for this use, because the bees will chew up regular cardboard.

Control

Strong colonies are the best defense against serious infestations. Nevertheless, a heavy infestation of adult small hive beetles and their offspring can readily overwhelm even a strong colony in short order. Worker bees can be seen trying to sting or pull at the beetles, but the beetles' hard exoskeleton and rounded body are effective defenses. Often bees will coral or surround adult SHBs, preventing the beetles from obtaining food; but instead of starving while entrapped, these beetles rub the mandibles of the surrounding bees with their clubbed antenna, causing the bees to regurgitate honey, which the beetles readily consume.

A number of small hive beetle traps are available, or you can create your own traps. These traps will capture adult beetles and help to reduce their populations. However, you must remember to remove the beetles from the traps; they will not be effective if too many beetles are caught. As long as you maintain them, the traps will help the colony help itself. Check bee journals for current information on SHB traps.

Reducing the hive entrance may offer a limited measure of control over incoming SHBs. But while reducing the entrance will let the guard bees do their job better, it will not work in apiaries that are heavily infested. And constricting the entrances is impracticable during the warm months, as it will hinder the bees from ventilating their hives to maintain proper colony temperatures, evaporating water from incoming nectar, and hamper them from carrying out evaporative cooling.

In the honey house, fans and a dehumidifier may help keep beetles under control, but freezing the supers for 24 hour at 23°F (–12°C) is reported to kill the beetles, eggs, larvae, and adults.

If small hive beetles are detected and you are in an area where they are known to be present, the following **preventive** measures are recommended:

- Keep your honey house **clean**, and do not store supers containing honey for any extended period. Extract as soon as possible. Wax capping coated with honey should be processed as soon as possible. Dead colonies that still contain frames of honey must be removed from the apiary and stored in a proper manner (see below).
- Store honey supers, if you must, in a cool room, as you would to prevent wax moths; temperatures as low as 40°F (4.4°C) will keep SHB out.
- Keep the apiary clean; that means do not throw burr comb on the ground. Collect it to later melt down to sell.
- If possible, minimize the time spent inspecting each colony; during inspections, bees will release their alarm odor, which adult SHBs can detect; they will home in on these colonies.
- Do not brush the SHB larvae off the bottom board onto the surrounding ground, because if any of these larvae are ready to depart the colony to pupate in the soil, you will have unwittingly assisted their departure.

If you discover that you have SHB, take the following measures:

- Destroy beetles in stored honey supers as soon as they are detected.
- Maintain only strong and healthy colonies and do not add infested equipment to uninfested colonies.
- Keep apiaries clean of **all** unused equipment; do not store empty supers on colonies.
- If beetle larvae have contaminated honey frames,

and most, if not all, of the cappings have been removed by power washing the honey out and then freezing the frames, be sure all the eggs, larvae, and adults have been killed; only then can the frames be returned to colonies for further cleaning and reuse by the bees.

- Experiment with trapping beetles or other cultural controls. Check websites and journals for SHB traps.
- If you find a colony that appears to keep SHBs at managed levels (inflicting little or no damage to the colony), you may consider using that colony to raise queens. This assumes that this colony might have a genetic predisposition for managing this parasite. However, since mating of queens with drones is random, such an undertaking may not yield a successful outcome.
- Moving infested colonies to another site will keep the soil in a given apiary from further infestations once all the SHB pupae emerge as adults. Nevertheless, those SHB that do emerge will fly to other sites and infest other apiaries that are nearby. SHB larvae pupate best in damp, well-drained soil. If you can find sites other than well-drained damp soil to locate your apiary, then to some degree, you can help eliminate the eventual buildup of large adult populations of SHBs.
- Free-range chickens will consume the beetle larvae, especially if the soil is disturbed to expose the beetles.

Again, the best defense against small hive beetles is a strong, healthy colony. Check the "Diseases and Pests" section in the References for more information and websites.

Hive Treatment for SHB

Some chemicals used to control the SHB have become less effective and also contaminate hive furniture and wax. **Do not treat** colonies that are storing honey, because the honey can become contaminated. If you are going to treat with chemicals, you must remove honey supers first. A better management strategy is to keep colonies strong and use in-hive bait traps.

Soil Drench

If you have too many beetles in an apiary, you could use a soil drench. GuardStar is a liquid soil treatment (40% permethrin—see note below) and has been approved in controlling the SHB around honey bee colonies. Hive beetles must pupate in the soil to complete their life cycle, and this insecticide will kill them in the soil. This pesticide provides treatment for the beetles, while minimizing contact with bees and honey. **Read, understand, and follow labels directions**.

Plastic strips, such as CheckMite+, can be used to kill adult SHB in colonies.

Note: Permethrin is highly toxic to bees and other insects, and extreme caution must be taken to avoid contact by spray or spray drift with the bees, hive equipment, or any other surfaces that bees may contact. Do not contaminate any water or food source that may be in the area. **Do not** apply during windy conditions. For better soil penetration and improved efficacy, cut the grass around your hives prior to application.

Other controls that are being tested are nontoxic in-hive baited traps and lures. As with all bee pests, keep current with the literature and check with your state's apiary inspectors or cooperative extension office. Consider IPM and biological or genetic bee lines to help control this pest.

Africanized Honey Bees

In 1956, a researcher in Brazil (W.E. Kerr) imported over 50 queens of *Apis mellifera scutellata* from South Africa into South America to improve bee stock in tropical regions. Their volatile, defensive nature, however, is still a problem. As a result, the Africanized honey bee has become a serious problem in some areas. First reported in Texas in 1990, colonies of Africanized honey bees (AHBs) have since taken up residency in several states in the US, including southern California, southern Nevada, Arizona, Texas, New Mexico, Oklahoma, western Louisiana, southern Arkansas, and central and southern Florida. If living in an area known to have AHBs, it is possible that one or more of your colonies may contain these bees. When you begin to open a colony for inspection and notice that the bees appear overly aggressive, it may be occupied by AHBs. Reassemble and close the colony immediately; however, if it appears that the situation is going to escalate, leave the colony and find a protective enclosure as soon as possible. Then call your local or state bee inspector to assist you in evaluating whether or not the colony has indeed been Africanized. **Remember:** Africanized bees are known to have killed people who have disturbed their colonies; this includes beekeepers as well as unsuspecting landscapers weed whacking within the vicinity of one of the colonies. In that the movement of AHBs into new areas is fluid, it is best to ascertain the situation in your location. For updates, contact your state's Department of Agriculture, the state bee inspector, or other beekeepers residing in your area.

If living in an area that is subject to this invasion, make sure you can differentiate European honey bees (EHBs) from the AHBs bees. In general, AHBs are smaller than EHBs, faster moving, more defensive, and more apt to abscond and swarm, especially if food becomes hard to find or the nest is threatened. This could happen several times a year in areas that have no harsh winters (see Appendix E).

AHBs can usurp weak colonies, such as a mating nuc, a colony with a caged queen, or a colony recently stressed by beekeeper manipulation. A small cluster or swarm, about the size of a grapefruit, may contain one or more queens; it will land on the outside of a weakened colony. The workers will slowly infiltrate the weaker colony, killing guard bees and eventually the resident queen. Once the resident queen is dead, the AHB queen, protected by a "ball" of bees, will enter and assume her duties.

AHBs are extraordinarily successful, and many beekeepers who have converted to keeping them (out of necessity to stay in the bee business) admire them for their low disease and mite incidence and their adaptability.

AHBs begin foraging earlier in the morning and remain at this task for extended periods of the day. They build up large populations in short periods of time and swarm frequently when food sources are abundant. (One author reports seeing bees working cactus flowers at night under a streetlamp.)

They work hard and build large populations quickly, then cast many swarms (or abscond if food is scarce) to take advantage of new resources. These behaviors are the reasons for the rapid expansion of AHBs since their introduction in South America in 1956 to their arrival in the United States (Brownsville, Texas) in 1990.

Beekeepers that manage colonies of AHB in areas where they are endemic have learned to work with them and are able to gather surplus honey as well as collect pollen to sell.

EHB queens that mate with AHB drones produce an *Africanized* colony. New research suggests that AHB queens mate mostly with African drones, and thus are *African*. Such colonies, which can be fine for a season, should be requeened with an EHB queen before the defensiveness becomes too extreme. By keeping good records, you can quickly tell if your colonies have been overtaken; it is **imperative** to mark all your queens in each hive and requeen with non-AHB stock when the marked queen that was there is absent. Keep in mind, however, that the unmarked queen may be either the resident queen who has lost her mark or she was replaced by a daughter, and therefore is not an Africanized queen.

If the behavior of a colony suddenly changes and the queen has no marking, suspect a takeover and requeen with a new, marked European queen. You may have to kill over half of the adult bees (pest control specialists often use 3% soapy water solution) and introduce the new queen on the emerging brood; the colony can be very aggressive to any newly introduced queen. Cover the hive entrance with a piece of queen excluder to keep the new, marked queen in and the AHB queens out. Having volatile AHBs in your apiary is not to your advantage—legally, socially, or economically.

MINOR INSECT ENEMIES

Although bees are often preyed on by other insects and spiders, these predators usually do not have any appreciable effect on a colony's well being. In some areas, however, any of these minor predators might become a serious problem. Spiders (Araneae) also prey on adult bees; some species, like crab spiders (Thomisidae) even wait for bees to arrive at a flower before attacking them. The most common types of spiders that would catch a bee are the orb weaver (Araneidae), grass (Agelenopsis), and "house" spiders; check with local agencies for spiders in your area.

In some southern areas, brown recluse (*Loxoscele recluse*), jumping (Salticidae), and black widow (*Latrodectus*) spiders are commonly found under inner covers or on the sides of hive furniture, especially in equipment that is stored outside. Scorpions (Scorpiones) can also be found in these regions, so be careful.

While not a problem in the temperate climates, in the subtropical areas ants (Hymenoptera: Formicidae) are a serious pest, and hives have to be placed on top of greased posts or oiled cans to keep out marauding ants. The more harmful ones in North America include Argentine ants (*Iridomyrmex humilis*), fire ants (*Solenopsis invicta* Bunen), and carpenter ants (*Camponotus* spp.). Ants can be controlled by keeping the apiary free of weeds, debris, and rotting wood, and by placing hives on stands. For more serious infestations, ant baits can be used (those containing boric acid and ground corncobs are nontoxic to bees and pets); contact your local extension agents as to what ants are in your area and what control measure to use that won't affect your bees.

Other ants (sweet-attracted species), earwigs, and cockroaches may use various hive parts, especially the inner cover, as a shelter or nest. Earwigs (Dermaptera) are found on top of the inner cover and may annoy bees. Keep tall vegetation mowed around hives. Nematodes or roundworms (Nematoda) are small worms that live in many different habitats. While some also live on bees, they are not really a serious pest. Termites (Isoptera) can damage hive parts especially if they are resting on the ground; and in some states (southeastern states and the Southwest) they can be a serious problem by eating wooden hive parts. Keep your apiary mowed and your hives on stands, and many of these problems will disappear.

Ant lion (Myrmeleontidae) larvae (also called doodlebugs) live in sandy soil and dig the typical cone-shaped pits for trapping ants or other insects that fall into them. They dig their pit traps by throwing out sand by means of upward jerks of the head; the long mandibles serve as a shovel. The small pits they form are about 1 inch deep (2.5 cm) with sloping sides (as steep as the soil will allow). At the bottom of the pit, the ant lion will hide in the soil, with its head just below the bottom of the crater, waiting for some hapless insect to fall in. While not a serious problem, if there are many in an apiary, the larvae can feast on crawling bees (and queens), and the ants they attract (which may be beneficial).

Some insect predators eat bees but do not typically pose a serious threat to strong colonies, and no control measures will be needed. Your field bees may be caught by:

- True bugs (Hemiptera) such as assassin bugs (Reduviidae) and ambush bugs (Phymatidae), which are particularly voracious.
- Robber flies (Diptera: Asilidae).
- Mantids (Mantodea).
- Hornets and wasps (Hymenoptera: Vespidae). These may be a problem in the fall or if colonies

have died from pesticides or mite predation. Wasps, hornets, and yellow jackets will clean out dead hives, feeding on dead insects, brood, pollen, honey, and even wax moths.
- Dragonflies (Odonata: Antisoptera) and damselflies (Zygoptera).

Other insects prey on the stored products in a colony or on the insects that eat the stored products; they can be a problem if there are many dead colonies. The most common insects found in a hive, or living in stored equipment, include:

- Moths—dried fruit moth (*Vitula edmandsae*), and Indianmeal moth (*Plodia interpunctella*).
- Beetles (Coleoptera), which may live inside hives, eating debris and the litter found there. The most common ones are dermestid beetles (Dermestidae), weevils (Curculionoidea), sap beetles (Nitidulidae), and scarab beetles (Scarabaeidae); the last two eat stored pollen. Some beetles are predators, such as ground beetles (Carabidae), or parasites, like the blister beetles (Meloidae), which eat or parasitize live bees.

Certain flies (Diptera) bother bees at times but are mostly considered a minor nuisance unless their natural prey is unavailable. Some flies are predators, but others are opportunists, found in colonies that died of other causes. Others will parasitize bees, but these are found mainly in tropical climates.

The following have been noted in the literature as being pests of bees:

- Humpbacked flies (Phoridae), blowflies (Calliphoridae), thick-headed flies (Conopidae), flesh flies (Sarcophagidae), and tachinid flies (Tachinidae).
- The bee louse (*Braula coeca*) eats food at the bee's mouth. This fly, which looks like a varroa mite (except that it has six instead of eight legs), may reach damaging levels in some regions. However, because bees are treated with acaricides for mite control, these flies are difficult to find now in regions where varroa mites are common.

To control these insect pests, store your empty equipment in cold or freezing temperatures. **Never** use insecticides or pest strips in stored hive bodies and comb: these will also kill bees and can be absorbed into the wax.

Other pests include:

- The pollen mite (Acari), which is found in deserted comb or in hive debris.
- Nematodes or roundworms (Nematoda), which are small worms that live in many different habitats. While some also live on bees, they are not really a serious pest.

ANIMAL PESTS

Skunks and Raccoons

Skunks (family Mustelidae), which include weasels and badgers, and raccoons (family Procyonidae) are serious pests to bees, often visiting hives in the early evening, as well as during the day. They can cause damage to both equipment and bees and dig up the beeyard looking for food to eat. By scratching at the entrance of a hive, skunks entice bees to expose themselves at the entrance's landing platform. Once the bees leave the safety of their hive, the skunks readily scoop them up and devour them. Repeated forays to the colony night after night will slowly decimate the adult population of a colony. Look for scratch marks in the ground near the hive entrance, as well as scratches on the surface of the bottom board; look for skunk "cuds" containing bee parts. These are telltale signs that skunks are at work. These daily forays also alter the behavior of a colony; it becomes highly defensive when opened for inspection. Even young skunks quickly learn bee colonies are a good food source. Skunks also feed on bumble bee and yellow jacket colonies.

Raccoons (Procyonidae) often take and scatter anything loose in the apiary, including feeder jars and frames of brood or honey that have been left out. Some raccoons are strong enough to lift the covers off hives and feed on bees, honey, or brood. Depending on the severity of winter in your area, these pests could be nearly a year-round problem, especially in queen-rearing yards where the colonies are smaller. In the Southwest and South and Central America, the coati (*Nasua* and *Nasuella* spp.), a member of the raccoon family, and sometimes even coyotes, can also be a problem.

Signs of mammalian visitors are:

- Defensive bees.
- Grass near hive entrance torn up.
- Scratch marks on the hive front or on earth at hive entrance.
- Outer covers off or skewed.
- Weak colony for no other apparent reason.
- Area near entrance muddy after a rain and tracks and scat (feces) can be seen.

Discouraging and eliminating these pests may be accomplished by:

- Using hive stands at least 18 inches high (45.7 cm), to keep bees out of reach (see "Hive Stands" in Chapter 4). This is the best and easiest way to eliminate small mammal predation, especially skunks.
- Sprinkling rock salt crystals on the ground around the hive. Although this method may deter these pests, until it rains, it will also kill the vegetation around the hives.
- Trapping skunks, which may be illegal in your area, may cause the skunks to discharge (which will not make you popular with your neighbors).
- Alternatively, live trapping skunks and raccoons (e.g. with a Havahart trap) can be done successfully (use cat food or marshmallows as bait). Rules and regulations permitting live trapping, or other ways of getting rid of skunk and raccoon problems, require that you conform to existing regulations on such matters. Transporting trapped animals to another location for release may not be permissible by law. Skunks may still be able to release liquid from their scent glands; covering the trap with a blanket before transporting it may deter the skunk from squirting the contents of its scent glands.
- Using poison baits; this method is not recommended, because it is not selective enough and can harm other animals and pets. Before using poison bait traps, contact your state game and wildlife departments and comply with regulations for controlling fur bearers.
- Extending a piece of hardware cloth in front of the hive entrance, which will allow bees to sting the skunk's belly; this is a good but temporary measure. Make sure it is fastened to the bottom board, or the skunk will tear it off.
- Placing a strip of carpet tacking, nail points up, on the landing board or anchored in the ground in front of the entrance, may deter skunks from revisiting the colony.

Bears

Bears (Ursidae) eat brood, adult bees, and honey and can do extensive damage to equipment, especially in Canada, where large bear populations exist. However, bears are now found in almost all states in the continental United States, and they are capable of destroying apiaries. Signs of bear damage are overturned hives, smashed hive bodies, frames scattered over the apiary, and entire supers that have been removed from the apiary and scattered 30 to 50 yards (27–46 m) away.

An electric fence around the apiary is the most effective control against this animal. Fences are available in most rural feed or farm supply stores; solar chargers to provide the electricity for the fence are now readily available and can also be purchased from most bee supply catalogs.

If there are bears in your state, your apiary will sooner or later be attacked by a bear, unless the colonies are enclosed within an electrically charged fence. It is not a question of if, but when. If you procrastinate as to whether or not you should install one, you will come to regret it. However, a word of caution: **never touch an electric fence**. Although the charge should not be fatal to humans, a two-year-old child crawling in wet grass could be electrocuted upon contacting an electric fence. An electric fence tester, which tests if your fence is working, can be purchased. This devise will also tell you the strength of the current flowing through your fence.

Raised platforms are used in Alaska, where grizzly bears are a problem, but these are extremely difficult and expensive to build. Locating apiaries away from bear routes may help, because these animals keep to knolls, tree edges, and stream banks. Do not leave combs or hive debris around an apiary, as this will attract not only bears but other pests. Alternative ways to reduce bear damage include moving bees to a new location and seeking the assistance of local conservation departments, who may trap bears that are frequent visitors.

Mice

With the exception of bears, mice (genera *Mus, Micromys, Clethrionomys, Peromyscus*) inflict the greatest physical damage to honey bee colonies. They enter hives in the fall and winter and, although they appear not to harm the bees themselves, can cause extensive damage to comb and woodenware. They may destroy weak colonies by feeding on bee bread, honey, wax moth larvae, and cocoons, bee brood, and bees. Their droppings and urine are another irritation that often disrupts cluster behavior, especially if the colony is weak. If you wrap or otherwise winterize your hives, mice can invade and make nests in the winter packing. Because mice carry viruses (such as hantavirus) and vermin (fleas) that may affect humans, keeping these pests out of bee hives can be important to your health. In the desert Southwest, pack rats (Cricetidae) can be a problem if they build their nests in abandoned equipment; their nests also contain kissing bugs (subfamily Triatominae), which can invade homes and vector Chagas disease.

Signs of mouse damage are:

- Chewed combs or wood; droppings on the bottom board.
- Holes chewed in entrance reducers, enlarging the opening to allow mice to enter.
- Nesting materials (grass, paper, straw, cloth, or such) in hives, usually in between frames.

Bee hives placed along forest edges and in fields of tall grasses are especially at risk. The following measures may help to control damage from mice:

- Place hives on stands (although mice can climb).
- Use entrance reducers; some beekeepers line them with metal to prevent mice from chewing through them. The exact time to add entrance reducers is a judgment call and will vary depending on latitude.
- In the fall, close the entrance with ½-inch mesh hardware cloth or metal mouse guards (can also be purchased from bee suppliers). Mice can squeeze through a space that measures ⅜ × 3 inches (1 × 8 cm).
- Keep weeds down around hives.
- Place mouse bait on bottom boards or around the base or under the hive. This measure is not recommended, because it is not selective. The only place to use poison bait and traps is in an enclosed space where extra equipment is stored, such as your honey house.

VANDALS (*HOMO SAPIENS*)

There has been an increase in recent years in the number of hives stolen or otherwise vandalized, which makes vandals the number one vertebrate pest of bees. The increasing demands for equipment, honey, bees, and hives for pollination services have all contributed to the prevalence of thieves. Furthermore, colonies are vandalized by the curious, or those who think they will be able to obtain free honey simply by opening up a colony. Those bent on mischief can overturn or otherwise damage hive furniture.

Vandals can be discouraged by placing apiaries near year-round dwellings. If it is not possible to place them near one's own residence, land can often be rented from a homeowner for a few pounds of honey a year. Branding your hive bodies and frames is good protection. If your hives are stolen, for example, and the bee inspector finds your brand on hives in some other yard, it is more likely the person responsible for the theft will be apprehended and your equipment returned to you. See Chapter 3 for other ideas on personalizing your hive furniture.

Instead of painting your hives white and placing them in open, highly visible areas, try going to the paint stores and getting cans of premixed colors that other consumers returned. Mixing these together often results in a nice, mud-colored paint that makes your hive boxes disappear into the background. Tree hedges or judiciously placed shrubs to create screens around the yard also help to discourage would-be thieves. So does fencing with a locked gate!

MISCELLANEOUS MINOR PESTS

Although many birds are insectivorous, few, if any, eat bees in large quantities in North America. Bee eaters (family Meropidae), common in Asia, Africa, and Europe, can decimate apiaries and eat virgin queens on mating flights. In North America, flycatchers and kingbirds (family Tyrannidae) feed on bees, and woodpeckers (family Picidae) can damage old abandoned or weak hives. But you should make no attempt to control birds by poisoning or shooting them, which is illegal.

Other minor pests, which could be major in some areas, include frogs, toads, lizards, squirrels, opossums, rats, and shrews. Livestock will knock over hives if they are not otherwise protected in pastures.

Bees could also bother livestock at watering containers or ponds, especially in desert regions. Additionally, coyotes (*Canis latrans*) have been known to eat brood and honey if they can get into a hive.

Snakes can be problematic if you live in areas where venomous snakes are common. These reptiles, while beneficial in keeping down mice and other rodents, like to nest in the warmer inner covers in cold weather, or underneath stored equipment. While most snakes are harmless, be careful and mindful of them when opening abandoned equipment.

INTEGRATED PEST MANAGEMENT

Integrated pest management, or IPM, is a strategy that is environmentally sensitive and uses **multiple tactics**, such as requeening, biological controls, and cultural, physical, or other nonchemical controls to "manage" pests (such as mites) and diseases instead of "controlling" or attempting to eliminate them, provided they do not damage the colony. Chemical acaricides or other products should be used as a last resort, and the least toxic chemical should be chosen. This emphasis on the least toxic chemical is even more critical for beekeeping than other situations where pests need to be managed, because bees are extremely sensitive to pesticides (and this includes the acaricides we use to control mites). The key words here are "pest management," not pest eradication (see illustration of IPM pyramid).

Most beekeepers have great respect for the environment surrounding their apiary, as well as the one inside a bee colony. Until the advent of the varroa mite, beekeepers kept to a bare minimum the number of chemicals administered inside a bee colony. The varroa mite changed all that, and at this time, it is extremely difficult to keep your bees alive without the use of miticides, which are placed inside bee colonies to reduce varroa mite numbers. However, the varroa mite is a pest that is extremely difficult to control, even with the use of miticides.

Although there are beekeepers who valiantly attempt to keep their bees healthy without miticide treatments, their efforts have yet to be proven successful enough to free beekeepers from using miticides. Even when miticides are used, none can kill all the mites in a colony. In short, beekeepers worldwide face a pest that has challenged both them and bee researchers as they have never been challenged

Pyramid of IPM Tactics for of Honey Bee Mites

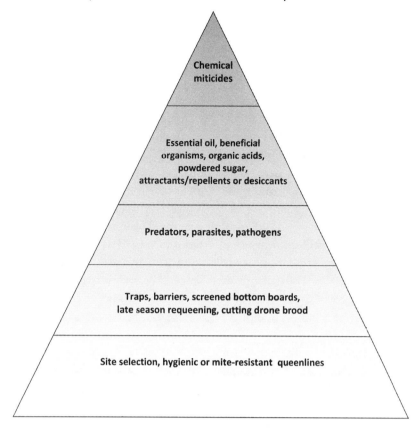

before. With the discovery of the varroa mite in Europe, South America, and the United States, many researchers have turned their attention to this mite, and yet, at this writing, we are still far from reaching a breakthrough (see "Diseases and Pests" and "Mites" in the References).

The discovery of varroa and tracheal mites, small hive beetles, and now a new *Nosema* pathogen has altered beekeeping practices forever. Gone are the days of the laissez-faire beekeeping of our grandfathers, when all they had to do was put bees in a hive and harvest honey. Now we, as beekeepers, face challenges that have changed beekeeping from a hobby requiring minimal knowledge or skill into a complex occupation with some risks of losing our bees. It has also focused attention to how important bees, and in fact all pollinators, are to the survival of our world. Without pollinating insects, we would be eating mostly rice and wheat.

That said, there are additional problems with keeping our bees healthy. Before mites, we needed to use only a few chemical antibiotics and could, with a little effort, avoid even those. Now we are not only in-troducing many different kinds of chemicals into bee colonies (not all of them are legal), but are stressing bees by moving them around to pollinate crops and thus exposing them to other bees that have more serious problems (resistant mites and diseases). Meanwhile, evidence is also mounting that the chemicals we are using to eliminate parasites and pathogens are also harming our bees. We are in a Catch-22 situation: If we don't use the chemicals, the bees can die. If we use the chemicals, we are increasing the chance that the bees will die from multiple chemical exposures or be more susceptible to pathogens because of the stress caused by the chemical exposure. What can we do?

Bees are not domesticated, nor do we really need to domesticate them; they work with us but should not be totally reliant on us. This total reliance on keeping mites and disease at bay artificially has led to weakened bees, not to mention the arms race of keeping ahead of the mites as they develop resistance to our chemical arsenal. Studies have shown that intensive reliance on a chemical to control pests or diseases will last about a decade. A new chemical, with a different way of killing the pest or disease, needs to be developed before resistance occurs, or you are back where you started, with the pest or disease killing your bees. There has to be another way to approach the problem, other than the serial use of chemicals—and fortunately there is.

Because there are now chemical-resistant mites and diseases, as part-time beekeepers, we need to develop an alternative action plan. We need to rely less on chemical crutches and more on augmenting the natural survival tactics and instincts of bees, such as breeding survivor bees and creating conditions in the hive to decrease the survivorship of pests and pathogens.

Remember, it is not possible to kill all mites in a colony or all mites in a region. The presence of some mites in the colony does not necessarily detract from hive health, provided the colony is strong, well fed, has nutritious food, and the mite numbers do not get out of hand.

Dr. Nancy Ostiguy, part of the bee team at Penn State University, has provided important additional information about IPM here. An IPM agenda for beekeeping is not a one-size-fits-all program. What you do will depend on your beekeeping goals, the number of colonies you maintain, the amount of time you have, and the environmental conditions. Here are some things that are common to all IPM programs, including:

- Pest and pathogen identification and monitoring.
- Preventing and controlling the spread of pests/disease.
- Designing your operation to minimize pests and pathogens.
  - Cultural control (site selection, resistant queen lines).
  - Hygienic or other resistant queens.
- Colony management.
- Biological controls (nematodes, fungus, predators).
- Mechanical controls (sticky boards, requeening, screened hive bottoms, drone trapping).

When prevention measures do not keep pest (mite) populations below the level where economic damage occurs, consider using biopesticides or bio-rational pesticides (essential oils, organic acids, powdered sugar, repellents, etc.). Apply conventional pesticides/controls only when mite levels are too high and the biopesticides are not effective.

Pest and Pathogen Identification

To implement an integrated pest management program you need to begin at the beginning. The first step is to make a list of the pests and pathogens that occur in your area and learn to identify them. For example, if you are living in bear territory, an important pest you should not ignore is bears. You could do an excellent job of keeping mite populations low, but a bear can wreck all your excellent work in a single evening. If you include a well-maintained electrified fence in your IPM program, the bears may still want to visit your hives, but they will be deterred by the fence. If bears are in your area, your bees will experience a bear attack; it is not a question of **if**, but **when**!

Use the resources available to you (your county or state agriculture department, your apiary inspectors) or take courses and workshops to increase your knowledge of bees and bee diseases and pests. Develop a "watch list" of problems in your area.

Once you have your list, and you have talked with other beekeepers in your area, you will be able to quickly determine which pests and pathogens are the biggest threats and about which you need to have the most knowledge. Start with these: memorize the life cycles of bee mites and learn to recognize the different disease symptoms. Then determine when to be especially vigilant.

For example, mite populations are the lowest in summer for tracheal mites, and winter/spring for varroa mites. Varroa mites are more likely to be found in brood during the summer and fall (in drone brood more than worker brood); during the winter the mites are more likely to be found on adult bees. These bits of information are incredibly useful for selecting varroa mite control tactics.

It is critical that you are able to recognize disease symptoms (see Chapter 13, on diseases of honey bees). Additional information is easily accessible on the web. Classes, workshops, and going through your colonies with your apiary inspector are other good ways to learn to recognize common diseases. Catching diseases early will not only improve the health of your bees, but you will lose fewer colonies and spend less money to keep them healthy. For example, American foulbrood found early (in a single colony at an early stage) is much less costly than detecting the disease after it has spread to your whole apiary. If the disease is confined to a single colony, it is easier to use a nonchemical tactic—shake the colony onto new, undrawn foundation and burn the old equipment.

Monitoring for Diseases and Pests

The next step is to learn how to monitor for pests and diseases. Most beekeepers will need to monitor for varroa mites, American foulbrood, sacbrood, chalkbrood, and deformed-wing bees (a marker for deformed wing virus). Tracheal mites, small hive beetles, and other pests/pathogens may or may not be found in your area; check with other beekeepers and your apiary inspector. Some pests and pathogens are difficult for beekeepers to identify. Tracheal mites are microscopic and cannot be seen with the naked eye, and most of the viruses have no overt symptoms and must be identified using laboratory techniques. Sacbrood and deformed wing viruses are exceptions in that they have clearly identifiable symptoms.

Unfortunately, reliable monitoring techniques that allow you to look at one colony (a sentinel colony) and predict the health status of other colonies do not exist; monitoring sentinel colonies should be used only when you have a high tolerance for significant colony loss. Because there is no reliable substitute for monitoring every colony, you need to determine how much time you have to dedicate to monitoring and how you are going to go about inspecting your bees. The most reliable way to monitor for American foulbrood, sacbrood, and chalkbrood is to inspect the brood area of your colony. You should consider doing this inspection whenever you are examining brood frames, cleaning bottom and/or sticky boards, or when examining a colony for other reasons.

Doing these inspections when there is a good nectar flow is an example of an IPM tactic, because during a nectar flow there is no likelihood of robbing, which eliminates the potential for disease transmission.

There are a number of ways to monitor for mites. The sticky board method requires that you have access to the bottom board. It is best to initiate this monitoring technique in the spring when you are reversing the hive bodies and can replace the solid bottom board with a screened bottom. This screened bottom should have a groove on each side of the wood, under the screen. This will allow you to insert and remove a sticky board under the screen from the back of the hive without disturbing your colony. (Check bee supply catalogs for equipment options.)

Another mite monitoring technique is to use a powdered sugar shake (see "Powdered Sugar Roll," this chapter). As with the sticky board monitoring method, the threshold could vary by location and season. In central Pennsylvania, the threshold for treatment is 1 mite in the spring and 10 mites in August. One standard suggested that if you find 3 mites per 100 bees from an alcohol wash, then treatment should be started. Recent information (2018) informs us that we may have to reevaluate all the procedures used to obtain varroa mite counts; you should be alert to new research as to how to obtain an accurate assessment of a colony's varroa mite population.

Try several different monitoring techniques to determine which one you are most likely to do on a regular basis—and do it!

Note Taking

While most of us are not thrilled with paperwork, doing a little bit each day will increase the effectiveness of your other activities. You may already keep track of some things such as the flowering times of local honey plants, or weather data, such as rainfall and temperature. This information is useful because it will tell you when certain pathogens will be a greater problem. Chalkbrood and sacbrood tend to be found more often in damp weather. Research is under way to determine if there is a relationship between weather and pathogen occurrence. If, in the course of your pest and pathogen monitoring efforts, you also record local weather, you may see a relationship that will be useful for predicting when you should be watching for problems. Temperature data can also be downloaded from websites covering your area. Also, if you record the results from your mite monitoring, you will know when it is important to step up your efforts to reduce mite populations beyond prevention measures.

When mite populations initially begin to exceed the ability of the colony to tolerate them, you should begin to implement soft chemical control tactics, such as formic acid and thymol. (Check for new treatments with your state agriculture department.) When mite populations have reached or exceeded the threshold needed to consider treatment options, consider treatments carefully: will they harm colony health? Without the treatment, will the colony survive? Because we only have some threshold data at this time (and not for all regions of the country), if you do not keep records, you will not know when it is time to implement soft chemical controls and when more serious mite-reducing treatments are necessary.

Design your operation to minimize pests and pathogens. The best offense is a good defense. Human public health specialists discovered many years ago that the most effective and least costly way to improve human health is **to prevent** diseases and parasites. The same is true for honey bees. You can do a number of things to reduce the risks caused by pathogens and parasites by creating less than ideal conditions for them. Many of the ways to create poor conditions for disease and good conditions for honey bees are known to beekeepers, but we have been able to ignore some or all of them in the past because the number of bee pathogens and pests were not as numerous, nor did they lead to the colony's demise.

Apiary Location

The first thing you may be able to control is **where** to locate your apiary.

Elsewhere in this book it was stated that you should choose a location with a good airflow; you don't want to be at the bottom of a hill where the air may be stagnant. Several diseases, including sacbrood and chalkbrood, thrive in moist conditions. Good air flow reduces moisture. If you could choose between a location with little air flow and too much air flow, the windy location would be better because you can put up windbreaks to protect your colonies from excessive wind speeds. One advantage of windier locations (approximately 10–20 miles/hour) is that the summer varroa mite numbers will likely be lower there than in colonies located where the air is stagnant. The problem with windy locations is higher colony losses in the winter. In order to survive in such locations, colony populations should be high, around 20,000 bees or more, and colonies should have more food to sustain them (at least one deep super filled with capped honey). You can also wrap colonies or provide windbreaks.

Another desirable apiary characteristic, if you are in the northern temperate regions, is to have a location in full sun (as long as there is an adequate and nearby water source). Full sun locations when daytime temperatures are in excess of 100°F (38°C) are not desirable. In addition, locations in full sun will increase the number of hours bees can fly each day. Research has shown that those colonies that fly more hours have lower mite levels; thus you are helping your bees help themselves. Another advantage of full sun is that your colonies will be warmer on sunny winter days. If you are concerned about the amount of heat absorbed by your hives, paint them white, or you can place shade cloth on the west side of your hives to decrease the sun's intensity during the hottest part of the day (see "Beekeeping in the Southwest," in Chapter 9).

Water and forage

As has been mentioned, your bees need an adequate supply of water. (Bees also will go to "dirty" or muddy water, which is OK if the water is not polluted). If your bees are in or near irrigated agricultural fields where pesticides are used, you may want to consider providing another source of water. If you do this, you'll want to make sure that the water you provide does not become a breeding ground for mosquitoes or for bacteria that could compromise the health of your bees. Use a water container such as a portable cattle waterer with a flow meter, or a small plastic kiddy pool, with rocks for the bees to land on. Empty and refill the pool every few days to reduce the problems that may be associated with the standing water. A Boardman feeder containing water is also a good way to provide your colonies with a source of water; that way you can monitor how much water the bees need by how fast the water is used up. This technique is especially good in arid areas where bees could be in nearby swimming pools or otherwise become a nuisance.

You should know the kinds and amount of forage that surrounds your apiary. It is ideal, not typical, if your bees have adequate forage for all the months of the year that they are active. Never place more colonies in an apiary than the surrounding forage can support. In many parts of the United States this means that no more than 15 colonies should be located in a given apiary, and the nearest adjacent apiary should be six miles from the center of the first one. When sufficient forage is available, bees will generally confine their foraging activity within a two-mile radius from their colony; by placing an adjoining apiary six miles from the first one, there will be little or no overlap of foraging pastures. There is no fixed rule on the number of colonies on a given site: the availability and abundance of nectar-producing plants will play a role in determining how many colonies work best in any given location.

If colony density is higher than the surrounding forage can support, the bees will have to be fed, resulting in lower productivity of these colonies. It is also likely that the bees will be stressed and more susceptible to diseases and pests. Another cause of colony stress occurs when there is a dearth of forage during a portion of the year. You may want to consider feeding your colonies if the dearth is longer or more intense than normal or if you have noticed colony losses in the summer or early fall resulting from depleted food stores. Be careful when feeding to watch out for robbing bees.

Apiary Layout

The arrangement of your colonies in your apiary is critical. Your goal is to improve the ability of the bees to find their home colony. S. Cameron Jay found high levels (40–50%) of drifting when six or more colonies where placed in the same row.

To avoid drifting due to the linear arrangement of hives, stagger your hives. Jay found drifting could be reduced (8–15%) if hives were placed in a circle or horseshoe configuration, with entrances facing outward; but more recent research has found extremely low (0.1–2%) drifting when hive entrances face south and hives are arranged in concentric circles. It is possible that south-facing hives have the lowest number of mites because of the amount of time sunlight bathes the hive entrance, causing the bees to fly for more hours each day; see "Hive Orientation," in Chapter 4.

A common arrangement used by beekeepers who have pollination contracts is to place four colonies on a pallet, with their hive entrances facing outward in four different directions. However, this arrangement gives the beekeeper limited options as to the direction of the entrances for the four hives. Although drifting is high under these conditions, there are few alternatives, given the fact there are four colonies per pallet. The advantage to the pallet is that monitoring one of the colonies on the pallet could be used as a predictor of the health status of the other three colonies. Conversely, if one colony is sick and dies, the other three colonies on the pallet are likely to get sick and die.

Honey bees, when they select a hive location on their own, tend to choose locations that are elevated more than one or two feet off the ground. It is possible that this characteristic preference is to avoid predation by skunks, raccoons, and other mammals. Beekeepers who have inspected colonies following a night of harassment by a skunk or raccoon can attest to the irritability of a colony subsequent to the disturbance. Preventing this source of bee stress can be easy to accomplish. High hive stands, or a properly maintained electrified fence, if the first live wire is approximately six inches off the ground, can reduce the likelihood of skunks and other mammals wreaking havoc in your apiary. If you install an electrified fence, make sure you put signage on the fence to warn individuals who might come in contact with the fencing. Sources for instructions on building a fence can be found online or from your county or state agriculture agency; or consult sources under "Diseases and Pests" in the References.

Hive Design

Historically the Langstroth and other hive designs have included solid bottom boards. These boards need to be scraped of debris each spring or they can become a resource for pests and pathogens. One method to decrease the buildup of debris and reduce your workload in the spring is to replace the solid bottom board with a screened bottom. A screen at the bottom of a hive is estimated to reduce varroa mite populations by 10 percent. This open bottom can also improve hive ventilation. A significant cause of colony death, prior to the introduction of mites, was inadequate hive ventilation. Thus a beekeeper is assisting in reducing the mite load in a colony while reducing the amount of work the bees must do to eliminate moisture in the colony. Make sure if you decide to install screen bottoms that you don't continue to use a solid bottom board beneath the screen. If the solid bottom is not removed, you will have provided an excellent location for large populations of wax moths, mice, and other pests.

If you are concerned about heat loss in the winter with an open hive bottom, choose a screened bottom design that includes grooves under the screen (used to insert sticky boards). A clean sticky board should be inserted below the screen before the onset of winter. Some beekeepers have had success leaving the screened bottom board open during the winter. Make sure to remove the sticky board in the spring when added ventilation may be required, or replace the winter sticky board with a new one in order to begin monitoring for mites again.

Before you go to the effort of inserting a solid bottom, remember that heat rises, and heat from a colony will escape from the bottom of the colony only when the apiary is windy. Avoid an apiary that is windy in the winter, because higher heat losses will occur, irrespective of the specifics of the hive design.

Today, there are other hive styles on the market that could work; before investing in other than the standard equipment, try a few first to see how they work in your area. Better yet, work with other beekeepers who are using different hive styles, and decide if you want to switch. These new hive types may be more expensive and, in some states, illegal. Work with state or local bee organizations first.

Apiary Maintenance

Apiary cleanliness is important. Leaving wax removed from hives strewn around the apiary will attract pests that are likely to move into any weak colony, as well as be a source of disease transmission (e.g., foulbrood). Keeping weeds and grass under control is critical for the operation of electrified fences; locating the first strand of wire approximately six inches off the ground can reduce predation by skunks and other mammals, particularly bears.

Weeds blocking the hive entrance can impair the ability of a colony to adequately vent the hive; vegetation should be kept down. Hive equipment that is broken or provides additional openings other than the front entrance increases the number of worker bees needed to guard the hive, as well as perform other tasks to maintain their home. This extra work has energy costs that will reduce a colony's productivity and increase its susceptibility to predators, robbing, and disease.

Colony Management

There are a number of things you can do to reduce pathogens and pests in colonies by your choice of management techniques. Many of these tactics are similar to what humans do to reduce the spread of pathogens and pests among each other. A significant number of these behaviors are easy to implement and, once they become second nature, will not increase the amount of time necessary to maintain your colonies.

Sanitation

Beekeepers are frequently told during disease management courses that hive tools should be cleaned between apiaries, to reduce the spread of disease. It would be a good idea to clean hive tools after each use (between colonies) in the same way that physicians wash their hands between patients (or at least we hope they do!). This hygienic practice will reduce the spread of pathogens within an apiary and increase the chances that a colony will be able to eliminate pathogens prior to a disease outbreak. An alcohol dip or wipe will eliminate many pathogens, but even better would be to use a propane torch to burn off any wax and propolis (first place the hive tools in a fireproof container). You can also put hive tools in your smoker and puff it until it flames up. **Do not burn yourself**. The ability of a pathogen to survive the heat or to remain once the wax and propolis are removed is very low; see Chapter 12 for other ideas.

Sanitation is also important for the control of the wax moth and the small hive beetle. If you don't want to rear wax moths for bait (see below), then drawn comb needs to be kept in a cool (below 40°F or 4°C) location or in active colonies that can defend against adult moths laying eggs. A solar wax melter is a useful devise to melt wax removed from colonies or wax from frames. Blocks of wax can then be stored for later uses, including coating plastic foundation or making candles, soap, or other value-added products.

Small hive beetles have created a new sanitation problem for beekeepers. If small hive beetle larvae are in your equipment, any supers removed for honey extraction cannot be stored; the honey must be immediately extracted, or the mess created by the larvae will be significant. As long as honey supers and brood frames are cold (below 50°F or 10°C) or patrolled by a sufficient number of honey bees, the "melt-down" caused by the beetle larvae will not occur. You will be tempted to give up beekeeping if you are faced with the mess that is created by this destructive beetle.

Equipment Replacement

American foulbrood spores are known to lie dormant in the comb for 80 or more years; also varroa mites prefer to reproduce in darker comb. This knowledge has led to the recommendation that beekeepers remove any comb older than three years and replace it with new wax and sterilized woodenware. A way to implement this practice and reduce the cost of replacing frames is to burn (or otherwise safely dispose of) frames that have been removed from weak colonies when you are moving frames of bees and brood from a strong colony to a weak colony (this practice is common when beekeepers are trying to equalize the strength of colonies in an apiary to prevent robbing). What you don't want to do is put the frame that is removed from the weak colony (to make room for the frame of bees and brood from the strong colony) into the strong colony. You don't know why the weak colony is weak. If it is from pathogens or parasites, you are moving disease to the strong colony. Most of us are not very thrilled to be in a room with a stranger who is coughing and/or sneezing; for that very reason you don't want to move that frame from the weak colony to the strong colony. Discard it.

Assuming the weak colony is free of any diseases and other serious problems, it may be safe to place the removed frame in a freezer to kill any problem that was undetected when strengthening the weak colony. Once it has been determined that the weak colony recovered by adding the frame of brood and bees, the frame(s) placed in the freezer can be reclaimed for use, where needed.

There are several possible ways to turn old comb into a salable product rather than having it become a "housekeeping" problem. Wax moth larvae are popular fish bait. You might be able to establish a relationship with a local sporting goods or pet store to supply them with bait or pet food (see Appendix F). Various beekeepers have developed devices, which a reasonably handy beekeeper can build, to remove the debris from old wax to produce clean wax that can be used for a variety of non-hive purposes. You can have your wax tested to ensure it is free from (or has very low levels of) contaminants. Nowadays, the closest you can come to clean wax is to have the bees construct their own wax combs and remove and render it as soon as possible.

Equalizing Colony Strength

Another management technique that was benign, prior to the multifarious new pests and pathogens that have been introduced over the last 30 years, but that is best to be avoided today, is the installation of new bees into hive equipment from "dead-outs." Several researchers have found a higher level of disease and death in colonies that were installed in equipment containing food left after a colony has died. If you are going to use equipment from dead colonies, then you need to remove or sterilize the remaining food. Irradiation has been successful in sterilizing food; another option is to power wash comb to remove contaminated food. (Do **not** take equipment to a self-serve carwash, as you will clog up the drainage system with wax and other debris. Rent or buy your own power washer.)

Genetics and Healthy Bees

All beekeepers need to pay more attention to the genetics of their colonies. If you have colonies that have survived longer than three years, you might want to consider raising queens from them. You may also want to purchase queens that have been reared from mite-resistant or tolerant stock. Bees that are more hygienic than the average colony are also desirable. Not only will these colonies keep mite populations lower, but they are also adept at removing diseased brood, and may even lower the rates of American foulbrood, chalkbrood, and sacbrood diseases.

Given all of the above, we must point out that our ability to retain identical genetic stock is extremely limited because of haplodipoidy, polyandry, and panmixa. In *haplodiploidy*, members of the order Hymenoptera, including honey bees, exhibit a chromosomal condition where some individuals (females) are diploid (having two sets of chromosomes), while others are haploid (males) having only one set of chromosomes. *Polyandry* is defined as many men. The queen mates with on average 14 or more drones, and none of them may have an identical set of genes. Hence, when the queen lays eggs using the semen of drone 5, all the female offspring (workers and queens) will belong to a subfamily produced from drone 5. Workers produced from the same mother (the queen) and the same father (drone) are called *super-sisters*.

As a consequence, a colony of bees may have **on average** 14 subfamilies. *Panmixa* refers to totally random mating. The queen flies to a drone congregating area to mate; assembled there are drones from many different colonies, many with their own set of genetic traits. As a consequence of haplodipoidy, polyandry, and paramixa, even if you have a colony that exhibits an abundance of favorable traits, the replacement of the existing queen with a new one to some degree alters the genetics of that particular colony.

Requeening

Replacing an existing queen with a new queen is an effective way to reduce some pathogen and pest problems. A new queen, if she is from a hygienic line, is the key to a colony being more effective at controlling American foulbrood, chalkbrood, and sacbrood. Requeening operations also interrupt brood rearing, which will slow the growth of the varroa mite population. If you are pleased with the genetic characteristics of your bees, you can requeen your colony by removing the old queen when the colony has larvae less than three days old. By removing the old queen and allowing the bees to replace her, no additional eggs and subsequent larvae will be available for mites to carry on their reproductive activities, curbing the growth of the mite population.

Splitting colonies, and introducing new queens or allowing the colony to rear its own queen, is an effective method to reduce mite populations. Some beekeepers have successfully controlled varroa mite populations this way; see Chapter 10, "Queens."

Physically Removing Pests

If your hives have screened bottoms, you can use sticky boards to remove mites from colonies. Make sure you use petroleum jelly or something equivalent on the sticky boards to capture the mites. Less viscous material *will not* prevent the mites from walking off the board and back into the colony. This mechanical control tactic requires that you remove the debris-filled sticky boards every three to five days. If you wait longer, the debris will build up on the board and the mites will be able to return to the colony; wax moths love this debris and can begin to breed here too. In the same way that you do not want to leave wax around your apiary, leaving the scrapings from the sticky boards in the apiary will attract pests. Take along a five-gallon bucket or other container and collect this debris so you can discard it later.

Drone Comb

Varroa mite populations can be reduced significantly by removing drone brood. Beekeepers using this technique report that mite levels remain low enough that other problems, such as deformed-wing bees, are not observed, and mite populations never reach the threshold that triggers the necessity of using a miticide. The timing of this tactic is important.

A good time to begin inserting a frame or two of drawn drone comb is the spring; this is the time when the queen begins laying drone eggs again. Mites have a preference for drone brood. Why? Drone development from eggs to adults takes 24 days, whereas workers develop in 21 days. More developing mites reach maturity on drone brood than on worker brood; hence if given a choice, reproductive mites will opt for drone brood to lay their eggs. By removing the drone frames before the drones emerge, you can remove a high percentage of the reproductive mites from the colony. If mite populations can be reduced to extremely low levels at the beginning of the season, the risk of the population growing to a number that can harm your colony is low. However, you must continue to monitor the mite levels through the season. Remember, with this technique you cannot forget to remove the capped drone comb. If you forget and the adult drones emerge, you have just increased the colony's mite population significantly.

Once the drone comb is filled with capped drone cells, it needs to be removed before the drones emerge. The drone comb can be fed to chickens that will eat the brood; you can also freeze combs to kill both the drone brood and the mites. In either case, the drone comb can then be returned to the colony, allowing the cycle to begin again. Pet stores may have a need for drone brood to feed to various pets, such as reptiles.

It may be prudent not to use drone comb removal (or drone zapping) during the swarming season, in that drones need to be available during the swarm season for mating purposes. Alternatively, you may wish to allow drone development to continue in your more desirable colonies.

And finally, by eliminating the use of miticides, your rendered beeswax will have less chemical residues.

Biological Controls

At this time effective lures and biological agents to control bee pests are extremely limited. There is some work looking into controlling varroa mites using fungi. A fungus for controlling the varroa mite is, at this writing, soon to be cleared for use in bee colonies. Work to develop an attractant (lure) for small hive beetles is in progress. Nematodes (microscopic roundworms) to control small hive beetles are now available. The nematodes are applied as a soil drench under and around the vicinity of the hive, where the hive beetles pupate; they parasitize the beetle pupae. In that they are soil inhabiting, they pose no danger to honey bees. Sources for purchasing these nematodes can be found online.

When Prevention Fails

There are times when prevention alone will not be sufficient. This is particularly the case when dealing with the varroa mite. One can say that the varroa mite is the most formidable foe ever encountered by honey bees, and, for the beekeeper, the most difficult to control. Evidence for this statement is the fact that we continue to try new strategies for mite control, and each time the mite seems to regain the upper hand.

There are times when the bees will need a bit of additional assistance beyond prevention measures or tactics such as drone brood removal. You should try to use the least toxic chemicals for treatment. Again, keep current with all the new work on IPM and bees, as researchers are finding new treatment options all the time.

Additional Practices to Keep Bees Healthy

The following are good practices to keep bees healthy and should be a part of any IPM program:

- Make sure all your colonies have ample, even large, stores of bee bread and honey. Starving colonies can succumb more easily. Feed your bees sugar syrup if there are insufficient quantities of honey.
- Assuming you are using two 10-frame brood boxes, one approach is to replace 20 percent (4 combs) each year. It will take five years for a complete turnover; in the sixth year, you have to start the process again. If you do not want to use foundation (because it might contain pesticides), consider letting the bees draw out their own comb (such as leaving frames with strips of foundation), or converting to top-bar hives. If you are not using pesticides yourself, consider coating plastic foundation with your own wax.
- Sample colonies for *Nosema*, mites, and other diseases before you consider any chemical treatments. If you treat colonies without knowing if they have a disease or parasite, you are not only wasting money, but you may be helping mites (or disease pathogens) become resistant to the treatments.
- Keep colonies headed by certain queen lines (specially bred lines of bees, such as hygienic) that are mite or disease resistant; you can also select those survivor queens in your own apiary.
- Emphasize cultural, mechanical, or other non-chemical control techniques in your management of honey bee pests and pathogens. There are no magic bullets, and all medications (including miticides) have side effects. Researchers are finding that the side effects of at least some of our treatments may be worse or equal to the problem we are treating.

We cannot give you an exact integrated pest management program to follow, because the specific tactics depend greatly on your goals, the number of colonies in your apiary, the amount of time you have available, and your level of tolerance for colony loss. You need to try various approaches and find the ones that work best for you. Keep reading and going to workshops and classes, because researchers are constantly looking for additional ways to improve bee health and longevity. They may be able to provide you with options that will help you reach your goals more effectively and inexpensively. Below are some additional tactics you might want to consider including in your IPM program to limit the growth of varroa and tracheal mites.

Reducing Varroa Mites

Here are some different techniques that you can "mix and match" to fit your particular situation for reducing varroa mites:

Pollinator Protection

Registration #	Product name	Active ingredient
2724-406	Zoecon rf-318 apistan strip	Fluvalinate (10.25%)
2724-406-62042	Apistan anti-varroa mite strips	
11556-138	Checkmite+ bee hive pest control strip	Coumaphos (10%)
11556-138-61671	Checkmite+ bee hive pest control strip	
61671-3	For-mite	Formic acid (65.9%)
70950-2	Avachem sucrose octanoate (40.0%)	Sucrose octanoate (40%)
73291-1	ApiLife Var	Thymol (74.09%), oil of eucalyptus (16%), menthol (3.73%)
75710-2	Mite Away Quick Strips	Formic acid (46.7%)
75710-3	Formic pro	Formic acid (42.25%)
79671-1	Apiguard	Thymol (25%)
83623-2	Hopguard 3	Hop beta acids resin (16%)
87243-1	Apivar	Amitraz (3.33%)
91266-1	Oxalic acid dihydrate	Oxalic acid (100%)
91266-1-73291	Oxalic acid dihydrate	
91266-1-91832	Oxalic acid dihydrate	

Source: https://www.epa.gov/pollinator-protection/

- Restrict brood rearing by caging the queen and removing capped brood.
- Treat adults (shake them into a screened box and dust with powdered sugar) to knock off the mites.
- Trap mites in drone brood (using drone foundation) and freeze.
- Scrape mites off returning foragers with pollen traps.
- Use botanical oils (such as Apilife Var) or organic acids.

Management Strategies for Both Mites

If you want to reduce the levels of both mites, try this:

- Split colonies, kill older foragers, requeen, and treat with bio-rational remedies.
- Keep bees healthy with plenty of bee bread/honey stores.
- Provide food supplements (protein and syrup) if needed.
- Use drone brood to trap varroa; cut out or freeze frames after cells are capped.
- Use oil patties for tracheal mites.
- Early in fall, harvest honey supers.

- Check mite levels.
- Requeen colonies in fall by using resistant bee stock, e.g. Buckfast or Russian for tracheal mites. Hygienic and Russian lines are beneficial for varroa mites and some diseases.
- Place shortening or sugar patties, or treat with menthol for tracheal mites. For varroa, use essential oils or other measures.
- If colony is highly infested, split the colony, requeen, and treat with bio-rational controls.
- Keep bees healthy with plenty of pollen and honey stores; provide food supplements if needed or when going into the winter.

Other Treatment Options for Varroa

Powdered (confectioner's) sugar sprinkled on bees knocks off some mites. Use this if you have honey supers on the colony and can't use other chemicals. Smoke the colony heavily to knock off mites, or dust bees with flour or powdered sugar. Insert sticky boards to catch the dislodged mites and keep them from crawling back up into the colony. You can also install a screened bottom board so mites will fall outside the colony entirely (check bee supply catalogs).

Dusting can be repeated weekly, but remember, it only removes mites temporarily clinging to adult bees but not the mites feeding on these adult bees. This technique may not keep varroa populations from spiking in the fall, so be prepared to use another control measure.

Cultural Practices for Disease

If a colony has been diagnosed with AFB or even EFB, shake all the adult bees off the combs. Then:

- Place bees into new equipment, including all new foundation, and requeen with a resistant queen.
- Feed sugar syrup and pollen supplement until the disease spores carried by the adult bees are flushed.

If this colony still develops the same disease, destroy it. As a last resort, use antibiotics if needed.

To prevent diseases from developing or spreading, use these techniques:

- Rotate combs every three years (date or color code the frames).
- Clean all tools (hive tools, smokers, suits) regularly.
- Learn to recognize and identify bee diseases; send in samples if in doubt.
- Requeen with resistant queen lines.

We all need to keep our bees healthy and able to develop natural immunities and resistance mechanisms. **Let's be part of the solution and not part of the problem.**

 Notes

CHAPTER 15

Pollination

The diet of honey bees consists primarily of honey and bee bread, which are produced by flowering plants in the form of nectar and pollen. The bees convert the nectar into honey and the pollen into bee bread. Flowering plants belong to the most dominant group of plants, the angiosperms, with at least 300,000 known species. With their appearance on the planet, an insect, the honey bee (which evolved from a carnivorous sphecoid wasp), coevolved with a vegetarian diet. The earliest known fossil record of a bee is about 87 million years old. While there are over 20,000 known bee species, found everywhere except Antarctica, they live in many, diverse habitats. Although honey bees are primarily vegetarians, during periods of dearth, they will resort to a carnivorous diet and eat the eggs, larvae, and possibly the pupae in their colony.

The relationship between flowering plants and honey bees can be looked at as a quid pro quo (something for something). The plant provides the honey bee with nectar (the primary source of carbohydrates) and pollen (the primary source of amino acids, fats, and minerals). In "recognition" for these essential products, the bee inadvertently assist the flowering plant by transferring pollen (sperm) from the male part of the plant to the female part (ovary); by so doing the bee assists the flower's reproductive activities (called pollination).

Growers are well aware of the essential role honey bees play in ensuring that their crops are well pollinated thus maximizing their yields of fruits and seeds. In addition to honey bees, there are a host of other pollinators, including bumble bees, solitary bees, and bats, to name a few (see "Non-*Apis* Bee Pollinators" in References). But in most cases, honey bees are selected to be the major crop pollinators because each colony can contain large populations, with a reasonably high percentage of foragers ready to visit flowers that require their services.

SOME DEFINITIONS

Flowers are the reproductive parts of the angiosperm plants where seeds are formed and from which fruits and vegetables develop. The earliest angiosperm in the fossil record found so far dates back 125 to 250 million years. The flower is the action site for pollinators, offering not only food (nectar and pollen) but a landing platform (petals), landing lights (flower color), and directional signals (nectar guides) to point the way to the food. Some of these guides have special, contrasting colors (even ultraviolet markings) to point out landing platforms and position the bee correctly. Some petals even have special conical cells (which are bumpy) to help bees get a firm foothold while pollinating the flower. A few flower species provide not food but wax and resins for nest material, and some even mimic the female of certain bees, to attract "pollinating" males. The orchids are especially diverse in their pollinating mechanisms.

The male part of the flower, called the *anther*, is located on the tip of the stamen and produces a powdery substance, called *pollen grains*, which contain the sperm cells. The female structure is called the *pistil* and contains the *stigma* (usually sticky) and the *ovule* (where the seeds are formed). Each species of flowering plants has its unique shape and form of

Fertilization of a Flower

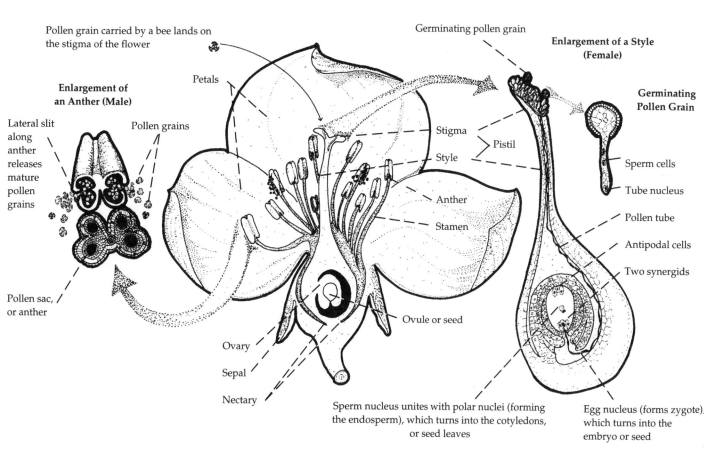

Pollen grain carried by a bee lands on the stigma of the flower

Enlargement of an Anther (Male)

Lateral slit along anther releases mature pollen grains

Pollen grains

Pollen sac, or anther

Petals

Stigma

Style

Anther

Stamen

Ovule or seed

Ovary

Sepal

Nectary

Germinating pollen grain

Enlargement of a Style (Female)

Germinating Pollen Grain

Pistil

Sperm cells

Tube nucleus

Pollen tube

Antipodal cells

Two synergids

Sperm nucleus unites with polar nuclei (forming the endosperm), which turns into the cotyledons, or seed leaves

Egg nucleus (forms zygote), which turns into the embryo or seed

pollen, which enables paleobotanists (who study ancient plants) to identify 10,000-year-old pollen from the mud of bogs and lake bottoms.

Most flowers that we recognize are those that are *perfect*; that is, the flower has both female and male parts. Fertilization takes place in the ovary, where pollen unites with a female ovule of the same flower species (see illustration above), creating seeds and sometimes fruit. The transfer of pollen from male to female sex organs is called *pollination*.

Some plants species have "imperfect" flowers, where the male and female reproductive organs are on separate flowers, which are located on separate plants. One plant carries the male part, and the other plant carries the female part; these plants are referred to as *dioecious*. The holly (*Ilex* spp.) is a good example of this; you must have a male plant (to produce pollen) and, nearby, a female plant to receive the pollen. Once fertilization occurs, the ovary enlarges and produces those familiar red berries.

Monoecious plants present a different scenario.

Both male and female reproductive parts are on the *same* plant but not together; they are located on separate female flowers and male flowers. The cucurbitaceae (e.g., squash, pumpkin, gourds, and cucumbers) are excellent examples of monoecious plants.

A large and familiar family that has many florets on a single flower head is the sunflower; they are called *composite* flowers. Besides sunflowers, these include asters and goldenrod.

All plants must be pollinated before seed (or fruit) will set. Pollen is transferred from the anthers to the stigma either abiotically (by wind, water, gravity) or biotically (by mammals, birds, humans, and insects). If the transfer takes place on the same blossom or on another blossom on the same plant, it is called *self-pollination*. Beans, for example, are self-pollinating. Though many kinds of beans and other plants do not need insect visitors, they do benefit from the extra pollen carried by them and may even set better or more fruit. Other examples are soybeans, peaches, and lima beans.

Fruit Proportion

Fruit	Proportion (honey bees)[a]
Almond	1.0 (100%)
Apple	1.0 (0.9)
Cherry	0.9 (0.9)
Citrus	0.8 (0.9)
Cranberry	1.0 (0.9)
Kiwi	0.9 (0.9)
Peach	0.6 (0.8)
Pear	0.7 (0.9)
Plum/prune	0.7 (0.9)
Strawberry	0.2 (0.1)

Field crops

Alfalfa	1.0 (0.6)
Canola	1.0 (0.9)
Sunflower	1.0 (0.9)

Vegetables and melons

Asparagus	1.0 (0.9)
Cucumber	0.9 (0.9)
Melons	0.8 (0.9)
Squash	0.9 (0.1)
Vegetable seed	1.0 (0.9)
Watermelon	0.7 (0.9)

Source: R.A. Morse and N.W. Calderone. 2000. The value of honey bee pollination in the United States. Bee Culture 128:1–15.

[a]The first number is the proportion dependent on insects for pollination; the number in parentheses is the percentage for which honey bees are the principle pollinator.

Apples and many fruits have a further complication—many are *self-sterile* or *self-incompatible*, which means the pollen from a flower of a Red Delicious apple tree will not result in fertilization when transferred to the stigma of the same flower or the flower of other Red Delicious trees.

But if the pollen goes from a Red Delicious apple tree flower to a Granny Smith apple tree blossom, this is called *cross-pollination*. The pollen must have another variety of apple to set fruit. The placement of apple varieties, the size of the orchard blocks (groups of the same variety of cultivar) of *pollinizers*, and length of rows may be factors in getting good fruit set in the orchard. It is important not to have too big a block of any one variety in any single area of the orchard. To correct this, some growers graft a limb of an appropriate pollenizer (usually a variety of crabapple tree, since crabapples are excellent pollen producers) onto every six apple trees. When grafting limbs onto self-sterile cultivars, it is imperative to select a variety that will bloom at the same time as the variety of apple trees in the orchard block.

Most angiosperm plant pollination, about 80 percent, is biotic, with insects being the most common carriers of pollen. In plants that are pollinated abiotically, about 98 percent of them are pollinated by wind and 2 percent by water. Many angiosperms (where the ovules are enclosed in an ovary) are wind pollinated, including all the grasses and their cultivated cousins (corn, oats, wheat, rice) and most trees, including hickory, walnuts, butternuts, maples, poplars, and some fruit trees. Wind-pollinated plants must release enormous quantities of pollen in order to guarantee successful pollination. The other major plant group, the gymnosperms (whose ovules are **not** enclosed within an ovary), are the cone-bearing trees and shrubs, which include the conifers, such as pines, fir, cedar, and junipers. The "flowers" of gymnosperms are inconspicuous, and the male and female reproductive parts are borne on separate parts of the plant.

Of course, wind pollination is a hit-or-miss situation; plants that rely on wind pollination have to produce excessive amounts of pollen to achieve reproduction success (hence one of the main reason that pollen allergies come from wind-borne pollen). Often, goldenrod (*Solidago* spp.) is blamed for fall allergies, when in fact it is likely the pollen from the wind-pollinated ragweed (*Artemisia* spp.) that is the culprit.

THE MECHANICS OF FERTILIZATION

When a bee visits a blossom on a pollen foraging trip, she will actively gather pollen from the anthers of flowers, and some grains of pollen will adhere to her hairy body by means of static electricity. While on that flower, or on subsequent flowers, or while hovering in the air, she will use her first two pairs of legs to remove the pollen adhering to her hairy body. She will add a little glue in the form of nectar or honey to the pollen and move the pollen grains to her pollen baskets (the corbiculae). Usually, a foraging bee is out to collect only pollen or only nectar; however, the bee will often accomplish getting both on any given foraging trip. While collecting pollen, the bee inadvertently comes in contact with the sticky tip of the flower's stigma, facilitating transferring the sperm contained in the pollen over to the stigma (a part of the female reproductive organ). Following this transfer from the anthers to the stigmas, the ultimate production of the next generation of an apple tree (as but one example) has been set in motion.

Once the pollen grains contact the tip of the female pistil, each will grow a thin *pollen tube* down the elongated stigma, or neck of the pistil, which delivers two sperm to the rounded base of the pistil (the ovary). Each ovary houses one or more small round bodies called *ovules*, in which the fertilization actually occurs. In the ovule, an egg nucleus is fertilized by one of the sperm cells, which will then develop into the embryo (early tissue of the new plant). The two other nuclei, called *polar nuclei*, when fertilized by the second sperm nucleus, develop into tissue called *endosperm*, which nourishes the developing embryo. As the ovules mature within the ovary, they become the seeds, which are protected within the surrounding ovary wall, now the mature fruit. In some cases, the fruit in the ovary is edible (for example, cucumbers, apples, beans), but in other cases, the ovary becomes a thick shell (for example, walnuts, almonds).

When the seeds are planted and then germinate, the first leaves to emerge from the seed (seed leaves or cotyledons) continue to nourish the growing plant until it is able to produce its own true leaves. The quantity of pollen that a plant produces (as a result of favorable growing conditions) may influence the pollinated plant to produce more or fewer seeds, thus stabilizing the next generation of plants produced (see "Pollination" in References section).

From this complex, double fertilization, almost all flowering plants are fertilized. Even "seedless" varieties of some crops need to be pollinated for fruit development. As a seed starts to form, it is aborted early in its development; the fruit is thus not truly seedless, but has undeveloped seeds.

BEES AS POLLINATORS

The most efficient pollinators—highly motile, small, and plentiful—are the insects. Major insect pollinators include beetles, flies, butterflies, moths, and bees. Bees are probably the principal pollinating agents of plants whose flowers have colors within the bee's visual range of blue, yellow, green, and ultraviolet.

Honey bees are the go-to insect for pollinating a vast number of crops, especially where there are vast acres of a specific crop (or monoculture), for example, apples, almonds, strawberries, and soybeans, to name a few. However, there are always exceptions, and alfalfa is one of them. When a honey bee alights on an alfalfa blossom and moves toward the nectar, a tripping mechanism releases the reproductive structures of the flower and the anthers hit the bee on the underside of the head (see illustration on tripping an alfalfa flower on p. 290). As a consequence, bees avoid working alfalfa blossoms, or work the flower from the side to avoid setting off the triggering mechanism. The result is a low yield of alfalfa seed. In order to increase the seed yields, alfalfa farmers have resorted to using two kinds of solitary bees to pollinate this plant, namely leafcutter and alkali bees.

Remember, over **$24 billion** worth of crops are pollinated by bees in the United States each year (the estimate worldwide was $235–$577 billion of global food product in 2020); the almond crop alone (restricted primarily to the Central Valley of California) costs growers an estimated $300 million for pollination from 1.5 million colonies (2019).

For the most part, honey bees are the most efficient pollinators. It often takes several trips by many bees to adequately pollinate a single apple or cucumber blossom. This is because there are many seeds in the ovaries in these blossoms, and each ovule requires a separate sperm. If the ovules do not receive the requisite number of sperm, the fruit may be lopsided, or the cucumber curled, making them less valuable. For example, the female cucumber blossom requires 20 or more visits by bees to be adequately pollinated. Since the cucurbits (which include cucumbers, pumpkins, melons, squash, and gourds) have both female and male blossoms on separate parts of the plant, the pollen needs to be gathered from the male flower and then transported to the female flower. To further complicate this activity, the blossoms of most cucurbits are opened for only part of a **single day**. Often bees, and particularly bumble bees, fail to exit the blossom before it closes and become trapped within the closed flower. These set of circumstances are reasons why honey bees are required to adequately pollinate large cucurbit fields.

A Honey Bee Tripping an Alfalfa Flower

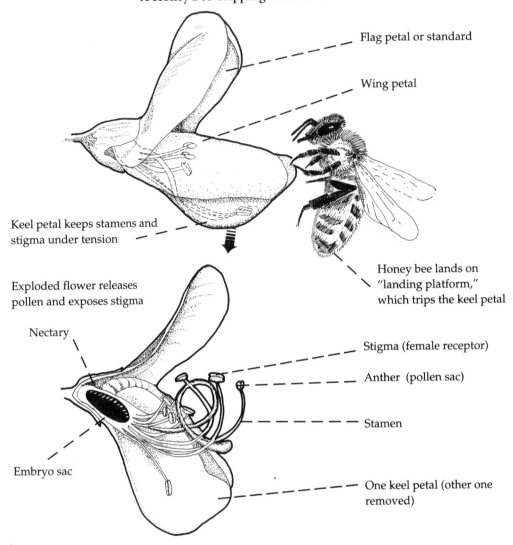

Flag petal or standard

Wing petal

Keel petal keeps stamens and stigma under tension

Honey bee lands on "landing platform," which trips the keel petal

Exploded flower releases pollen and exposes stigma

Nectary

Stigma (female receptor)

Anther (pollen sac)

Stamen

Embryo sac

One keel petal (other one removed)

When a honey bee lands on the keel petal of the alfalfa flower, its weight trips the spring-loaded petal, and the central column of the flower (containing the style and anthers) is released; as the anthers hit the bee on the underside of its head, they deposit pollen on the bee. Honey bees "learn" to avoid springing the central column while still taking the flower's nectar. For this reason honey bees are not the best pollinators of alfalfa fields. Diagram courtesy of the Xerces Society.

Almost all large commercial growers and orchardists plant large blocks of the same crop, necessitating the need to contract beekeepers to place honey bee colonies in these blocks in order to guarantee that their crops receive adequate pollination. Colonies of bees are moved to various crops needing pollination over the course of the spring and summer, depending on when the crops bloom. Honey bees are desirable pollinating agents for these reasons:

- Each colony contains large populations of foragers to work crops within a narrow pollinating window.
- Bees will usually work only one type of flower on each trip (*flower fidelity*), not mixing pollen types. For example, when the honey bee flies out to gather nectar and pollen from an apple blossom, this is the only type of flower she will visit on this flight.
- Crops can be sprayed with certain attraction pheromones, which assure that the bees will work only the target crop.

Commercial Crops That Benefit from (or Require) Insect Pollination

Common name	Latin name	Pollinator	Commercial product of pollination	Pollinator impact	Number of honey bee hives per acre
Acerola	*Malpighia glabra*	Honey bees, solitary bees	fruit (minor commercial value)		
Alfalfa	*Medicago sativa*	Alfalfa leafcutter bee, alkali bee, honey bees	seed		1
Allspice	*Pimenta dioica*	Honey bees, solitary bees (*Halictus* spp., *Exomalopsis* spp., *Ceratina* spp.)		3-great	
Almond	*Prunus dulcis, Prunus amygdalus,* or *Amygdalus communis*	Honey bees, bumble bees, solitary bees (*Osmia cornuta*), flies	nut	3-great	2–3
Apple	*Malus domestica* or *Malus sylvestris*	Honey bees, orchard mason bee, bumble bees, solitary bees (*Andrena* spp., *Halictus* spp., *Osmia* spp., *Anthophora* spp.), hover flies (*Eristalis cerealis, Eristalis tenax*)	fruit	3-great	1, 2 semi dwarf, 3 dwarf
Apricot	*Prunus armeniaca*	Honey bees, bumble bees, solitary bees, flies	fruit	3-great	1
Avocado	*Persea americana*	Stingless bees, solitary bees, honey bees	fruit	3-great	
Azarole	*Crataegus azarolus*	Honey bees, solitary bees	fruit	1-little	
Beet	*Beta vulgaris*	Hover flies, honey bees, solitary bees	seed	1-little	
Blackberry	*Rubus fruticosus*	Honey bees, bumble bees, solitary bees, hover flies (*Eristalis* spp.)	fruit	3-great	
Black currant, red currant	*Ribes nigrum, Ribes rubrum*	Honey bees, bumble bees, solitary bees	fruit	2-modest	
Blueberry	*Vaccinium* spp.	Alfalfa leafcutter bees, southeastern blueberry bee, bumble bees (*Bombus impatiens*), solitary bees (*Anthophora pilipes, Colletes* spp., *Osmia ribifloris, Osmia lignaria*), honey bees	fruit	3-great	3–4
Boysenberry	*Rubus* spp.	Honey bees, bumble bees, solitary bees	fruit		
Broad bean	*Vicia faba*	Honey bees, bumble bees, solitary bees	seed	2-modest	
Broccoli	*Brassica oleracea* cultivar	Honey bees, solitary bees	seed	1-little	
Brussels sprouts	*Brassica oleracea* Gemmifera group	Honey bees, solitary bees	seed	1-little	
Buckwheat	*Fagopyrum esculentum*	Honey bees, solitary bees	seed	3-great	1
Cabbage	*Brassica oleracea* Capitata group	Honey bees, solitary bees	seed	1-little	
Cantaloupe, melon	*Cucumis melo* L.	Honey bees, squash bees, bumble bees, solitary bees (*Ceratina* spp.)	fruit	4-essential	2–4
Carambola, starfruit	*Averrhoa carambola*	Honey bees, stingless bees	fruit	3-great	
Caraway	*Carum carvi*	Honey bees, solitary bees, flies	seed	2-modest	
Cardamom	*Elettaria cardamomum*	Honey bees, solitary bees		3-great	
Cashew	*Anacardium occidentale*	Honey bees, stingless bees, bumble bees, solitary bees (*Centris tarsata*), butterflies, flies, hummingbirds	nut	3-great	
Cauliflower	*Brassica oleracea* Botrytis group	Honey bees, solitary bees	seed	1-little	
Celery	*Apium graveolens*	Honey bees, solitary bees, flies	seed		

Commercial Crops That Benefit from (or Require) Insect Pollination *(continued)*

Common name	Latin name	Pollinator	Commercial product of pollination	Pollinator impact	Number of honey bee hives per acre
Cherry, sour	*Prunus cerasus*	Honey bees, bumble bees, solitary bees, flies	fruit	3-great	
Cherry, sweet	*Prunus avium* spp.	Honey bees, bumble bees, solitary bees, flies	fruit	3-great	
Chestnut	*Castanea sativa*	Honey bees, solitary bees	nut	2-modest	
Chili pepper, red pepper, bell pepper, green pepper	*Capsicum annuum, Capsicum frutescens*	Honey bees, stingless bees (*Melipona* spp.), bumble bees, solitary bees, hover flies	fruit	1-little (pollinators important in greenhouses to increase fruit weight, but less in open fields)	
Chinese cabbage	*Brassica rapa*	Honey bees, solitary bees	seed	1-little	
Clover (not all species)	*Trifolium* spp.	Honey bees, bumble bees, solitary bees	seed		1
Clover, alsike	*Trifolium hybridum* L.	Honey bees, bumble bees, solitary bees	seed		1
Clover, arrowleaf	*Trifolium vesiculosum* Savi	Honey bees, bumble bees, solitary bees	seed		1
Clover, crimson	*Trifolium incarnatum*	Honey bees, bumble bees, solitary bees	seed		1
Clover, red	*Trifolium pratense*	Honey bees, bumble bees, solitary bees	seed		1
Clover, white	*Trifolium alba*	Honey bees, bumble bees, solitary bees	seed		1
Coconut	*Cocos nucifera*	Honey bees, stingless bees	nut	2-modest	
Coffea spp. *Coffea arabica, Coffea canephora*	*Coffea* spp.	Honey bees, stingless bees, solitary bees	fruit	2-modest	
Coriander	*Coriandrum sativum*	Honey bees, solitary bees	seed	3-great	
Cotton	*Gossypium* spp.	Honey bees, bumble bees, solitary bees	seed, fiber	2-modest	
Cowpea, blackeyed pea, blackeye bean	*Vigna unguiculata*	Honey bees, bumble bees, solitary bees	seed	1-little	
Cranberry	*Vaccinium oxycoccus, Vaccinium macrocarpon*	Bumble bees (*Bombus affinis*), solitary bees (*Megachile addenda*, alfalfa leafcutter bees), honey bees	fruit		3
Crown vetch	*Coronilla varia* L.	Honey bees, bumble bees, solitary bees	seed (increased yield from pollinators)		
Cucumber	*Cucumis sativus*	Honey bees, squash bees, bumble bees, leafcutter bee (in greenhouse pollination), solitary bees (for some parthenocarpicgynoecious green-house varieties pollination is detrimental to fruit quality)	fruit	3-great	1–2
Elderberry	*Sambucus nigra*	Honey bees, solitary bees, flies, longhorn beetles	fruit	2-modest	

Commercial Crops That Benefit from (or Require) Insect Pollination *(continued)*

Common name	Latin name	Pollinator	Commercial product of pollination	Pollinator impact	Number of honey bee hives per acre
Feijoa	*Feijoa sellowiana*	Honey bees, solitary bees	fruit	3-great	
Fennel	*Foeniculum vulgare*	Honey bees, solitary bees, flies	seed	3-great	
Flax	*Linum usitatissimum*	Honey bees, bumble bees, solitary bees	seed	1-little	
Grape	*Vitis* spp.	Honey bees, solitary bees, flies	fruit	0-no increase	
Guar bean, Goa bean	*Cyamopsis tetragonoloba*	Honey bees	seed	1-little	
Guava	*Psidium guajava*	Honey bees, stingless bees, bumble bees, solitary bees (*Lasioglossum* spp.)	fruit	2-modest	
Hog plum	*Spondias* spp.	Honey bees, stingless bees (*Melipona* spp.)	fruit	1-little	
Hyacinth bean	*Dolichos* spp.	Honey bees, solitary bees	seed	2-modest	
Jujube	*Zizyphus jujuba*	Honey bees, solitary bees, flies, beetles, wasps	fruit	2-modest	
Karite	*Vitellaria paradoxa*	Honey bees	nut	2-modest	
Kiwifruit	*Actinidia deliciosa*	Honey bees, bumble bees, solitary bees	fruit	4-essential	
Lemon	*Citrus limon*	Honey bees (also will often self-pollinate)	fruit		
Lima bean, kidney bean, haricot bean, adzuki bean, mungo bean, string bean, green bean	*Phaseolus* spp.	Honey bees, solitary bees	fruit, seed	1-little	
Lime	*Citrus limetta*	Honey bees (also will often self-pollinate)	fruit		
Longan	*Dimocarpus longan*	Honey bees, stingless bees		1-little	
Loquat	*Eriobotrya japonica*	Honey bees, bumble bees	fruit	3-great	
Lupine	*Lupinus angustifolius* L.	Honey bees, bumble bees, solitary bees	seed		
Lychee	*Litchi chinensis*	Honey bees, flies	fruit	1-little	
Macadamia	*Macadamia ternifolia*	Honey bees, stingless bees (*Trigona carbonaria*), solitary bees (*Homalictus* spp.), wasps, butterflies	nut	4-essential	
Mango	*Mangifera indica*	Honey bees, stingless bees, flies, ants, wasps	fruit	3-great	
Mustard	*Brassica alba, Brassica hirta, Brassica nigra*	Honey bees, solitary bees (*Osmia cornifrons, Osmia lignaria*)	seed	2-modest	
Okra	*Abelmoschus esculentus*	Honey bees (incl. *Apis cerana*), solitary bees (*Halictus* spp.)	fruit	2-modest	
Onion	*Allium cepa*	Honey bees, solitary bees, blowflies	seed	1-little	
Orange, grapefruit, tangelo	*Citrus* spp.	Honey bees, bumble bees	fruit	1-little	
Papaya	*Carica papaya*	Honey bees, thrips, large sphinx moths, moths, butterflies	fruit	1-little	
Peach, nectarine	*Prunus persica*	Honey bees, bumble bees, solitary bees, flies	fruit	3-great	1
Pear	*Pyrus communis*	Honey bees, bumble bees, solitary bees, hover flies (*Eristalis* spp.)	fruit	3-great	1
Persimmon	*Diospyros kaki, Diospyros virginiana*	Honey bees, bumble bees, solitary bees	fruit	1-little	
Pigeon pea, Cajan pea, Congo bean	*Cajanus cajan*	Honey bees, solitary bees (*Megachile* spp.), carpenter bees	seed	1-little	

Commercial Crops That Benefit from (or Require) Insect Pollination *(continued)*

Common name	Latin name	Pollinator	Commercial product of pollination	Pollinator impact	Number of honey bee hives per acre
Plum, greengage, mirabelle, sloe	*Prunus domestica, Prunus spinosa*	Honey bees, bumble bees, solitary bees, flies	fruit	3-great	1
Pomegranate	*Punica granatum*	Honey bees, solitary bees, beetles	fruit	2-modest	
Quince	*Cydonia oblonga* Mill.	Honey bees	fruit		
Rambutan	*Nephelium lappaceum*	Honey bees, stingless bees, flies	fruit	1-little	
Rapeseed	*Brassica napus*	Honey bees, solitary bees	seed	2-modest	
Raspberry	*Rubus idaeus*	Honey bees, bumble bees, solitary bees, hover flies (*Eristalis* spp.)	fruit	3-great	1
Rose hips, dog roses	*Rosa* spp.	Honey bees, bumble bees, carpenter bees, solitary bees, hover flies		3-great	
Rowanberry	*Sorbus aucuparia*	Honey bees, solitary bees, bumble bees, hover flies	fruit	4-essential	
Safflower	*Carthamus tinctorius*	Honey bees, solitary bees	seed	1-little	
Scarlet runner bean	*Phaseolus coccineus* L.	Bumble bees, honey bees, solitary bees, thrips	seed		
Service tree	*Sorbus domestica*	Bees, flies	fruit	2-modest	
Sesame	*Sesamum indicum*	Honey bees, solitary bees, wasps, flies	seed	2-modest	
Squash, pumpkin, gourd, marrow, zucchini	*Cucurbita* spp.	Honey bees, squash bees, bumble bees, solitary bees	fruit	4-essential	1
Strawberry	*Fragaria* spp.	Honey bees, stingless bees, bumble bees, solitary bees (*Halictus* spp.), hover flies	fruit	2-modest	1
Strawberry tree	*Arbutus unedo*	Honey bees, bumble bees	fruit	2-modest	
Sunflower	*Helianthus annuus*	Bumble bees, solitary bees, honey bees	seed	2-modest	1
Tamarind	*Tamarindus indica*	Honey bees (incl. *Apis dorsata*)	fruit	1-little	
Tangerine	*Citrus tangerina*	Honey bees, bumble bees	fruit	1-little	
Turnip, canola	*Brassica rapa*	Honey bees, solitary bees (*Andrena ilerda, Osmia cornifrons, Osmia lignaria, Halictus* spp.), flies	seed	3-great	1
Vetch	*Vicia* spp.	Honey bees, bumble bees, solitary bees	seed		
Watermelon	*Citrullus lanatus*	Honey bees, bumble bees, solitary bees	fruit	4-essential	1–3

Sources: K.S. Delaplane and D.F. Mayer. 2000. Crop pollination by bees. New York: CABI. S.E.; McGregor. 1976. Insect pollination of crops. Washington, DC: USDA-ARS; see "Pollination" in the References.

RECOMMENDATIONS FOR GROWERS

Many growers and orchardists have planted large blocks of crops that require migratory honey bee pollinators. A good rule of thumb is to place one or two colonies per acre of crop; and more would be better. During periods of inclement weather, the placement of bee colonies in large blocks of monoculture crops will help ensure the likelihood that adequate pollination will take place.

Honey bee colonies in the spring of the year have lost a considerable number of adult bees by way of attrition during the winter months. Growers rightfully expect that each colony brought to their farm will have an adequate bee population in order to carry out the work of pollinating their crops. For example, almond growers can inspect each colony that arrives in their orchard and have the right to reject colonies with a low adult population of bees. The number of adult bees per colony varies, but as a general rule, growers expect each colony to have four to eight frames covered with adult bees. The USDA estimates there were approximately 1.39 million acres of almonds in 2018, requiring over 1.9 million bee colonies to ensure adequate pollination of this crop. In order to grasp the importance of honey bees as pollinators, consider the fact that the California Department of Agriculture estimated that the almond orchards in 2017 required approximately 73 percent of all the bee colonies in the United States.

If growers are not diligent, bees can be killed by pesticides that are sprayed on the crops or on non-target plants and weeds, or that drifts over hives or into the water supply. Beekeepers and growers must understand each other's obligations while the colonies are on the growers' property. The beekeeper expects that the growers will avoid any spraying of insecticides, herbicides, and fungicides, when possible, while the bees are working their crops. Fungicides, for example, while labeled "safe for bees," have been shown to adversely affect the beneficial fungi in bee colonies, which are mostly responsible for converting pollen into bee bread. Additionally, many pesticides (as well as some used by beekeepers for disease or mite control) can be found in beeswax, which can last for many years.

There are many lists published on bee-toxic pesticides (see "Pesticides" in the References). It has been reported that in areas where many pesticides have been used, and where all the pollinators are dead, some plants must be pollinated by hand.

In many cases, native pollinators have been killed by destructive farming practices or pesticide use. Many non–honey bee pollinators are valuable to growers and need to be protected and cultivated. You, as beekeepers, can also raise alternate pollinators as a sideline (see "Non-*Apis* Bee Pollinators" in the References section, and the website for the USDA-ARS lab in Logan, Utah). These native bees may become increasingly important to future growers as the number of feral honey bee colonies is diminished by mite infestation.

Both non-*Apis* bees and honey bees require adequate amounts of pollen and nectar during flowering periods to enable them to continue brood rearing and, particularly in times of dearth, to keep them from depleting their stores. It may be necessary to move colonies to other areas with blooming flowers. This can more easily be accomplished if you have only a few colonies. Alternatively, if you have suffi-cient property or access to fallow land that you can use, plant or sow seeds of flowering plants that bees can forage on (see list of common bee plants).

LEASING BEES

Many beekeepers lease their hives to fruit and vegetable growers whose crops benefit from or require bees for pollination. The need for bee pollination is increasing, in part because of declining bee populations (especially feral or wild honey bee nests) caused by urbanization of natural foraging land, pesticide use, mites, and pollution. Some factors to consider when leasing or renting bees are:

- Number of hives: If other factors are favorable, count on one colony per acre of fruit crops, more for other crops.
- Weather: Optimum flying conditions for bees include temperatures between 60° and 90°F (15.6–32.2°C), winds of less than 15 mph (24 km/hr), and fair, sunny days.
- Colony strength: Each colony should have at least four frames of brood and bees, and a laying queen. Each of these frames must be covered with adult bees.
- Timing: Set out bees just as the crop comes into bloom (about 10% bloom); if set out too early, bees may work other blooming plants and may not switch to the target crop.
- Leasing fees: Although there is no flat fee for leasing bees, some factors that may affect the price include time of year; pesticide hazard; loss of queen, bees, and/or honey; and the difficulty of getting into and out of the field.

Some beekeepers remove frames of bee bread from colonies to stimulate bees to collect more pollen. Others use pollen traps for the same reason. The use of pollen traps reduces the amount of pollen that bees can collect to keep pollen stores at adequate levels; as a consequence of this restriction, the bees recognize that their stores are deficient and send out more pollen foragers. You can also replace queens with queens whose genetics produce worker bees that are excessive pollen gatherers.

All these techniques require close attention to the condition of the bees and the crop, as well as the weather.

Poisonous Plants

Name	Toxic part
Abies alba, silver fir	Honeydew reported
Aconitum spp., monkshood[a]	Honey?/Pollen
Aesculus californica, California buckeye	Nectar/Honey/Pollen
Andromeda spp., andromeda	Honey/nectar?
Arbutus unedo, strawberry tree	Nectar
Astragalus spp., locoweed, tragacanth	Nectar
A. miser v. *serotinus*, timber milk vetch	Nectar
Caltha vulgaris, march marigold	Pollen occasionally
Camellia reticulate, netvein camellia	Nectar
Coriaria arborea, New Zealand tutu[a]	Honeydew from passion vine hopper *Scolypupa australis* feeding on tutu
C. japonica[a]	Honeydew
Corynocarpus laevigatus, New Zealand laurel or karaka, summer titi	Nectar
Cuscuta spp., dodder	Nectar
Cyrilla racemiflora, southern leatherwood, titi	Nectar
Datura stramonium, jimsonweed[a]	Honey toxic to human
D. metel, Egyptian henbane	Honey
Delphinium consoida, forking larkspur	Plant toxic to bees, mammals
Digitalis purpurea, foxglove[a]	Pollen?
Euphorbia sequeirana, spurge	Honey/pollen
E. marginata, snow on the mountain	Plant poison to bees
Gelsemium sempervirens, yellow jessamine[a]	Nectar/pollen?
Helenium hoopwaii, orange sneezewood	Toxic to mammals/bees
Hyoscyamus niger, black henbane	Nectar/pollen, plant toxic to bees
Kalmia latifolia, mountain laurel[a]	Nectar/plant toxic to mammals
Ledum palustre, wild rosemary	Honey/pollen?
Macadamia integrifolia, macadamia	Cyanide gas from bloom
Nerium oleander, oleander[a]	Honey
Papaver somniferum, opium poppy	Pollen
Ranunculus spp., buttercup	Nectar/pollen?
Rhododendron spp., azalea, and *R. anthopogon, R. lutea, R. occidentalis, R. ponticum, R. prattii, R. thomsonii*[a]	Nectar/pollen
Sapindus spp., soapberry	Nectar
Senecio jacobaea, tansy ragwort	Nectar bitter, plant toxic to mammals
Solanum nigrum, black nightshade	Toxic plant
Sophora microphylla, yellow kowhai	Nectar
Stachys arvensis, nettle betony, staggerweed	Nectar or yeasts in nectar
Taxus spp., yew	Pollen
Tilia spp., basswood	Occasional nectar toxicity under drought conditions
Tulipa spp., tulip	Nectar, occasional toxicity
Tripetaleia paniculata, an azalea	Nectar
Tulipa spp., tulips	Stigmatic nectar?
Veratrum spp., Western false hellebore	Nectar
Zygadensis (= *Zygadenus=Zigadenus*) *venenosus*, death camas	Nectar/pollen

Sources: J. Skinner. 1997. In: Honey bee pests, predators and diseases, 3rd edn. Medina, OH: A.I. Root. J. Ramsay. 1987. Plants for beekeeping in Canada and northern USA. Cardiff, UK: IBRA.
[a] Plant parts also toxic to humans and animals.

Common Bee Plants

Common name	Latin name	Common name	Latin name
Acacias, esp. catclaw acacia	*Acacia* spp., *A. greggii*	Common buckthorn	*Rhamnus cathartica*
Alfalfa	*Medicago sativa*	Common hackberry	*Celtis occidentalis*
Almond, apricot	*Prunus dulcis, A. armeniaca*	Common vetch	*Vicia sativa*
Alsike clover	*Trifolium hybridum*	Cotton	*Gossypium* spp. Extrafloral nectaries
American holly	*Ilex opaca*		
Anise hyssop	*Agastache foeniculum*	Crab apple	*Malus sylvestris*
Appalachian Mountain mint	*Pycnanthemum flexuosum*	Cranberry	*Vaccinium macrocarpon*
Apple	*Malus domestica*	Creeping thistle	*Cirsium arvense*
Asparagus	*Asparagus officinalis*	Creosote bush	*Larrea tridenta*
Aster spp.	Asteraceae family	Crimson clover	*Trifolium incarnatum*
Basswood, American	*Tilia americana* and other *Tilia* spp.	Currant	*Ribes* spp.
		Dandelion	*Taraxicum officinale*
Bee bee tree	*Evodia danelli* (=*Tetradium*)	Dead nettle	*Lamium* spp.
Birdsfoot trefoil	*Lotus corniculatus*	Devil's walking stick	*Aralia spinosa*
Black cherry	*Prunus serotina*	Dogbane	*Apocynum* spp.
Black chokeberry	*Aronia melanocarpa*	Dogwood	*Cornus* spp.
Black gum or tupelo	*Nyssa* spp.	Elm	*Ulmus* spp.
Black locust	*Robinia pseudoacacia*	Eucalyptus	*Eucalyptus* spp.
Blackberry	*Rubus* spp.	Fairy duster	*Calliandra eriophylla*
Blackhaw	*Viburnum prunifolium*	Figwort	*Scrophularia californica* and other species
Bladderpod	*Lesquerella gordonii*		
Blue curls	*Trichostema* spp.	Fireweed	*Epilobium angustifolium*
Blue thistle, viper's bugloss, blue weed	*Echium vulgare*	Gallberry	*Ilex glabra*
		Garlic chives	*Allium tuberosa*
Blue vervain	*Verbena hastata* L.	Germander, thyme	*Teucrium canadense*
Blue vine	*Cynanchum leave* (*Gonolobus laevis*)	Globe thistle	*Echinops ritro*
		Golden honey plant, wingstem	*Verbesina alternifolia* (=*Actinomeris*)
Blueberry	*Vaccinium* spp.		
Bonset	*Eupatorium* spp.	Goldenrod	*Solidago* spp.
Borage	*Borago officinalis*	Hawthorn	*Crataegus* spp.
Brazilian pepper tree	*Schinus terebinthfolius*	Heath	*Erica* spp.
Buckeye or horsechestnut	*Aesculus* spp.	Heathers	*Calluna vulgaris*
Buckthorn	*Rhamnus* spp.	Henbit deadnettle	*Lamium* sp.
Buckwheat	*Fagopyrum esculentum*	Holly	*Ilex* and various species
Butterfly weed	*Asclepias tuberosa*	Honey locust	*Gleditsia triancanthos*
Buttonbush	*Cephalanthus occidentalis*	Honeysuckle	*Lonicera* spp.
Cactus (giant saguaro)	*Carnegiea gigantea*	Horehound	*Marrubium vulgare*
Cajeput, tea tree	*Melaleuca quinquenervia*	Horsemints	*Monarda* spp.
Camphorweed	*Heterotheca subaxillaris*	Joe-Pye weed (=*Eupatorium*)	*Eutrochium* spp.
Canola, rapeseed	*Brassica rapa* and other *Brassica* spp.		
		Knapweed	*Centaurea* spp.
Catalpa	*Catalpa speciosa*	Knotweed or smartweed	*Polygonum* spp.
Catnip	*Nepeta mussinii*	Korean evodia, bee bee tree	*Tetradium daniellii* (=*Evodia*)
Chaste tree	*Vitex agnus-casteum*		
Cherry	*Prunus cerasus*	Lavender	*Lavandula angustifolia*
Chickweed	*Stellaria media*	Leadwort	*Amorpha fruticosa*
Chives	*Allium schoenoprasum*	Leopardsbane	*Doronicum cordatum*
Citrus (orange, lemon, etc.)	*Citrus* spp.	Lespedeza	*Lespedeza* spp.
Cleome or spider plant	*Cleome* spp.	Lima bean	*Phaseolus limensis*
Clethra summersweet	*Clethra alnifolia*	Linden, basswood, lime tree	*Tilia* spp.
Clover, sweet and white Dutch	*Melilotus* spp. and *Trifolium* spp.	Loosestrife, purple	*Lythrum salicaria*
		Lungwort	*Pulmonaria* spp.

Common Bee Plants *(continued)*

Common name	Latin name	Common name	Latin name
Mangrove, black	*Avicennia nitida, A. germinans*, and other spp.	Saw palmetto	*Serenoa repens* and other *Serenoa* spp.
Manzanita	*Arctostaphylos* spp.	Selfheal	*Prunella vulgaris*
Maple	*Acer* spp.	Shad, shadblow or	*Amelanchier arborea* and
Marigold	*Calendula officinalis*	serviceberry	other spp.
Mesquite	*Prosopis* spp.	Smartweed	*Polygonum* spp.
Milk vetch	*Astragalus* spp.	Snakeweed	*Gutierrezia* spp.
Milkweed, common, swamp, whorled	*Asclepias* spp.	Sourwood	*Oxydendrum arboreum*
		Soybean	*Glycine max*
Mints	*Mentha* spp.	Spanish needles, beggar's ticks	*Bidens* spp.
Mountain bluet	*Centaurea montana* (Knapweed)	Speedwell	*Veronica spicata*
		Spurge	*Euphorbia* spp.
Mountain mint	*Pycnanthemum flexuosum*	Star thistle	*Centaurea* spp.
Mustard	*Brassica arvenisi* (L.) and others	Sumac; African sumac	*Rhus* spp., *R. lancia*
		Sunflower	*Helianthus annuus* and other spp.
N.J. tea	*Ceanothus* spp.		
Ohio buckeye	*Aesculus glabra*	Sweet autumn clematis	*Clematis terniflora*
Oilseed rape (canola)	*Brassica napus* L., *Brassica rapa*	Sweet corn	*Zea mays* (pollen only)
		Tall ironweed	*Vernonia altissima*
Oregano	*Origanum vulgare*	Tallow tree	*Sapium sebiferum*
Palms	*Sabal* spp.	Thistles	*Cirsium* spp. and other genera
Palo verde	*Cercidium* spp.		
Plum	*Prunus*	Thyme	*Thymus* spp.
Privet	*Ligustrum* spp.	Tufted vetch	*Vicia cracca* and other *Vicia* spp.
Pussy willow	*Salix discolor*		
Rabbitbrush	*Chrysothamnus* spp.	Tulip-tree or tulip poplar	*Liriodendron tulipifera*
Raspberry	*Rubus* spp.	Tupelo	*Nyssa* spp.
Red chokeberry	*Aronia arbutifolia, Photinia pyrifolia*	Vetch	*Vicia* spp.
		Virgin's bower	*Clematis virginiana*
Red clover	*Trifolium pratense*	White clover, lawn clover	*Trifolium repens*
Red maple	*Acer rubrum*	White sweet clover	*Melilotus alba*
Redbud	*Cercis canadensis*	Wild buckwheat	*Eriogonum* spp.
Red-flowering thyme	*Thymus praecox*	Wild carrot	*Daucus carota*
Russian olive, autumn olive	*Eleagnus angustifolia, E. umbellata*	Willow	*Salix* spp.
		Woundwort	*Stachys byzantina*
Russian sage	*Perovskia atriplicifolia*	Yellow sweet clover	*Melilotus officinalis*
Sage	*Salvia* spp.	Yellow rocket	Cruciferae family (Mustards)
Salt cedar	*Tamarix* spp.		

Source: J.M. Graham (ed.). 1992. Hive and the honey bee. Hamilton, IL: Dadant and Sons.

Notes: Some species of honey plants can be invasive weeds. Check your state agriculture department and websites. Some good websites are http://ag.arizona.edu/pima/gardening/aridplants/; http://www.aces.edu/department/ipm/npt1 .htm, http://www.aces.edu/department/ipm/npt2.htm; http://uvalde.tamu.edu/herbarium//index.html; http://www .vftn.org/projects/bryant/navbar_pages/texas_regions.htm; and http://MAAREC.cas.psu.edu.

Also see References under "Honey Plants."

POLLINATION CONTRACTS

To be fair, the beekeeper and the grower should draw up a pollination contract or agreement. This document will help prevent misunderstandings while at the same time detailing the expectations of all parties. Key points should include:

- Location of crop.
- Date of placement of bees into the crop and their removal (relative to bloom time and condition).
- Number and strength of colonies.
- Pattern of colony placement (on pallets or spread throughout the orchard).
- Rental fee and date on which it is paid.
- Agreement by the grower to not apply bee-toxic pesticides while bees are pollinating their crop, or give the beekeeper a 48-hour heads-up before any application that may be toxic to bees.
- Agreement by grower to warn beekeeper of other spraying in the area.
- Agreement by grower to reimburse beekeeper for any additional movement of colonies in, out of, or around the crop.
- Agreement by grower to provide right-of-entry to beekeeper for management of bee colonies.

Search online for sample contracts; see, for example, https://beeaware.org.au/ pollination/pollination -agreements/.

PROBLEM PLANTS

Some plants are harmful, even poisonous to bees. For example, the sundew (Droseraceae), Venus fly-trap (*Dionaea musipula*), and pitcher plants (Sarraceniaceae) are insect-eating plants that attract their victims by secreting a sweet sap, odor, or both. These plants grow in wet areas and are not usually attractive to bees; the number of bees lost to them is minimal.

Plants like *Rhododendron* spp. and laurel (*Kalmia*) are said to produce a toxic nectar that is ultimately poisonous to insects that retrieve it. If some of this nectar is converted to honey back at the colony, this honey may also be toxic to humans. (While rare, if your apiary is near an area where these plants are growing, you may want to get suspect honey tested or be alert for bee kills.) Certain environmental conditions, such as abnormally cold or dry weather, may cause otherwise nontoxic plants, such as linden trees (*Tilia* spp.) to yield toxic nectar or pollen. In some instances, wild bees, such as bumble bees, can collect these substances that are toxic to honey bees without adverse effects. The list in the sidebar on p. 296 summarizes information on toxic plants.

HONEY PLANTS

If you want to secure a good honey crop, you need to have your apiaries adjacent to or nearby some good bee forage. Although you cannot control the weather, you can look over an area before locating an apiary, to see if there will be ample forage during most of the growing season. In general, industrial and residential land, or agricultural land that has tree farms (such as Christmas trees) will be areas where bees may struggle to survive; check online maps of your location to see what the land is used for.

There are always exceptions, of course; for instance. New York City would be a cement desert but for the fact that a park of almost one thousand acres (Central Park) is located there. Many New York City beekeepers keep their colonies on rooftops; check sites online for more information on urban honey plants.

Bees located in such areas may forage for any form of liquid containing sugar, especially during dearths; these liquids may contain contaminants, such as pesticides, as well as industrial and/or commercial waste products. For example, if you see honey of a strange color (or flavor), you may want to investigate further for a nearby factory or waste site (e.g., bright orange honey was seen in a colony located near a soda factory that was throwing sugary orange waste in a landfill).

Lists of honey plants are available from local bee groups, newsletters, books, and on websites, so please look them up. Major honey plants are those that will give a surplus crop of honey, while minor honey plants provide the day-to-day source of food used for brood rearing or for building up a colony coming out of winter or during a dearth period. Remember that soil type, moisture, sunlight, temperature, and humidity will affect the amount of food (nectar and pollen) plants produce.

In some instances, you can sow seeds of crops (like buckwheat or sunflowers) or locate your apiary near honey-yielding plants (including trees) if you have enough land. With the current interest in the decline in pollinators, planting a "pollinator garden" in your area may be a good way to change fallow or other unused land into a field that will be inviting to many pollinators. Check the list of honey plants in the sidebar and "Pollination" in the References section of this book, as well as other sources online, for honey plants in your area.

 **Notes**

Anatomy of Honey Bees

The anatomy of the honey bee is similar to that of other insects except for the specialization of certain organs and structures needed by bees to carry out functions peculiar to them. Parts common to other insects include the three basic insect parts—head, thorax, and abdomen; the exoskeleton (the hard, waxy protein [chitin] covering); the free respiratory system (no lungs); the ventral or bottom spinal cord; and the free circulatory system.

EXTERNAL ANATOMY OF DRONE

Drones are larger than both the queen and the workers and can be recognized by their large eyes, which meet at the top of the head, and the blunt abdomen. The reproductive organs are enclosed within the body and remain inside the body until the mature drone flies to mate with a virgin queen. Drones mature after emerging from the cell in about 12 days and can be seen flying back to the colony in the afternoon after orientation flights. During the copulatory act, the organs are everted, except for the testes (which remain inside); after coupling, the entire organ will break off, and the drone dies. Queen breeders who instrumentally inseminate queens collect semen from mature drones by squeezing the thorax, which causes a drone to evert its sex organs.

External Anatomy of the Drone Honey Bee

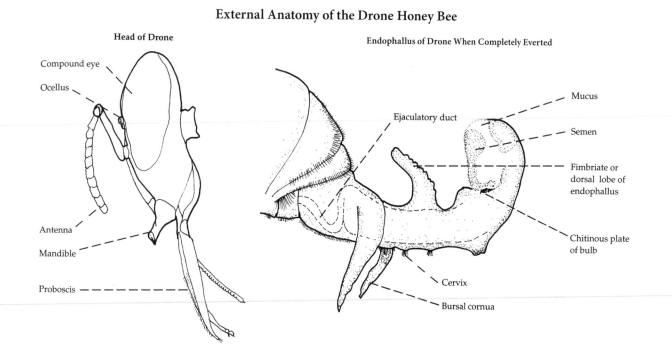

Head of Drone

Compound eye
Ocellus
Antenna
Mandible
Proboscis

Endophallus of Drone When Completely Everted

Ejaculatory duct
Mucus
Semen
Fimbriate or dorsal lobe of endophallus
Chitinous plate of bulb
Cervix
Bursal cornua

What Makes Bees Exceptional Pollinators

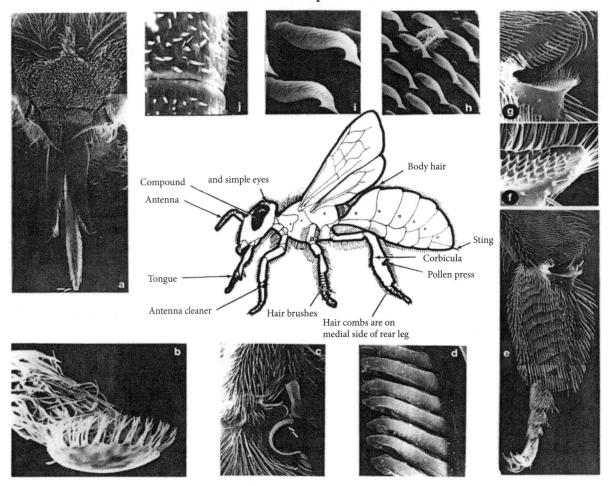

WORKER: WHAT MAKES BEES EXCEPTIONAL POLLINATORS

On page 12, the diagram showing the worker anatomy gives the names of the parts of a worker honey bee. The diagram here includes the parts of the worker blown up using a scanning electron microscope (SEM) to give the fascinating details of how this pollinating machine is put together.
Images include:

a. The head of a honey bee, showing the mouthparts.
b. The tip of the proboscis which has sensory hairs to help guide the proboscis to the nectary as well as to provide information on the food being ingested.
c. Close-up of antenna cleaner, showing modified hairs that act as a comb. The antennae are important sensory organs of the bee and must be kept clean and free of debris.
d. Close-up of the modified hair comb of antenna cleaner in *c*.

e. Medial view of the hair combs on the hind leg; the rake apparatus on the opposite hind leg removes the pollen from the hair combs and transfers it to the press, also on the hind leg. From there the pollen is transferred to the corbicula (pollen basket).
f. Close-up of base of pollen press.
g. Pollen press showing modified hairs to collect pollen grains.
h. Close-up of modified hairs on hind leg of bee with a captured pollen grain from a sunflower.
i. Modified hair in *h*, at higher magnification; the peg-like hair aid in collecting pollen grains.
j. Close-up of antenna showing the sensory hairs and pegs that enable the bee to smell.

For more information and photos, see Erickson et al. 1986 under "Bees, Beekeeping, and Bee Management" in the References. Photos by D. Sammataro and M. Garment.

Morphological Differences between Honey Bee Castes and Drones

Item	Queen	Drone	Worker
Larval diet	Royal jelly	Drone jelly?	Worker jelly
Cell orientation	Vertical	Horizontal	Horizontal
Development time	16 days	24 days	21 days
Eyes	3920–4920 facets	13,000 facets	400–6300 facets
Brain	Small	Large	Large
Mandibular gland products[a]	9-Oxodec-*trans*-2-enoic acid[b]	Absent	10-Hydroxydecenoic acid
	9-Hydroxydec-*trans*-2-enoic acid[c]		
Mandibles	Unmodified, cutting only		Small and notched, modified for comb building, flat center ridge
Tongue	Short	Short	Short
Honey stomach	Small	Small	Well developed
Wings	Appear reduced in length	Extend over abdomen	Extend over abdomen
Legs	Middle tibia = no spine	Middle tibia = no spine	Middle tibia = spine Hind leg = pollen- collecting apparatus
Wax glands	Absent	Absent	Present, 4 pairs
Nasonov gland	Absent	Absent	Present
Ovaries (n = ovarioles)	Well developed; n = 300	Absent	Undeveloped; n = 2–12
Spermatheca	Present	Absent	Rudimentary
Sting	Lightly barbed, curved, waxy	Absent	Strongly barbed, straight, no wax
Antennae	1600 chemoreceptors	?	2400 chemoreceptors
Reproductive role	Egg laying	Mating	Maternal care
Number per colony	1 or 2	100–2500	30,000–60,000
Food	Fed royal jelly	Begs/eats honey	Honey, pollen
Life span	1–8 years	Up to 60 days	Up to 4–6 months

[a] Brood food comes from the mandibular (white liquid) and the hypopharyngeal (clear) glands of young, nurse bees.
[b](9-ODA).
[c](9-HDA).

INTERNAL ANATOMY

Compare the internal anatomy of the queen and drone to that of the worker. The queen has larger ovaries than the worker, a sting, but no wax glands or pollen-carrying structures on her hind legs. See the figures showing the external anatomy and the table on morphological differences between honey bee castes and drones.

Sting Anatomy

The sting structure of bees is a modified egg-laying device, or *ovipositor*, developed into a defensive structure. Made up of two barbed lancets, these are supported by the quadrate plates and extensive musculature in the last abdominal segment (called VIIT, for the seventh tergite; see the illustration showing the sting anatomy of the worker on p. 305). When the bee stings, the two lancets move into the skin with a sawing action and anchor themselves by their barbs. Once implanted, the sting sac can pump venom into its vic-

Internal Digestive Organs of a Worker Honey Bee

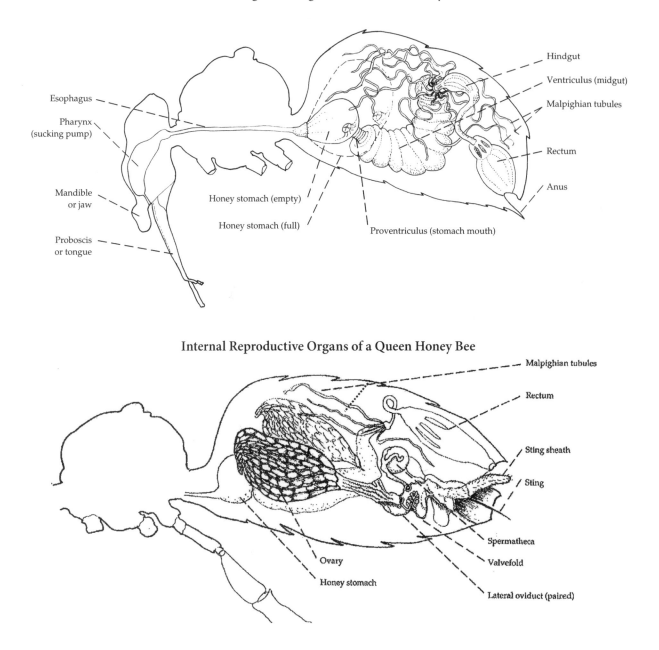

Internal Reproductive Organs of a Queen Honey Bee

tim unless it is scraped off. Queens usually reserve their stinging apparatus to kill rival queens, which, because queens' lancets are only slightly barbed, they can sting repeatedly. Worker bees' lancets are markedly barbed. Drones have no sting.

The poison gland secretes venom into an attached poison sac, which holds the venom until it is pumped through the sting. This gland was originally labeled the acid gland and is found as such in the older bee literature. Another smaller gland, the Dufour (or alkali) gland, also opens in the sting chamber and may secrete lubricants for the sting apparatus or for the eggs. It may also be involved in pheromone production.

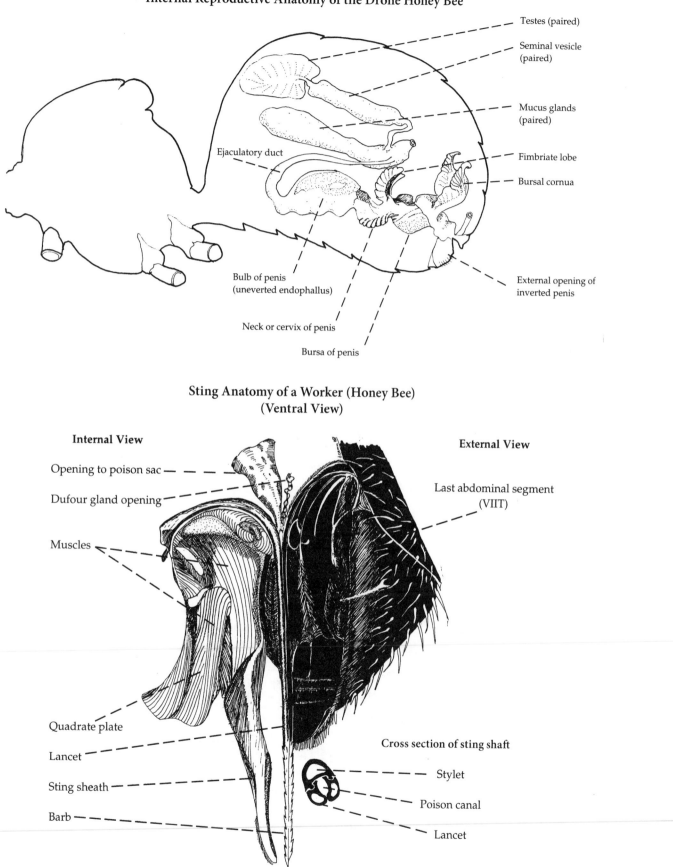

Internal Reproductive Anatomy of the Drone Honey Bee

Testes (paired)

Seminal vesicle (paired)

Mucus glands (paired)

Fimbriate lobe

Bursal cornua

Ejaculatory duct

External opening of inverted penis

Bulb of penis (uneverted endophallus)

Neck or cervix of penis

Bursa of penis

Sting Anatomy of a Worker (Honey Bee) (Ventral View)

Internal View

External View

Opening to poison sac

Dufour gland opening

Muscles

Last abdominal segment (VIIT)

Quadrate plate

Lancet

Sting sheath

Barb

Cross section of sting shaft

Stylet

Poison canal

Lancet

APPENDIX B

Pheromones

Honey bee behavior both inside and outside the hive is regulated to a large extent by chemical substances called pheromones. Pheromones are secreted by glands and trigger in other bees certain behavioral responses or physiological activities. Refer to the table on glands of honey bees for more information.

QUEEN PHEROMONES

Located in the mandibles of a queen's head are the mandibular glands, which produce and secrete the pheromones called queen substance, which include the chemicals *(E)*-9-hydroxydec-*trans*-2-enoic acid (10-HDA), (E)-9-oxo-2-decenoic acid (9-ODA), methyl *p*-hydroxybenzoate (HOB), and 4-hydroxy-3-methoxyphenylethanol diacetate (HVA). These compounds elicit various responses in worker and drone honey bees (see the table on the glands of honey bees). Inside the hive these substances have been shown to inhibit ovary development in workers and deter them from making new queens. Their absence evokes the opposite response: Queen cup construction (which leads to queen cells) is undertaken.

A swarm—either flying out of the hive to a new

Glands of Honey Bees

Gland location	Number	Substance secreted	Worker	Queen	Drone
Head					
Hypopharyngeal	2	Brood food and royal jelly, enzymes, glucose oxidase	X		
Salivary	2	Fatty substance, enzymes	X	X	X
Mandibular	2	Lipid component of larval food + 10-HDA	X		
		Alarm, alerts other workers (2-Heptanone)	X		
		Enzymes	X		
		Queen substance[a] contains over 20 pheromones including 9-HDA,[b] 9-ODA,[c] HOB,[d] and HDA + HVA		X	
		Congregating pheromone			X
Thorax					
Salivary	2	Saliva, enzymes (derived from larval silk gland)	X	X	X
Legs					
Arnhart, tarsal "gland"	?	Worker pheromone: attract workers to nest or to forage	X		
		Footprint + pheromone + HDA mandibular from queen inhibits queen cup construction		X	
		Drone pheromone: ?			X

Gland location	Number	Substance secreted	Worker	Queen	Drone
Abdomen					
Wax	4 × 2	Beeswax: nest construction	X		
Nasonov		Scent pheromone[e]: to attract other workers	X		
Abdominal tergites (Ab. Terg. 3–5)		Pheromone: recognition of queen by worker; stabilize queen retinue, inhibit worker ovary; attract drone to queen (precopulation)		X	
Rectal	6	Enzyme: catalase to break down starch	X	?	?
Spermathecal		Polysaccharide + proteins		X	
Drone accessory	2	Peptides + mucus			X
Sting apparatus		Venom	X	X	
Dufour		Egg coating, distinguishes eggs laid by queen and those laid by workers		X	
Kozhevnikov	2	Pheromone: queen attracts and recognized by workers		X	
Sting sheath	2	Alarm pheromone (acetates, isopentyl, etc.)	X		
Alkaline		Lubricant for sting?	?	X	
Sting shaft		Waxy esters? (lubricant?)	?	X	
Larvae and pupae					
Larval silk	2	Silk: spun into cocoon	X	X	X
?		Brood pheromones: incubation, attracts and recognized by workers	X	X	X
?		Inhibits queen rearing, workers' ovary development		X	
?		Stimulates foraging, especially for pollen	?	?	?
? Diploid drone larva		Cannibalism pheromone		?	

Sources: E. Crane. 1990. Bees and beekeeping: Science, practice, and world resources. Ithaca, NY: Comstock. M.L. Winston. 1987. Biology of the honey bee. Cambridge, MA: Harvard University Press.

Note: 2-Hep = 2-heptanone; 9-HDA = 9-hydroxy-2-enoic acid; 10-HDA = 10-hydroxy-2-decenoic acid; 9-ODA = *(E)*-9-oxodec-2-enoic acid; HOB = methyl *p*-hydroxybenzoate; HVA = 4-hydroxy-3-methoxy phenylethanol.

[a] There are many pheromones in queen mandibular glands, one or more of which (1) inhibit worker ovary development, (2) stimulate workers to release Nasonov pheromone, (3) stimulate workers to forage, (4) regulate workers clustering in swarms, and (5) attract workers.

[b] 9-HDA inhibits queen cup construction, allows workers to recognize queen (virgin), and stabilizes a swarm.

[c] 9-ODA attracts drone to queen, inhibits queen rearing and queen recognition by workers, and suppresses ovary development in workers.

[d] HOB + 9-ODA + two isomers of 9-HDA + HVA induces attendant behavior of workers to the queen.

[e] Scent pheromones: citral, nerol, geraniol, (E,E)-farnesol; 9-HDA, 9-hydroxy-2-decenoic acid; HOB, methyl p-hydroxybenzoate; HVA, 4-hydroxy-3-methyoxyphenylethanol; 9-ODA, 9-oxo-2-decenoic acid.

homesite or settled in a cluster—is aware of its queen's presence by means of these and other substances. Queen substance also guides drones toward queens who are on mating flights. Researchers are still investigating other pheromones or other volatile compounds, and with the more sophisticated instrumentation, new compounds are being discovered, such as brood pheromones and volatiles from mites.

WORKER PHEROMONES

Several different chemical pheromones are produced by workers. Two of these are alarm pheromones. One identified alarm odor (isopentyl acetate or isoamyl acetate) is released from a membrane at the base of the sting. It smells like banana oil and stimulates bees to sting or fly at intruders. Other alarm com-

Glands of a Worker Honey Bee

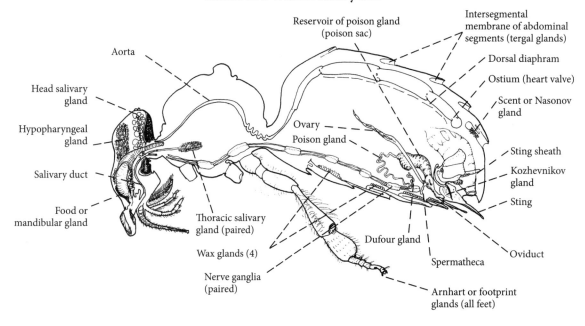

pounds from the sting have recently been isolated and are currently being studied. Another alarm odor (2-heptanone) is released by the mandibular glands of workers. Items anointed with this odor are attacked by bees; it currently is being investigated as a mite and wax moth deterrent.

A third pheromone is the scenting or orientation odor and is composed mainly of four chemicals (nerolic, geranic, and citrol acids, and geraniol). This chemical complex is produced by the Nasonov or scent gland near the dorsal tip of the abdomen. Dispersal of these substances is aided by the fanning action of bees secreting it. Upon smelling these chemicals, bees move toward the source (as in a swarm). Honey bees produce many other pheromones such as brood ester pheromone (BEP), which is emitted by developing larvae and may have internal and external components that alter a worker bee's physiology and behavior. Researchers are continuing to study drone and brood pheromones, and it is likely that ad-

ditional pheromones will be discovered along with their targeted function(s).

Honey bee pheromones are divided into two groups, the *primer* and the *releaser*. Releaser pheromones (short-term effect) trigger a rapid, immediate response from the receiving bees, usually outside the colony; primer pheromones (long-term effect) trigger or suppress developmental events inside the hive. The pheromones are liquids but may change into gas, solids, or remain liquid after secretion. The Nasonov gland produces a releaser pheromone; queen substance is both a primer and a releaser pheromone. In the latter case, the queen substance suppresses queen cup (and queen cell) construction, which is a long-term effect; but on her nuptial flight, the pheromone has a short-term effect on drones. Our knowledge of pheromones is increasing thanks to new technological tools; we can now identify and trace these chemicals within the bee colony as well as from individual bees.

Bee Sting Reaction Physiology

LOCAL REACTION

What happens in your body when you are stung by a bee? As in reaction to any invader, the body's natural defenses are called on to help. Basically, bee venom is a foreign protein (called an *antigen*) that stimulates the production of the body's defense proteins (called *antibodies*). Antibodies belong to a family of proteins known as gamma globulin, also called immunoglobulins. The bee sting antigens appear to stimulate a specific class of immunoglobulins known as immunoglobulin E, or IgE (see illustration on bee sting reaction physiology).

Because the bee venom antigen causes the production of specific antibodies, in this case IgE, people not otherwise exposed to honey bee proteins must be stung more than once before any type of reaction will occur. After the initial injection of venom, the body seems to "remember" that particular antigen and will likely react faster to subsequent stings, with further antibody production.

In a *local reaction*, the antigen of the bee venom appears to cause the production of the IgE bodies, which are attached to tissue cells (called *mast cells*). Mast cells contain numerous vesicles filled with histamine and other substances, which when released promote inflammation. The action of the antigen with the IgE/mast cell complex causes the histamine-filled vesicles to empty. Once released into the body, histamine has several effects, which include the expansion of blood vessels (vasodilation), the increased permeability of capillary cell walls to proteins and fluids, and the constriction of the respiratory passages. The first two may be responsible for the in-flammation, swelling, and itching associated with bee stings. Most beekeepers are reported to have this kind of local reaction. With repeated stings, the body may become immune to the venom, thereby reducing the localized swelling (except it always hurts).

If you are being stung repeatedly by bees, cover your eyes and mouth (carbon dioxide [CO_2] that you exhale attracts bees) and run as fast and as far as possible from the stinging insects (or go into a car or shed). The lethal dose reported by researchers is 5 to 10 bee stings per pound of body weight, which means an average male, weighing 160 pounds (72.6 kg), can receive more than 1500 stings and still live; the critical factor is that the kidneys may shut down, due to the overload of venom in the bloodstream.

SYSTEMIC REACTION

In a systemic reaction, the same mechanisms come into play as in the local reaction, with one big difference: the antigen/IgE/mast cell complex reaction can cause death. This allergic reaction, called hypersensitivity, appears to be a result of the large amounts of histamine released from the mast cells.

In some people, the second bee sting may be enough to be fatal. Because the body "remembers" the bee venom antigen, the subsequent inoculations usually cause a faster reaction, which means more histamine is released each time a person is stung. Usually, a systemic reaction builds up gradually, with the victim showing greater distress, such as difficulty breathing, after each stinging incident.

An antihistamine and adrenaline (epinephrine) should be immediately administered to people experiencing a systemic reaction, to counteract the effects of the released histamine and give relief for breathlessness. In these cases, medical advice must be sought immediately to treat allergies.

Desensitization or Immunity

People who develop hypersensitivity to bee stings can become desensitized to bee venom. Most beekeepers become immune to venom after repeated exposure. Desensitization can also be undertaken by an allergist. What an allergist does is to incrementally increase the amount of venom that the victim receives, thus allowing the body to form enough of these blocking antibodies to combat the allergic reaction. In either case, the immune processes or desensitization are probably the same; frequent injections of the specific venom induce the body to manufacture a "blocking" antibody, IgG. The IgG competes with the IgE in its reaction activities to bee venom antigens. Because the IgG antibodies are not fixed to the mast cells, but float freely, they are better able to combine with the venom antigens. Less histamine is therefore released, and the discomfort or allergic response is prevented.

Some beekeepers can, over long periods of time and exposure, become allergic to bee venom as well as beeswax, honey, propolis, bee debris, and bee bodies. The percentage of beekeepers who do become allergic to bee stings, although hard to assess because individual body chemistry, allergy history, exposure, and genetic predisposition are so varied, is small. Most people will probably already know if they are allergic, but those who do not can contact local allergy clinics or hospital outpatient facilities for testing. See "Venom" in the References for more information.

Bee Sting Reaction Physiology

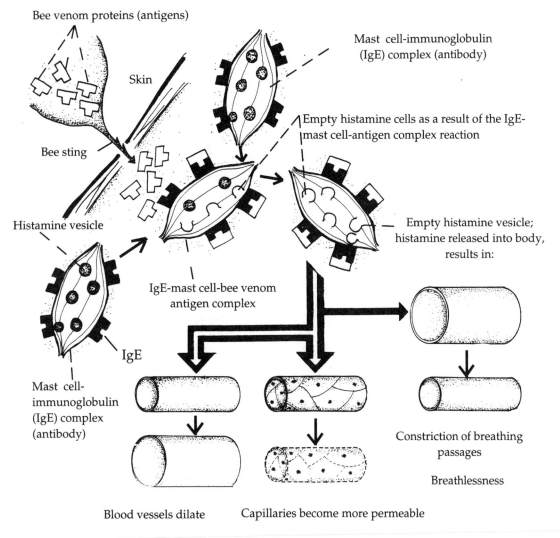

Bee venom proteins (antigens)

Mast cell-immunoglobulin (IgE) complex (antibody)

Skin

Bee sting

Empty histamine cells as a result of the IgE-mast cell-antigen complex reaction

Histamine vesicle

Empty histamine vesicle; histamine released into body, results in:

IgE-mast cell-bee venom antigen complex

IgE

Mast cell-immunoglobulin (IgE) complex (antibody)

Constriction of breathing passages

Breathlessness

Blood vessels dilate

Capillaries become more permeable

Swelling, inflammation

Paraffin Dipping

Today, some wood preservatives are contaminating the beeswax comb in bee colonies, and beekeepers are reluctant to use them to extend the life of wooden hive furniture. A better option to preserve hive woodenware is to dip it in boiling paraffin. This process is used in many countries and is very inexpensive if there is a lot of equipment to treat. It will clean up older boxes or can be used on new, fresh wood. It is also used to sterilize diseased equipment. Use a vat for dipping that has these features:

- Made of metal.
- An enclosure to provide a "double skin" around the vat.
- An electric heat source.
- Deep enough to hold two shallow supers or one deep, with some wiggle room.
- All joints welded from both sides to reduce the risk of bursting when in use; hot wax is dangerous.

You can dip: outer covers (without metal tops), supers and bottom boards, nucs.

Paraffin wax blocks can be purchased from commercial warehouses or online. The grade with a melting point of 140°F (60°C) is best. Heat the wax to 320°F (160°C); this can take up to two hours. Use a thermometer attached to a stick—**do not** rely on guesswork here; a thermostat would be better, one that can regulate the heat automatically. Some beekeepers add linseed oil and beeswax to the mixture.

Clean old equipment first by pressure washing or sanding to remove old paint and wax. If water is used, **make sure** wood is dry before immersing; otherwise you could cause the wax to boil over. Additionally, the woodenware must be **dry**, as dry as possible, as the wax will seal any water in the wood and could lead to rot from the inside out.

Immerse the boxes into the boiling wax for about 2 minutes, using a heavy weight to keep them below the surface. Diseased boxes must be treated for 10 to 15 minutes. Then lift the supers from the wax onto a rack or other place to drain. Boxes previously painted with oil-based paints will blister and need to be scraped.

Boxes can be used straight after dipping without being painted; paraffin wax is a good preservative on its own. You can still paint the boxes while they are hot, with water-based latex. Two coats are best; the paint should be pulled into the wood by the drying wax.

The secret is to immerse the boxes long enough in the hot wax so the wood is thoroughly heated and good penetration by the wax can be obtained. The boxes will be very hot when they come out of the dipper, so use thick fireproof gloves and tongs.

Finally, some commonsense safety precautions:

- Locate the wax dipper away from anything flammable; outdoors is best.
- Wear eye protection in case the wax splashes.
- Have a metal cover ready to put over the vat in case the wax boils over.
- Wear protective footwear.
- Wear an apron and face mask.
- Have on hand a first aid kit, for burns.
- Have available a fire extinguisher rated for oil/wax fires (dry chemical powder).

For more information, see "Paraffin Dipping" in the References.

Differences between European and Africanized Honey Bees

Differences in Physical, Behavioral, and Developmental Characteristics

	AHB	EHB
Physical characteristics		
Worker length	12.73 mm	13.89 mm
Cell size[a]	4.6–5.0 mm	5.2–5.4 mm
Wing beat[b]	290 Hz	243–270 Hz
Wing position	Wings elevated off abdomen	On abdomen
Fresh weight[c]	2 g	2.5 g
Nest size, volume	22 liters	45 liters
Comb area (1000 cm^2)	8–11	23.4
Honey area	0.9	2.8
Queen's egg laying		
Maximum/day	4000, 2–4 times brood area	2500
Total/year	105,000	58,000
Behavioral differences		
Swarming	3–16 times/year	3–6 times/year
Absconding	30% of colonies	Not common
Supersedure	> 40% in tropics	?
Pollen	Much more collected	Normal
Mating time	Late afternoon	Early afternoon
Defensiveness/stinging	Bounce off veil, sting in vicinity	Sting in apiary
Time before first sting	14 seconds	229 seconds
No. stings/minute	35	< 2
Pursuit distance	160 m	22 m
Recovery time	28 minutes	3 minutes
Entrance	Attack other bees at entrance	Not so prevalent
Drone rearing	Stimulated, more drones, mate more frequently with EHB queens	Suppressed
Returning foragers	Returning bees fly into entrance	Land on landing board
Foraging		
Unassisted spread/year	200–500 km	4 km

Differences in Physical, Behavioral, and Developmental Characteristics *(continued)*

Maximum distance to new nest site	75 km	5 km
	AHB	**EHB**
Running on comb	Excessive, runs off comb, festoons at lower edge; parades around inside box or inner cover	Normal, not usual to run much
Reaction to smoke	More bees leave colony, more bees eat honey	Fans to expel smoke, some eat honey
Cell caps	Thin, hard to uncap	Normal
Usurpation rates	20–30%/year in fall (Arizona)	
Requeening	Can be very difficult, if requeening with EHB about 40% loss	Normally 10% loss
Life expectancy	Shorter than EHB, 12–18 days as foragers	Longer than AHB
Developmental stage		
Egg to adult worker	18.5 days	21 days
Egg to adult queen	15 days	16 days
Worker's first flight	3 days old	10–14 days old
Queen's mating flight	5–6 days	7–10 days
Drone's mating flight	7.5 days	13 days
Brood production	Almost 2 times as much when resources good	Slower

Sources: D.M. Caron. 2001. Africanized honey bees in the Americas. Medina, OH: A.I. Root. E. Crane. 1990. Bees and beekeeping: Science, practice, and world resources. Ithaca, NY: Comstock. S.S. Schneider, G. DeGrandi-Hoffman, and D.R. Smith. 2004. The African honey bee: Factors contributing to a successful biological invasion. Annu. Rev. Entomol. 49:351–376.

[a] Cell size: measure 10 cells three times; if average ≥ 49 mm, then bee is AHB; if 52 mm, then bee is EHB.
[b] Wingbeat: measure free-flying, undisturbed bees.
[c] Fresh weight of 30 callow bees, average.

Weight Differences between All Developmental Stages

Worker bee stage	Age (days)	EHB weight (mg)	AHB weight (mg)
Egg[a]	1	0.1435	0.1064
	2	0.1196	0.0723
	3	0.0961	0.1122
Larva	4	0.1692	0.2218
	5	1.6493	1.1640
	6	9.8635	4.9440
	7	54.547	30.060
Capped over	8	109.10	93.16
Capped and spinning	9	160.60	113.66
Spinning	10	148.35	116.24
Prepupa (immobile)	11	142.74	113.48
White pupa	12	145.69	114.10
	13	144.48	104.88
Light pink compound eye	14	142.89	111.20
Dark pink compound eye	15	136.30	109.88
Dark pink-purple compound eye and ocelli	16	141.05	107.26
Dark purple compound eye, light brown wing bases	17	135.34	107.78

Weight Differences between All Developmental Stages *(continued)*

Worker bee stage	Age (days)	EHB weight (mg)	AHB weight (mg)
Dark purple compound eye, light brown head, and thorax	18	136.11	101.38
Black compound eyes, light gray wing pad, dark brown head, and thorax	19	133.12	92.34 Worker emerges
Black antennae, medium gray wing pads	20	122.37	
Imago/emerging	21	116.11	

Source: D. Sammataro and J. Finley. 2007. The developmental changes of immature African and four lines of European honey bee workers. Apiacta 42:64–72.

[a] Egg dimensions: eggs are 1.3–1.8 mm long and take 48–144 hours to hatch, the average being 72 hours.

Rearing Wax Moths (*Galleria mellonela*)

Wax moth larvae, while a pest in the beeyard, can be a lucrative sideline operation, sold as bait for fishermen or as pet food for exotics, such as turtles, lizards, snakes, and even birds. Look around for potential markets in your area.

DIETS FOR REARING WAX MOTH LARVAE

There are two basic diets you can use. Here they are:

1. Boil together for several minutes ⅓ cup sugar (or honey), ⅓ cup glycerol, ⅓ cup water. Cool, and then add ¼ teaspoon of a vitamin supplement, such as you would use to feed to your chickens. (Do **not** add antibiotics.) Then stir in 5 cups dried cereal, such as a mixed-grain baby cereal. Mix together and place wax moth eggs on top of the mixture in a glass jar.
2. A better recipe is to warm 200 ml glycerin, 200 ml honey. Add to 2 boxes complete or fortified baby cereal. Mix together and store in the refrigerator until used (or add wax moth eggs right away). This will be a dry mix.

BEGINNING YOUR CULTURE

Select a glass (or plastic) jar with a wide top. A peanut butter or 2-quart fruit or mason jar works fine. Fill container 2 inches (5 cm) deep with prepared diet. Cover the jar with a paper towel, a 20-mesh wire screen disc, and a jar lid with a large hole cut in the center (or you could use a mason jar rim). The paper towel is to keep foreign material from entering and the young wax moth larvae from leaving; the larger

larvae will eat through the paper if it is not protected by the wire.

Now add eggs, larvae, pupae, or adults to your culture. Whichever you add, they should all be the same stage (i.e., don't mix eggs and pupae). Add 10 to 50 larvae, or whatever stage you are starting it with, to each jar, depending on the size. Do **not** crowd too many larvae into one jar, as you will get smaller worms. It is best to start with eggs or adults that will lay eggs, since field-collected larvae may have parasites.

Now place your culture jars in an enclosed room, **out** of the sun, where it is warm, around 80° to 93°F (27–34°C). The larvae can tolerate a range of temperatures, from 77° to 99°F (25–37°C). But remember, the warmer the room, the faster they will grow; and the opposite is also true. Date the jar on the outside when you started the culture, so you can time when the wax moth larvae will mature.

HARVESTING AND INCREASING

If you plan on harvesting larvae to eat, collect them as they begin to spin their cocoons, since they void all fecal matter just before spinning. Try not to disturb the growing medium (diet) when collecting them. If you wish to hold the larvae for a long period, you can store them at 59°F (15°C) and 60 percent humidity for up to a year. This will keep them from pupating.

To increase your colony, you will need to rear some larvae to adult moths. Either place some adults, or around 50 cocoons, in a separate jar (you need no food in these jars, since adults do not feed). Now add a piece of paper, folded like an accordion, in the jar to give the female moths a place to lay eggs. Cover the jar with paper towels tied in place, and keep the moths at room temperature.

By putting in 50 cocoons, you will get a good mix of males and females. Within 10 days, the adult moths will emerge. Each female moth will mate in the jar and begin to lay eggs in the folds of the paper. Each female can lay 300 to 600 eggs over a period of five days.

Once you see the eggs (laid in strips, they are round and slightly tan in color), place a 2-inch strip of eggs in separate jars on top of fresh diet. Don't wait too long to transfer the eggs or the larvae could crawl out and starve to death (they are very tiny and **very** mobile at this stage).

BAIT WORMS

If you are using your worms just for bait, harvest them before they spin their cocoons. Put them in cold storage for four or five hours at 32° to 33°F (0–0.6°C). To ensure an even temperature of 32°F, put ice cubes and some water in a large bowl. Now put your larvae in a smaller jar and place the jar in the ice bath. Keep them in the bath until all the ice has melted.

After that they will not spin, and you can store them at 40° to 45°F (4.4–7.2°C) until ready to sell as bait. Place the larvae in a separate container and keep them moist with paper toweling, wood shavings, or other absorbent material.

HELPFUL TIPS

If your worms start dying suddenly, are attacked by fungi, virus, or change to a dark color, you could have a disease problem. Burn that batch and any other batch that may have been contaminated by it, wash the jars in hot, soapy water, and pour boiling water in them. Then start with a fresh batch.

TASTY INSECT TREATS

It has been suggested that the growing larvae can be delicately flavored by incorporating small amounts of various spices into their diet. Wax worms are great as snack items. Fried in hot oil, they pop like popcorn, and if lightly salted are reported to have good or better flavor than potato chips or corn puffs.

REFERENCES

See "Wax Moth Rearing" in the References.

Pointers for Extreme Urban Beekeeping
by James Fischer (NYC) and Toni Burnham (DC)

1. Urban "Location" Measured in Blocks, Not Miles: Despite the well-known "three-mile radius" of a bee colony's immediate foraging area, distances seem more daunting to urban bees. New packages established on rooftops and installed on foundation within what should be "easy flying range" of large areas of parkland (such as Central Park in Manhattan and Prospect Park in Brooklyn) had trouble drawing out even two mediums of foundation from mid-April to late July and required extensive feeding. Hives closer to major parks did much better. (But, per a 2015 study by David Tarpy in PLOS One, the researchers observed that urban colonies in smaller NC cities share patches of forage and may travel longer distances around their hives to find it: "poor habitat quality within this radius would impose additional stressful, above-average flight distances.")

2. Urban Winds Are Not the Beekeeper's Friend: Rectilinear street grids tend to increase wind velocities between tall buildings. This effect is even more pronounced along rivers. The result is that some locations have winds that make it difficult for bees returning to the hive. Worker bees weigh only a tenth of a gram, so careful site evaluation via regular weighing of hives, or using a recording anemometer and a bee counter might be required. Bees from hives above 15 stories might not be able to make it back, due to wind turbulence along the sides of the buildings. A good resource for data is https://www.wunderground.com (click "find nearest station" for a station near your hives).

3. Rooftop Hives Are Often Warmed by the Rooftops over Winter: Heat rises, and many roofs are black, so hives on older roofs may be kept unexpectedly warm, fail to cluster, and eat themselves into starvation when other hives have clustered and enjoy reduced consumption of stores. Putting one's hives on bricks, cinder blocks, or a hive stand is a very good idea. Wrapping that stand with pallet wrap is also a good idea, to create a dead-air space below the hive, and to block the wind.

4. Mouse Guards Are Not Just for Country Bees: While mice are rare in cities, and rats tend to be far too large to fit into a hive entrance, there are many species of roaches attracted to a hive. For example, the Bronx Zoo is famous for having a significant number of species of roaches who live well off bird seed and other animal feed, living on bottom boards and above inner covers in the Bronx Zoo hives. Blocking roach access to the hives was solved with a 4-inch-wide strip of ¼-inch hardware cloth or mesh, cut to the width of the hive entrance. Fold the mesh into a V shape and push it into the entrance with a hive tool; this should stay in place by friction alone.

5. Urban Swarming and Supersedure Can Result in Unmated Queens: Because there are a smaller number of honey bee colonies in urban areas and fewer feral colonies, hives that swarm or replace their queen by means of supersedure may result in unmated queens. In New York City, we have seen naturally mated queens only in the Greenpoint and the Red Hook neighborhoods of Brooklyn. No other area within city limits has yet produced its own mated queens. In Washington, DC, with a greater density of hives to the acre and a 13-story building height limit, queens mate successfully all

over town (including at the most industrial sites). Be sure to connect with your local beekeeping community to learn which particular challenges apply most to your urban location.

6. Trees Are the Primary Urban Nectar Sources: New York City has 5.2 million trees, nearly 600,000 of them lining streets, according to the Million Trees NYC organization. More than 44,509 acres of the city are shaded by trees, 24 percent of the total land area. This means that urban beekeepers should "Super Early, Harvest Often." Maples begin blooming in mid-February, and one cannot super early enough to exploit all the blooms. The question becomes: How early do the bees start flying? Regular hive weighing can detect the first nectar being brought in and prompt the beekeeper to get supers on. In Washington, DC, trees are also a primary nectar source, but beekeepers harvest only once, in midsummer, if they are able to harvest at all. In London, England, the urban tree canopy is not a reliable nectar source. Because of the small hive beetle, supering in advance or after the nectar flow can threaten the long-term viability of a SHB-infested colony.

7. What the Neighbors Don't Know Won't Scare Them (or, Keep It Quiet): While beekeepers expect their neighbors to share a deep, abiding concern for the well-publicized problems facing bees, and to comprehend that one cannot have local food without also having local bees, many neighbors seem to expect locally grown, biodynamically managed, organically certified veggies to appear by magic. A generation has grown up without ever being stung, and they are now raising their own children, so there are many otherwise rational people who are absolutely certain that an insect sting is dangerous. Outreach can only go so far, and one cannot give away 300 bottles of honey to residents in each of the high-rises that surround one's hives. The astute urban beekeeper purchases paint in colors that match the most common brands of rooftop air-conditioning equipment to use on hives, and may consider wearing workmen's coveralls and a veil rather than the usual bee suit if beehives are exposed to view. If the coveralls say "Joe's Air Conditioning" in large letters on the back, so much the better. Subtlety is an art. And don't expect those law-citing and law-abiding neighbors **not** to stoop to vandalism or spraying insecticide even if beekeeping is legal in your community and your hives meet all regulations. Especially for ground-level hives, best practices in subtlety include "out of sight / out of mind" locations, access controls, and painting of hives to be unobtrusive (for example, the same color as commercially available plastic garden sheds).

8. That being said, swarm control is crucial to continued tolerance of beekeeping in urban areas. Swarms are still treated by police, fire, and emergency management as if they were unexploded bombs. The prudent beekeeper utilizes splits, swarm traps, and constant vigilance in spring to control swarming. Many beekeeping communities have an organized swarm response team: consider joining it or sharing information that there is a reliable, not scary response to swarms one call away. In Washington, DC, we also reach out proactively to every police district to alert them to our presence and availability to help.

9. Decontaminate Packages and Quarantine Nucs Entering Your City: With only a small number of feral hives in the larger parks prior to urban beekeeping becoming popular, much of New York City was essentially disease- and pest-free since at least 2006. Both the Hudson and East Rivers are too wide and too windy for bees to fly across, so we are keeping bees on small islands off the coast of North America. Other urban areas may have similar barriers in the form of a solid ring of auto dealers, big-box retailers, and other suburban sprawl surrounding the urban core. All packages purchased by the community of beekeepers are decontaminated by spraying with oxalic acid at pickup locations, and purchased nucs are treated with formic acid before being hived. The result has been a significant number of hives without detectable varroa for most of the year, a real pleasure. Sadly, some opportunists refuse to decontaminate the packages they sell, which has resulted in some areas having varroa infestations and higher *Nosema* counts. Outreach and education are ongoing in this regard, as one's bees can be no healthier than one's neighbors' bees. Although you may not be your brother's keeper, in cities, you are your neighbor's beekeeper, like it or not. This matters more in cities without natural barriers to block bee movement.

10. Welcome Back, My Friends, to the Class That Never Ends: Apologies to Emerson, Lake, and Palmer, but education is the best solution to most beekeeping problems. And there is **always** something you do not know.

11. Logistics Are a Pain in the "Beehind": When most urban beekeepers in NYC live in small apartments and do not own cars, simple things like assembling and painting woodenware, extracting honey, and storing honey supers in winter become daunting logistical challenges. Shared facilities and organized work days were the only possible way to help new beekeepers get established. Even mundane things like glassware must be planned, ordered, and provisioned in advance of need. When choosing an out-apiary, ask about on-site storage, availability of water, and off-hours access (some bee work must be done after sunset or before dawn, or on weekends/holidays). Join a club to get access to tools you will only need from time to time, like refractometers, chest freezers, extractors, and so on.

Notes: Jim Fischer pollinated apples and produced Demeter-certified honey in the mountains of Virginia for over a decade of "retirement," and when he moved to New York City in 2006 he found that beekeeping had been prohibited by Rudolph Giuliani in the 1990s. Legalizing beekeeping required supporters, so he taught a free beekeeping course in Manhattan's Central Park, creating a constituency to legalize in 2010. Jim invented Fischer's Bee-Quick (http://bee-quick.com), an organic-certified repellent he distills to clear bees from honey supers; the Nectar Detector (http://netardeteor.com); and other ubiquitous beekeeping tools. He and his wife Joanne produce "Hi-Rise Hive" honey, sold at the better stores on the Upper East Side.

Toni Burnham is the original driving force behind the DC Beekeeper Alliance (https://dcbeekeepers.org) and is a regular contributor to *Bee Culture* magazine. Her twitter bio states "Dotcom dropout who has gone to the bees," and she famously produces honey at the US Congressional Cemetery under the "Rest in Bees" label.

Varroa Mite Infestation

Mite Density Chart

Number of Mites per 300 Adult Bees	Colony Infestation (%)	Number of Mites per 8 Samples of 300 Adult Bees	Apiary Infestation (%)
1	1	8	1
2	1	16	1
3	2	24	2
4	3	32	3
5	3	40	3
6	4	48	4
7	5	56	5
8	5	64	5
9	6	72	6
10	7	80	7
11	7	88	7
12	8	96	8
13	9	104	9
14	9	112	9
15	10	120	10
16	11	128	11
17	11	136	11
18	12	144	12

Notes: Count the number of mites in a sample of 300 adult bees. Use that number and this simple table to determine percentage of infestation. For example, if 12 mites were counted on 300 bees, check the number 12 on the table: that would be an 8% colony infestation. If the colony (or apiary) infestation is more than 10 to 12%, treatment of some sort is recommended. After Lee et al. 2010 and by permission of the author.

Chemotherapy for Control of Varroa Mites

A list of some of the current chemical miticides (acaracides) used to control *Varroa*; some of these may also work for *Tropilaelaps*.

Product Trade Name ®	Active Ingredient	Chemical Class
Apiguard	thymol	essential oil
Apilife VAR	thymol, eucalyptol, menthol, camphor	essential oils
Apistan **	fluvalinate	synthetic pyrethroid
Amitraz, Miticur, Api-warol (tablets)	formamidine	Formetanate, Methanimidamide
Apitol	cymiazole	iminophenyl thiazolidine derivative
Apivar **	amitraz	amadine
Bayvarol **	flumethrin	synthetic pyrethroid
Check-Mite+ **	Perizin, coumaphos	organophosphate
Folbex	bromopropylate	chlorinated hydrocarbon
Sucrocide	sucrose octanoate	sugar esters
Hivestan	fenpyroximate	pyrazole (alkaloid)
generic	formic acid	organic acid
generic	lactic acid	organic acid
generic	oxalic acid	organic acid

** No longer effective in some areas

Sources: Chemical controls for varroa mites compiled from Rosenkranz et al. 2010 and http://www.maf.govt.nz /biosecurity/pests-diseases/ animals/varroa/guidelines/control-of-varroa-guide.pdf. In: *Recent Developments Focused on the Problems of Honey Bee Pollinators*, edited by D. Sammataro and J.A. Yoder, in press.

Glossary

abdomen The posterior or third region of the body of the bee that encloses the honey stomach, stomach proper, intestines, sting, and reproductive organs.

abdomen of a drone bee The posterior or third region of the body that contains the stomach, intestine, and reproductive organs.

abdomen of a queen bee The posterior or third region of the body that contains the stomach, intestine, reproductive organs, sting, and Dufour gland.

abdomen of a worker bee The posterior or third region of the body that contains the honey stomach, stomach proper, intestines, sting, reproductive organs, wax glands, and pheromone gland (Nasonov gland).

abscond Bees abandoning a hive because of wax moth, excessive heat or water, mites, lack of food, or other unfavorable conditions.

acarine disease The name of the disease later known to be caused by the mite *Acarapis woodi* (Rennie) infesting the thoracic trachea of adult bees; see **tracheal mite**.

acarology The study of mites.

Africanized honey bee The common name for *Apis mellifera scutellata* hybrids inhabiting the Americas that have retained much of their defensive characteristics.

afterswarms Swarms that leave the hive after the first, or prime, swarm has departed, accompanied by one or more virgin queens. Such swarms in the order of their departure are often referred to as secondary, tertiary, etc., swarms.

alarm odor or pheromone A chemical substance released from the vicinity of a worker bee's sting that alerts other bees to danger. Isopentyl acetate is the principal alarm odor; 2-heptanone, found in the worker bee's mandibular glands, is a secondary alarm odor.

allergic reaction A systemic or general reaction to bee venom.

American foulbrood (AFB) disease A brood disease of bees that affects the late larval and prepupal stages and is caused by the bacterium *Paenibacillus* (=*Bacillus*) *larvae* spp. *larvae*.

antenna (*pl.* antennae) One of two long, segmented, sensory filaments extending from the head of insects; includes taste and odor receptors.

apiary The location and sum total of honey bee colonies in a single location; a bee yard.

apiculture The science and art of raising honey bees for economic benefit.

Apis cerana The scientific name for Indian or Asian honey bees. This species is found throughout Asia, and in certain countries it is the honey bee of commerce. It is the natural host of the varroa mite.

Apis dorsata The scientific name for the largest of the species of honey bees, often referred to as the giant bee. This species is found only in Southeast Asia.

Apis florea The scientific name for the smallest of the species of honey bees, often called the dwarf bee. This species is found in Asia, though it is more Western in distribution than the other Asian species.

Apis mellifera The scientific name for the European honey bee, which is found throughout the Western world, though originally it is thought to be Near Eastern in origin. This bee has been carried from Europe to all areas of the world except the Arctic and Antarctica.

Apistan strips Plastic strips impregnated with the pesticide tau-fluvalinate, used to control varroa mites; however, it is currently not effective in controlling varroa in most areas because the mites have become resistant to this pesticide.

Bacillus larvae (now *Paenibacillus larvae* spp. *larvae*)　The old scientific name of the organism (bacterium) that causes American foulbrood disease.

ball a queen　An attack on a queen by a number of worker bees intent on killing her by pulling at her legs and wings and then suffocating and overheating her. In this process the bees form a small cluster or ball of bees around the queen.

basitarsus　Part of the tarsus, with rows of hairs referred to as *combs*, located on its medial surface that receives the pollen grains from the first two pairs of legs.

bee bread　Pollen gathered by bees, mixed with various bacteria, enzymes, and fungi, is converted into a nutritious substance. It is an essential food for larvae and young bees. Bee bread is stored in honeycomb and is often covered with a layer of honey for preservation.

bee cellar　A portion of a dwelling or building made especially for overwintering bee colonies in climates where winters are severe.

bee escape　A device that permits bees to pass only one way, preventing their return. It is used to remove bees from honey supers.

Bee Go　A chemical (a butyrate) repellent used with a fume board to remove bees from honey supers.

beehive　A box or receptacle for housing a colony of bees. Modern beehives are made of wood or plastic and adhere to the bee space dimensions.

bee louse　The common name for *Braula coeca* (Diptera: Braulidae), a flightless fly found only on honey bees. The fly steals food from a bee's mouth but is not considered a serious problem. Since the introduction of the varroa mite and the chemicals used to control it, the bee louse is rarely seen.

bee paralysis　The condition which affects adult bees that are unable to fly or work normally. It is now known to be induced by a virus or carried by parasite bee mites, such as the varroa mite.

bee space　A space that permits safe passage for bees on opposite sides of facing combs but is too narrow to encourage comb building and too large to induce propolization by bees. It measures from ¼ to ⅜ inches (6.4–9.5 mm).

bee suit　Coveralls, usually white, that fit over normal clothing with or without an attached veil. This apparel is worn when working with bees.

beeswax　A complex mixture of organic compounds secreted by four pairs of glands on the ventral or underside of the worker bee's abdomen and used by bees for building comb. Its melting point is from 143.6° to 147.2°F (62–64°C).

bee veil　A cloth or wire netting worn with a hat or helmet for protecting the head and neck from stinging insects.

bottom board　The floor of a beehive.

brace comb or burr comb　Small pieces of comb made as connecting links between two frames or between a comb and the hive itself. Burr comb often does not connect to any other part but is an extension of the comb, often built during a honeyflow.

brood　Immature stages of the bees not yet emerged from their cells (eggs, larvae, and pupae).

brood chamber　The part of the hive interior in which brood is reared; the brood chamber includes one or more hive bodies and the combs within.

brood foundation　Heavier wax foundation sheets, usually wired with vertical wire and used in brood-nest frames; designed for frames to be placed in the broodnest.

broodnest　The part of the hive interior in which brood is reared; this is the warmest part of a colony, since the eggs and larvae must be incubated at around 95°F (35°C).

brood rearing　The raising of young bees from eggs to adults.

Buckfast bee　A strain of bees developed by Brother Adam in England and bred for resistance to tracheal mites, disinclination to swarm, hardiness, honey production, and good temperament.

candied honey　Crystallized, granulated, or solidified honey; see **Dyce process**.

capped brood　Larval cells that have been capped over with a brown covering consisting of wax and propolis. Once the cells are capped, the larvae spin their cocoons and turn into pupae, the third stage of complete metamorphosis.

cappings　The thin, light-colored wax covering cells full of honey. Once cut from honey frames they are referred to as cappings and make the finest grade of beeswax.

cappings scratcher　A forklike device used to scrape the wax cappings off the sealed honey in order to extract the honey. Also used to remove drone brood to test for the presence of varroa mites.

Carniolan bee　The common name for *Apis mellifera carnica*, a race of honey bee named for Carniola, Austria, but originating in the Balkan region. It is gray-black in color, very gentle, and conserves honey well.

caste　Worker and queen bees are both members of the female gender but differ in both form and function. Drones are male bees with the same form and function, and therefore do not belong to a caste.

Caucasian bee　The common name for *Apis mellifera caucasica*, a race of honey bee originating in the Caucasus Mountains. Grayish in color, they use more propolis than other bee races.

cell　The hexagonal unit-compartment of a comb.

cellar wintering   The placing of bees in an unheated cellar or special building during the winter months. In some cases these sites are temperature controlled.

cell bar   A wooden strip on which queen cups are placed for rearing queens.

chalkbrood   A disease caused by the fungus *Ascosphera apis*, which turns larvae into white then gray and black chalklike mummies.

chilled brood   Immature bees that have died from the cold because of insufficient numbers of adult bees to maintain proper incubation temperatures. This may also result from beekeepers inspecting hives during low temperatures or when the forager numbers have been diminished by pesticides.

chorion   Membrane or shell covering a bee egg.

chunk comb honey   Honey cut from the comb and placed with liquid honey in glass containers.

clarifying   The removal of any foreign material or wax from honey usually by allowing the honey to settle, at either room or a higher temperature. Wax and most other material will float to the top and can be skimmed off.

cleansing flight   Bees flying out of the hive during winter and early spring when temperatures are sufficient to permit flight. During these flights bees relieve themselves of waste products. It is common when snow is on the ground to find yellow spots over the snow.

cluster   The form or arrangement in which bees cling together as a swarm assembles. Most swarm clusters are found on the branches of trees and shrubs. In the winter bees form a cluster inside the hive in order to conserve heat.

colony   An aggregate of worker bees, drones, and a queen living together in a hive or in some other dwelling (usually a cavity) as one social unit.

comb   A back-to-back arrangement of two series of hexagonal cells made of beeswax to house/store eggs, brood, bee bread, or honey. There are approximately five worker cells to the linear inch and about four drone cells to the inch.

comb honey   Honey in the comb, produced in small, square wooden sections about 4¼ × 4¼ inches (11 × 11 cm) in size, in plastic rings (Ross Rounds), or plastic boxes.

compound eyes   The two large lateral eyes of the adult honey bee composed of many lens elements called ommatidia.

conical escape   An escape board made with cones set in a frame to permit a one-way exit for bees, to clear honey supers of bees.

corbicula (*pl.* corbiculae)   An anatomical structure located on the lateral side of the hind tibia that functions as a pollen-carrying basket.

cremed honey   See **Dyce process**.

crystallization   Granulation of honey; when honey as a supersaturated solution has candied or become solid instead of liquid. Cremed honey is a commercially made soft form of granulated honey.

cut comb honey   Honeycomb cut to fit in small plastic boxes or wrapped individually for retail sale.

deep super   Hive furniture that holds standard, full-depth frames; the usual depth is 9½ inches or 9⅝ inches (24–25 cm).

Demaree   The beekeeper G. Demaree, who invented a popular method of swarm control in 1884; also used as a verb "to demaree," in describing this method. It consists of separating the queen from most of the brood.

dequeen   Removal of a queen from a colony.

dextrose   See **glucose**.

diastase   An enzyme in honey that assists in the conversion of starch to sugar.

disease resistance   When strains of bees selected from stock show high survival despite the presence of diseases.

dividing   Separating a colony in such a way as to form two or more colonies.

division board   A thin vertical board of the same dimensions as a frame; also called dummy board or follower board. It is used to reduce the size of the brood chamber (to a few frames) or to fill up the gap in a hive body using only nine frames. It is also used to permanently divide the hive into two or more parts.

division board feeder   A plastic or wood container the shape of a frame, hung in a hive and filled with syrup to feed bees.

drawn combs   Honeycombs having the cell walls built up by honey bees from a sheet of foundation base.

drifting   The movement of bees into hives other than their own. Young bees who have not yet learned the location of their hive tend to drift more frequently than do older bees. This phenomenon is common when hives are placed in long rows; in such cases bees tend to drift into hives near the end of rows. On days when wind is a factor, drifting also becomes a problem.

drone   The male honey bee, developing from unfertilized eggs, which are haploid, or have half the chromosome numbers. The development of individuals from unfertilized eggs is known technically as *parthenogenesis*.

drone brood   Brood that is reared in larger cells and produces drone bees. When drone cells are sealed, the cappings have the appearance of bullet heads and can therefore be easily distinguished from the cappings over worker brood and those covering honey.

drone comb   Comb having cells measuring about 4 cells to the linear inch or about 18.5 cells to the square inch. Such comb is used for the specific purpose of raising drones or is placed in honey supers to facilitate the extraction of honey and to control varroa.

drone congregating area (DCA)   A specific area to which drones fly year after year waiting for an opportunity to mate with virgin queens.

drone layer   A queen that lays only unfertilized eggs, resulting in only drone offspring, because she is old, is low on sperm, was improperly mated, was not mated at all, or is diseased or injured.

drumming   Pounding on the sides of a hive or other bee dwelling to drive the bees upward. This method is used to transfer bees from bee trees into a bee hive or to drive bees from one hive body into another or completely out of a given hive body.

dwindling   The rapid dying off of old bees in the spring (often referred to as spring dwindling). It can be caused by nosema and viral diseases or mite infestation.

Dyce process   A patented process involving pasteurization and controlled granulation to produce a finely crystallized or granulated honey product that spreads easily at room temperature; also sometimes called "cremed" honey.

dysentery   A condition of adult bees resulting from an accumulation of feces on the inner or outer surface of hive furniture. Dysentery is first detected by finding small spots of feces around the entrance and within the hive, and its presence usually occurs during winter. It is caused by unfavorable wintering conditions that restrict cleansing flights and/or low-quality food; can be confused with nosema disease, caused by a microsporidian fungus.

egg   The first stage in the bee's life cycle, usually laid by the queen. It is cylindrical, 1/16 inch (1.6 mm) long, and enclosed with a flexible shell or chorion.

embed   To force wire into wax foundation by heat, pressure, or both in order to strengthen the foundation. Such additional support to the foundation keeps it intact, especially when honey is being extracted from the comb.

emerging brood   Young or teneral bees chewing their way out of the capped brood cells.

entrance reducers   A strip of wood notched with different-size holes to regulate the size of the hive entrance and hence the flow of bees into or out of a hive. Reducers are useful when robbing in an apiary becomes prevalent. Reducers also restrict mice from entering hives; metal reducers are now available for this purpose.

escape board   A board having one or more bee escapes in it that is used to remove bees from honey supers.

European foulbrood   A bacterial brood disease of bees caused by *Melissococcus* (=*Streptococcus*) *plutonius*.

extender patties   Vegetable shortening and sugar patties with antibiotics added (e.g., Terramycin). This patty has been used to suppress American foulbrood disease but is now not recommended as it can lead to resistance of the bacteria to the antibiotic.

extracted honey   Honey removed from combs by means of a centrifugal extractor.

extractor   A machine used for removing honey from combs by spinning the frames and throwing the honey against the extractor walls; the combs remain intact.

eyelets   Metal pieces that fit in the holes of frame end bars; used to prevent reinforcing wires from cutting into the wood and thus slackening the taut wires.

feeders   Various types of appliances and containers for feeding bees sugar syrup.

fermentation   A chemical breakdown of high-moisture honey by yeasts; in honey, fermentation is caused only by the genus *Zygosaccharomyces*, a yeast able to tolerate the high sugar content of honey. Fermented honey cannot be reversed and is unsalable.

fertile queen   A queen inseminated instrumentally or naturally with drone spermatozoa. By either method, the sperm is stored in her spermatheca. Such a queen is capable of laying fertilized eggs.

fertilization   Usually refers to eggs laid by queen bees, which become fertilized when sperm stored in the spermatheca unites with an egg while it is being laid. Can also refer to the process where a pollen grain grows down a flower's female stigma to fertilize eggs in the ovary, producing seed and fruit.

festooning   The activity of young bees engorged with honey and clinging to one another in a suspended form while secreting wax scales. The function of festooning is unknown.

field bees or foragers   Worker bees, usually at least 16 days old, that work in the field to collect pollen, nectar (or rob honey from other hives), honeydew, water, and propolis.

flight path   Refers to the direction bees fly leaving their hive. If obstructed by a beekeeper standing in front of the hive entrance, bees will collect behind the obstruction and may become defensive.

food chamber   A hive body filled with honey; used as extra food or for winter stores.

foulbrood   Malignant, contagious bacterial diseases of honey bee brood; see **American foulbrood** and **European foulbrood**.

foundation   A thin sheet of beeswax or plastic, embossed or stamped with the base of a normal worker or drone cell on which the bees will construct a complete or drawn comb.

frame Four pieces of wood (or preformed plastic) that form a rectangle, designed to hold wax or plastic foundation / honey comb. A frame consists of one top bar, a bottom bar (of one or two pieces), and two end bars.

frass Excreta of insects, used especially in reference to moths and butterflies; the black droppings found on comb infested with wax moth larvae.

fructose A simple sugar (or monosaccharide); formerly called levulose (fruit sugar). It is a disaccharide found in honey.

fumagillin An antibiotic made from a fungus used for the treatment of nosema disease.

fume board A cloth-coated wooden frame with a metal cover, sprinkled with a bee repellent, such as Bee Go, used to remove bees from honey supers.

Fumidil-B The older trade name (U.S.) for fumagillin; now made in Canada and labeled Fumagilin-B.

Galleria mellonella **L.** The scientific name of the greater wax moth. The larvae of this moth chew and destroy honeycomb. They are used for fish bait and for scientific research.

glucose A simple sugar; formerly called dextrose (grape sugar). It is one of the two main sugars found in honey and forms most of the solid phase in granulated honey.

grafting A process of removing newly hatched worker larvae from their cells and placing them in artificial queen cups, for the purpose of rearing queens.

granulation A term applied to crystallized, candied, creamed, or solidified honey.

grease patty A patty made with vegetable shortening and sugar; used to control tracheal mites.

gynandromorph Having both male and female organs and/or body parts.

head The front (anterior) part of an insect containing the eyes, antennae, and mouthparts.

2-heptanone A chemical substance produced in the mandibular glands of worker honey bees that elicits an alarm reaction.

hive (*n*.) A home for bees provided by humans, that is, a hive box. (*v*.) To place a swarm into a hive box.

hive body A wooden (or plastic) box or rim that contains the frames.

hive stand A structure that serves as a base support for a hive. Such a stand keeps the bottom board off the damp ground.

hive tool A metal device with a curled scraping surface at one end and a flat blade at the other end; used to separate hive furniture when inspecting bees, to scrape frames, and to remove frames from the hive.

honey A sweet, viscous liquid composed of sugars. It is made from flower nectar gathered by the bees, ripened or evaporated into honey, and stored in the combs. Well-cured honey contains about 17 percent water.

honey bee The common name for *Apis mellifera* (honey bearer). The word is written as two words (in American publications) and as one word in Europe. An arthropod in the class Insecta, order Hymenoptera, and superfamily Apoidea that is a social, honey-collecting insect living in perennial colonies. Also known as *A. mellifica* (honey maker).

honeyflow Loosely, a time of year when there is a plentiful supply of nectar that bees can collect. Its signs include fresh, white wax and combs filled with liquid. It is a time when bees produce and store surplus honey.

honey house A building used for extracting and packaging honey, storing supers, and so on.

honey stomach, crop, or sac An enlargement of the back or posterior end of the bee's esophagus that lies in the front part of the abdomen. This organ can hold a large amount of liquid due to its invaginated walls; now found to contain *Lactobacillus* and other beneficial bacteria.

hybrid queen The result of crossing different bee races or lines to produce a queen of superior qualities.

hypopharyngeal glands A pair of organs located in the head of a worker bee that produce brood food and royal jelly.

increase To add to the number of existing colonies in an apiary, by dividing those already on hand or in hiving swarms or packages.

infertile As in worker bees, not able to lay fertile eggs. In plants, it means unable to reproduce.

inner cover A lightweight cover with an oblong hole in its center; used under a standard telescoping outer cover on a bee hive.

instrumental insemination The introduction of drone spermatozoa into the genital organs of a virgin queen by means of special instruments; sometimes called artificial insemination.

introducing or queen cage A small box made of wire screen and wood or plastic, used in shipping queens or introducing a new queen to a colony.

invertase An enzyme that speeds the transformation of sucrose (a complex sugar) into the monosaccharides (or simple sugars) fructose and glucose.

Isle of Wight disease An early name for acarine disease, the infestation of bees by tracheal mites. It is named after the place where these mites were first discovered in the early 1900s.

isopentyl acetate The alarm odor in honey bees produced in the sting chamber.

Italian bee The common name for *Apis mellifera ligustica*, the most common race of European bees used commercially. Introduced from Italy in the 1860s; workers have brown and yellow bands on their abdomen; queens have brown or orange abdomens with few or no stripes.

K-wing The appearance of the bee's wing in the shape of the letter K, in which the hind wing is held in front or over the fore wing; caused by an infestation of tracheal mites or by a virus, or both.

lactic acid bacteria (LAB) Several species of bacteria able to tolerate acidic conditions and responsible for fermentation, especially of sugars; special strains are commonly used for probiotics (such as in yogurt). New research has discovered LAB living in the honey stomachs of bees and other pollinators, which could help protect the bees from pathogens and help in the preservation and conversion of their food.

larva (*pl.* larvae) The second stage in the development of an insect, such as the honey bee, that has complete metamorphosis. It is comparable to the caterpillar (or eating) stage of a moth or butterfly.

laying worker A worker bee that lays unfertilized eggs, which will develop into drones. Laying workers develop usually in colonies that have been queenless for a long time.

levulose See fructose.

local reaction Bee stings that elicit a brief, sharp pain at the site(s) of the sting(s). The sting site may swell, turn red, and itch, but such symptoms will usually subside in a matter of minutes or hours.

mandibles The jaws of an insect. In the honey bee and most insects, the mandibles move in a horizontal rather than in a vertical plane. The bees use the jaws to form honeycomb, to scrape pollen, and to pick up hive debris.

mating flight The flight taken by a virgin queen, during which she mates in the air with one or more drones. Normal queens mate ten to twenty times in two or more mating flights.

mead A wine made with honey as the main source of food for the yeast.

menthol crystals Crystalline form of the essential oil of the mint plant *Mentha avensis*; used to control tracheal mites.

metamorphosis The developing process of most insects, in four stages: egg, larva, pupa, and adult; also called complete metamorphosis.

migratory beekeeping The moving of bee colonies from one locality to another during a single season to pollinate different crops or to take advantage of more than one honeyflow.

mite An eight-legged creature or acarine (like the tick and spider). At least three species are currently parasitic on bees: *Tropilaelaps*, tracheal, and varroa mites.

movable frame A wooden or plastic frame containing honeycomb constructed to provide the bee space between frames. When placed in a hive it remains essentially unattached by brace comb and heavy deposits of propolis, thus permitting its removal with ease.

Nasonov gland The gland associated with the seventh abdominal tergite of the worker honey bee. This gland is commonly called the scent gland because its contents attract bees to gather in a cluster.

nectar A sweet plant exudation secreted by special nectary glands containing sugars. It is found chiefly in the flowers or reproductive organ of plants. Extrafloral nectaries are found outside of flowers (on leaves or stems).

nectaries Specialized tissues contained in organs or glands of plants that secrete nectar.

Nosema apis The scientific name of a microsporidian parasite (now considered a fungus) of honey bees, causing nosema disease. Since the 1990s, a new species, *N. ceranae*, has been identified and is reported to be more virulent that *N. apis* and may have replaced it in many areas. The parasite *Nosema* spp. is also found in other insects, such as locusts and bumble bees.

nosema disease An abnormal condition of adult bees resulting from the presence of nosema spores in their intestines; often treated with the antibiotic fumagillin.

nucleus A small colony of bees often used in queen rearing and called a nuc; comes in three to five frame sizes.

nurse bees Young worker bees with fully functional food glands whose duty is to feed larvae and the queen and to perform particular hive duties. Generally nurse bees are 3 to 10 days old.

observation hive A hive made largely of glass or Plexiglas with an outside entrance to permit observation of the bees at work from inside a building.

ocellus (*pl.* ocelli) A simple eye with a single lens. The honey bee has three ocelli on the top of its head that distinguish light from dark.

ommatidium (*pl.* ommatidia) One of the visual units or lenses that make up the compound eye.

osmophilic yeasts Yeasts that occur naturally in honey and are responsible for fermentation in honey that has more than 18 percent water. These yeasts belong to the genus *Zygosaccharomyces*.

out apiary An apiary kept at a distance from the home of the beekeeper; also called an outyard.

outer cover The top cover that fits over a hive to protect it from the weather. The two most common covers are migratory and telescoping.

ovary The egg-producing female organ of a plant or animal.

oxytetracycline The antibiotic sold as Terramycin, registered to control American foulbrood (AFB). Some report that AFB is resistant to this antibiotic and are using tylosin instead.

package A special wire-screened shipping box containing a quantity of bees (2–5 lb.) with or without a queen.

Paenibacillus The current scientific genus name of the organism (bacterium) that causes American foulbrood disease; formerly known as *Bacillus*.

paradichlorobenzene (PDB) A crystalline chemical used to control wax moths; also called moth crystals. It is not recommended anymore because of its carcinogenic nature.

parthenogenesis The development of young from unfertilized eggs. In honey bees, the unfertilized eggs are laid by virgin queens, laying workers, or mated queens and produce only drone bees.

pheromone A chemical substance that is released externally by one insect or animal and stimulates a response in other insects (or animals) of the same species.

piping A series of sounds, a loud shrill tone followed by shorter ones, made by queens. These sounds are usually made by newly emerged virgin queens to illicit quacking from queens still in their cells, which enables her to locate and destroy them.

play flights Short flights taken in front of the hive and in its vicinity to acquaint the young bees with their immediate surroundings and hive location; also called orientation flights. These may also have other functions and are sometimes mistaken for robbing or preparation for swarming. They are common during late afternoon.

pollen The dustlike male reproductive cell bodies of flowers formed on the anthers and collected by bees. Bees collect pollen and turn it into bee bread by a complex fermentation process; it provides the protein part of the honey bee's diet.

pollen basket or corbicula A flattened depression surrounded by curved spines located on the outside of the tibiae of the bee's third set of legs. It is used to carry pollen gathered from flowers back to the hive, where the pollen pellets are deposited into cells and packed together as the bees ram their heads against the pellets. This same basket is also used by bees to collect and transport propolis back to the hive.

pollen insert A device inserted into the entrance of a colony into which hand-collected pollen is placed. As the bees leave the hive and pass through the trap, some of the pollen adheres to their bodies. This allows the pollen to be carried to the target blossoms, resulting in cross-pollination.

pollenizer The plant that furnishes pollen. Crab apple trees are often pollenizers in apple orchards.

pollen patty A cake or patty made of pollen pellets and sugar syrup. These patties are fed to stimulate brood rearing.

pollen substitute A food material used to substitute wholly for pollen in the bees' diet; commonly made from soy flour or other products.

pollen supplement A bee food that is mixed with pollen to augment the bees' diet. It can contain brewer's yeast (distiller's soluble from a brewer), soybean flour, natural pollen, and other ingredients formulated in different ways to be digestible to bees.

pollen trap A device for removing pollen from the pollen baskets of bees as they return to their hives.

pollination The transfer of pollen from the anthers (male part) to the stigma (female) of flowers.

pollinator The agent that transfers the pollen.

prime swarm The first swarm to issue from the parent colony; usually contains the old queen.

proboscis A structure formed from the free parts of the bees' maxillae and labium, forming a tube for ingesting nectar, honey, honeydew, and water.

propodeum The first abdominal segment fused to the bee's thorax, which connects the thorax to the abdomen and is typical "wasp waist" of the Apocrita suborder that includes bees, ants, and wasps. The propodeum has a pair of spiracles.

propolis or bee glue A gluey or resinous material that bees collect from trees and plants and use to strengthen the comb or seal cracks. It has antimicrobial properties.

protein Naturally occurring, complex organic macromolecule that contains amino acids. Pollen contains protein, which is an important nutrient to developing bees.

prothoracic/mesothorax spiracle This spiracle lies between the pro- and mesothorax. It is the largest of the spiracles; by way of this spiracle the parasitic mite *Acarapis woodi* enters the tracheal system of the bees.

pupa (*pl.* pupae) The third stage in the development of an insect that is encapsulated in a cocoon. In this stage, the organs of the larva are replaced by those that will be used as an adult.

queen   The reproductive member of a colony capable of laying both fertilized and unfertilized eggs, once she has been properly mated. A fully developed mated female bee. The queen is larger and longer than a worker bee and is recognized by workers because of special pheromones.

queen cage   A small cage in which a queen and five or six worker bees may be confined; used for shipping queens and usually contains a candy plug.

queen cell   A special, elongated cell suspended vertically from honeycomb and resembling a peanut shell in which a queen bee is being raised. It is usually an inch or more in length when fully developed and capped.

queen cup   A cup-shaped cell produced by bees and suspended vertically from the honey comb that may eventually develop into a queen cell. These cups can also be obtained commercially or produced by individuals using a mold. Commercial cups are made either of beeswax or plastic. These cups become queen cells when a queen deposits an egg in them or when a queen breeder transfers a young larva into the cup. These cups are also suspended in a vertical position.

queen excluder   A device made of wire, wood and wire, plastic, or punched plastic, having openings of about 0.16 to 0.17 inch (0.41–0.42 cm). This permits workers to pass through but excludes queens and drones. It is used to confine the queen to a specific part of the hive, usually the brood chamber.

queenright   A bee colony having a laying queen.

rabbet   A narrow ledge cut into the top ends of hive bodies on which the frames hang. Some rabbets are cut so that resting frames will be at the right bee space to the top of the box; others are lower, requiring a metal strip to correct the bee space.

rendering wax   The process of melting combs and cappings to separate wax from its impurities and thus refining the beeswax. Wax is usually melted using a hot water tank or solar wax melter.

requeen   To place a new queen into a hive made queenless.

reversing   To exchange places of different hive bodies of the same colony, usually to expand the nest. Done in the spring, the upper hive box full of bees and brood is reversed with the lower, emptier box on the bottom board.

robbing   Applied to bees stealing honey/nectar from other colonies.

round sections   Section comb honey made in plastic rings instead of square wooden or plastic boxes. Also called Ross Rounds.

royal jelly   A highly nutritious glandular secretion of young bees; used to feed the queen and the young queen larvae.

sacbrood   A brood disease of bees caused by a filterable virus.

scent gland   See **Nasonov gland**.

sex attractant   A chemical substance that attracts an animal of the same species, male or female, for the purpose of mating. The sex attractant in *Apis mellifera* is [t]9-oxo-2-enoic acid (9-ODA) from a queen's mandibular gland.

shallow super   Any one of several super sizes less than the depth of a deep super. Commonly, shallow supers vary from 4¼ to 6¼ inches (11–16 cm) in depth.

skep   A dome-shaped beehive without movable frames, usually made of twisted straw. The use of skeps for keeping bees is illegal in most states in the United States.

slumgum   The refuse from melted comb and cappings after the wax has been rendered or removed.

small hive beetle   The common name for *Aethina tumida*, the beetle first identified in Florida in 1998. Before its discovery in the United States, the beetle was known to exist only in tropical or subtropical areas of Africa. How it found its way to North America is not known.

smoker   A metal container with attached bellows that burns organic fuels to generate smoke; used to control the defensive behavior of bees during routine colony inspections.

solar wax melter   A glass-covered insulated box used to melt wax from combs and cappings, by using the heat of the sun.

spermatheca   A small sac connected with the oviduct of the queen and in which is stored the spermatozoa received during mating with drones or by instrumental insemination.

spermatozoa   Male reproduction cells (gametes) that fertilize eggs.

spiracles   Openings in the body wall connected to the tracheal tubes through which insects breathe.

split   (*v.*) To divide the components of a hive and its population of bees to form a new colony. (*n.*) A colony divided, thus increasing the number of hives.

Starline hybrid   An Italian strain, crossbred for vigor, honey production, and prolific populations of bees. Not commonly found anymore.

sting   An organ of defense of workers and queen bees. It is an egg-laying device (or ovipositor) modified to form a piercing shaft, through which painful organic venom is delivered.

super   A piece of hive furniture in which bees store surplus honey; so called because it is placed over or above the brood chamber.

supering   The act of placing supers on a colony in anticipation of a honey crop.

supersede   To rear a young queen that will replace the mother queen in the same hive. Shortly after the young queen starts to lay eggs, the old queen usually disappears.

surplus honey The honey removed from a hive, in excess of what bees need for their own use, such as winter food stores.

swarm The aggregate of worker bees, drones, and a queen that leave the original colony to establish a new one. Swarming is the natural propagation method to form a new bee colony.

swarm cells A special type of queen cells constructed from fresh wax that is light yellow in color; they can be distinguished from supersedure and emergence queen cells because they are found in colonies during only the two swarm seasons. Another way of identifying that they are indeed swarm cells is that there are many in a colony during the swarm season, possibly ten to forty. Robust in appearance, they are commonly located near the lower edges of combs.

swarming season Swarming is bimodal: the most intense swarm period is during the spring and early summer; a lesser swarming period occurs during late summer and early fall.

systemic reaction A reaction from a bee sting or stings that can be life-threatening and require immediate medical attention. Such a reaction is far more serious than stings that elicit pain at the site of the sting, and symptoms include urticaria (hives), throat tightness, difficulty breathing, and a drop in blood pressure. An EpiPen is often used when someone has a systemic reaction, followed by a trip to the hospital.

Terramycin The trade name of an antibiotic used to combat European or American foulbrood disease; generic name is oxytetracyline. Research has shown that AFB is now resistant to this antibiotic in many areas.

thin comb-honey foundation A thin, wireless wax foundation that is used for section comb or chunk honey production.

thorax The central or second region, between the head and abdomen, of the bee's body that supports the wings and legs.

top bar The top part of a frame.

trachea (*pl.* tracheae) The breathing tubes of an insect opening into the spiracles.

tracheal mite The common name for *Acarapis woodi* (Rennie) (Acari: Tarsonemidae), a mite that inhabits the trachea of adult honey bees. It can be controlled by grease patties and menthol crystals.

travel stain The darkened or stained surface of comb honey, which is caused by bees walking on its surface over a long time; usually from propolis.

***Tropilaelaps* spp.** The scientific name for one of two common mites associated with Asian honey bees;

the other is *Euvarroa* spp. These mites are not currently present in the New World.

tylosin Newer antibiotic used for treating colonies with American foulbrood disease, sold as Tylan.

uncapping knife A knife with a sharp blade used to shave or remove the cappings from combs of sealed honey before extracting. The knives are usually heated by steam, hot water, or electricity.

unfertilized An ovum (egg) that has not been united with the sperm. Insect eggs not fertilized usually become males.

uniting Combining two or more weak colonies to form a large one. To prevent fighting between colonies, a sheet of newspaper is placed between the colonies to separate them until they become familiar with each other's scent by means of tearing down the paper.

unripe honey Honey that is more like nectar, containing over 18 percent water.

unsealed brood or open brood Immature or larval bees not yet capped over with wax; the term can include cells containing eggs.

***Varroa destructor* Truman and Anderson 2000** The scientific name for the varroa mite; a destructive parasitic mite (eight-legged) that feeds on brood but is carried by adult bees. First discovered in Florida in 1986 and formerly called *Varroa jacobsoni* Oudemans, this mite is now found worldwide, most recently in Kenya and Uganda.

varroosis A disease due to the actions of the varroa mites feeding on honey bees.

ventriculus The stomach of the bee, located in the abdomen behind the honey stomach but before the hindgut; also called the midgut.

virgin queen An unmated queen. If she remains unmated due to a variety of circumstances, she will be capable of laying only unfertilized eggs. These eggs will then yield only drone honey bees.

virus (Latin for *toxin*) An infectious agent (100 times smaller than bacteria) that must grow and reproduce in a host cell; viruses infect all cellular life. Tobacco mosaic virus was the first one discovered, in 1899. Currently 5000 viruses have been described; honey bees have about 18 known viruses.

wasp A close relative of a honey bee, usually in the genus *Vespula*. Wasps are carnivorous, and some species prey on bees.

wax bloom A powdery coating forming on the surface of beeswax. It is composed of volatile components of beeswax.

wax glands The eight glands that secrete beeswax. They are located in pairs on the last four visible ventral abdominal segments (sternites) of young workers.

wax moth The common name for *Galleria mellonella* L., a moth whose larvae eat comb, pollen, and pupae.

wax scale A drop of liquid beeswax that hardens into a scale; so named because it has the appearance of a fish scale. These scales, once extruded by the bee's wax glands, harden on contact with the air and serve as the building blocks of honeycomb.

windbreaks Specifically constructed fences or natural barriers to reduce the force of wind in an apiary during cold weather as well as to reduce drifting in areas of frequent prevailing winds or breezes.

winter cluster The arrangement or organization of adult bees within the hive during the winter period.

winter hardiness The ability of some strains of bees to survive long winters by frugal use of honey stores and low bee populations.

wired foundation Wax foundation containing embedded vertical wires to prevent the finished drawn comb from sagging.

worker bee A female bee whose organs of reproduction are only partially developed. Workers are responsible for carrying on all the routine tasks of a bee colony.

References

The books and articles listed below are only a small representation of what is available and are just to get you started. Some information is now out of date in many of these books, and many are out of print. With internet technology, many of these citations can be accessed through your computer or at internet libraries. You can find a plethora of information on each of the topics listed here by searching the internet; **but beware:** some of the information may not be accurate. **Read with discrimination.**

The references are divided into the following sections:

ABBREVIATIONS

ABJ = American Bee Journal
Ag. = Agriculture
Ann. = Annals
Annu. Rev. Entomol. = Annual Review of Entomology
ARS = Agriculture Research Service
CES = Cooperative Extension Service
Dept. Entomol. = Department of Entomology

(ed.) = editor(s)
edn. = edition
Entomol. = Entomology
Ext. Serv. = Extension Service
IBRA = International Bee Research Association
J. = Journal
JAR = Journal of Apicultural Research
n.d. = no date
Pub. = Publication

BEES, BEEKEEPING, AND BEE MANAGEMENT

Belknap, C. 2018. The complete guide to beekeeping for fun and profit: Everything you need to know explained simply. Atlantic.

Berrevoets, E. 2009. Wisdom of the bees: Principles for biodynamic beekeeping. Steiner Books.

Biggle, J. 2012. The Biggle bee book: A swarm of facts on practical beekeeping, carefully hived. Skyhorse Publishing.

Blackiston, H., and D.M. Caron. 2017. Beekeeping for dummies. Wiley.

Bonney, R.E. 1990. Hive Management: A seasonal guide for beekeepers. Garden Way Publishing.

Brown, R. 1998. Honey bees: A guide to management. Northern Bee Books.

Buchmann, S.L., and B. Repplier. 2005. Letters from the hive: An intimate history of bees, honey, and humankind. New York: Bantam Books.

Bush, M. 2011. The practical beekeeper. Vols. 1–3: Beekeeping naturally. X-Star Publishing.

Caron, D.M. 1999. Honey bee biology and beekeeping. Kalamazoo, MI: Wicwas Press.

Caron, D.M., and L.J. Connor. 2013. Honey bee biology and beekeeping. Rev. edn. Kalamazoo, MI: Wicwas Press.

Carreck, N., and T. Johnson. 2007. Aspects of sociality in insects. Central Association of Beekeepers.

Carter, G.A.J. 2004. Beekeeping: A guide to the better understanding of bees, their diseases and the chemistry of beekeeping. Delhi: Biotech Books.

Chambers, J. 2018. Beekeeping for West Virginia. CreateSpace Independent Publishing.

Chandler, P.J. 2014. The barefoot beekeeper: A simple sustainable approach to small-scale beekeeping using top bar hives. Chandler.

Collison, C.H. 2003. What do you know? Everything you've ever wanted to know about honey bees, beekeeping. Medina, OH: A.I. Root Co.

Conrad, R. 2007. Natural beekeeping: Organic approaches to modern apiculture. White River Junction, VT: Chelsea Green.

Conrad, R. 2013. Natural beekeeping: Organic approaches to modern apiculture. White River Junction, VT: Chelsea Green.

Corona, M., et al. 2007. Vitellogenin, juvenile hormone, insulin signaling, and queen honey bee longevity. Proc. Natl. Acad. Sci. 104: 7128–7133.

Cramp, D. 2013. The complete step-by-step book of beekeeping: A practical guide to beekeeping, from setting up a colony to hive management and harvesting the honey, shown in 400 photographs. Lorenz Books.

Cramp, D. 2017. The practical book of beekeeping: A complete how-to manual on the satisfying art of keeping bees and their day to day care. Lorenz Books.

Cramp, D., and J. Fleetwood. 2013. Bees and honey: A hive of knowledge and practical inspiration for budding beekeepers. Lorenz Books.

Crane, E. 1999. The world history of beekeeping and honey hunting. Routledge.

Crane, E. 1990. Bees and beekeeping: Science, practice and world resources. Ithaca, NY: Cornell University Press.

Crowder, L., and H. Harrell. 2012. Top-bar beekeeping: Organic practices for honeybee health. Chelsea Green.

Cushman, D.A. "Estimation of Bees on a Frame," http://www.dave-cushman.net/bee/beesest.html.

Cummings, D., and S. Wilson. 2016. The good living guide to beekeeping: Secrets of the hive, stories from the field, and a practical guide that explains it all. Good Books.

Dadant, D. 2018. First lessons in beekeeping. Dover.

Dade, H.A. 1961. Anatomy and dissection of the honeybee. Cardiff, Wales, UK: Bee Research Assoc.

Danforth, B.N., et al. 2006. The history of early bee diversification based on five genes plus morphology. Proc. Natl. Acad. Sci. 103: 15118–15123.

Dartington, R. 2018. New beekeeping in a long deep hive. Northern Bee Books.

Davies, A. 2017. Beekeeping: Inspiration and practical advice for beginners. National Trust Books.

Davis, I., and R. Cullum-Kenyon. 2018. The BBKA guide to beekeeping. Bloomsbury Wildlife.

Delaplane, K.S. 2006. Honey bees and beekeeping: A year in the life of an apiary. University of Georgia, Georgia Center for Continuing Education.

Delaplane, K.S. 2007. First lessons in beekeeping. Hamilton, IL: Dadant & Sons.

De Vito, D. 2012. Beekeeping: A primer on starting and keeping a hive. GMC Distribution.

Drake, J. 2009. The complete book of beekeeping: A complete step by step guide. CreateSpace.

Dullas, W. 2008. ABC's of beekeeping problems and problem beekeepers. IUniverse.

Eckert, J.E. 1933. The flight range of the honeybee. *J Agric Res* 47: 257–285, https://naldc.nal.usda.gov/download/IND43968380/PDF.

Ellis, H. 2004. Sweetness and light: The mysterious history of the honeybee. New York: Three Rivers.

Erickson, Eric H., Jr., S.D. Carlson, and M.B. Garment. 1986. A scanning electron microscope atlas of the honey bee. Ames: Iowa State University Press.

Ferrari, S., M. Silva, M. Guarino, and D. Berckmans. 2008. Monitoring of swarming sounds in bee hives for early detection of the swarming period. Computers and Electronics in Agriculture 64 (1): 72–77.

Flottum, K. 2002. Honey as a crop. Small Farm Today 19 (2): 48–49.

Flottum, K. 2008. The backyard beekeeper: An absolute beginner's guide to keeping bees in your yard and garden. Medina, OH: A.I. Root Co.

Flottum, K. 2011. Better beekeeping: The ultimate guide to keeping stronger colonies healthier and more productive bees. Beverly, MA: Quarry Books.

Flottum, K. 2014. The backyard beekeeper: An absolute beginner's guide to keeping bees in your yard and garden. Beverly, MA: Quarry Books.

Fry, S.N., and R. Wehner. 2002. Honey bees store landmarks in an egocentric frame of reference. J. Comparative Physiology A: Sensory, Neural, and Behavioral Physiology 187 (12): 1009–1016.

Goodman, L. 2003. Form and function in the honey bee. Cardiff, Wales, UK: International Bee Res. Assoc.

Gould, J.L., and C. Gould. 1995. The honey bee (2nd edn.). New York: Scientific American Library, W.H. Freeman.

Graham, J.M. (ed.). 1992. The hive and the honey bee. Hamilton, IL: Dadant & Sons.

Graham, J., and L.L. Langstroth. 2015. The hive and the

honey bee: A new book on beekeeping which continues the tradition of Langstroth on the hive and the honeybee. Hamilton, IL: Dadant.

Guler, A. 2008. The effects of the shook swarm technique on honey bee (*Apis mellifera* L.) colony productivity and honey quality. JAR 47 (1): 27–34.

Hanson-Harding, A. 2014. Beekeeping: Harvest your own honey. Rosen.

Head, V. 2012. The beekeeping handbook: A practical apiary guide for the yard, garden, and rooftop. Fox Chapel.

Hemenway, C. 2013. The thinking beekeeper: A guide to natural beekeeping in top bar hives. New Society Publishers.

Hemenway, C. 2017. Advanced top bar beekeeping: Next steps for the thinking beekeeper. New Society Publishers.

Hooper, T. 1997. Guide to bees and honey. Somerset, UK: Marston House, Yeovil.

Horn, T. 2005. Bees in America: How the honey bee shaped a nation. Lexington: University Press of Kentucky.

Hughes, C. 2010. Urban beekeeping: A guide to keeping bees in the city. Good Life Press.

Hunter, J. 2015. The classic guide to beekeeping: From hives to honey. Amberley.

Jacobson, A. 2016. Beekeeping: The essential guide: A step-by-step guide to beekeeping for beginners and advanced. CreateSpace Independent Publishing.

Johnson, S. 2018. Beginner's guide to beekeeping: Everything you need to know. Hal Leonard.

Jones, J.C., et al. 2005. The effects of rearing temperature on developmental stability and learning and memory in the honey bee, *Apis mellifera*. J. of Comparative Physiology A: Neuroethology, Sensory, Neural, and Behavioral Physiology 191: 1121–1129.

Jones, J.C., M.R. Myerscough, S. Graham, and B.P. Oldroyd. 2004. Honey bee nest thermoregulation: Diversity promotes stability. Science 305 (5682): 402–404.

Kidd, S.M. 2002. The secret life of bees. New York: Penguin Books.

Langstroth, L.L. 2004. Langstroth's hive and the honeybee: The classic beekeeper's manual. Mineola, NY: Dover Publications.

Lusby, E., and D. Lusby. 2018. Biological beekeeping: The way back. X-Star Publishing.

Lynn, R.C., and T. Cooney. 2003. Raising healthy honey bees. Seattle: Christian Veterinary Mission.

MacFawn, D. 2017. Beekeeping tips and techniques for the Southeast United States, beekeeping finance. Outskirts Press.

Mackenzie, A. 2013. Beekeeping: An easy step-by-step guide to setting up and maintaining a hive. Hobble Creek Press.

Mahindru, S.N. 2018. Beekeeping. APH Publishing.

Mcfarland, R. 2015. Better beekeeping: The natural, treatment-free and easy way to raise healthy bees in your backyard. Page Street.

Melathopoulos, A. "The Biology and Management of Colonies in Winter." http://www.capabees.com/shared/2013/02/winteringpdf.pdf.

Menzel, R., et al. 2005. Honey bees navigate according to a map-like spatial memory. Proc. Natl. Acad. Sci. 102 (8): 3040–3045.

Menzel, R., and M. Giurfa. 2006. Dimension of cognition in an insect, the honeybee. Behavioral and Cognitive Neuroscience Reviews 5: 24–40.

Milius, S. 2009. Fossil shows first all-American honeybee. ScienceNews. Web edition: 7/23/09. https://www.sciencenews.org/article/fossil-shows-first-all-american-honeybee.

Mitchener, C.D. 2000. The bees of the world. Baltimore: Johns Hopkins University Press.

Mitcher, C.D., and D. Grimaldi. 1988. The oldest fossil bee: Apoid history, evolution stasis, and antiquity of social behavior. Proc. Natl. Acad. Sci. 85: 6424–6426.

Moore, A. 2014. Beekeeping secrets: The safe way to raise bees. Speedy Publishing.

Moore, L. 2013. Buzz: Urban beekeeping and the power of the bee. New York: NYU Press.

Morrison, A., and M. Vilaubi. 2013. Homegrown honey bees: An absolute beginner's guide to beekeeping; Your first year, from hiving to honey harvest. Storey.

Oldroyd, B.P., and S. Wongsiri. 2006. Asian honey bees: Biology, conservation, and human interactions. Cambridge, MA: Harvard University Press.

Patterson, R. 2012. Beekeeping: A practical guide. Right Way.

Peacock, P. 2008. Keeping bees: A complete practical guide. T.F.H. Publications.

Petterson, J., R. Persson, et al. 2016. Beekeeping: A handbook on honey, hives and helping the bees. Weldon Owen.

Phillips, T. 2017. Beginning beekeeping: Everything you need to make your hive thrive. DK Publishing.

Pierce, A.L., L.A. Lewis, and S.S. Scheider. 2007. The use of the vibration signal and worker piping to influence queen behavior during swarming in honey bees, *Apis mellifera*. Ethology 113: 267–275.

Pisano, T. 2013. Build your own beekeeping equipment. Storey.

Poinar, G.O., Jr., and B.N. Danforth. 2006. A fossil bee from early cretaceous Burmese amber. Science 314: 614.

Remolina, S.C., et al. 2007. Senescence in the worker honey bee *Apis mellifera*. J. Insect Physiology 53: 1027–1033.

Rich, M.I. 2018. Business plan: Beekeeping. Lulu Com.

Ryde, J. 2010. Beekeeping: Self-sufficiency. Skyhorse.

Ryde, J. 2016. Beekeeping. IMM Lifestyle Books.

Schneider, S.S. 2015. The honey bee colony: Life history. In: J.M. Graham (ed.), The hive and the honey bee, pp. 73–109. Dadant & Sons.

Scott, H. 1999. Bee lessons: Think bees, thank natural life, and bee happy. Chapel Hill, NC: Professional Press.

Scott-Dupree, C.D. 1998. The complete beekeeper. Guelph, ON: University of Guelph.

Seeley, T.D. 1996. The wisdom of the hive: The social physiology of bee colonies. Cambridge, MA: Harvard University Press.

Sexton, C.A. 2007. Honey bees. Minneapolis: Bellwether Media.

Shimanuki, H. K., K. Flottum, and A. Harman (eds.). 2006. The ABC and XYZ of bee culture: An encyclopedia of beekeeping. 41st edn. Medina, OH: A.I. Root Co.

Skinner, J.A. 2004. Beekeeping in Tennessee. Knoxville: University of Tennessee, Agricultural Ext. Serv.

Smithers, C.N. 2011. Backyard beekeeping. Rosenberg Publishing.

Specht, S. 2009. Secrets of the Hive. Resource: Engineering and Technology for Sustainable World 16 (2): 13–15.

Spencer, H. 2007. The honeybee: The most profitable thing you can have. Small Farm Today 24 (1): 39–40.

Strange, J.P., R.P. Cicciaarelli, and N.W. Calderone. 2008. What's in that package? An evaluation of quality of package honey bee (Hymenoptera: Apidae) shipments in the United States. J. Economic Entomol. 101 (3): 668–673.

Tautz, J. 2008. The buzz about bees: Biology of a superorganism. Berlin: Springer-Verlag.

Tew, J.E. 2001. Beekeeping principles: A manual for the beginner; A guide for the gardener. Clarkson, KY: Walter T. Kelley Co.

Tew, J.E. 2004. Backyard beekeeping. Auburn: Alabama Cooperative Extension System, Alabama A&M University and Auburn University.

Tew, J.E. 2014. Wisdom for beekeepers: 500 tips for successful beekeeping. Bloomsbury.

Thomas, T.D. 2002. Beeing. Guilford, CT: Lyons Press.

Tourneret, E., S. Saint Pierre, and M. Pettus. 2018. Honey from the earth: Beekeeping and honey hunting on six continents. Deep Snow Press.

Towne, K. 2017. Beekeeping: Step-by-step guide for beginner and advanced beekeepers. Spruce.

Towne, W.F. 1994. Frequency discrimination in the hearing of honey bee (Hymenoptera: Apidae). J. Insect Behavior 8: 281–286.

Traynor, J. 1993. Almond pollination handbook. Bakersfield, CA: Kovak Books.

Traynor, K. 2006. The honey bee's contribution to medicine. ABJ 146 (10): 859–860.

Trood, J. 2018. World of beekeeping. Blurb.

Tsujiuchi, S., et al. 2007. Dynamic range compression in honey bee auditory system toward waggle dance sounds. PLOS One 2: e234.

Virtual atlas of the honeybee. On the website of Freie University Berlin. http://www.neurobiologie.fu-berlin.de/beebrain/. Accessed July 26, 2008.

Waller, G.D. 1980. Honey bee life history. Beekeeping in the United States, Agriculture Handbook 335, https://beesource.com/resources/usda/honey-bee-life-history/.

Waring, A., and C. Waring. 2006. Teach yourself beekeeping. New York: McGraw-Hill.

Waring, A., and C. Waring. 2016. Get started in urban beekeeping. John Murray Learning.

Warner, A. 2017. Beekeeping: Building your backyard beekeeping business. CreateSpace Independent Publishing.

Whitaker, J.M. 2018. Ethics of beekeeping. Northern Bee Books.

Whitfield, C.W., et al. 2006. Thrice out of Africa: Ancient and recent expansions of the honey bee, *Apis mellifera*. Science 314: 642–645.

Beekeeping Websites

Search online for additional websites.
http://apisenterprises.com/
http://www.badbeekeeping.com/weblinks.htm
https://dcbeekeepers.org/
http://www.masterbeekeeper.org/aapa/

BEESWAX

Ahnert, P. 2018. Beehive alchemy: Projects and recipes using honey, beeswax, propolis, and pollen to make soap, candles, creams, salves, and more. Beverly, MA: Quarry Books.

Ahnert, Petra. 2015. Beeswax alchemy: How to make your own candles, soap, balms, salves, and home décor from the hive. Beverly, MA: Quarry Books.

Battershill, N., and D. Constable. 1996. Beeswax crafts: Candlemaking, modelling, beauty creams, soaps and polishes, encaustic art, wax crayons. Tunbridge Wells, UK: Search Press.

Beeswax from the Apiary. 1971. Advisory Leaflet #347. Edinburgh, Scotland: H.M.S.O. Press.

Berthold, R., Jr. 1993. Beeswax crafting. Cheshire, CT: Wicwas Press.

Brown, R.H. 1981. Beeswax (2nd edn.). Burrowbridge, Somerset, UK: Bee Books New and Old.

Cliff, A. 2010. The bee book: Beekeeping basics, harvesting honey, beeswax, candles and other bee business. Melbourne: Manna Press.

Coggshall, W.L., and R.A. Morse. 1984. Beeswax—

production, harvesting, processing and products. Kalamazoo, MI: Wicwas Press.

Dalziel, C. 2017. The beeswax workshop: How to make you own natural candles, cosmetics, cleaners, soaps, healing balms and more. Berkeley, CA: Ulysses Press.

Edmonds, K. 1994. Candles, naturally: A complete guide to rolling beeswax candles. Illume Press.

Flottum, K. 2018. In business with bees: How to expand, sell, and market honey bee products and services including pollination, bees and queens, beeswax honey, and more. Beverly, MA: Quarry Books.

Hepburn, H.R. 1986. Honeybees and wax: An experimental natural history. New York: Springer-Verlag.

Ittner, R. 2011. Candlemaking the natural way: 31 projects made with soy, palm and beeswax. Lark Books.

Munn, P. 1998. Beeswax and propolis for pleasure and profit. IBRA.

Root, H.H. 1951. Beeswax, its properties, testing, production and applications. Brooklyn, NY: Chemical Pub.

Super formulas, arts and crafts: How to make more than 360 useful products that contain honey and beeswax. 1993. Valley Hills Press.

White, E.C. 1995. Soap recipes: Seventy tried-and-true ways to make modern soap with herbs, beeswax and vegetable oils. Valley Hills Press.

BOOKS ON BEES—HISTORIC

This list is for collectors of old, out-of-print books.

Alley, H. 1883. The bee-keeper's handy book. Boston: A. Wenham.

Atkins, W., and K. Hawkins. 1924. How to succeed with bees. Watertown, WI: G.B. Lewis.

Bee and honey patents of the world. 1969. Arlington, VA.

Bonsels, W. 1922. Adventures of Maya the bee. New York: T. Seltzer.

Casteel, D.B. 1912. The manipulation of the wax scales of the honey bee. Bull. #161. Washington, DC: USDA.

Coleman, M.L. 1939. Bees in the garden and honey in the larder. New York: Doubleday/Doran.

Comstock, A.B. 1905. How to keep bees. NY: Doubleday/Page.

Cook, A.J. 1888. The beekeeper's guide. East Lansing, MI: Agricultural College.

Cowan, T.W. 1908. Wax craft. London: Sampson, Low, Marston.

Dadant, C.P. 1917. First lessons in beekeeping. Hamilton, IL: ABJ.

Dadant, C.P. 1920. Dadant system of beekeeping. Hamilton, IL: ABJ.

Dadant, C.P. 1926. Huber's observations on bees. Hamilton, IL: ABJ.

Dietz, H.F. 1925. Pollination and the honey bee. Indianapolis: W.B. Burford.

Edwards, T. 1923. The lore of the honey-bee. London: Methuen.

Flower, A.B. 1925. Beekeeping up to date. London: Cassell.

Gilman, A. 1929. Practical bee breeding. New York: J.P. Putnam's Sons.

Harrison, C. 1903. The book of the honey bee. London: John Lane / Bodley Head.

Hawkins, K. 1920. Beekeeping in the South. Hamilton, IL: ABJ.

Herrod-Hempsall, W. 1930. Beekeeping new and old. Vols. 1 and 2. London: British Bee J.

Langstroth, L.L. 1888. Langstroth on the honey-bee, revised by Dadant. Hamilton, IL: Dadant.

Latham, A. 1949. Allen Latham's bee book. Hapeville, GA: Hale.

Lyon, D.E. 1910. How to keep bees for profit. New York: Grosset & Dunlap.

Maeterlinck, M. 1924. Life of the bee. New York: Dodd-Mead.

Manley, R.O.B. 1946. Honey farming. London: Faber & Faber.

Miller, C.C. 1911. Fifty years among the bees. Medina, OH: A.I. Root.

Miner, T.B. 1850. American bee keeper's manual. New York: C.M. Saxton.

Naile, F. 1942. Life of Langstroth. Ithaca, NY: Cornell University Press.

Pellett, F.C. 1931. The romance of the hive. New York: Abingdon.

Pellett, F.C. 1938. History of American beekeeping. Ames, IA: Collegiate Press.

Phillips, E.F. 1915. Beekeeping. Norwood, MA: Norwood Press.

Phillips, E.F. 1943. Beekeeping. New York: Macmillan.

Root, A.I. 1877. The ABC of bee culture. Medina, OH: A.I. Root.

Sechrist, E.L. 1944. Honey getting. Hamilton, IL: Dadant & Sons.

Sechrist, E.L. 1947. The bee master. Roscoe, CA: Earth-master.

Sechrist, E.L. 1955. Amateur beekeeping. New York: Devin-Adair.

Snelgrove, L.E. 1946. Swarming: Its control and prevention. Weston-Super-Mare, UK: Snelgrove.

Spinner, J. 1952. The house apiary. Exeter, UK: Cornwell.

Stuart, F.S. 1947. City of bees. New York: McGraw Hill.

Sturges, A.M. 1924. Practical beekeeping. Philadelphia: David McKay.

Teale, E.W. 1945. The golden throng. New York: Dodd-Mead.

Webb, A. 1944. Beekeeping for profit and pleasure. New York: Macmillan.

Wedmore, E.B. 1932. A manual of beekeeping for English-speaking beekeepers. London: Longmans, Green.

CANADIAN BEE ASSOCIATIONS AND CONTACTS

Please check online for updated information.

Boucher, Claude, provincial apiarist, Ministère de l'Agriculture, des pêcheries et de l'alimentation du Québec, 675, route Cameron, bureau 101, Ste-Marie, Québec, G6E 3V7. (418) 386-8191, poste 302; (418) 386-8099, télécopieur. Email: claude.boucher@mapaq.gouv.qc.ca.

Currie, Rob, university professor, Dept. Entomol., University of Manitoba, Winnipeg, MB, R3T 2N2. (204) 474-6022; (204) 474-7628. Email: rob_currie@umanitoba.ca.

Foster, Leonard J., university professor, UBC Centre for Proteomics, Dept. of Biochemistry and Molecular Biology, University of British Columbia, Vancouver, BC. (604) 822-8311; (604) 822-2114. Email: ljfoster@interchange.ubc.ca.

Giovenazzo, Pierre, university professor, Département de biologie Faculté des sciences et de genie, Université Laval, Québec, G1K 7P4. (418) 656-2131-8081; (418) 656-2043. Email: pierre.giovenazzo@bio.ulaval.ca.

Gruszka, John, provincial apiarist, Saskatchewan Agriculture and Food, box 3003, 800 Central Avenue, Prince Albert, SK, S6V 6G1. (306) 953-2790; (306) 953-2440. Email: jgruszka@agr.gov.sk.ca.

Guzman, Ernesto, university professor, Dept. Environmental Biology, University of Guelph, Guelph, ON, N1G 2W1. (519) 837-0442; (519) 824-4120 x53609. Email: eguzman@uoguelph.ca.

Houle, Émile, research technician, CRSAD, 120a Chemin du roy, Deschambeault, Québec, G0A 1S0. (418) 286-3353, 227; (418) 286-3597. Email: emile.houle@crsad.qc.ca.

Jordan, Chris, provincial apiarist, PEI Department of Agriculture, box 1600, Charlottetown, PEI, C1A 7N3. (902) 569-7638; (902) 368-5729. Email: cwjordan@gov.pe.ca.

Kelly, Paul, research technician, Dept. of Environmental Biology, University of Guelph, Guelph, ON, N1G 2W1. (519) 836-8897; (519) 837-0442. Email: pgkelly@uoguelph.ca.

Kevan, Peter, university professor, Environmental Biology, University of Guelph, Guelph, ON, N1G 2W1. (519) 824-4120 x52479; (510) 837-0442. Email: pkevan@uoguelph.ca.

Lafrenière, Rhéal, provincial apiarist, MAFRI, 204-545 University Crescent, Winnipeg, MB, R3T 5S6. (204) 945-4825; (204) 945-4327. Email: Rheal.Lafreniere@gov.mb.ca.

Marceau, Jocelyn, research technician, MAPAQ, édifice 2, RC-22, 1665 Blvd. Hamel Ouest, Québec, PQ, G1N 3Y7. (418) 643-7255; (418) 644-8263. Email: jmarceau@mapaq.gouv.qc.cal.

Maund, Christopher, provincial apiarist, Crop Development, New Brunswick Dept. of Agriculture & Aquaculture, PO Box 6000, Fredericton, NB, E3B 5H1. (506) 453-3477; (506) 453-7978. Email: chris.maund@gnb.ca.

McRory, Doug, provincial apiarist, Ontario Beekeeping Association, Foods of Plant Origin, Food Inspection Branch, 1 Stone Road West, Guelph, ON, N1G 4Y2. (519) 826-3595; (519) 826-4375 fax; (888) 466-2372 6-3595. Email: doug.mcrory@ontario.ca.

Melathopoulos, Andony, research technician, AAFC Research Station, PO Box 29, melathopoulosa@agr.gc.ca, Beaverlodge, AB, T0H 0C0. (780) 354-5130; (780) 354-8171.

Moran, Joanne, provincial apiarist, Nova Scotia Agriculture, Kentville Agriculture Centre, Kentville, NS, B4N 1J5. (902) 679-8998; (902) 679-6062. Email: jmoran@gov.ns.ca.

Nasr, Medhat, provincial apiarist, Crop Diversification Centre, North Agriculture Research Division, Alberta Agriculture and Rural Development, 17507 Fort Road, Edmonton, AB, T5Y 6H3. (780) 415-2314; (780) 422-6096. Email: medhat.nasr@gov.ab.ca.

Ostermann, David, provincial apiarist, MAFRI, 204–545 University Crescent, Winnipeg, MB, R3T 5S6. (204) 945-3861; (204) 945-4327. Email: dostermann@gov.mb.ca.

Otis, Gard, university professor, Dept. of Environmental Biology, University of Guelph, Guelph, ON, N1G 2W1. (519) 824-4170 x52478; (519) 837-0442. Email: gotis@uoguelph.ca.

Pernal, Stephen, research scientist, AAFC Research Station, Box 29, Beaverlodge, AB, T0H 0C0. (780) 354-5135; (780) 354-8171. Email: pernals@agr.gc.ca.

Rogers, Richard, consultant, Wildwood Labs Inc., 53 Blossom Dr., Kentville, NS, B4N 3Z1. (902) 679-2818. Email: drogers@wildwoodlabs.com.

Scott-Dupree, Cynthia, university professor, Dept. of Environmental Biology, University of Guelph, Guelph, ON, N1G 2W1. (519) 824-4120, 52477; (519) 837-0442. Email: csdupree@uoguelph.ca.

Skinner, Alison, tech transfer specialist, Ontario Beekeepers' Association Research Office, Orchard Park Office Centre, 5420 Hwy 6 North, Guelph, ON, N1H 6J2. (519) 836-3609. Email: alison_bee@yahoo.com.

Tam, Janet, tech transfer specialist, Ontario Beekeepers' Association Research Office, Orchard Park Office Centre, 5420 Hwy 6 North, Guelph, ON, N1H 6J2. (519) 836-3609. Email: shrewless@yahoo.com.

van Westendorp, Paul, provincial apiarist, British Columbia Ministry of Agriculture and Lands, 1767 Angus Campbell Road, Abbotsford, BC, V3G 2M3. (604) 556-3129; (604) 556-3030. Email: paul.van westendorp@gov.bc.ca; vanwestendorp@telus.net.

White, Jane, natural resources specialist, Dept. of Natural Resources, Newfoundland and Labrador, Agrifoods Development Branch, PO Box 2006, Corner Brook, NL, A2H 6J8. (709) 637-2662; (709) 637-2591. Email: janewhite@gov.nl.ca.

Winston, Mark, university professor, Morris J. Wosk Centre for Dialogue, Simon Fraser University, Harbour Centre, 3309–515 W. Hastings St., Vancouver, BC, V5B 5K3. (778) 782-7894; (778) 782-7892. Email: winston@sfu.ca.

CHILDREN'S BOOKS ON BEES

Claybourne, A. How do bees make honey? 1995. Starting Point Science Series. Sagebrush.

Cole, J., and B. Degen. 1996. The magic school bus: Inside a beehive. New York: Scholastic.

Hodgson, N.B. Guide to children's literature on bees and beekeeping. Bibliography, Bee Research Association. Astley Abbotts, Bridgnorth, Shropshire, UK. This bibliography contains 271 titles in English and 75 in other languages.

Houghton, G. 2004. Bees, inside and out. New York: PowerKids Press.

Kalman, B., A. Larin, and N. Walker. 1998. Hooray for beekeeping! New York: Crabtree.

McDonnell, J. How we get honey (step-by-step). Children's Press.

Rustad, M.E.H. Honey bees. 2003. Mankato, MN: Pebble Books.

Schaefer, L.M. 1999. Honey bees and flowers. Mankato, MN: Pebble Books.

Schaefer, L.M., and G. Saunders-Smith. 1999. Honey bees and hives. Mankato, MN: Pebble Books.

COLONY COLLAPSE DISORDER

CCD Working Group. 2009. MAAREC, Topics: Colony collapse disorder: http://maarec.cas.psu.edu/bkCD /Bee_Diseases/disease_index.html.

Chauzat, M.-P., M. Higes, R. Martín-Hernández, A. Meana, N. Cougoule, and J.-P. Faucon. 2007. Presence of *Nosema ceranae* in French honey bee colonies. JAR 46 (2): 127–128.

Chen, Yanping. 2002. Phylogenetic analysis of acute bee paralysis virus strains. Appl. Environ Microbiol. 68 (12): 6446–6450.

Chen, Y., J.D. Evans, I.B. Smith, and J.S. Pettis. 2008. *Nosema ceranae* is a long-present and wide-spread microsporidian infection of the European honey bee

(*Apis mellifera*) in the United States. J. Invertebrate Pathology 97 (2): 186–188.

Dobbelaere W., D.C. de Graaf, W. Reybroeck, E. Desmedt, J.E. Peeters, and F.J. Jacobs. 2001. Disinfection of wooden structures contaminated with *Paenibacillus larvae* subsp. larvae spores. J. Applied Microbiology 91 (2): 212–216. https://doi.org/10.1046 /j.1365-2672.2001.01376.x.

Ellis, J. 2007. Colony collapse disorder (CCD) in honey bees. University of Florida, Dept. Entomol. and Nematology. Florida CES, Gainesville.

Encyclopaedia Britannica. Colony collapse disorder. https://www.britannica.com/science/colony -collapse-disorder.

Fries, I., R. Martín, A. Meana, P. García-Palencia, and M. Higes. 2006. Natural infections of *Nosema ceranae* in European honey bees. JAR 45 (4): 230–233.

Gill, C. 2007. Honey bees in crisis. PennState Agriculture (Summer–Fall): 17–19.

Gliński, Z., and J. Jarosz. 2001. Infection and immunity in the honey bee *Apis mellifera*. Apiacta 36 (1): 12–24.

Higes, M., F. Esperón, and J.M. Sánchez-Vizcaíno. 2007. Short communication: First report of black queen-cell virus detection in honey bees (*Apis mellifera*) in Spain. Spanish J. Ag. Research 5 (3): 322–325.

Higes, M., R. Martín, and A. Meana. 2006. *Nosema ceranae*, a new microsporidian parasite in honeybees in Europe. J. Invertebrate Pathology 92 (2): 81–83.

Higes, M., R. Martín-Hernández, C. Botías, E.G. Bailón, A.V. González-Porto, L. Barrios, M.J. Del Nozal, J.L. Bernal, J.J. Jiménez, P.G. Palencia, and A. Meana. 2008. How natural infection by *Nosema ceranae* causes honeybee colony collapse. Environmental Microbiology 10 (10): 2659–2669.

Higes, M., R. Martín-Hernández, E. Garrido-Bailón, P. García-Palencia, and A. Meana. 2008. Detection of infective *Nosema ceranae* (microsporidia) spores in corbicular pollen of forager honeybees. J. Invertebrate Pathology 97 (1): 76–78.

Huang, W.-F., M. Bocquet, K.-C. Lee, I.-H. Sung, J.-H. Jiang, Y.-W. Chen, and C.-H. Wang. 2008. The comparison of rDNA spacer regions of *Nosema ceranae* isolates from different hosts and locations. J. Invertebrate Pathology 97 (1): 9–13.

Huang, W.-F., J.-H. Jiang, Y.-W. Chen, and C.-H. Wang. 2007. A *Nosema ceranae* isolate from the honeybee *Apis mellifera*. Apidologie 38 (1): 30–37.

Jacobsen, R. 2008. Fruitless fall: The collapse of the honey bee and the coming agricultural crisis. New York: Bloomsbury.

Kashmir Bee Virus. 2004. Apiculture Factsheet #230, Ministry of Agriculture and Lands, Government of British Columbia, Canada.

Klee, J., A.M. Besana, E. Genersch, S. Gisder, A. Nanetti, D.Q. Tam, T.X. Chinh, et al. 2007. Widespread dispersal of the microsporidian *Nosema ceranae*, an emergent pathogen of the western honey bee, *Apis mellifera*. J. Invertebrate Pathology 96 (1): 1–10.

Klee, J., W. Tek Tay, and R.J. Paxton. 2006. Specific and sensitive detection of *Nosema bombi* (Microsporidia: Nosematidae) in bumble bees (*Bombus* spp.; Hymenoptera: Apidae) by PCR of partial rRNA gene sequences. J. Invertebrate Pathology 91 (2): 98–104.

National Pesticide Information Center. Bee colony collapse disorder. 2017. http://npic.orst.edu/envir/ccd.html.

Pajuelo, A.G., C. Torres, and Fco. J. Orantes-Bermejo, 2008. Colony losses: A double blind trial on the influence of supplementary protein nutrition and preventative treatment with fumagillin against *Nosema ceranae*. JAR 47 (1): 84–86.

Paxton, R.J., J. Klee, S. Korpela, and I. Fries. 2007. *Nosema ceranae* has infected *Apis mellifera* in Europe since at least 1998 and may be more virulent than *Nosema apis*. Apidologie 38 (6): 558–565.

Ribière, M., Jean-Paul Faucon, and M. Pépin. 2000. Detection of chronic honey bee (*Apis mellifera* L.) paralysis virus infection: Application to a field survey. Apidologie 31: 567–577.

Rodriguez-Saona, C. 2007. Honey bees, colony collapse disorder, and cranberries. Plant & Pest Advisory 13 (2). http://njaes.rutgers.edu/pubs/plantandpest advisory/2007/cb0426.pdf.

US Environmental Protection Agency. Pollinator protection. https://www.epa.gov/pollinator-protection/colony-collapse-disorder.

van Engelsdorp, D., D. Cox-Foster, M. Frazier, N. Ostiguy, and J. Hayes. 2006. Colony collapse disorder preliminary report. Mid-Atlantic Apiculture Research and Extension Consortium (MAAREC), CCD Working Group. http://maarec.cas.psu.edu/bkCD/Bee_Diseases/disease_index.html.

Williams, G.R., M.A. Sampson, D. Shutler, and R.E.L. Rogers. 2008. Does fumagillin control the recently detected invasive parasite *Nosema ceranae* in western honey bees (*Apis mellifera*)? J. Invertebrate Pathology 99 (3): 342–344.

Williams, G.R., A.B.A. Shafer, R.E.L. Rogers, D. Shutler, and D.T. Stewart. 2008. First detection of *Nosema ceranae*, a microsporidian parasite of European honey bees (*Apis mellifera*), in Canada and central USA. J. Invertebrate Pathology 97 (2): 189–192.

Zoet, K. 2008. Dutch bee breeding and *Nosema ceranae*. ABJ 148 (3): 185–186.

COURSES

Search online for the many beekeeping courses offered by individuals, organizations, and private companies.

DISEASES AND PESTS

Akratanakul, P. 1987. Honeybee diseases and enemies in Asia: A practical guide. Rome: FAO.

Atkins, J., and B. Atkins. 2016. The business of bees: An integrated approach to bee decline and corporate responsibility. Greenleaf.

Bailey, L., and B.V. Ball. 1991. Honey bee pathology. 2nd edn. London: Academic Press.

Benjamin, A., and B. McCallum. 2009. A world without bees. W.W. Norton.

Chen, Y.P., J.D. Evans, I.B. Smith, and J.S. Pettis. 2008. *Nosema ceranae* is a long-present and wide-spread microsporidian infection of the European honey bee (*Apis mellifera*) in the United States. J. Invertebrate Pathology 97 (2): 186–188.

Chen, Y.P., J.S. Pettis, A. Collins, and M.F. Feldlaufer. 2006. Prevalence and transmission of honeybee viruses. Applied and Environ. Microbiology 72 (1): 606–611.

Chen, Y.P., and R. Siede. 2007. Honey bee viruses. Advances in Virus Research 70: 33–80.

Control of parasitic mites in honey bees: Final report. 2001. University of Manitoba. Saskatchewan Agri-Food Innovation Fund.

Currie, R.W. 2001. Chalkbrood and disease control in honey bees. University of Manitoba. Saskatchewan Agri-Food Innovation Fund.

Devillers, J., and Minh-Hà Pham-Delègue. 2002. Honey bees: Estimating the environmental impact of chemicals. London: Taylor & Francis.

Dobbelaere, D.C. de Graaf, W. Reybroeck, E. Desmedt, J.E. Peeters, and F.J. Jacobs. 2001. Disinfection of wooden structures contaminated with *Paenibacillus larvae* subsp. *larvae* spores. J. Applied Microbiology 91 (2): 212–216. https://doi.org/10.1046/j.1365-2672.2001.01376.x.

Donahue, J.S. 2016. 50 ways to save the honey bees (and change the world). Cider Mill Book Publishers.

Eade, K., and V. Eade. 2013. Bless the bees: The pending extinction of our pollinators and what you can do to stop it. Createspace.

Embry, P. 2018. Our native bees: America's endangered pollinators and the fight to save them. Timber Press.

Erickson, E.H., R.E. Page, and A.A. Hanna. 2002. Proceedings of the 2nd international conference on Africanized honey bees and bee mites. Medina, OH: A.I. Root Co.

Evans, C. 2016. Improving the health of honey bees and other pollinators: National strategy and research action plan. Nova Science Publishers.

Flores, A. 2007. Honey bee genetics vital in disease resistance. Agri. Research 5 (1): 14. http://www.ars.usda.gov/is/AR/archive/jan07/bee0107.htm.

Foulbrood disease of honey bees: Recognition and control. Central Science Laboratory National Bee Unit, Department for Environment, Food and Rural Affairs (DEFRA), UK.

Frazier, M., D. Caron, and D. VanEngelsdorp. 2011. A field guide to honey bees and their maladies. College Park: Pennsylvania State University.

Fries, I., F. Feng, A. Da Silva, S.B. Slemenda, and N.J. Pieniazek, 1996. *Nosema ceranae* n. sp. (Microspora, Nosematidae), morphological and molecular characterization of a microsporidian parasite of the Asian honey bee *Apis cerana* (Hymenoptera, Apidae). European J. Protistology 32 (3): 356–365.

Fries, I., R. Martín, A. Meana, P. García-Palencia, and M. Higes. 2006. Natural infections of *Nosema ceranae* in European honey bees. JAR 45 (4): 230–233.

Fuselli, S.R., et al. 2017. Control and prevention of American foulbrood in honey bees. Nova Science Publishers.

Gillespie, A. 2014. Hives in the city: Keeping bees alive in an urban world. Croyden Hill, UK.

Higes, M., R. Martín, and A. Meana. 2006. *Nosema ceranae*, a new microsporidian parasite in honeybees in Europe. J. of Invertebrate Pathology 92 (2): 93–95.

Higes, M., R. Martín-Hernández, C. Botías, E.G. Bailón, A.V. González-Porto, L. Barrios, et al. 2008. How natural infection by *Nosema ceranae* causes honeybee colony collapse. Environmental Microbiology 10 (10): 2659–2669.

Hive Equipment Sterilization. 2009. Pennsylvania Beekeeper Newsletter 1 (January): 13. www.pastatebeekeepers.org.

Honey bee diseases and pests: A practical guide. FAO Agricultural and Food Engineering Technical Report No. 4. http://www.fao.org/3/a-a0849e.pdf.

Hornitzky, M. 2008. Nosema disease: Literature review and three year survey of beekeepers, part 2. Rural Industries Research and Development Corporation.

Imhoof, M., and C. Lieckfelf. 2015. More than honey: The survival of bees and the future of our world. Greystone Books.

Kashmir Bee Virus. 2004. Apiculture Factsheet #230. Ministry of Agriculture and Lands, Government of British Columbia. July.

Matheson, A. 1996. The conservation of bees. IBRA. London: Academic Press.

Miksha, R. 2004. Bad beekeeping. Victoria, BC: Trafford.

Monck, M., and D. Pearce. 2007. Mite pests of honey bees in the Asia-Pacific region. Australian Centre for International Agricultural Research, Canberra.

Morse, R.A., and R. Nowogrodzki (eds.). 1990. Honey bee pests, predators, and diseases. 2nd ed. Ithaca, NY: Cornell University Press.

Oldroyd, B.P., and S. Wongsiri. 2006. Asian honey bees: Biology, conservation, and human interactions. Cambridge, MA: Harvard University Press.

Powell, G. 2006. Cleaning up American foulbrood. Iowa Honey Producers Association, Buzz Newsletter. January.

Protecting honey bees from pesticides. 2005. Oklahoma City: Oklahoma Dept. of Agriculture, Food and Forestry.

Ritter, W., and P. Akratanakul. 2006. Honey bee diseases and pests: A practical guide. Rome: Food and Agriculture Organization of the United Nations.

Ritter, W., and A. Pongthep. 2006. Honey bee diseases and pests: A practical guide. New York: Food and Agricultural Organization of the United Nations.

Shimanuki, H., and D. Knox. 1991. Diagnosis of honey bee diseases. USDA Handbook AH-690.

Suryanarayanan, S., and D.L. Kleinman. 2017. Vanishing bees: Science, politics, and honeybee health. New Brunswick, NJ: Rutgers University Press. https://www.rutgersuniversitypress.org/.

Suszkiw, J. 2005. New antibiotic approved for treating bacterial honey bee disease. US Dept. of Agriculture, Agricultural Research Service. http://ars.usda.gov/IS/pr/2005/051219.htm.

Szabo, T.I., and D.C. Szabo. 2006. Honey bees in bear territory. Am. Bee Journal 146 (6): 512–514.

Tentcheva, D., and L. Gauthier. 2004. Prevalence and seasonal variations of six bee viruses in *Apis mellifera* L. and *Varroa destructor* mite populations in France. Applied Environmental Microbiology 70 (12): 7185–7191.

Tropilaelaps: Parasitic mites of honey bees. 2005. Department for Environment, Food and Rural Affairs. London.

Diseases and Pests Websites

Acari: The Mites: http://tolweb.org/tree?group=Acari.

North American Bee Associated Mites: http://insects.ummz.lsa.umich.edu/beemites/.

Ontario Beekeepers' Association: https://www.ontariobee.com/outreach/honey-bee-pests-and-diseases.

Penn State Extension: https://extension.psu.edu/a-quick-reference-guide-to-honey-bee-parasites-pests-predators-and-diseases.

University of Georgia: http://bees.caes.uga.edu/bees-beekeeping-pollination/honey-bee-disorders/honey-bee-disorders-bacterial-diseases.html.

Wikipedia: https://en.wikipedia.org/wiki/List_of_diseases_of_the_honey_bee.

USDA: https://www.ars.usda.gov/is/np/honeybeediseases/honeybeediseases.pdf.

EQUIPMENT

Celia, C. A livestock trailer for bees. 2006. ABJ 146 (6): 507–508.

Free hive beetle DIY trap. http://www.greenbeehives.com/.

Goodwin, M., and H. Haine. 1998. Using paraffin wax and steam chests to sterilise hive parts that have been in contact with colonies with American Foulbrood Disease. New Zealand Beekeeper 5 (4): 21.

Great Lakes IPM. www.greatlakesipm.com. Varroa mite (sticky) boards and bee lures.

Griffiths, G.L. 1992. Protocol for wax dipping bee equipment. In: Preservation of wooden hive equipment. Miscellaneous Publication 4/92, March 1992. Compiled by L. Allan, senior apiculturalist. Department of Agriculture, Western Australia.

Hot wax dipping of beehive components for preservation and sterilization. Publication No. 01/051, Project No. DAV-167A. 2001 Rural Industries Research and Development Corporation (RIDC) and the Department of Natural Resources and Environment. Agri-Futures Australia. http://www.rirdc.gov.au.

Kalnins, M.A., and B.F. Detroy. 1984. Effects of wood preservative treatment on honey bees and hive products. J. Ag. and Food Chemistry 32: 1176–1180. http://dx.doi.org/10.1021/jf00125a060.

Northern Tool and Equipment, Tools Bench Scale #19333. NorthernTool.com.

Plans and dimensions for a 10-frame bee hive. Pub.#CA 33-24. Beltsville, MD: USDA-ARS, Ent. Res. Div., n.d.

Robinson, P.J., and J.R.J. French. 1984. Beekeeping and wood preservation in Australia. Proceedings of the 21st forest products research conference, vol. 1.

Robinson, P.J., and J.R.J. French. 1986. Beekeeping and wood preservation in Australia. Australian Bee Journal 67 (1): 8–10.

Sanford, M. T. 1998. Preserving woodenware in beekeeping operations. ENY-125, Florida CES, Institute of Food and Agricultural Sciences, University of Florida. http://edis.ifas.ufl.edu.

Tew, J. 1984. Paraffin wax dipping of hive equipment. Gleanings in Bee Culture 112 (8): 422, 426.

Weatherhead, T.F., and M.J. Kennedy. 1980. Adhesives and wood preservatives for the beekeeper in the 1980s. Australasian Beekeeper 82 (4): 88–94.

FEEDING BEES

Avni, D., A. Dag, and S. Shafir. 2009. The effect of surface area of pollen patties fed to honey bee (*Apis mellifera*) colonies on their consumption, brood production and honey yields. JAR 48 (1): 23–28.

DeGrandi-Hoffman, G., G. Wardell, F. Ahumada-Segura, T. Rinderer, R. Danka, and J. Pettis. 2008. Comparisons of pollen substitute diets for honey bees: Consumption rates by colonies and effects on brood and adult populations. JAR 47 (4): 265–270.

De Jong, D., E.J. Da Silva, P.G. Kevan, and J.L. Atkinson. 2009. Pollen substitutes increase honey bee haemolymph protein levels as much as or more than does pollen. JAR 48 (1): 34–37.

Dufault, R., B. LeBlanc, R. Schnoll, C. Cornett, et al. 2009. Mercury from chlor-alkali plants: Measured concentrations in food product sugar. Environmental Health 8: 2. https://doi.org/10.1186/1476-069X-8-2.

Feeding bees. 2011. National Bee Unit, Best Practice Guideline No. 7. Food and Environment Research Agency (UK). http://www.nationalbeeunit.com/downloadNews.cfm?id=121.

Huang, Z. n.d. Feeding honey bees. Michigan Pollinator Initiative, Michigan State University. https://pollinators.msu.edu/resources/beekeepers/feeding-honey-bees/.

Mattila, H.R., and B.H. Smith. 2008. Learning and memory in workers reared by nutritionally stressed honey bee (*Apis mellifera* L.) colonies. Physiology and Behavior 95 (5): 609–616.

Rosendale, D.I., I.S. Maddox, M.C. Miles, M. Rodier, M. Skinner, and J. Sutherland. 2008. High-throughput microbial bioassays to screen potential New Zealand functional food ingredients intended to manage the growth of probiotic and pathogenic gut bacteria. International J. of Food Science and Technology 43 (12): 2257–2267.

Somerville, D. 2005. Fat bees, skinny bees: A manual on honey bee nutrition for beekeepers. Rural Industries Research and Development Corporation, Barton, Australia.

HONEY AND HONEY PRODUCTS

Balayiannis, G., and P. Balayiannis. 2008. Bee honey as an environmental bioindicator of pesticides' occurrence in six agricultural areas of Greece. Archives of Environmental Contamination and Toxicology 55 (3): 462–470.

Bishop, H. 2005. Robbing the bees: A biography of honey. New York: Free Press.

Bishop, H. 2006. Robbing the bees: A biography of honey, the sweet liquid gold that seduced the world. New York: Free Press.

Blasa, M., M. Candiracci, A. Accorsi, M.P. Piacentini, and E. Piatti. 2007. Honey flavonoids as protection agents against oxidative damage to human red blood cells. Food Chemistry 104 (4): 1635–1640. http://dx.doi.org/10.1016/j.foodchem.2007.03.014.

Bogdanov, S. 2006. Contaminants of bee products. Apidologie 37 (1): 1–18.

Bogdanov, S., T. Jurendic, R. Sieber, and P. Gallmann.

2008. Honey for nutrition and health: A review. J. of the American College of Nutrition 27 (6): 677–689.

Briggs, Margaret. 2007. Honey and its many health benefits. Gardners Books.

Conti, M.E., and F. Botre. 2001. Honeybees and their products as potential bioindicators of heavy metals contamination. Environmental Monitoring and Assessment 69 (3): 267–282. http://www.kluweronline.com/issn/1420-2026/contents.

Coughlin, H. 2018. The goodness of honey: 40 sweet and savoury recipes. London: Kyle Books.

Crane, E. 1999. The world history of beekeeping and honey hunting. New York: Routledge.

Cutting, K.F. 2007. Honey and contemporary wound care: An overview. Wound Management & Prevention 53 (11): 49–54. http://www.o-wm.com/article/8058.

Dyce, E.J. 1961. Finely crystallized or granulated honey, general information. Ithaca, NY: Cornell University Press.

Edgar, J.A., E. Roeder, and R.J. Molyneux. 2002. Honey from plants containing pyrrolizidine alkaloids: A potential threat to health. J. Ag. and Food Chemistry 50 (10): 2719–2730.

Fessenden, R., and M. McInnes. 2009. The honey revolution: Restoring the health of future generations. World Class Emprise. www.thehoneyrevolution.com.

Fleetwood, J. 2014. Honey: Nature's wonder ingredient; 100 amazing uses from traditional cures to food and beauty, with tips, hints and 40 tempting recipes. https://www.thriftbooks.com/w/honey_jenni-fleetwood/2437809/#isbn=1846813743&idiq=32053296.

Forte, G., S. D'Ilio, and S. Caroli, 2001. Honey as a candidate reference material for trace elements. J. of AOAC International 84 (6): 1972–1975.

Frankel, S., G.E. Robinson, and M.R. Berenbaum. 1998. Antioxidant capacity and correlated characteristics of 14 unifloral honeys. JAR 37: 27–31.

Gheldof, N., and N.J. Engeseth. 2002. Antioxidant capacity of honeys from various floral sources based on the determination of oxygen radical absorbance capacity of inhibition of in vitro lipoprotein oxidation in human serum samples. J. Ag. and Food Chemistry 50: 3050–3055.

Gheldorf, N., et al. 2003. Buckwheat honey increases serum antioxidant capacity in humans. J. Ag. and Food Chemistry 51: 1500–1505.

Goltz, L. 2001. Honey color: How important? ABJ 141 (1): 33–35.

Guzman-Novoa, E. and J.L. Uribe-Rubio. 2004. Honey production by European, Africanized and hybrid honey bee (Apis mellifera) colonies in Mexico. ABJ 144 (4): 318–320.

Honey Market News. Fruit & Vegetable Division, Agr. Market Serv., Washington, DC. Ongoing.

Killion, C.E. (1951) 1981. Honey in the comb. Paris, IL: Killion & Sons.

Loveridge, J. Honeycomb. On Chemistry of Bees website. http://www.chemsoc.org/exemplarchem/entries/2001/loveridge.index-page4.html.

Masterton, L. 2013. The fresh honey cookbook. North Adams, MA: Storey Publishing.

McInnes, M., and S. McInnes. 2009. The hibernation diet. World Class Emprise. www.thehoneyrevolution.com.

Molan, P.C. 2006. The evidence supporting the use of honey as a wound dressing. Lower Extremity Wounds 5: 40–54.

Morse, R.A., and Mary Lou Morse. 2015. Honey shows: Guidelines for exhibitors, superintendents, and judges. Kalamazoo, MI: Wicwas Press.

Nicholls, J., and A.M. Miraglio. 2003. Honey and healthy diets. Cereal Foods World 48 (3): 116–119.

Orey, C. 2018. The healing powers of honey: A complete guide to nature's remarkable nectar. New York: Kensington Books.

Pascoe, T. 1996. Beekeeping and back pain. Bee Biz 2: 10–11.

Przybylowski, P., and A. Wilczynska. 2001. Honey as an environmental marker. Food Chemistry 74 (3): 289–291.

Subrahmanyam, M. 2007. Topical application of honey for burn wound treatment—an overview. Annals of Burns and Fire Disasters 20: 137–139.

Tananaki, C., A. Thrasyvoulou, E. Karazafiris, and A. Zotou. 2006. Contamination of honey by chemicals applied to protect honeybee combs from wax-moth (Galleria mellonela L.). Food Additives and Contaminants 23 (2): 159–163.

Terry's Apiaries, https://www.heirloomfm.org/vendor/terrys-apiaries/.

Tosi, E., M. Ciappini, E. Re, and H. Lucero. 2002. Honey thermal treatment effects on hydroxymethylfurfural content. Food Chemistry 77 (1): 71–74.

HONEY COOKBOOKS

Bass, L.L. 1983. Honey and spice. Ashland, OR: Coriander Press.

Berto, H. 1972. Cooking with honey. New York: Gramercy Pub. Co.

Charlton, J., and J. Newdick. 1995. A taste of honey. Edison, NJ: Chartwell Books.

Davenport, M. 1992. Cooking with honey! Tigard, OR: Paddlewheel Press.

Geiskopf, S. 1979. Putting it up with honey: A natural foods canning and preserving cookbook. Ashland OR: Quicksilver Productions.

National Honey Board. https://www.honey.com/recipes.

Opton, G., and N. Highes. 2000. Honey: A connoisseur's guide with recipes. Berkeley, CA: Ten Speed Press.

Savannah Bee Company. https://savannahbee.com/the-fresh-honey-cookbook.

HONEY PLANTS

Ayers, G.S. 2002. Honey quality from a unifloral source. ABJ 142 (9): 657–660. Ayers is a regular contributor to ABJ on honey plants.

Burgett, D.M., B.A. Stringer, and L.D. Johnston. 1989. Nectar and pollen plants of Oregon and the Pacific Northwest. Blogett, OR: Honeystone Press.

Chittka, L., and J. D Thomson (eds.). 2001. Cognitive ecology of pollination: Animal behavior and floral evolution. New York: Cambridge University Press.

Crompton, C.W., and W.A. Wojtas. 1993. Pollen grains of Canadian honey plants. Pub. 1892. Ottawa: Agriculture and Agri-Food Canada.

Dalby, R. 2004. A honey of a tree: Black locust. ABJ 144 (5): 382–384.

Delaplane, K. 1998. Bee conservation in the Southeast. University of Georgia, College of Agricultural and Environmental Sciences. Cooperative Extension, Bulletin 1164.

Hooper, T., and M. Taylor. 1988. The beekeepers' garden. London: Alphabooks.

Tew, J. 1998. Nectar and pollen producing plants of Alabama: A guide for beekeepers. Wooster: Ohio State University, OARDC.

Tew, J. 2000. Some Ohio nectar and pollen producing plants. Fact sheet. Wooster: Ohio State University Extension.

Honey Plants Websites

Search online for additional websites.

MAAREC (Mid-Atlantic Apiculture Research and Extension Consortium): http://MAAREC.cas.psu.edu.

Wikipedia (crop plants pollinated): http://en.wikipedia.org/wiki/List_of_crop_plants_pollinated_by_bees.

Wikipedia (North American nectar sources): https://en.wikipedia.org/wiki/List_of_Northern_American_nectar_sources_for_honey_bees.

INTERNATIONAL BEEKEEPING

Adjare, S. 1990. Beekeeping in Africa. Rome. FAO Ag. Serv. Bull. 68/6.

Bradbear, N. 2004. Beekeeping and sustainable livelihoods. Rome: Agricultural Support Systems Division, Food and Agriculture Organization of the United Nations.

Crane, E. 1999. The world history of beekeeping and honey hunting. New York: Routledge.

International Bee Research Association. 16 North Road, Cardiff, Wales, UK, CF1 3DY. www.ibra.org.uk.

Ransome, H.M. 2004. The sacred bee in ancient times and folklore. Mineola, NY: Dover Publications.

Sammataro, D. 1980. Lesson plans for beekeeping. Washington, DC: US Peace Corps.

MEAD

Morse, R.A. 1980. Making mead. Ithaca, NY: Wicwas Press.

Schramm, K. 2003. The complete meadmaker: Home production of honey wine from your first batch to award-winning fruit and herb variations. Boulder, CO: Brewers Publications.

MITES

Bienefeld, K., J. Radtke, and F. Zautke. 1995. Influence of thermoregulation within honeybee colonies on the reproduction success of *Varroa jacobsoni* Oud. Apidologie 26: 329–330.

Currie, R.W. 2001. Management of varroa mites in honey bees. Saskatchewan Agri-Food Innovation Fund Project 97000002.

Currie, R.W., and P. Gatien. 2006. Timing acaricide treatments to prevent *Varroa destructor* (Acari: Varroidae) from causing economic damage to honey bee colonies. Canadian Entomologist 138 (2): 238–252.

Danka, R.G., and J.D. Villa. 2004. Bees that resist mites are busy groomers. Agricultural Research Magazine 52: 21.

Danka, R.G., J.D. Villa, et al. 1995. Field test of resistance to *Acarapis woodi* (Acari: Tarsonemidae) and of colony production by four stocks of honey bees (Hymenoptera: Apidae). J. Economic Entomol. 88: 584–591.

Dillier, F.-X., P. Fluri, and A. Imdorf. 2006. Review of the orientation behaviour in the bee parasitic mite *Varroa destructor*: Sensory equipment and cell invasion behavior. Revue Suisse de Zoologie 113 (4): 857–877.

Donzé, G., et al. 1996. Effect of mating frequency and brood cell infestation rates on the reproductive success of the honeybee parasite *Varroa jacobsoni*. Ecol. Ent. 21: 17–26.

Donzé, G., and P.M. Guerin. 1994. Behavioral attributes and parental care of *Varroa* mites parasitizing honeybee brood. Behav. Ecol. & Sociobiol. 34: 305–319.

Halliday, B., D.E. Walter, H.C. Proctor, R.A. Norton, and M.J. Colloff. 2001. Acarology, Proceedings of the 10th International Congress [July 5–10, 1998]. 960 pp.

Imdorf, A., and J-D. Charrière. Alternate varroa control: Swiss Bee Research Center, Dairy Research Station Liebefed, CH-3003 Bern.

Kober, T. 2001. Honey double sieve used for determining varroa infestation levels. ABJ 141 (1): 37.

Mangum, W.A. 2007. Honey bee biology: A coexistence between varroa mites and bees, without any miticide treatments. ABJ 147 (6): 489–491.

Martin, S.J. (ed.). 2007. Apicultural research on varroa: Original research in the 21st century. IBRA.

Pettis, J.S., and W.T. Wilson. 1995. Life history of the honey bee tracheal mite (Acari: Tarsonemidae). Ann. Ento. Soc. Am. 89: 368–374.

Ramsey, S.D., C. Gulbronson, J. Mowery, R. Ochoa, D. vanEngelsdorp, and G.R. Bauchan. 2018. A multi-microscopy approach to discover the feeding site and host tissue consumed by *Varroa destructor* on host honey bees. Microscopy Microanal. 24 (Suppl 1), https://doi.org/10.1017/S1431927618006773.

Sammataro, D. 2004. *Tropilaelaps* infestation of honey bees (*Troplilaelaps clareae*, T. koenigerum). In OIE Manual of Diagnostic Tests and Vaccines for Terrestrial Animals. 5th ed., vol. 2: 992–995. Paris: *Office International* des Epizooties.

Sammataro, D. 2006. An easy dissection technique for finding tracheal mites (Acari: Tarsonemidae) in honey bees (with video link). International J. Acarology 32: 339–343. Video: http://www.ars.usda.gov/pandp/docs.htm?docid=14370.

Sammataro, D., S. Cobey, B.H. Smith, and G.R. Needham. 1994. Controlling tracheal mites (Acari: Tarsonemidae) in honey bees (Hymenoptera: Apidae) with vegetable oil. J. Economic Entomol. 57(4): 910–916.

Sammataro, D., U. Gerson, and G.R. Needham. 2000. Parasitic mites of honey bees: Life history, implications and impact. Ann. Rev. Entomology 45: 517–546.

Sammataro, D., and G.R. Needham. 1996. Host-seeking behaviour of tracheal mites (Acari: Tarsonemida) on honey bees (Hymenoptera: Apidae). Experimental and Applied Acarology 20: 121–136.

Satta, A., I. Floris, M. Eguaras, P. Cabras, V. L. Garau, and M. Melis. 2005. Formic acid–based treatments for control of *Varroa destructor* in a Mediterranean area. J. Economic Entomol. 98 (2): 267–273.

Underwood, R.M., and R.W. Currie. 2003. The effects of temperature and dose of formic acid on treatment efficacy against *Varroa destructor* (Acari: Varroidae), a parasite of *Apis mellifera* (Hymenoptera: Apidae). Experimental and Applied Acarology 29: 303–313. 2003.

Walter, D.E. 2006. Invasive mite identification: Tools for quarantine and plant protection. Fort Collins: Colorado State University, and Raleigh, NC: USDA/APHIS/PPQ Center for Plant Health Science and Technology. http://www.lucidcentral.org/keys/v3/mites/Invasive_Mite_Identification/key/Whole_site/Home_whole_key.html.

Walter, D.E., Gerald Krantz, and Evert Lindquist. Acari: The mites. Tree of Life Web Project. http://tolweb.org/Acari/2554.

NON-*APIS* BEE POLLINATORS

Canto-Aguilar, M.A., and V. Parra-Tabla. 2000. Importance of conserving alternative pollinators: Assessing the pollination efficiency of the squash bee, *Peponapis limitaris*, in *Cucurbita moschata* (Cucurbitaceae). J. Insect Conservation 4 (3): 203–210.

Costa, L., and G.C. Venturieri. 2009. Diet impacts on *Melipona flavolineata* workers (Apidae, Meliponini). JAR 48 (1): 38–45.

Dogterom, M. 2002. Pollination with mason bees: A gardener and naturalists' guide to managing mason bees for fruit production. Beediverse Publishing.

dos Santos, C.G., F.L. Megiolaro, J.E., Serrão, and B. Blochtein. 2009. Morphology of the head salivary and intramandibular glands of the stingless bee *Plebeia emerina* (Hymenoptera: Meliponini) workers associated with propolis. Ann. Entomol. Soc. America 102 (1): 137–143.

Kremen, C., N.M. Williams, and R.W. Thorp. 2002. Crop pollination from native bees at risk from agricultural intensification. Proc. Natl. Acad. Sci. 99 (26) 16812–16816.

Richards, K.W. 1996. Effect of environment and equipment on productivity of alfalfa leafcutter bees (Hymenoptera: Megachilidae) in southern Alberta, Canada. Canadian Entomologist 128 (1): 47–56.

Sampson, B.J., S.J. Stringer, J.H. Cane, and J.M. Spiers. 2004. Screenhouse evaluations of a mason bee *Osmia ribifloris* (Hymenoptera: Megachilidae) as a pollinator for blueberries in the southeastern United States. Fruits Review 3 (3–4): 381–392.

Shuler, R.E., T'A.H. Roulston, and G.E. Farris. 2005. Farming practices influence wild pollinator populations on squash and pumpkin. J. Economic Entomol. 98 (3): 790–795.

Stephen, W.P., and S. Rao. 2005. Unscented color traps for non-Apis bees (Hymenoptera: Apiformes). J. Kansas Entomol. Soc. 78 (4): 373–380.

Stubbs, C.S., F.A. Drummond, and E.A. Osgood. 1994. *Osmia ribifloris biedermannii* and *Megachile rotundata* (Hymenoptera: Megachilidae) introduced into the lowbush blueberry agroecosystem in Maine. J. Kansas Entomol. Soc. 67: 173–185.

Tuell, J.K., J.S. Ascher, and R. Isaacs. 2009. Wild bees (Hymenoptera: Apoidea: Anthophila) of the Michigan highbush blueberry agroecosystem. Ann. Entomol. Soc. America 102 (2): 275–287.

Weinberg, D., and C.M.S. Plowright. 2006. Pollen collection by bumblebees (*Bombus impatiens*): The effects of resource manipulation, foraging experience and colony size. JAR 45 (2): 22–27. ISSN 0021-8839.

PARAFFIN DIPPING (APPENDIX D)

Goodwin, M., and H. Haine. 1998. Using paraffin wax and steam chests to sterilise hive parts that have been in contact with colonies with American foulbrood disease. New Zealand Beekeeper 5 (4): 21.

Griffiths, G.L. 1992. Protocol for wax dipping bee equipment. In: Preservation of wooden hive equipment. Miscellaneous Publication 4/92. March 1992. Compiled by L. Allan, senior apiculturalist, Department of Agriculture, Western Australia.

Hot wax dipping of beehive components for preservation and sterilization. 2001. Publication No. 01/051, Project No. DAV-167A. Kingston, Australia: Rural Industries Research and Development Corporation and the Department of Natural Resources and Environment; http://www.rirdc.gov.au.

Kalnins, M.A., and B.F. Detroy. 1984. Effects of wood preservative treatment on honey bees and hive products. J. Agricultural Food Chemistry 32: 1176–1180.

Matheson, A. 1980. Easily constructed paraffin wax dipper. New Zealand Beekeeper 41 (4): 11–12.

Robinson, P.J., and J.R.J. French. 1986. Beekeeping and wood preservation in Australia. Australian Bee Journal 67 (1): 8–10.

Tew, J. 1984. Paraffin wax dipping of hive equipment. Gleanings in Bee Culture 112 (8): 422, 426.

Weatherhead, T.F., and M.J. Kennedy. 1980. Adhesives and wood preservatives for the beekeeper in the 1980s. Australasian Beekeeper 82 (4): 88–94.

Williams, D. 1980. Preserving beehive timber. New Zealand Beekeeper 41 (4): 13.

PESTICIDES

Alix, A., and C. Vergnet. 2007. Risk assessment to honey bees: A scheme developed in France for nonsprayed systemic compounds. Pest Management Science 63 (11): 1069–1080.

Chauzat, M.-P., and J.-P. Faucon. 2007. Pesticide residues in beeswax samples collected from honey bee colonies (*Apis mellifera* L.) in France. Pest Management Science 63 (11): 1100–1106.

Chauzat, M.-P., J.-P. Faucon, A.-C. Martel, J. Lachaize, N. Cougoule, and M. Aubert. 2006. A survey of pesticide residues in pollen loads collected by honey bees in France. J. Economic Entomol. 99 (2): 253–262.

Choudhary, A., and D.C. Sharma. 2008. Dynamics of pesticide residues in nectar and pollen of mustard (*Brassica juncea* (L.) Czern.) grown in Himachal Pradesh (India). Environmental Monitoring and Assessment 144 (1–3): 143–150.

Colin, M.E., J.M. Bonmatin, I. Moineau, C. Gaimon, S. Brun, and J.P. Vermandere. 2004. A method to quantify and analyze the foraging activity of honey bees: Relevance to the sublethal effects induced by systemic insecticides. Archives of Environmental Contamination and Toxicology 47 (3): 387–395.

Cutler, G.C., and C.D. Scott-Dupree. 2007. Exposure to clothianidin seed-treated canola has no long-term impact on honey bees. J. of Economic Entomol. 100 (3): 765–772.

Desneux, N., A. Decourtye, and J.-M. Delpuech. 2007. The sublethal effects of pesticides on beneficial arthropods. Annual Review of Entomol. 52: 81–106.

Frazier, M., C. Mullin, J. Frazier, and S. Ashcraft. 2008. What have pesticides got to do with it? ABJ 148 (6): 521–523.

Karazafiris, E., C. Tananaki, U. Menkissoglu-Spiroudi, and A. Thrasyvoulou. 2008. Residue distribution of the acaricide coumaphos in honey following application of a new slow-release formulation. Pest Management Science 64 (2): 165–171.

Martel, A.-C., S. Zeggane, C. Aurières, P. Drajnudel, J.-P. Faucon, and M. Aubert. 2007. Acaricide residues in honey and wax after treatment of honey bee colonies with Apivar® or Asuntol® 50. Apidologie 38 (6): 534–544.

Tew, J.E. 1996. Protecting honey bees from pesticides. Extension fact sheet. Wooster: Ohio State University Extension, OARDC.

Thompson, H.M., and C. Maus. 2007. The relevance of sublethal effects in honey bee testing for pesticide risk assessment. Pest Management Science 63 (11): 1058–1061.

POLLEN

Hodges, D. (1974) 1994. The pollen loads of the honeybee. London: IBRA.

Kesseler, R., and M. Harley. 2005. Pollen: The hidden sexuality of flowers. 2nd edn. Papadakis.

Kirk, W.D.J. 2006. A colour guide to pollen loads of the honey bee. 2nd rev. edn. Cardiff, Wales, UK: IBRA.

LeBlanc, B.W., O.K. Davis, S. Boue, A. DeLucca, and T. Deeby. 2009. Antioxidant activity of Sonoran Desert bee pollen. Food Chemistry 115 (4): 1299–1305.

Pleasants, J.M., R.L. Hellmich, G.P. Dively, M.K. Sears, D.E. Stanley-Horn, H.R. Mattila, J.E. Foster, P. Clark, and G.D. Jones. 2001. Corn pollen deposition on milkweeds in and near cornfields. Proc. Natl. Acad. Sci. 98 (21): 11919–24. https://doi.org/10.1073/pnas.211287498. PMID 11559840. http://www.pnas.org/cgi/pmidlookup?view=long&pmid=11559840.

Price, L.D., F. Chukwuma, and J.J. Adamczyk Jr. 2004. Honey bee (Hymenoptera: Apidae) pollen load rate based on pollen grain size. J. Entomol. Science 39 (4): 677–678.

Wenning, C.J. 2003. Pollen and the honey bee. ABJ 143 (5): 394–397.

POLLINATION

Barth, F.G. 1991. Insects and flowers. Princeton, NJ: Princeton University Press.

Chittka, L., and N.E. Raine. 2006. Recognition of flowers by pollinators. Current Opinion in Plant Biology 9: 428–435.

Dafni, A., P.G. Kevan, and B.C. Husband (eds.). 2005. Practical pollination biology. Ontario: Enviroquest.

Dag, A., R.A. Stern, and S. Shafir. 2005. Honey bee (Apis mellifera) strains differ in apple (Malus domestica) pollen foraging preference. JAR 44 (1): 15–20.

David, A., and S. Bucknall. 2004. Plants and honey bees: An introduction to their relationships. West Yorkshire, UK: Northern Bee Books, Mytholmroyd.

Delaplane, K., and D.F. Mayer. 2000. Crop pollination by bees. New York: CABI Publishing.

Dryer, A.G., M.G.P. Rosa, and D.H. Reser. 2008. Honeybees can recognize images of complex natural scenes for use as potential landmarks. J. Experimental Biology 211: 1180–1186.

Ellis, A., and K.S. Delaplane. 2008. Effects of nest invaders on honey bee (Apis mellifera) pollination efficacy. Agriculture, Ecosystems and Environment 127 (3–4): 201–206.

Free, J.B. (1970) 1993. Insect Pollination of Crops. 2nd ed. New York: Academic Press.

Gould, J. 2004. Animal navigation. Current Biology 14: R221–R224.

Gruter, C., M.S. Balbuena, and W.M. Farnia. 2008. Informational conflicts created by the waggle dance. Proceedings of the Royal Society B, Biological Sciences 275: 1321–1327.

James, R.R., and T.L. Pitts-Singer. 2008. Bee pollination in agricultural ecosystems. New York: Oxford University Press.

Kerns, C.A., and D.W. Inouye. 1993. Techniques for pollination biologists. Niwot: University Press of Colorado.

Klein, B.A. 2006. Caste-dependent sleep of worker honey bees. J. Experimental Biology 211: 3028–3040.

Ledford, H. 2007. Plant biology: The flower of seduction. Nature 445: 816–817.

McGregor, S.E. 1976. Insect pollination of crops. Washington, DC: USDA, ARS. Online.

National Research Council. 2007. Status of pollinators in North America. Washington, DC: National Academies Press.

Nicolson, S.W., M. Nepi, and E. Pacini (eds.). 2007. Nectaries and nectar. Dordrecht, Netherlands: Springer.

O'Neal, R.J., and G.D. Waller. 1984. On the pollen harvest by the honey bee (Apis mellifera L.) near Tucson, Arizona (1976–1981). Desert Plants 6 (2).

Pierce, A.L., L.A. Lewis, and S.S. Schneider. 2007. The use of the vibration signal and worker piping to influence queen behavior during swarming in honey bees, Apis mellifera. Ethology 113: 267–276.

Pollination. 2000. Mid-Atlantic Apicultural Research and Extension Consortium, MAAREC Publication 5.2.

Seeley, T.D., and P.K. Visscher. 2003. Choosing a home: How the scouts in a honey bee swarm perceive the completion of their group decision making. Behavioral Ecology and Sociobiology 54: 511–520.

Skinner, J.A. n.d. Making a pollination contract. Ag. Ext. Serv., University of Tennessee.

Whynott, D. 1991. Following the bloom: Across America with the migratory beekeepers. New York: Tarcher/Penguin.

Pollination Websites

The Great Pollinator Project: On NYC Beewatchers website, www.nycbeewatchers.org.

North American Pollinator Protection Campaign: http://www.nappc.org/.

Pollinator Conservation Digital Library: http://libraryportals.com/PCDL/plants.

Pollinator Conservation Program, Xerces Society: http://www.xerces.org/Pollinaotr_Insect_Conservation/.

Pollinator Partnership: www.pollinator.org.

PRODUCTS OF THE HIVE (OTHER THAN HONEY)

Bankova, V. 2005. Recent trends and important developments in propolis research. Compl. and Alt. Medicine 2 (1): 29–32. https://doi.org/10.1093/ecam/neh059. http://ecam.oxfordJ.s.org/cgi/content/full/2/1/29.

Fearnley, J. 2001. Bee propolis: Natural healing from the hive. Souvenir Press.

Krell, R. 1996. Value-added products from beekeeping.

FAO Ag Serv. Bull. No. 124. Food and Agriculture Organization of the United Nations, Rome.

Munstedt, K., and M. Zygmunt, 2001. Propolis: Current and future medical uses. ABJ 141(7): 507–510.

Quiroga, E.N., D.A. Sampietro, et al. 2006. Propolis from the northwest of Argentina as a source of antifungal principles. J. Applied Microbiology 101 (1): 103–110. http://dx.doi.org/10.1111/j.1365-2672 .2006.02904.x.

Sforcin, J.M. 2007. Propolis and the immune system: A review. J. Ethnopharmacology 113 (1): 1–14. http:// dx.doi.org/10.1016/j.jep.2007.05.012.

QUEENS

Al-Lawati, H., G. Kamp, and K. Bienefeld. 2009. Characteristics of the spermathecal contents of old and young honeybee queens. J. Insect Phys. 55(2): 117–122. http://dx.doi.org/10.1016/j.jinsphys.2008.10.010.

The development of honey bee artificial insemination. Apiculture Program, Dept. Entomol., North Carolina State University. http://www.cals.ncsu.edu /entomology/apiculture/PDF%20files/2.14.pdf.

Frake, A.M., L.I. De Guzman, and T.E. Rinderer. 2009. Comparative resistance of Russian and Italian honey bees (Hymenoptera: Apidae) to small hive beetles (Coleoptera: Nitidulidae). J. Economic Entomol. 102 (1): 13–19.

Harris, J.L. 2008. Effect of requeening on fall populations of honey bees on the northern Great Plains of North America. JAR 47 (4): 271–280.

Honey bee artificial insemination instrument. Chung Jin Biotech Co. Ltd. http://younan99.en.ecplaza.net /catalog.asp?DirectoryID=81469&CatalogID =197772. Accessed April 5, 2008.

Jurica, C.A. 1985. Practical queen production in the north. Johnstown, NY: Jurica Apiaries.

Mattila, H.R., and T.D. Seeley. 2007. Genetic diversity in honey bee colonies enhances productivity and fitness. Science 317: 362–364.

Pettis, J.S., A.M. Collins, R. Wilbanks, and M.F. Feldlaufer. 2004. Effects of coumaphos on queen rearing in the honey bee, *Apis mellifera*. Apidologie 35 (6): 605–610.

Rinderer T.E. (ed.). Bee genetics and breeding. (1986) 2009. Northern Bee Books.

Villa, J.D., and T.E. Rinderer. 2008. Inheritance of resistance to *Acarapis woodi* (Acari: Tarsonemidae) in crosses between selected resistant Russian and selected susceptible U.S. honey bees (Hymenoptera: Apidae). J. Economic Entomol. 101 (6): 1756–1759.

SMALL HIVE BEETLES

Arbogast, R. T., B. Torto, and P.E.A. Teal. 2009. Monitoring the small hive beetle *Aethina tumida* (Coleoptera: Nitidulidae) with baited flight traps: Effect of distance from bee hives and shade on the numbers of beetles captured. Florida Entomologist 92(1): 165–166.

Caron, D.M., A. Park, J. Hubner, R. Mitchell, and I.B. Smith. 2001. Small hive beetle in the Mid-Atlantic states. ABJ 141 (11): 776–777.

Ellis, J.D., and K. S. Delaplane. 2008. Small hive beetle (*Aethina tumida*) oviposition behaviour in sealed brood cells with notes on the removal of the cell contents by European honey bees (*Apis mellifera*). JAR 47 (3): 210–215.

Hood, M.W. 2004. The small hive beetle, *Aethina tumida*: A review. Bee World 85 (3): 51–59.

Neumann, P., and D. Hoffmann. 2008. Small hive beetle diagnosis and control in naturally infested honeybee colonies using bottom board traps and CheckMite + strips. J. Pest Science 81 (1): 43–48. http://dx.doi .org/10.1007/s10340-007-0183-8.

Neumann, P., and J.D. Ellis. 2008. The small hive beetle (*Aethina tumida* Murray, Coleoptera: Nitidulidae) distribution, biology and control of an invasive species. JAR 47 (3): 181–183.

Nolan, M.P., IV, and W.M. Hood. 2008. Comparison of two attractants to small hive beetles, *Aethina tumida*, in honey bee colonies. JAR 47 (3): 229–233.

Sanford, M.T. 2005. Small hive beetle, *Aethina tumida* (Murray). Featured Creatures, University of Florida. EENY-94. http://entomology.ifas.ufl.edu/creatures /misc/bees/small_hive_beetle.htm.

Small Hive Beetle Websites

MAAREC (Mid-Atlantic Apiculture Research and Extension Consortium): http://maarec.psu.edu/pdfs /Small_Hive_Beetle_-_PMP.pdf.

SOCIAL INSECTS

Beshers, S.N., and J.H. Fewell. 2001. Models of division of labor in social insects. Ann. Review Entomol. 46: 413–440.

Moritz, R.F.A., and E.E. Southwick. 1992. Bees as superorganisms: An evolutionary reality. New York: Springer-Verlag.

Reeve, H.K., and B. Hölldobler. 2007. The emergence of a superorganism through intergroup competition. Proc. Natl. Acad. Sci. 104: 9736–9740.

Wilson, E.O., and B. Hölldobler. 2005. Eusociality: Origin and consequences. Proc. Natl. Acad. Sci. 102 (38): 13367–13371. https://doi.org/10.1073/pnas

.0505858102. http://www.pnas.org/content/102/38/13367.full.pdf+html.

SUPPLIERS, FOREIGN

B.J. Sherriff, Carclew Road, Mylor Downs, Falmouth, Cornwall, TR11 5UN, UK. www.bjsherriff.co.uk.

Maisemore Apiaries, Old Road, Maisemore, Gloucester, GL2 8HT, UK. www.bees-online.co.uk.

National Bee Supplies, Merrivale Road, Exeter Road Industrial Estate, Okehampton, Devon, EX20 1UD, UK. www.beekeeping.co.uk.

Northern Bee Books. http://www.northernbeebooks.co.uk/booklist/all-available-new-books/.

Saf Natura s.r.l., 36015 Schio (VI) Italia, Via Lago di Misurina, 26 Z.I. Email: safnatura@witcom.com.

Swienty A/S, Hortoftvej 16, Ragebol, DK-6400 Sonderborg, Denmark. www.swienty.com.

Thomas, 86 rue Abbé Thomas, BP 2, F-45450, Fay aux Loges, France. www.thomas-apiculture.com.

Thorne Beehive Works, Wragby, Market Rasen LN8 5LA, UK. www.thorne.co.uk.

SUPPLIERS, MAJOR, U.S.

Betterbee Inc., 8 Meader Road, Greenwich, NY 12834. www.betterbee.com.

Dadant and Sons, 51 S. 2nd St., Hamilton, IL 62341. www.dadant.com.

Mann Lake Ltd., 501 1st St., S. Hackensack, MN 56452. www.mannlakeltd.com.

Walter T. Kelley Co., 807 West Main St., Clarkson, KY 42726. www.kelleybees.com.

Note: Check bee magazines for other suppliers.

VENOM

Allan, S. 2008. Allergy: Bee-keepers hold clues for T-cell tolerance. Nature Reviews Immunology 8 (12): 910.

Balit, C., G. Isbister, and N. Buckley. 2003. Randomized controlled trial of topical aspirin in the treatment of bee and wasp stings. J. Toxicol. Clin. Toxicol. 41 (6): 801–808. https://doi.org/10.1081/CLT-120025345.

Bee venom collector devices. Apitronic Services. http://www.beevenom.com/collectordevices.htm#COLL.

Bowling, A.C. 2006. Complementary and alternative medicine and multiple sclerosis. New York: Demos Medical Publishing.

de Graaf, D.C., M. Aerts, E. Danneels, and B. Devreese. 2009. Bee, wasp and ant venomics pave the way for a component-resolved diagnosis of sting allergy. J. Proteomics 72 (2): 145–154.

Hellner, M., D. Winter, R. Von Georgi, and K. Münstedt. 2008. Apitherapy: Usage and experience in German beekeepers. Evidence-Based Complementary and Alternative Medicine 5 (4): 475–479.

Kokot, Z.J., and J. Matysiak. 2009. Simultaneous determination of major constituents of honeybee venom by LC-DAD. Chromatographia 69 (11–12): 1401–1405.

Mraz, C. 1995. Health and the honeybee. Burlington, VT: Queen City.

Sanford, M. 2003. Bee stings and allergic reaction. ENY122, one of a series of the Entomology and Nematology Department, Florida CES, Institute of Food and Agricultural Sciences. Gainesville: University of Florida. http://edis.ifas.ufl.edu.

Sturm, G.J., C. Schuster, B. Kranzelbinder, M. Wiednig, A. Groselj-Strele, and W. Aberer. 2009. Asymptomatic sensitization to Hymenoptera venom is related to total immunoglobulin E levels. International Archives of Allergy and Immunology 148 (3): 261–264.

Visscher P., R. Vetter, and S. Camazine. 1996. Removing bee stings. Lancet 348 (9023): 301–302. https://doi.org/10.1016/S0140-6736(96)01367-0.

WAX MOTH REARING (APPENDIX F)

Gojmerac, W. 1978. Grow your own wax worms. Fin and Feathers. January.

Hocking, B., and F. Matsumura. 1960. Bee brood as food. Bee World 41 (5): 113–120.

How to grow a wax worm farm. wikiHow. https://m.wikihow.com/Grow-a-Wax-Worm-Farm.

Keeney, G. Ohio State University Insectary, Columbus.

Taylor, R.L. Butterflies in my stomach. Santa Barbara, CA: Woodbridge Press.

Taylor, R.L., and B.J. Carter. 1976. Entertaining with insects. Santa Barbara, CA: Woodbridge Press.

Terry, S. How to raise waxworms. Pets on mom.com. https://animals.mom.me/how-to-raise-wax-worms-3047967.html.

WINTERING

Melathopoulos, A. "The Biology and Management of Colonies in Winter." http://www.capabees.com/shared/2013/02/winteringpdf.pdf.

Wahl, C., L. Mizer, and D. Sammataro. Wintering bees in cold climates: Fall management, preparing for winter, how to be a good beekeeper in January, and diagnosing spring colony deadouts. Address correspondence to C. Wahl at CWL5@cornell.edu, or Department of Biomedical Sciences, College of Veterinary Medicine, Cornell University, Ithaca, NY 14853.

Index

abandonment, 143

ABC and XYZ of Bee Culture, The (Shimanuki, Flottum, and Harma), 209, 213, 235

abdomen, 11–12

absconding, 188

Acarapis woodi, 250–255

acaricides, 262–263

Acarine disease, 250–255

acid gland, 15

acute bee paralysis virus (ABPV), 247, 248, 249

adenosine diphosphates (ADT), 18

adenosine triphosphates (ATP), 18

aerosol treatments for stings, 87

Africanized honey bees, 2, 8, 61, 86, 146, 148, 150–151, 166, 269–270, 313–315

Africanized Honey Bees and Bee Mites (Eickwort), 250

afterswarms, 150

age polyethism, 27

age progression, labor and, 26–27

air flow/wind, 278

alarm odors, 15, 37, 74, 77–78, 86

alcohol washes, 260

Alexander-type veils, 52

alfalfa, 289, 290*fig*

alighting boards, 45

alkali gland, 15

allergies, 3, 87

ambush bugs, 271

American foulbrood (AFB), 111, 124, 178, 197, 238–243, 239*fig*, 240*t*, 244, 281, 285

amoeba disease, 232–233

anaphylactic shock, 3

anatomy of bee, 11–16, 12*fig*, 301–305

angiosperms, 286–288

animals, 272–273

ant lions, 271

antenna, 13

anther, 286

antihistamines, 87

ants, 68, 266, 270–271

aphid lethal paralysis virus, 247

apiary
  identification on, 65–66
  layout of, 72*fig*, 279
  maintenance of, 280
  preparing to go to, 75
  site for, 64–65, 64*fig*, 236, 278–279
  See also bees; colonies; hives

apiary inspection laws, 2

Apiary Inspectors of America, 228

Apis iridescent virus, 247

apitherapy, 223

apples, 288

Argentine ants, 271

Arkansas bee virus, 247

Ascophaera apis, 245

assassin bugs, 271

attendant bees, 31, 154

auger-hole pollen traps, 113*fig*

back care, 143

bait boxes/hives, 62, 63, 195–196

balling the queen, 85–86, 89

banking frames, 155, 155*fig*

banking queens, 154, 155

basitarsus, 13–14

basket-type extractors, 210, 210*fig*

"battery box," 154

bears, 65, 180, 273

bee bearding, 40

bee blowers, 142–143

bee bread, 14, 16, 17, 19, 35, 111–112

bee brood, 220–221

bee brushes, 54

bee eaters, 274

bee inspectors, 58, 59

bee lice, 271

bee parasitic mite syndrome. *See* parasitic mite syndrome (PMS)

bee space, 33, 42, 45–46, 45*fig*

bee stings. *See* stings

bee suits, 52

bee venom, 221–223

bee venom therapy, 223

bee virus X, 247

bee virus Y, 247

bee wear, 51–53, 76–77

beekeeping equipment
  accessory equipment, 54–55
  basic hive parts, 42–49
  bee wear, 51–53, 76–77
  for beginners, 50–51
  hardware, 53–54
  replacement of, 281
  supers and hive bodies, 49–50

bees
  anatomy of, 11–16, 12*fig*, 301–305
  ancestors of, 5–6
  being chased by, 86
  colony activities of, 26–40

developmental stages of, 18*fig*

dissecting, 253–254, 254*fig*

feeding, 103–116

glands of, 306–307*t*, 308*fig*

hybrid, 10–11

importation of, 7–8

leasing, 295, 299

life stages of, 16–18

obtaining, 56–64

package, 56–57, 88–102, 89*fig*

pheromones of, 7

preparing for, 64–73

races of, 7–10

removing from honey supers, 139–143

smoking, 77–79

social structure of, 6–7

taxonomy for, 6, 6*fig*

temperament of, 84–85

understanding, 5–25

in veil, 85

working, 74–87

See also apiary; colonies; drones; hives; queens; workers

beeswax, 34, 217–220. *See also* comb building; foundations

beetles, 271. *See also* small hive beetles (SHB)

beginners, equipment for, 50–51

beneficial microbes, 35, 35*fig*

Benton mailing cage, 153–154, 153*fig*

Berkeley bee virus, 247

bicyclohexylammonium fumagillin, 232

Big Sioux virus, 247